T0331247

Scientific Basis for Nuclear Waste Management XXXII

MATERIALS RESEARCH SOCIETY
SYMPOSIUM PROCEEDINGS VOLUME 1124

Scientific Basis for Nuclear Waste Management XXXII

Symposium held December 1–5, 2008, Boston, Massachusetts, U.S.A.

EDITORS:

Neil C. Hyatt
The University of Sheffield
Sheffield, United Kingdom

David A. Pickett
Southwest Research Institute
San Antonio, Texas, U.S.A.

Raul B. Rebak
GE Global Research
Schenectady, New York, U.S.A.

Materials Research Society
Warrendale, Pennsylvania

CAMBRIDGE
UNIVERSITY PRESS

University Printing House, Cambridge CB2 8BS, United Kingdom

One Liberty Plaza, 20th Floor, New York, NY 10006, USA

477 Williamstown Road, Port Melbourne, VIC 3207, Australia

314-321, 3rd Floor, Plot 3, Splendor Forum, Jasola District Centre, New Delhi - 110025, India

79 Anson Road, #06-04/06, Singapore 079906

Cambridge University Press is part of the University of Cambridge.

It furthers the University's mission by disseminating knowledge in the pursuit of education, learning and research at the highest international levels of excellence.

www.cambridge.org
Information on this title: www.cambridge.org/9781605110967

Materials Research Society
506 Keystone Drive, Warrendale, PA 15086
http://www.mrs.org

First published 2009
First paperback edition 2012

Single article reprints from this publication are available through University Microfilms Inc., 300 North Zeeb Road, Ann Arbor, MI 48106

CODEN: MRSPDH

A catalogue record for this publication is available from the British Library

ISBN 978-1-605-11096-7 Hardback
ISBN 978-1-107-40845-6 Paperback

CONTENTS

SPENT NUCLEAR FUEL

*Invited Paper

NUCLEAR WASTE GLASSES
AND VITRIFICATION

CERAMIC WASTEFORMS

*Invited Paper

ENGINEERED BARRIER SYSTEMS AND THE NEAR FIELD

x

CEMENT-BASED SYSTEMS

GEOLOGICAL DISPOSAL AND
WASTE TREATMENT

CONTAINER CORROSION

*Invited Paper

MIGRATION AND COLLOIDS

*Invited Paper

PREFACE

Symposium Q, "Scientific Basis for Nuclear Waste Management XXXII," was held December 1–5 at the 2008 MRS Fall Meeting in Boston, Massachusetts. The scientific program featured 72 oral and 34 poster presentations, from 15 countries, covering: national research programs; advanced fuel cycles; behavior of spent nuclear fuel; nuclear waste glasses and vitrification; ceramic wasteforms; engineered barrier systems and the near field; cementitious wasteforms; geological disposal; container corrosion; wasteform performance and natural analogues; migration and colloids. The symposium featured a special focus session on Yucca Mountain, and a joint session on actinide fuels and wasteforms with Symposium O, "Structure/Property Relationships in Fluorite-Derivative Compounds."

The success of this 32nd Scientific Basis for Nuclear Waste Management symposium and the publication of this volume is the result of contributions from many people. We thank the Session Chairs for their assistance in delivering the symposium: Graham Fairhall, Virginia Oversby, Michael Ojovan, Karl Whittle, Nicholas Collier, Paul Bingham, Ricardo Carranza, Xihua He, Fraser King, Melody Carter, Martin Stennett, Edgar Buck, and Joe Small. We are grateful to our Symposium Assistant, Daniel Reid, for his help in ensuring that the symposium ran smoothly. We also wish to thank the many participants who reviewed the manuscripts published in this proceedings volume. We also thank the entire MRS staff for their impeccable support in organizing the symposium and in the production of this volume.

On behalf of the symposium participants, we wish to record our particular thanks to Lars Werme and Virginia Oversby as Chair and Secretary, respectively, of the international advisory committee, for their dedication to the Symposium on the Scientific Basis for Nuclear Waste Management. The new Chair, Neil Hyatt, was elected by the international advisory committee during the 2008 MRS Fall Meeting.

We gratefully acknowledge the generous support of our symposium sponsors:

CEA Marcoule (France)
GE Global Research (USA)
National Nuclear Laboratory (UK)
Southwest Research Institute (USA)

We look forward to meeting again in St. Petersburg, Russia, for the 33rd symposium which begins May 24, 2009; in San Francisco, USA, for the 34th symposium which will be held at the 2010 MRS Spring Meeting which starts April 5, 2010; and in Buenos Aires, Argentina, for the 35th symposium which commences on October 2, 2011.

Neil C. Hyatt
David A. Pickett
Raul B. Rebak

March 2009

MATERIALS RESEARCH SOCIETY SYMPOSIUM PROCEEDINGS

MATERIALS RESEARCH SOCIETY SYMPOSIUM PROCEEDINGS

Prior Materials Research Society Symposium Proceedings available by contacting Materials Research Society

National Programs and
Advanced Fuel Cycle

Mater. Res. Soc. Symp. Proc. Vol. 1124 © 2009 Materials Research Society
1124-Q01-01

Long-Term Performance of the Proposed Yucca Mountain Repository, USA

Peter N. Swift[1], Kathryn Knowles[1], Jerry McNeish[1], Clifford W. Hansen[1], Robert L. Howard[2], Robert MacKinnon[1], S. David Sevougian[1]

[1]Sandia National Laboratories, Department of Energy Office of Civilian Radioactive Waste Management Lead Laboratory for Repository Systems, 1180 North Town Center Drive, Mail Stop LL423, Las Vegas, NV 89144
[2]University of Nevada Las Vegas National Supercomputing Center for Energy and the Environment, 4505 Maryland Parkway, Box 454028, Las Vegas, NV 89154

ABSTRACT

This paper summarizes the historical development of the United States Department of Energy's (DOE's) 2008 performance assessment for the proposed high-level radioactive waste repository at Yucca Mountain, Nevada, and explains how the methods and results meet regulatory requirements specified by the United States Nuclear Regulatory Commission (NRC) and the United States Environmental Protection Agency (EPA). Topics covered include (i) screening of features, events and processes, (ii) development of scenario classes, (iii) descriptions of barrier capability, and (iv) compliance with applicable quantitative standards for individual protection, individual protection following human intrusion, and ground water protection.

INTRODUCTION

The United States Department of Energy Office of Civilian Radioactive Waste Management (DOE-OCRWM) submitted a license application on 3 June 2008 to the United States Nuclear Regulatory Commission (NRC) seeking authorization to construct a repository at Yucca Mountain, Nevada, for the permanent disposal of spent nuclear fuel (SNF) and high-level radioactive waste (HLW) [1]. On 8 September 2008, the NRC accepted the DOE's application for technical review and docketed it to begin the formal regulatory process of review and hearings. The three- to four-year regulatory process is expected to lead to a licensing decision regarding construction authorization no sooner than September 2011.

The total system performance assessment (TSPA) model developed to support the DOE's license application provides quantitative estimates of long-term performance of the repository, including estimates of future radiation doses to a hypothetical "reasonably maximally exposed individual" (RMEI) [2], and is based on more than two decades of scientific investigations. Modeling work performed for this TSPA builds on prior iterations of system-level analyses [3-8] beginning in the early 1990s and continuing through the 2002 Yucca Mountain Site Recommendation [9] and Final Environmental Impact Statement [10].

REGULATORY FRAMEWORK

The management of SNF and HLW is governed in the United States by the provisions of the Nuclear Waste Policy Act of 1982 (NWPA), as amended [11]. As required by the NWPA, the United States Environmental Protection Agency (EPA) has issued public health and environmental radiation protection standards for Yucca Mountain [12], and the NRC has issued final and proposed regulatory requirements [13,14] at 10 Code of Federal Regulations (CFR) Part 63 that establish criteria for the implementation of the EPA standards. Additional guidance relevant to evaluating compliance with the NRC regulations is provided in the *Yucca Mountain Review Plan* [15].

The EPA and NRC regulations define the overall framework for the TSPA as an analysis that identifies and evaluates relevant features, events, and processes (FEPs) that could affect repository performance, and then estimates performance taking into account uncertainties associated with significant FEPs and weighting their consequences by their probabilities of occurrence. In practice, this regulatory direction is implemented as a probabilistic uncertainty analysis using numerical models for components of the repository and the geologic setting, linked into a system-level model suitable for Monte Carlo uncertainty analysis [16-21].

In addition to defining the overall method to be used for the performance assessment, regulations also define the quantitative limits on long-term performance that are to be met for demonstrations of regulatory compliance. Specifically, 10 CFR 63.311 defines an individual protection standard that limits the mean annual dose to the RMEI during 10,000 years after disposal to 0.15 mSv (15 mrem). The rule also provides separate standards for allowable releases of radiation to ground water (10 CFR 63.331), and a limit for mean annual doses to the RMEI following an assumed and stylized human intrusion event (10 CFR 63.321). The final EPA rule for implementing a dose standard after 10,000 years [12] extends the individual protection and human intrusion standards to 1,000,000 years, setting a limit on the mean annual dose to the RMEI of 1 mSv (100 mrem) for the period between 10,000 and 1,000,000 years.

The EPA and NRC regulations define the scope of the long-term analysis by defining criteria for FEPs. Specifically, performance assessments shall not consider features, events, or processes "that are estimated to have less than one chance in 100,000,000 per year of occurring." Impacts of FEPs that have a higher probability of occurrence need not be evaluated if overall performance in the initial 10,000 years "would not be changed significantly" by their occurrence (40 CFR 197.36(a)(1)). Regulatory requirements call for the use of "multiple barriers, consisting of both natural barriers and an engineered barrier system" (10 CFR 63.113(a)). Although no quantitative limits apply to the performance of components of the multiple barrier system, the DOE is required to describe the capability of barriers and to provide the technical basis for that description, consistent with the performance assessments used to demonstrate compliance with the system-level standards.

As required by 10 CFR 63 Subpart G, all scientific and engineering work that directly supports the performance assessment for the license application is performed and documented in accordance with appropriate quality assurance standards [22].

THE TSPA MODEL

The TSPA is a system-level model that integrates submodels for each of the various components of the natural and engineered barriers, simulating the various processes that can affect repository performance, including water movement, material degradation, and radionuclide transport [17-19]. Because of the complexity of the individual processes and the need for a large number of system-level simulations to support the Monte Carlo uncertainty analysis, the TSPA model relies on simplifications (abstractions) of some of the major processes. As described in Ref. 2, section 6, the TSPA model relies both on linked external process models that simulate specific processes as well as a set of abstractions that run within GoldSim software [23].

The Yucca Mountain TSPA identified the scenarios to be analyzed for the license application by following a five-step approach shown in Figure 1. A comprehensive list of potentially relevant FEPs was identified based on input from past Yucca Mountain performance assessments and input from other programs internationally, and the FEPs were evaluated and screened according to criteria specified by the NRC, e.g., at 10 CFR 63.114 [24,25]. Of the 374 FEPs identified for evaluation, 222 were excluded from the quantitative TSPA modeling on the basis of either low probability, low consequence, or incompatibility with regulatory requirements. The remaining 152 FEPs were included in one or more of the performance assessment analyses for evaluating compliance with the regulatory standards.

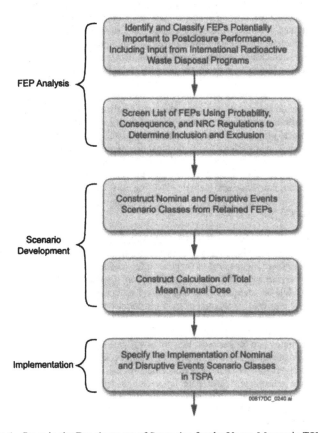

Figure 1. Steps in the Development of Scenarios for the Yucca Mountain TSPA (from Ref. 2, figure 6.1.1-1)

5

The large majority of FEPs included in the performance assessment analyses are features and processes that are relevant to essentially all possible future states of the disposal system. Some included FEPs, however, represent rare events that may or may not occur in the future, and therefore can be used to define scenario classes for analysis. The TSPA analyzes four discrete scenario classes that characterize the range of future states of the system, in addition to a stylized human intrusion scenario prescribed by regulation for evaluation under a separate standard (10 CFR 63.311) [13,15]. The four scenario classes that are treated probabilistically include an early failure scenario class, in which one or more waste packages or overlying drip shields fails prematurely due to undetected manufacturing or emplacement defects; an igneous disruption scenario class in which a volcanic event causes magma to intersect the emplacement region, with or without an accompanying eruption; a seismic disruption scenario class, in which ground motion or fault displacement damages waste packages and drip shields; and a nominal scenario class in which none of these three types of events occurs. Each of the event-based scenario classes is further subdivided into separate modeling cases that simulate consequences of specific events: i.e., early drip shield and waste package failure cases, an igneous intrusion case (without eruption), a volcanic eruption case, a seismic ground motion damage case, and a fault displacement damage case. The total mean annual dose for 10,000 years is developed by summing the mean annual doses for each modeling case, noting that calculated releases from nominal performance in this analysis are zero during the first 10,000 years, and taking into account the probability of occurrence of each type of event. Analyses include both aleatory uncertainty associated with the timing and characteristics of the events and epistemic uncertainty associated with the physical properties of the disposal system [16,20,21]. For analyses of 1,000,000-year performance, when releases from nominal processes (e.g., general corrosion) become significant and damage from seismic ground motion becomes relatively likely to occur simply because of the very long time under consideration, the seismic ground motion and nominal modeling cases are combined into a single modeling case.

BARRIER CAPABILITY

Consistent with the requirements of 10 CFR 63.113(a), the DOE has identified three barriers that are important to waste isolation at Yucca Mountain: an upper natural barrier that limits the amount of water reaching the drift environment; the engineered barrier system that both limits water flow and restricts radionuclide migration from the waste; and the lower natural barrier that significantly delays radionuclides during ground water transport away from the repository. Each of these barriers is composed of multiple components that contribute to the overall capability of the system.

The upper natural barrier includes the surface soils and topography that limit the amount of precipitation that infiltrates into the bedrock, the 200 to 400 meters of unsaturated rock that damp the episodic effects of weather-related infiltration, and the capillary barrier provided by the drift walls that tends to divert flow around the emplacement regions. The overall effect of the upper natural barrier is to reduce water flow (percolation flux) in the mountain to a small fraction (5 to 7 percent) of the mean annual precipitation above the repository, and to further reduce mean seepage (the amount of water that enters a drift) to a small fraction of percolation flux (2-11%) in intact drifts (Ref. 2, Section 8.3.3.1.1[a]).

The engineered barrier system includes: the drift environment; the titanium drip shield, the waste package, including the Alloy 22 outer corrosion barrier, the inner stainless steel

6

container; the waste forms and associated shipping containers; the emplacement pallet; and the invert material that forms the floor of the drift. The drip shield contributes to barrier performance by providing a robust barrier to the flow of water: modeled failures of the drip shield due to corrosion occur between 270,000 and 340,000 years after closure, and the drip shield effectively prevents flowing water from reaching the waste until then. Earlier failures of the drip shield may occur due to emplacement errors, igneous disruption, fault displacement, or extreme ground motion events, but all are anticipated to occur at very low probabilities. Similarly, the Alloy 22 outer corrosion barrier of the waste packages provides a very effective barrier to both water flow and radionuclide transport. Under nominal conditions (without ground motion damage), cracks in the closure welds of waste package, which would allow diffusive transport of radionuclides, are modeled to occur only rarely before approximately 170,000 years after closure. Taking ground motion damage into account, stress corrosion cracks can occur earlier in the repository history, but releases are still anticipated to occur by diffusion only. Larger penetrations of the waste package outer barrier due to general corrosion of the Alloy 22, which would allow advective transport of radionuclides in flowing water, rarely occur in the model before about 560,000 years. By 1,000,000 years, approximately 9 percent of the waste packages have penetrations due to general corrosion, regardless of damage due to seismic ground motion. Seismic damage that could result in advective transport of radionuclides is accounted for in the model, but rarely occurs (Ref. 2, Section 8.3.3.2 [a]).

The lower natural barrier contains 250 to 380 meters of unsaturated rocks (estimated to be up to 120 meters less during wetter future climates) between the repository and the water table, and approximately 18 km laterally along the potential transport pathway ground water from the repository to the location of the RMEI. Processes that will delay radionuclide transport through the lower natural barrier include the slow rate of water flow, diffusion of dissolved-phase radionuclides from fractures into the rock matrix where water moves more slowly, sorption of dissolved species, dispersion and dilution of radionuclides transported in colloidal phases, and reversible filtration of colloidal species. Estimated transport times through the unsaturated portion of the lower natural barrier vary considerably, depending on the degree of sorption and the extent to which transport occurs in fractures or in matrix porosity. For nonsorbing radionuclides (e.g., ^{99}Tc), median transport times through the unsaturated zone range from tens of years (for cases in which transport is largely in fractures) to many thousands of years (for cases in which transport is largely in matrix porosity) (Ref. 26., Figure 6.6.2-8[b]). Transport times for sorbing species are longer. Similar ranges of transport times are estimated for the saturated zone. For nonsorbing radionuclides (e.g., ^{99}Tc), median transport times range from 100 years to 22,000 years, and strongly sorbing species (e.g., ^{226}Ra) show median transport times from 18,000 years to more than 1,000,000 years (Ref. 27, section 6.7.1[a]). Overall, the lower natural barrier effectively delays transport of many radionuclides sufficiently that radioactive decay greatly reduces their concentrations by the time they reach the location of the RMEI.

Figure 2 summarizes overall barrier performance for the seismic ground motion modeling case (which is a major contributor to the total mean annual dose) by displaying the mean inventory through time that has been released from the disposal system for selected radionuclides, together with the total inventory that was initially present when the simulation began. All values are corrected for radioactive decay and ingrowth; thus, the total inventory curve drops through time as radioactive decay reduces the overall radioactivity of the waste.

(a)

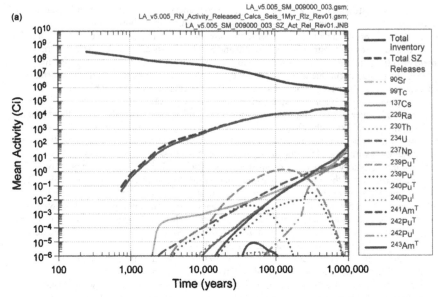

Figure 2. Mean activity released from the saturated zone at the location of the RMEI, seismic ground motion modeling case (from Ref.2, figure 8.3-26[a]a).

The release curves for individual radionuclides show a characteristic pattern of increasing as the probability of having an event that causes releases from the disposal system increases, and then eventually declining as radioactive decay removes the species from the overall inventory of material released. Interpretation of this figure shows that 1,000,000 years after closure the disposal system still contains, on average, approximately 95 percent of the total activity remaining at that time, and the large majority of the activity that has been released is in the form of ^{99}Tc, which transports essentially without retardation. Short-lived species (e.g., ^{90}Sr and ^{137}Cs) do not appear at all on the figure because they are fully contained within the disposal system until after decay has essentially removed them from the inventory. Radionuclides with intermediate half-lives (e.g., most isotopes of Pu and Am) show mean releases beginning after 10,000 years, but releases are small and radioactive decay results in declining mean release totals before 1,000,000 years. Radionuclides for which released activity is flat or still climbing at 1,000,000 years are those that have very long half lives (^{99}Tc, [242]Pu, [237]Np, [234]U), and [226]Ra, for which the release rate is determined by the concentration of its precursor species [234]U.

EVALUATION OF COMPLIANCE WITH QUANTITATIVE LONG-TERM STANDARDS

Figures 3 and 4 show estimated expected annual doses, summed over all scenario classes. Individual curves on these plots represent different samplings of epistemic uncertainty associated with the characteristics of the physical system, and are averaged over the aleatory uncertainty

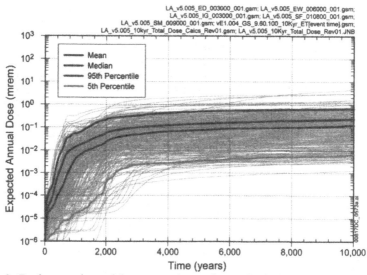

Figure 3. Total expected annual dose, summed over all scenario classes, for 10,000 years. (from Ref. 2, Figure 8.1-1[a])

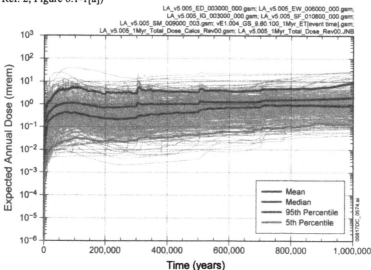

Figure 4. Total expected annual dose, summed over all scenario classes, for 1,000,000 years (from Ref. 2, Figure 8.1-2[a]).

9

associated with the time and characteristics of events. Summary curves show the mean, median, 5th, and 95th percentile values of the expected annual dose. For the 10,000-year individual protection standard (10 CFR 63.311), the mean curve shown on Figure 3 is the appropriate measure for comparison to the regulatory limit of 0.15 mSv/yr (15 mrem/yr). For the 1,000,000-year standard, the mean curve on Figure 4 is the appropriate curve for comparison to the EPA limit of 1 mSv/yr (100 mrem/yr). The maximum mean annual dose during 10,000 years is 0.24 mrem (0.0024 mSv), and the maximum mean annual dose during 1,000,000 years is 2.0 mrem (0.02 mSv); both values are well below the regulatory limits. Figure 5 shows the mean contribution to the total mean annual dose from each of the major modeling cases included in the analysis.

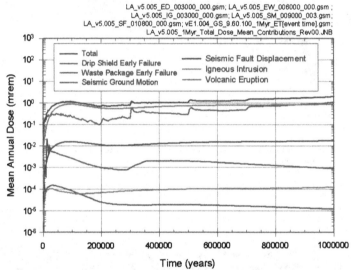

LA_v5.005_ED_003000_000.gsm; LA_v5.005_EW_006000_000.gsm ;
LA_v5.005_IG_003000_000.gsm; LA_v5.005_SM_009000_003.gsm;
LA_v5.005_SF_010800_000.gsm; vE1.004_GS_9.60.100_1Myr_ET[event time].gsm;
LA_v5.005_1Myr_Total_Dose_Mean_Contributions_Rev00.JNB

Figure 5. Mean annual dose contributions from individual modeling cases to the total mean annual dose for 1,000,000 years (from Ref. 2, figure 8.1-3b[a]).

Compliance with the 10,000-year ground water protection standards (10 CFR 63.331) is demonstrated by estimating concentrations of radionuclides in ground water at the location of the RMEI, excluding human intrusion and unlikely natural events (i.e., igneous disruption). Specifically, the combined activity of ^{226}Ra and ^{228}Ra in ground water must be less than 5 picocuries per liter, and the total gross alpha activity, including ^{226}Ra but excluding uranium and radon, must be less than 15 picocuries per liter, with both values including natural background. The maximum mean ground water concentration of combined radium derived from the proposed repository is estimated to be 1.3×10^{-7} pCi/L, far below the natural background level of 0.5 pCi/L, which in turn is well below the regulatory limit. Similarly, the mean maximum gross alpha activity resulting from releases from the proposed repository is estimated to be 6.7×10^{-5} pCi/L, far below the natural background of 0.5 pCi/L, which in turn is well below the regulatory limit. Estimated mean annual doses from beta and photon emitting radionuclides in 2 liters per day of ground water are 0.06 mrem (0.0006 mSv) for the whole body and 0.26 mrem (0.0026 mSv) for the thyroid, well below the regulatory limit of 0.04 mSv/yr (4 mrem/yr) (Ref. 2, Section 8.1.2[a]).

Compliance with the individual protection standard for human intrusion (10 CFR 63.321) is based on the consequences of a single penetration of a waste package by a borehole, creating a path for ground water transport of radionuclides from the repository to the saturated zone. Borehole penetration is assumed to occur at 200,000 years after repository closure, consistent with regulatory requirements to consider intrusion at the earliest time that the waste package would be sufficiently degraded that an intrusion could occur without recognition by the drillers. Estimated maximum median annual dose resulting from a single borehole penetration is 0.011 mrem (0.00011 mSv), well below the EPA limit of 1 mSv (100 mrem/yr) for the period after 10,000 years (Ref. 2, Section 8.1.3.2[a]).

CONCLUSIONS

The 2008 TSPA for the proposed Yucca Mountain repository provides one of the technical bases for the DOE's license application to the NRC for authorization to construct the repository. The TSPA is based on detailed site characterization information and process modeling of the key components, and on a systematic analysis of the potentially relevant features, events, and processes that that may affect long-term performance of the disposal system. Modeling results indicate that the largest contributions to the estimated maximum mean annual dose come from the igneous intrusion and seismic ground motion scenario classes, taking into account the probability of occurrence of these events. Nominal corrosion processes that lead to degradation and failure of the waste packages become the primary release mechanism late in the million-year period. All estimated performance measures are well below regulatory limits, both for the final NRC rule applicable for 10,000 years and for the final EPA rule applicable for 1,000,000 years.

ACKNOWLEDGMENTS

The authors thank the many hundreds of people whose work over more than two decades has contributed to this analysis. This manuscript has been adapted from a paper originally prepared for the 2008 International High-Level Radioactive Waste Conference in Las Vegas, Nevada [28]. It has been authored by Sandia National Laboratories under Contract DE-AC04-94AL85000 with the U.S. Department of Energy. The United States Government retains and the publisher, by accepting the article for publication, acknowledges that the United States Government retains, a non-exclusive, paid-up, irrevocable, world-wide license to publish or reproduce the published form of this manuscript, or allow others to do so, for United States Government purposes. The statements expressed in this article are those of the author and do not necessarily reflect the views or policies of the United States Department of Energy, Sandia National Laboratories, or the University of Nevada Las Vegas.

ACRONYMS

CFR	United States Code of Federal Regulations
DOE	United States Department of Energy

EPA	United States Environmental Protection Agency
FEP	feature, event, and/or process
HLW	high-level radioactive waste
mrem/yr	millirem per year
mSv/yr	millisievert per year
NRC	United States Nuclear Regulatory Commission
NWPA	United States Nuclear Waste Policy Act
OCRWM	Office of Civilian Radioactive Waste Management
pCi/L	picocurie per liter
RMEI	reasonably maximally exposed individual
SNF	spent nuclear fuel
TSPA	total system performance assessment

REFERENCES

1. US DOE (United States Department of Energy), *Yucca Mountain Repository License Application*, DOE/RW-0573, Rev 0 (2008).
2. SNL (Sandia National Laboratories), *Total System Performance Assessment Model/Analysis for the License Application*, MDL-WIS-PA-000005 Rev 00, AD 01, U.S. Department of Energy Office of Civilian Radioactive Waste Management, Las Vegas, Nevada (2008).
3. R.W. Barnard, M.L. Wilson, H.A. Dockery, J.H. Gauthier, P.G. Kaplan, R.R. Eaton, F.W. Bingham, and T.H. Robey, *TSPA 1991: An Initial Total-System Performance Assessment for Yucca Mountain.* SAND91-2795, Sandia National Laboratories, Albuquerque, New Mexico (1992).
4. CRWMS M&O (Civilian Radioactive Waste Management Systems Management and Operating Contractor), *Total System Performance Assessment - 1993: An Evaluation of the Potential Yucca Mountain Repository.* B00000000-01717-2200-00099 REV 01, CRWMS M&O, Las Vegas, Nevada (1994).
5. M.L. Wilson, J.H. Gauthier, R.W. Barnard, G.E. Barr, H.A. Dockery, E.Dunn, R.R. Eaton, D.C. Guerin, N. Lu, M.J. Martinez, R. Nilson, C.A. Rautman, T.H. Robey, B. Ross, E.E. Ryder, A.R. Schenker, S.A. Shannon, L.H. Skinner, W.G. Halsey, J.D. Gansemer, L.C. Lewis, A.D. Lamont, I.R. Triay, A. Meijer, and D.E. Morris, *Total-System Performance Assessment for Yucca Mountain – SNL Second Iteration (TSPA-1993).* SAND93-2675. Executive Summary and Two Volumes, Sandia National Laboratories, Albuquerque, New Mexico (1994).
6. CRWMS M&O (Civilian Radioactive Waste Management Systems Management and Operating Contractor), *Total System Performance Assessment - 1995: An Evaluation of the Potential Yucca Mountain Repository.* B00000000-01717-2200-00136 REV 01. Las Vegas, Nevada: CRWMS M&O (1995).
7. DOE (U.S. Department of Energy), *Viability Assessment of a Repository at Yucca Mountain: Total System Performance Assessment,* DOE/RW-0508 Volume 3, U.S. Department of Energy, Office of Civilian Radioactive Waste Management, Washington, D.C. (1998).
8. CRWMS M&O (Civilian Radioactive Waste Management Systems Management and Operating Contractor), *Total System Performance Assessment for the Site Recommendation.* TDR-WIS-PA-000001 REV 00 ICN 01. Las Vegas, Nevada: CRWMS M&O (2000).

9. DOE (U.S. Department of Energy), *Yucca Mountain Site Suitability Evaluation.* DOE/RW-0549 (2002).

10. DOE (U.S. Department of Energy), *Final Environmental Impact Statement for a Geologic Repository for the Disposal of Spent Nuclear Fuel and High-Level Radioactive Waste at Yucca Mountain, Nye County, Nevada.* DOE/EIS-0250 (2002).

11. 42 U.S.C. 10101 and following, The *Nuclear Waste Policy Act of 1982, as amended.*

12. EPA (United States Environmental Protection Agency), 40 Code of Federal Regulations Part 197, *Public Health and Environmental Radiation Standards for Yucca Mountain, NV* (2007), as amended in *Federal Register* vol. 73, p. 61256 (2008).

13. NRC (US Nuclear Regulatory Commission), *10 Code of Federal Regulations Part 63: Disposal of High-Level Radioactive Wastes in a Geologic Repository at Yucca Mountain, Nevada* (2008).

14. NRC (US Nuclear Regulatory Commission), "Implementation of a Dose Standard After 10,000 Years" (proposed rule), *Federal Register* V. 70, p. 53313 (2005).

15. NRC (US Nuclear Regulatory Commission), *Yucca Mountain Review Plan: Final Report,* NUREG-1804, Revision 2 (2003).

16. J.C. Helton, C.W. Hansen, and C.J. Sallaberry, "Yucca Mountain 2008 Performance Assessment: Conceptual Structure and Computational Organization," *Proceedings of the 2008 International High-Level Radioactive Waste Management Conference,* September 7-11, 2008, American Nuclear Society (2008).

17. R.J. Mackinnon, A. Behie, V. Chipman, Y. Chen, J. Lee, K. P. Lee, P. Mattie, S. Mehta, K. Mon, J. Schreiber, S.D. Sevougian, C. Stockman, and E. Zwahlen, "Yucca Mountain 2008 Performance Assessment: Modeling the Engineered Barrier System," *Proceedings of the 2008 International High-Level Radioactive Waste Management Conference,* September 7-11, 2008, American Nuclear Society (2008).

18. P.D. Mattie, T. Hadgu, B. Lester, A. Smith, M. Wasiolek, and E. Zwahlen, "Yucca Mountain 2008 Performance Assessment: Modeling the Natural System," *Proceedings of the 2008 International High-Level Radioactive Waste Management Conference,* September 7-11, 2008, American Nuclear Society (2008).

19. S.D. Sevougian, A. Behie, B. Bullard, V. Chipman, M. Gross, and W. Statham, "Yucca Mountain 2008 Performance Assessment: Modeling Disruptive Events and Early Failures," *Proceedings of the 2008 International High-Level Radioactive Waste Management Conference,* September 7-11, 2008, American Nuclear Society (2008).

20. C.J. Sallaberry, A. Aragon, A. Bier, Y. Chen, J. Groves, C.W. Hansen, J.C. Helton, S. Mehta, S. Miller, J. Min, and P. Vo, "Yucca Mountain 2008 Performance Assessment: Uncertainty and Sensitivity Analyses for Physical Processes," *Proceedings of the 2008 International High-Level Radioactive Waste Management Conference,* September 7-11, 2008, American Nuclear Society (2008).

21. C.W. Hansen, K. Brooks, J. Groves, J.C. Helton, K.P. Lee, C.J. Sallaberry, W. Statham, and C. Thom, "Yucca Mountain 2008 Performance Assessment: Uncertainty and Sensitivity Analysis for Expected Dose," *Proceedings of the 2008 International High-Level Radioactive Waste Management Conference,* September 7-11, 2008, American Nuclear Society (2008).

22. DOE (US Department of Energy), *Quality Assurance Requirements and Description.* DOE/RW-0333P (2008).

23. GoldSim Technology Group, *User's Guide, GoldSim Probabilistic Simulation Environment.* Version 9.60. Two volumes, GoldSim Technology Group, Issaquah, Washington (2007).

24. SNL (Sandia National Laboratories), *Features, Events, and Processes for the Total System Performance Assessment: Methods,* ANL-WIS-MD-000026 REV 00, U.S. Department of Energy Office of Civilian Radioactive Waste Management, Las Vegas, Nevada (2008).

25. SNL (Sandia National Laboratories), *Features, Events, and Processes for the Total System Performance Assessment: Analyses,* ANL-WIS-MD-000027 REV 00, U.S. Department of Energy Office of Civilian Radioactive Waste Management, Las Vegas, Nevada (2008).

26. SNL (Sandia National Laboratories), *Particle Tracking Model and Abstraction of Transport Processes,* MDL-NBS-HS-000020 REV 02 AD 02 (2008).

27. SNL (Sandia National Laboratories), *Saturated Zone Flow and Transport Model Abstraction,* MDL-NBS-HS-000021 REV 03 AD 02, U.S. Department of Energy Office of Civilian Radioactive Waste Management, Las Vegas, Nevada (2008).

28. P.N. Swift, K. Knowles, J. McNeish, C.W. Hansen, R.L. Howard, R. MacKinnon, and S.D. Sevougian, "Yucca Mountain 2008 Performance Assessment: Summary," SAND2008-3075C, *Proceedings of the 2008 International High-Level Radioactive Waste Management Conference,* September 7-11, 2008, American Nuclear Society (2008).

Mater. Res. Soc. Symp. Proc. Vol. 1124 © 2009 Materials Research Society   1124-Q08-01

Corrosion Issues Related to Disposal of High-Level Nuclear Waste in the Yucca Mountain Repository

David J. Duquette, Ronald M. Latanision, Carlos A. W. Di Bella, and Bruce E. Kirstein
United States Nuclear Waste Technical Review Board
2300 Clarendon Blvd., Suite 1300
Arlington, Virginia 22201, U.S.A.

ABSTRACT

The current policy of the United States is to dispose of spent nuclear fuel and high-level nuclear waste underground in geologic repositories. The U.S. Department of Energy (DOE) has been developing plans for a repository to be located at Yucca Mountain, in Nevada. In June 2008, DOE submitted an application to the U.S. Nuclear Regulatory Commission (NRC) for a construction license for that repository. NRC accepted the application for docketing in September 2008.

This paper discusses DOE's bases for and approach to modeling the localized and general corrosion aspects of the Alloy 22 outer shell of the container that DOE plans to use for encapsulating the waste in the repository. The modeling is necessary to predict the corrosion behavior for the container's extraordinarily long "service period" — more than a million years.

INTRODUCTION

The U. S. Nuclear Waste Technical Review Board (Board) was created by the Nuclear Waste Policy Amendments Act of 1987 as an independent federal agency within the Executive Branch. The duties of the Board are to evaluate the technical and scientific validity of activities undertaken by DOE within scope of the Nuclear Waste Policy Act (NWPA) of 1982, as amended, and advise the Secretary of Energy and Congress of the Board's findings, conclusions, and recommendations. The Board is composed of 11 members, who are nominated by the National Academy of Sciences and appointed by the President to 4-year terms. Board members, all of whom serve part-time, are chosen from a broad range of scientific and engineering disciplines, including geology, hydrogeology, geochemistry, risk analysis, transportation, chemical and nuclear engineering, and other relevant disciplines. The Board membership always has included 1 or 2 corrosion experts. The Board is supported by a small permanent staff. Two of the authors (DJD and RML) are Board members; the two other authors are part of the Board's staff.

The docketing of DOE's license application in September 2008 started an intensive review by NRC employees and contractors of supporting materials submitted with the license application. That review is part of NRC's process for determining from a radiological point of view whether the repository DOE proposes for Yucca Mountain can be constructed and operated safely and will protect the public and the environment before and after the repository is closed. The Board's ongoing peer review of the technical and scientific validity of DOE activities differs significantly from NRC's review. For example, many DOE activities within the scope of the NWPA, e.g., transportation, are not part of the license application. In addition, although safety is paramount, other attributes of an activity, such as cost, schedule, capacity, process integration, non-radiological safety, and resource availability can be and are among the important considerations of the Board's evaluation. Finally, as part of its review of technical and scientific validity, the Board evaluates the depth of technical understanding of key processes. In contrast, the level of technical understanding of a process required by safety regulations is confined to what is necessary to show either that the process is not important for safety or that assumptions about the physical behavior of the process are conservative and indicate that the effect of the process on safety is bracketed.

All waste to be disposed of in the proposed repository will be contained in double-shell, welded-metal cylinders — the "waste package." The waste package is described in DOE's license application materials;[1] weights and dimensions of waste packages may be found in a DOE information exchange document.[2] The outside diameters of the outer shells range from approximately 1.7m to approximately 2.0m and the lengths from approximately 3.5m to approximately 5.7m, depending on the form of the waste. The current design includes an outer shell of 25.4mm-thick Alloy 22. The current design for the inner shell is 50.8mm thick 316 stainless steel. The mass of a fully loaded waste package ranges from ~40,000 to ~74,000 kg. The Alloy-22 outer shell is the topic of this paper because its corrosion resistance accounts for most of the performance of the waste package in preventing or limiting releases. According to DOE's analysis, for the nominal scenario, i.e., for the scenario with igneous, seismic, and early-failure effects omitted, the outer shells of fewer than 10% for the waste packages would be penetrated by general corrosion after a million years, and less than 1% of the surface area of the penetrated outer shells would be gone.[3] Thus, the waste package is an important barrier for isolating waste from the human environment. In its license application, DOE takes the inner shell into account for structural support of the outer shell so long as the outer shell is intact but assumes that the inner shell has essentially no corrosion resistance. While such an assumption, which is very conservative from a safety point of view, may be acceptable in a license application, it masks the true behavior of the waste package. Liquid water, which is necessary for corrosion of the stainless steel to occur, cannot contact the inner shell until the Alloy-22 outer shell is penetrated. In almost all cases, such penetration should not occur until temperatures in the waste package are low, i.e., <40°C. At such temperatures, the inner shell is likely to be a formidable corrosion barrier.

Projected service conditions in the repository

Temperature: Ventilation (natural and forced) will remove essentially all of the decay heat of emplaced waste for the 50-100[+]-year period while the repository is open. When the repository is closed and sealed, decay heat will dissipate into the rock surrounding the tunnels rather than into the ventilation air, and the waste package and rock temperatures will achieve peak values approximately 60-80 years after closure. Then temperatures will begin to decrease. In the current design, DOE's calculations indicate that waste package surface temperatures will fall within the blue band shown in Figure 1. The roughly thousand-year period when temperatures are above 96°C is known as the "thermal pulse." The specific temperature of an individual waste package will depend on the thermal power of the waste in the package and the package's location in the repository. Figure 1 also incorporates uncertainties in thermal parameters, such as thermal conductivity of the rock, and is based on the assumption that the geology of the repository is stable for a million years. The calculations can be extrapolated beyond a million years as long as the geology remains stable.

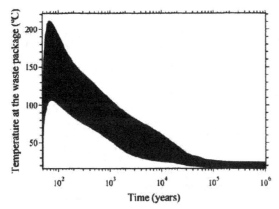

Figure 1. Approximate waste package surface temperature vs. time since repository closure.[4]

Pressure: Since the repository rock is permeable and the repository will be located well above the water table, the pressure of the gas phase in the tunnels will be essentially atmospheric during waste emplacement and after repository closure. The repository will be located approximately 300m below the surface at an elevation of approximately 1100m above sea level. Nominal atmospheric pressure at an elevation of 1100m is 89 kilopascal. The boiling point of pure water at this pressure is ~96°C.

Chemistry: Atmospheric (dry) corrosion of Alloy 22 is not a significant concern; however, aqueous corrosion of Alloy 22 will occur, albeit slowly, whenever and wherever liquid water resides on the surface of the waste package. The chemistry of the water will be influenced by soluble solids present on waste package surfaces, e.g., salts and soluble organic compounds in dust deposited on the waste package during the long ventilation period, and by dissolved solids in water that drips onto waste packages. Based on samples of dusts in the air outside the portal of the underground Exploratory Studies Facility at Yucca Mountain, the dusts that deposit on waste packages during the ventilation period are likely to contain the following: calcium, sodium, magnesium, potassium, silicon, ammonium, chloride, sulfate, bicarbonate, nitrate, fluoride, formate, acetate, and propionate, among others.[5] Since the concentration of oxygen in the gas phase will be approximately 20 volume % at all times except during a portion of the thermal pulse, the environment will be mildly oxidizing. Anions present in the water tend to make it neutral-to-basic, but local situations could exist where, due to concentration, hydrolysis, radiolysis, or other processes, low-pH environments exist temporarily.

Humidity: As the tunnel rock heats up after closure, water remaining in the fractures and pores of the rock will vaporize, which will drive air away from the tunnels, resulting in a nearly pure steam atmosphere. After the rock reaches its peak temperature and cools to 96°C, water

17

vapor will condense, drawing air back into the repository tunnels. The relative humidity in the tunnels will return to its natural state of greater than 90%.

Liquid Water: Corrosion is not a concern during the period before repository closure because the dry desert air used for ventilating the repository will preclude formation of liquid water on the waste package surfaces. After the thermal pulse, the saturation of the pores in the rock will return to ~90%, and the relative humidity will be greater than 90%. Under such high-relative humidity conditions, water films at least a few molecules thick are likely to be present on waste package surfaces, and general corrosion will occur. The film will exist at lower relative humidities, or the film will be thicker at the same relative humidity, if there are soluble salts on waste package surfaces, because the salts will dissolve in the film and depress the equilibrium vapor pressure of the resultant solution. Waste packages surfaces will be coated with a 50-100[+]-year accumulation of dust, which will contain some soluble salts. Due to deliquescence, liquid water in the form of concentrated brines also can exist on waste package surfaces at temperatures well above the boiling point of pure water if certain salts are present. For example, mixtures of sodium and potassium chlorides and nitrates can deliquesce at temperatures above 200°C at atmospheric pressure if the vapor phase over the salts is essentially pure steam. Whether other mixed salt combinations that might exist on waste package surfaces could form brines at high temperatures via deliquescence is still under debate. The possible formation of brines, the composition of the brines, and the presence or absence of solid dust particles or solids resulting from the precipitation of insoluble species, are relevant to the corrosion performance of the Alloy-22 outer barrier of the containers.

EXPERIMENTS AND MODELS

Reference 6 describes the Alloy-22 general and localized corrosion models used in DOE's license application and the analyses and experimental data underlying the models.[6] This paper deals with some of the issues the Board sees in some of the models, with particular emphasis on recent experimental work that became available too late to be incorporated into Reference 1. The recent work includes work sponsored by the State of Nevada at the Institute for Metal Research (IMR) of the Chinese Academy of Sciences [7-9], work sponsored by NRC at the Center for Nuclear Waste Regulatory Analyses (CNWRA)[10], and work sponsored by the DOE at Sandia National Laboratories (SNL).[11,12] We also touch briefly on the issue of "oxide wedging."

General Corrosion

Base general corrosion rate: Personnel at Lawrence Livermore National Laboratory (LLNL) produced much of the general corrosion data that DOE relied on to develop a model of the long-term general corrosion behavior of Alloy 22 in repository-relevant environments. LLNL personnel designed, built, and then operated for approximately 10 years baths approximately half-full of solutions[1] in which samples of Alloy 22 and other materials that were

---

[1] Four different solution compositions were used. All were based on the composition of the water in the aquifer below the proposed repository location, a sodium carbonate water. Each composition was adjusted to reflect concentration processes and possible acidification or alkalization. The compositions of the solutions in the baths are listed in Reference 6.

reference materials or alternate candidates for use in a repository at Yucca Mountain were exposed. Some samples were completely immersed; some were at the water line; and some were in the vapor space above the solutions. The corrosion rates for the immersed samples were higher than other samples. Therefore data from only the immersed samples were used to develop the model. Approximately half of the baths were maintained at 60°C; the other half were maintained at 90°C. Corrosion rates were determined by weight loss.

Two geometric sample configurations were used in the baths. One set (samples with crevice formers attached) showed greater average weight loss than the other set (boldly exposed samples). DOE conservatively decided to use only the set showing the greater average weight loss as the basis for its model of the Alloy-22 general corrosion rate. There were no discernible differences between average weight losses at 60°C and at 90°C. The samples used for the model had been immersed in the baths for approximately 5 years. The mean general corrosion rate determined from the data was approximately 8nm/yr. Assuming a general corrosion rate of 8nm/yr and no other forms of Alloy-22 degradation, the outer layer of the waste package would last an astounding 3 million years.

Weight loss data from samples immersed for 9½ years in the baths were not available when the model was frozen. Evaluation of the 9½-year samples just began earlier this year. The 5-year data, data from samples with less time in the baths, and shorter-term electrochemical data indicate that the Alloy-22 general corrosion rate decreases with time, as would be expected. It is anticipated that the 9½-year data will show that trend continuing, although eventually the corrosion rate would reach an equilibrium value corresponding to the limiting passive current density. In its modeling of general corrosion, DOE conservatively used a fixed general corrosion rate based on the corrosion rate averaged over the entire 5 year period rather than a corrosion rate that decreases with time or even the instantaneous corrosion rate at the end of the 5-year period. This is very conservative; in our opinion DOE's model for general corrosion significantly overpredicts recession rates at below-boiling temperatures.

Unfortunately, the 60°C and 90°C temperatures and the environments in the baths were too similar to discern variations in corrosion rate with temperature or environment. DOE's model assumes the base general corrosion rate for Alloy 22 applies for 60°C and any environment to be encountered in Yucca Mountain. To adjust the general corrosion rate for temperature DOE used short term, electrochemical (polarization resistance) experiments at different temperatures to determine the variation of general corrosion rate with temperature.

DOE performed polarization resistance measurements at 60, 80, and 100°C in solutions made by dissolving various amounts of sodium chloride and potassium nitrate to obtain solutions ranging between 1 and 6 molal chloride and 0.05 and 3.0 molal nitrate. A total of 360 polarization resistance measurements were made, and corrosion rates were calculated from the measurements.

The corrosion rates then were fit to a classical Arrhenius relationship,

$$\ln(R_T) = \ln(A) - E_a/RT$$

where R_T is the generalized corrosion rate, nm/yr,

                 A is the preexponential factor, nm/yr,

                 E_a is the apparent activation energy, J/mol,

                 R is the gas constant, 8.314 J/mol K, and

                 T is temperature, K.

The resultant fit yielded an apparent activation energy of 40.78 kJ/mol with a standard deviation of 11.75 kJ/mol.

Recent CNWRA data. CNWRA measured corrosion rates of Alloy 22 in highly concentrated solutions of $NaCl-NaNO_3-KNO_3$ at atmospheric pressure and temperatures ranging from 130 to 220°C.[10] The test durations ranged from 80 to 140 days. The results appear consistent with the temperature-dependent Alloy-22 general corrosion rate model developed by DOE. However, corrosion rates appear to vary with composition of the solutions, i.e., the samples immersed in low pH (pH = ~3) solution had higher corrosion rates than samples immersed in higher pH solutions (pH=~7) at the same temperature.[2]

Earlier, similar tests run at CNWRA in the same starting solution over the same temperature range gave different results.[13] Corrosion rates were significantly higher and showed no correlation with temperature. The major differences between the recent tests and the earlier tests were; (1) The earlier tests had shorter durations (30 to 80 days versus 80 to 140 days); and (2) Water captured in the vent bubbler (see Figure 2a) was recycled in the recent tests; in the earlier tests, deionized water was used to make up losses.

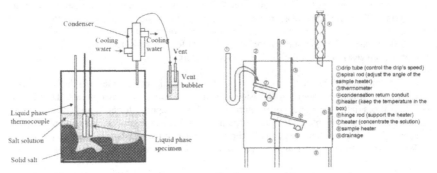

Figure 2a. CNWRA Apparatus   **Figure 2b.** IMR Apparatus
          Simplified Schematic   Simplified Schematic

Recent IMR Data. Personnel at IMR, using equipment with many apparent similarities to the equipment used in the CNWRA studies discussed above, studied corrosion of Alloy 22 at atmospheric pressure at ~160°C.[7,8] Instead of a concentrated solution of $NaCl-NaNO_3-KNO_3$ used by CNWRA, the IMR work used "simulated unsaturated zone pore water" of various

---

[2] The pHs reported above are for samples of solution that were diluted by a factor of 10 to prevent crystallization. Therefore the pHs of the solutions themselves are likely to be lower than reported above.

concentrations. Both CNWRA and IMR reported the amounts of salts used to make up their test solutions, but did not report the composition of the solutions. The IMR also used a test solution made up of magnesium chloride hexahydrate to which a small amount of magnesium nitrate hexahydrate had been added ($[Cl]/[NO_3]\sim50$). Instead of immersing the sample, IMR researchers dripped solutions onto an inclined plate maintained at $100°C$, from which the heated solution dripped onto an Alloy-22 sample heated to $160°C$. The duration of the IMR tests ranged from 10 to 30 days. Figure 2b shows a schematic of the IMR apparatus. The pHs of the solutions were measured before the solutions were introduced into the apparatus and as the solutions were draining from the samples. The pHs of the virgin solutions were essentially neutral (from ~6 to ~ 7.3); the pHs of the solutions draining from the samples were acidic (pHs from ~1 to ~3.3). No quantitative corrosion results were reported. However, photographs and photomicrographs of the test samples show severe corrosion, which the IMR investigators believe is crevice corrosion.

Discussion. The Board has little concern about general corrosion of Alloy 22 at temperatures below boiling because DOE's model appears very conservative. Weighing of the samples (at SNL) that resided in the LLNL baths for 9½ years has been underway for several months. No results have been reported yet. If corrosion rates determined by weighing the 9½-year samples do not show a continuing trend of a decrease in mean general corrosion rate (averaged over time) with residence time in the bath, we may be concerned with the validity of the data. We recognize that the general corrosion rate will eventually reach an equilibrium value. However, this point did not appear to have been reached after 5 years in the baths.

The new CNWRA and IMR data give reason for concern about general corrosion of Alloy 22 at temperatures above boiling. The CNWRA results indicate that the general corrosion rate of Alloy 22 may vary with composition of the environment, i.e., at lower pHs the general corrosion rate of Alloy 22 may be higher. This is not an unexpected finding, but it appears to contradict DOE's model, which predicts no variation in the general corrosion rate of Alloy 22, regardless of composition of the solution in contact with the metal (within the range of compositions that DOE expects to be present at Yucca Mountain). The IMR experiments indicate the possibility of severe corrosion for a scenario of no drip shield, large temperature differences between the tunnel walls and the waste package surface, and with acid gasses condensing within the brine on waste package surfaces. We do not know what the possibility of that scenario is. The likelihood of the scenario should be explored by DOE. Performing the experiments on titanium Grade 7, the material proposed for the drip shield, also would be worthwhile. (The DOE model for titanium predicts a higher general corrosion rate for titanium Grade 7 than for Alloy 22.) Both the new CNWRA data and the new IMR data raise the possibility of problematic general corrosion rates at above boiling temperatures.

The Arrhenius form of the DOE model for the general corrosion rate of Alloy 22 is appropriate, particularly since the temperature range of interest is so narrow, 25-200°C. The value for the apparent activation energy is reasonable. It is likely that the corrosion rates obtained from the polarization resistance measurements are greater than they will be at steady state, but since the purpose of the experiments was to determine the variation of corrosion rate with temperature and not to determine the rate itself, this does not seem to be a significant limitation. Limiting the solutions tested to chloride- and nitrate-containing solutions, not varying the pH of the solutions, and limiting the temperature range of the polarization resistance tests to 60-100°C do not seem appropriate. However, even assuming the most severe conditions that might be encountered in the repository, it appears — barring the evidence from the IMR tests — unlikely that the containers will fail due to general corrosion of the Alloy 22 for hundreds of millennia.

Localized Corrosion

DOE treats Alloy-22 localized corrosion differently depending on the environment in which the localized corrosion may occur. If the environment is due to liquid water dripping onto the waste package (seepage), then localized corrosion is modeled as described below. If the environment is due to the deliquescence of salts deposited on the waste package (e.g., during the ventilation phase) localized corrosion is screened out as inconsequential.

Localized Corrosion due to Seepage. DOE's model for localized corrosion due to seepage is based in part on the assumption that seepage onto the waste package cannot occur when the surface of the waste package is above 120°C. For waste package surface temperatures at and below 120°C, DOE assumes that localized corrosion will initiate on the portion of the waste package surface contacted by seepage if the composition, temperature, and pH of the seepage water are such that the corrosion potential, E_{corr}, of Alloy 22 immersed in a surrogate for the seepage water is equal to or greater than the critical potential, E_{crit}. The critical potential is also a function of the composition, temperature, and pH of the seepage water contacting the waste package. The Board concurs that this general approach is a reasonable *conceptual* model for initiation of localized corrosion. Initiation is particularly important because DOE assumes that seepage-based localized corrosion, once initiated, will penetrate rapidly and will not stifle.

The data for the determination of E_{corr} and E_{crit} were obtained in separate experiments. To determine E_{corr}, DOE measured open circuit potentials versus time for various Alloy-22 samples in model aqueous solutions over a temperature range of 25-90°C for periods of up to 3 years and then used multiple linear regression on the data to fit a model for E_{corr} as a function of nitrate and chloride concentrations, temperature, and pH, i.e., $E_{corr} = f_1([NO_3], [Cl], T, pH)$.

To determine E_{crit}, DOE used the ASTM G61-86 cyclic potentiodynamic polarization (CPP) technique to determine the repassivation potentials for a number of creviced Alloy-22 samples in a variety of aqueous solutions. E_{crit} was taken as the potential value where the reverse CPP scan intersected the forward CPP scan. The resultant data were fit to express critical potential as a function of nitrate concentration, chloride concentration, and temperature, i.e., $E_{crit} = f_2([NO_3], [Cl], T)$.

Discussion. DOE assumes that localized corrosion of Alloy 22 will initiate if E_{corr}, as represented by function f_1, is greater than E_{crit}, as represented by function f_2. The data on which function f_1 is based are long-term data — up to 3 years in many cases but at least 8 months in all cases. On the other hand, the data on which function f_2 is based are short-term data — a matter of approximately a day for each CPP curve. The question then must be asked if mixing long-term data and short-term data to form the basis for predicting localized corrosion initiation is appropriate. If so, is it likely to overpredict the occurrence of localized corrosion or underpredict it? In addition, despite the long term nature of the corrosion potential experiments, it is not clear that all the experiments had reached a stable value by the end of the tests, and some of the test results were "noisy" at the end of the tests, also making their stability questionable.

A problem with interpreting some of the data used to develop the localized corrosion model is that DOE's long-term corrosion tests in the baths at LLNL, many of which were on creviced samples, apparently do not corroborate the model DOE has developed by fitting E_{corr} and E_{crit} equations to the electrochemical corrosion potential and CPP data. For example, according to the localized corrosion model, a few of the tests used to develop the generalized corrosion model discussed above should have shown localized corrosion. However, none of them did. Similarly, other long-term corrosion tests that the model indicates should have developed localized corrosion did not. With a single exception, the only tests where the model predicted localized corrosion and localized corrosion did occur were in highly concentrated solutions of calcium chloride with minor amounts (or none) of calcium nitrate. Such solutions are not very likely to exist in the repository. In summary, the model used to predict localized corrosion under seepage conditions appears to be overly conservative in that it predicts that localized corrosion will occur in many instances where experimental data indicate that it will not occur — at least not in the time frame of the experiments.

Propagation. If seepage-based localized corrosion initiates, DOE's model assumes that it will propagate at a constant rate, chosen from a logarithmic linear distribution ranging from 12.7 µm/yr to 1270 µm/yr. Apparently, these values were selected from Alloy-22 uniform corrosion rates in highly aggressive solutions extracted from the product literature of one of the manufacturers of Alloy 22. In our opinion, such data only should be used to compare the relative corrosion resistance of different alloys, and localized corrosion rates, in general, are much higher than uniform corrosion rates. However, using the highest published corrosion rates results in penetration of the waste packages, once localized corrosion has been initiated, in a matter of less than 2,000 years, which, for a million-year (or more) period of concern, is essentially instantaneously.

It is not clear from currently available documentation how DOE models what happens after the waste package is penetrated by localized corrosion, i.e., what the area, morphology, and geometry of the penetration(s) are. One approach could be to assume that the entire area contacted by seepage disappears when penetration occurs. This would be an extremely conservative view and inconsistent with the very nature of localized corrosion. An even more extreme assumption would be to assume that the entire waste package disappears at the time of penetration. In either case, the corrosion resistance of the container alloys essentially becomes irrelevant, and the containment of radionuclides becomes the responsibility of the natural barriers.

Localized corrosion due to deliquescence. That brines can form on waste package surfaces at temperatures up to ~200°C due to deliquescence is well established. Therefore, the possibility of localized corrosion (and other forms of aqueous corrosion) during the thermal pulse period is a concern. Localized (crevice) corrosion has been observed in autoclave (pressurized) experiments performed on Alloy 22 in aqueous solutions of 2.5m and 6.4m Cl⁻ with [NO₃⁻]/[Cl⁻] ratios of 0.5 or 7.4 at temperatures of 160°C and 220°C.[14] Localized corrosion was observed in all cases, but was not anticipated in the solutions with nitrate to chloride ratios of 7.4.[3] The test solutions were made by dissolving sodium chloride, sodium nitrate, and potassium nitrate in water. Brines containing sodium, potassium, chloride, and nitrate cannot form at atmospheric pressure in the higher end of the temperature range of interest unless the nitrate to chloride ratio of the brine is well above 1. However, if the nitrate concentrations are lower than anticipated, or if the nitrate is removed by chemical, physical, or biological processes (e.g., by reacting with organic materials in the dust deposited during ventilation), whether the stable brine would decrease in amount, form a metastable brine or solidify (thus being rendered innocuous) has not been determined. In addition, whether other brines could form at high temperatures from other mixtures of salts that may exist at Yucca Mountain has not been explored systematically. In other words, in our view the possibility of other high-temperature corrosive environments has not been conclusively ruled out.

The Board sponsored a public workshop on localized corrosion in September 2006 to discuss deliquescence-based corrosion of Alloy 22 in repository-relevant environments.[15] At that workshop, DOE presented a strong case suggesting that the nitrate to chloride ratios in the repository would be sufficiently high that localized corrosion could not occur. DOE representatives also suggested that, if those assumptions were incorrect, the propagation of localized corrosion would effectively be stifled, because (a) migration rates for nitrate into occluded regions are higher than chloride migration rates, and repassivation would occur within the occluded regions; and/or (b) the amount of water and/or aggressive species are so low that the occluded regions would effectively be "starved" as the damage propagated and localized corrosion would essentially halt. The Board made two recommendations to DOE as a result of the workshop: (1) determine the level of nitrate needed to inhibit localized corrosion over the *entire* temperature range (i.e., up to ~200°C), and (2) determine the relative migration rates for the migration of nitrate and chloride ions into crevices. At this time it appears that that the first of these recommendations is in DOE's corrosion test plan for fiscal year 2009.[12]

At or below a waste package surface temperature of 120°C, i.e., for waste package surface temperatures for which DOE believes seepage-based localized corrosion is possible, thin films of brines may form due to deliquescence of salt mixtures in the dust deposited on the packages. There is essentially no difference between films caused by seepage and films caused by deliquescence, although the films caused by deliquescence could be thinner. Therefore the $E_{corr} \geq E_{crit}$ model DOE uses to determine whether localized corrosion due to seepage would initiate should apply equally well to the determination of whether deliquescence-based localized corrosion would initiate. Yet DOE seemingly has chosen to ignore the possibility of deliquescence-based localized corrosion at low temperatures, with no apparent explanation. This is a shortcoming that DOE needs to explain or rectify.

---

[3] The minimum nitrate to chloride ratio necessary to inhibit localized corrosion has not been determined over the full range of temperatures that waste package surfaces may experience. Moreover, the minimum ratio appears to increase with temperature and may be a function of other variables, also.

Oxide Wedging

The 2 cylinders that make up the waste package do not have an interference fit. Design specifications call for the outer diameter of the inner, 316 stainless steel cylinder to be 4 to 10 mm less than the inner diameter of the outer, Alloy-22 cylinder. Thus, there would be a small gap between the 2 cylinders. If the outer cylinder is penetrated, water could enter the gap space between the cylinders, causing corrosion of both the inner wall of the outer cylinder and the outer wall of the inner cylinder. Corrosion products of both metals have lower densities than the parent metals, and therefore the gap will eventually fill with corrosion products. Further corrosion could result in local stresses that might exacerbate waste package degradation, either by simple deformation or by introducing the possibility of stress corrosion cracking.

DOE screens out oxide wedging as inconsequential.[16] That is, DOE omits the phenomenon from the total system performance assessment (TSPA-LA) in its license application on the basis that oxide wedging would have an insignificant effect on dose. The rationale advanced by DOE for screening oxide wedging out is that oxide wedging would result in enhanced degradation of the inner vessel, preventing it from acting as an effective substrate from which the corrosion products could exert stresses on the outer vessel. Even if the corrosion products were to exert stresses on the outer vessel, DOE argues that the effects are sufficiently accounted for by conservative modeling — specifically by assuming a maximized stress corrosion crack area and by taking no credit for the decreased rate of transport of radionuclides through the corrosion products.

DOE commissioned an independent review of TSPA-LA. The results of the review were published earlier this year.[17] The reviewers agreed that screening out oxide wedging was appropriate. However, they found DOE's rationale deficient. The reviewers suggested that stresses caused by oxide wedging would be ameliorated by stretching of the Alloy 22, which is highly ductile. The reviewers also pointed out that any cracks (e.g., cracks caused by stress corrosion cracking) in the Alloy-22 vessel would fill with water and corrosion products, resulting in very slow transport of oxygen to the stainless steel inner vessel, thereby limiting the rate of corrosion of the inner vessel. The corrosion products in the gap between the inner and outer cylinder also would retard oxygen transport. The reviewers criticized DOE for not considering this scenario. We agree with the reviewers.

CONCLUSIONS

The waste package is an extremely important barrier. For the nominal scenario, few waste packages would have failed after a million years, and most of those failed would have relatively little area penetrated. The technical bases for the corrosion performance of the Alloy-22 outer shell of the waste package at below-boiling temperatures appear robust, although the model for localized corrosion at below-boiling temperatures appears overly conservative. There are fewer data at high temperatures, however, and it is not apparent that all of the above-boiling corrosion data are consistent. DOE has plans to address high-temperature corrosion issues. We encourage DOE to execute the plans soon. Treating deliquescence-based localized corrosion and seepage-based localized corrosion of Alloy 22 differently, as DOE does in the corrosion models in its license application, seems inappropriate since the mechanisms are likely to be the same.

The million-plus year service life for Alloy 22 is a unique and unprecedented materials engineering challenge. Even though the general corrosion data are of high quality and their extrapolation for 10^6+ years is based on conservative application of physical and chemical principles, one is uneasy predicting or bounding behavior for this long a period. The corrosion community has not had to deal with anything approaching such a long service life. Perhaps the single largest unknown in predicting the corrosion behavior of Alloy 22 in the repository environment is the specific chemistry of the environment in the repository horizon, in occluded regions such as metal to metal crevices, in deliquesced brines under dust, etc. A further complication is the evolution of the chemical environment with time, which could result in depletion of nitrates due to reaction with organic materials in deliquescent brines.

While the general corrosion of the containers may be reasonably predicted, or at least bounded, it appears that deliquescence of certain salts that can cause liquid water to be present at temperatures well above the boiling point of pure water introduces many uncertainties about localized corrosion (and, to a lesser extent, general corrosion) at elevated temperatures. It should not be concluded, however, that localized corrosion of the Alloy 22, even if it should result in a breach of the containers, will necessarily compromise the integrity of the repository. Even assuming complete penetration of the Alloy 22 by localized corrosion, before the first spent fuel material would contact the repository environment the inner stainless steel shell of the waste package would also have to corrode and aggressive species would have to be transported across the gas within that shell to the fuel rod assemblies followed by penetration of the zircaloy cladding on the assemblies. Even then, before radionuclides can reach the biosphere, degradation of the spent fuel must occur, followed by transport of the degraded spent fuel material across the gap between the assembly and the wall of the container, transport through a presumably limited localized corrosion penetration zone of the container, and transport across the repository environment into the walls of the repository with subsequent geological confinement (perhaps the ultimate isolation step).

Accordingly, because "defense in depth," a concept that has been adopted by virtually every national nuclear waste isolation program, must be followed, corrosion of the containers will not necessarily compromise the entire concept. Nevertheless, an understanding of the corrosion processes that may lead to damage to the containers is imperative. If it can be shown that the containers will contain the waste during the entire period of statutory emplacement, scientific and technical credibility may lead to public credence.

ACKNOWLEDGEMENTS

We appreciate the helpful technical and editorial comments and suggestions by Karyn D. Severson of the Board's staff. Figure 1 is based on data in reference 4. Figures 2a and 2b are slight modifications of figures in references 10 and 7, respectively. The views in this paper are those of the authors and not necessarily those of the U.S. Nuclear Waste Technical Review Board.

26

REFERENCES

1. U.S. Department of Energy Office of Civilian Radioactive Waste Management, *Yucca Mountain Repository License Application,* DOE/RW-0573, Rev. 0, June 2008 (see particularly Chap. 1.5 of the *Safety Analysis Report*)
2. U.S. Department of Energy Office of Civilian Radioactive Waste Management, IED Waste Package Configuration, 800-IED-WIS0-02101-000, May 2007
3. Sandia National Laboratories, *Total System Performance Assessment Model/Analysis for the License Application,* MDL-WIS-PA-000005 REV00, January 2008 (see particularly Appendices K and L)
4. Sandia National Laboratories, *Multiscale Thermohydrologic Model,* ANL-EBS-MD-000049 REV 03 Addendum 01, August 2007
5. T. Oliver, "Effects of Temperature on the Composition of Soluble Salts in Dust at Yucca Mountain, Nevada" (Presentation at the Geological Society of America 2007 Annual Meeting Session 1: Geochemistry, October 28, 2007, Denver, CO)
6. Sandia National Laboratories, *General Corrosion and Localized Corrosion of Waste Package Outer Barrier,* ANL-EBS-MD-000003 REV 03, July 2007
7. J. F. Yang, J. H. Dong, E.-H. Han, J. Q. Wang, and W. Ke [Environmental Corrosion Center, Institute of Metal Research, Chinese Academy of Sciences Shenyang, China], "C22 Corrosion in Dripped Pore Water," undated (posted at http://www.state.nv.us/nucwaste/news2008/pdf/cc2phase2a.pdf on October 15, 2008)
8. J. F. Yang, "Final Results for C22 Corrosion Test IMR April 16, 2008 [PowerPoint presentation]," (posted at http://www.state.nv.us/nucwaste/news2008/pdf/cc2phase2a.pdf on October 15, 2008)
9. Y. Z. Jia, J. Q. Wang, E. H. Han, J. H. Dong, W. Ke [Environmental Corrosion Center, Institute of Metal Research, Chinese Academy of Sciences Shenyang, China], undated (posted at http://www.state.nv.us/nucwaste/news2008/pdf/cc2phase2a.pdf on October 15, 2008)
10. L. Yang, K.T. Chiang, H. J. Gonzalez, J. Auguste, Y. M. Pan, and R. T. Pabalan [Center for Nuclear Waste Regulatory Analyses Southwest Research Institute San Antonio, Texas], "Corrosion Rates of Alloy 22 Materials in Deliquescent Salt Brines at Elevated Temperatures," November 2007
11. P. Russell, D. Wall, and N. Brown, "Plans for Long-Term Corrosion Testing and Recent Results," (Presentation to the U.S. Nuclear Technical Review Board, January 16, 2008, Las Vegas, NV)
12. F. D. Wall and N. R. Brown [Sandia National Laboratories], *Long-Term Corrosion Testing Plan,* SAND2008-4922, August 2008
13. L. Yang, D. Dunn, G. Cragnolino, X. He, Y. M. Pan [Center for Nuclear Waste Regulatory Analyses Southwest Research Institute San Antonio, Texas], A. Csontos, and T. Ahn [U.S. Nuclear Regulatory Commission Washington, DC], "Corrosion Behavior of Alloy 22 in Concentrated Nitrate and Chloride Salt Environments at Elevated Temperatures," paper presented at NACE International Corrosion Conference 2007, March 11-15, 2007, Nashville, Tennessee
14. Raùl B. Rebak, "Newer Alloy 22 Data and Their Relevance to High-Temperature Localized Corrosion," (Presentation to the U.S. Nuclear Waste Technical Review Board, September 26, 2006, Las Vegas, NV)

15. U.S. Nuclear Waste Technical Review Board, *Workshop on Localized Corrosion,* September 25-26, 2006, Las Vegas (see http://www.nwtrb.gov/meetings/meetings.html)

16. Sandia National Laboratories, *Features, Events, and Processes for the Total System Performance Assessment: Analyses,* ANL-WIS-MD-000027 REV 00, March 2008

17. G. E. Apostolakis, R. G. Ballinger, C. Fairhurst, M. Gascoyne, L. Smith, C. G. Whipple, "Report of the Independent Performance Assessment Review (IPAR) Panel," Prepared fro Sandia National Laboratories, March 31, 2008. See also R. G. Ballinger, "Review of the TSPA-LA by the Independent Performance Assessment Review (IPAR) Panel," presentation to the United States Nuclear Waste Technical Review Board, May 29, 2008

Mater. Res. Soc. Symp. Proc. Vol. 1124 © 2009 Materials Research Society 1124-Q08-02

Repository Design–Postclosure Safety Analyses Iterations–A Retrospective Evaluation of the Last Decade of Development of the Yucca Mountain Repository

Robert W. Andrews[1], Gerald Nieder-Westermann[1], Jack Bailey[1] and Robert Howard[2]

[1]Bechtel SAIC Company, LLC, 1180 N. Town Center Drive, Las Vegas, NV 89144
[2]University of Nevada Las Vegas National Supercomputing Center for Energy and the Environment, 4505 Maryland Parkway, Box 454028, Las Vegas, NV 89154

ABSTRACT

Over the past decade there have been several major programmatic decisions associated with the conceptual design and related pre- and post-closure safety analyses of the planned geologic repository for spent nuclear fuel (SNF) and high-level radioactive wastes (HLW) designated for Yucca Mountain, Nevada. These programmatic decisions have resulted from the ongoing evolution of the repository design and scientific understanding of the natural and engineered features that affect the safety analyses of the design. These decisions have also been affected by changes in the regulatory requirements as well as the tradeoffs between pre-closure hazard and safety analyses of near term risks, including worker safety, longer term risks analyzed in the postclosure performance assessment, and cost. The above factors have been considered by the U.S. Department of Energy (DOE) in the formal decision making process dictated by the U.S. Congress as well as the governing laws such as the Nuclear Waste Policy Act and the applicable U.S. Nuclear Regulatory Commission (NRC) regulation 10 CFR Part 63.

INTRODUCTION

The DOE has submitted a license application to the NRC to construct a geologic repository at Yucca Mountain, Nevada for the disposal of SNF and HLW [1]. In connection with the License Application, the DOE submitted to the NRC a Final Environmental Impact Statement (EIS) for the proposed repository and (shortly thereafter) a Final Supplemental EIS for the proposed repository [2]. The license application contains a description of the repository design and analyses of how the repository design functions to provide safety to workers and the public during construction and operations prior to closure of the facility and protects the public and the groundwater resources in the vicinity of the site following closure of the facility.

Major program decisions associated with the Yucca Mountain repository are affected by design decisions and the scientific knowledge available at the time of each design decision. The scientific understanding affects the uncertainty in the long term predicted behavior of the design in the postclosure performance assessment, which in turn impacts the acceptability of the design solution by the regulatory authorities and the public. Therefore, reviewing the basis for the design and the basis for confidence in the projections of the behavior of the design in the natural environment is an important element in the acceptability of the design and the analyses.

The following discussion summarizes the major evolutions in the repository design and postclosure performance assessments over the last decade with the goal of illustrating how the design presented in the license application has been refined to meet both postclosure performance objectives of isolating the radioactive wastes as well as preclosure operational considerations to minimize worker and public safety risks.

DESIGN – PERFORMANCE ASSESSMENT ANALYSES DETAILS

The planned Yucca Mountain repository is located on federal land adjacent to the Nevada Test Site about 90 miles northwest of Las Vegas (Figure 1). The repository design presented in the license application consists of surface facilities comprised of waste receipt, wet handling, aging, canister receipt and closure, and other related facilities and a geologic repository comprised of access and exhaust mains, ventilation shafts and emplacement drifts designed for the emplacement of waste packages and drip shields. The repository is located 200 to 500 m below the surface and about 300 m above the regional groundwater level.

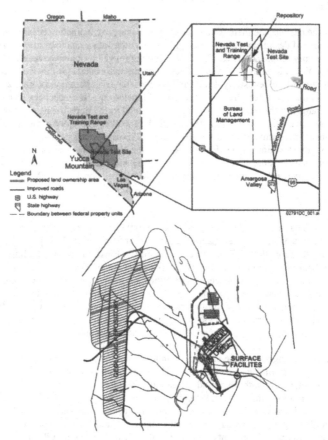

Figure 1. Location of planned surface facilities and geologic repository at Yucca Mountain, Nevada [1, Figures 1-2 and 1.3.1-2]

The subsurface facilities within the repository include emplacement drifts, drip shields and waste packages as illustrated in Figure 2. The waste packages consist of an inner vessel of stainless steel surrounded by an outer corrosion resistant layer of Alloy 22 as illustrated in Figure 2. The waste packages contain commercial SNF in transportation, aging and disposal (TAD) canisters, DOE HLW immobilized as a glass waste form contained in pour canisters, and DOE and naval SNF contained in canisters as illustrated in Figure 3.

The planned repository design has been analyzed for compliance with both pre-closure and post-closure requirements specified in 10 CFR Part 63. The results of these analyses are presented in the License Application [1]. There is a clear nexus between the pre-closure

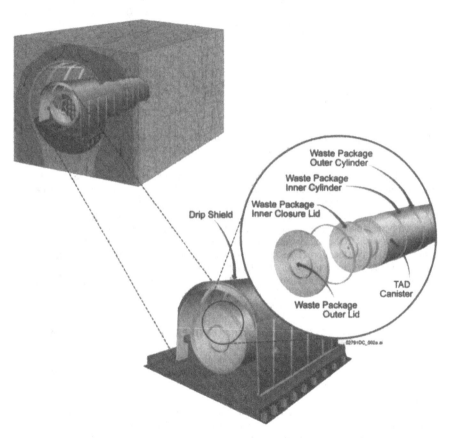

Figure 2. Schematic drawing of emplacement drift with associated drip shield and TAD canister containing 21 pressurized water reactor (PWR) commercial SNF assemblies [3, Figures 11 and 4]

31

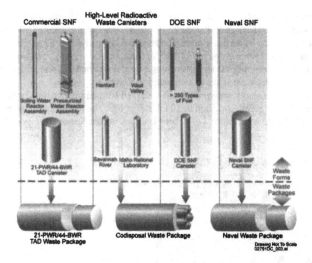

Figure 3. Spent nuclear fuel (SNF) and high-level radioactive waste waste forms and waste package types to be emplaced in the Yucca Mountain repository. [1, Figure 1.5.2-1]

construction and operations conditions and the post-closure performance, in the sense that the pre-closure conditions define the initial conditions for the post-closure analyses. This paper focuses on the integration of the repository design with the post-closure performance assessments. The post-closure performance is analyzed with respect to the expected dose to the reasonably maximally exposed individual (RMEI) located about 18 km downgradient from the repository as well as the concentration of radionuclides in a representative volume of groundwater at that location. These analyses are performed using the total system performance assessment (TSPA) model that analyzes the affect of features, events and processes on the expected behavior of the natural and engineered barriers. The predicted postclosure doses to the RMEI from expected releases from the repository are illustrated in Figure 4 and are significantly less than the levels specified in 10 CFR Part 63.

INTEGRATION OF REPOSITORY DESIGN AND PERFORMANCE ASSESSMENT

The repository design and performance assessment described in the Yucca Mountain License Application [1] have been developed through a series of iterations over the past decade. The last decade of iterations are represented by the major programmatic milestones of the Viability Assessment (VA) in 1998 [4], the Site Recommendation (SR) [5 and 6] and Final Environmental Impact Statement (FEIS) [7] in 2002 and the License Application (LA) [1] and Supplemental Environment Impact Statement (SEIS) [2] in 2008. These iterations have resulted from both evolution of the repository design (to improve constructability, operations and safety) and the evolution of scientific understanding associated with ongoing site characterization. The evolution of the design and scientific information from 1998 to 2008 is summarized in Tables I and II, respectively.

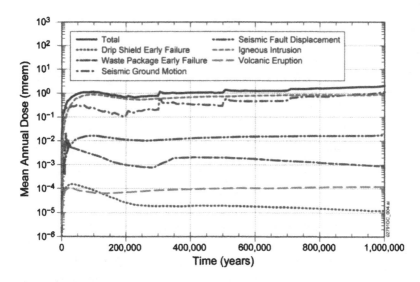

Figure 4. Expected total mean annual dose to the reasonably maximally exposed individual and contribution from different modeling cases [1, Figure 2.4-18]

Table I. Evolution of Key Repository Design Features and Characteristics (1998 – 2008).

Repository Design Feature Characteristic	VA Design (1998)[4]	SR-FEIS Design (2002) [5, 6, 7]	LA-SEIS Design (2008)[1 and 2]
Emplacement Drift Spacing	28 m	81 m	81 m
Emplacement Drift Length	107 km	56 km	68 km
Emplacement Drift – Waste package spacing	Point load – 2 m	Line load – 0.1 m	Line load – 0.1 m
Emplacement Drift Ventilation rate	0.1 m³/s	15 m³/s	15 m³/s
Emplacement Drift Invert material	Concrete	Steel with ballast	Steel with ballast
Emplacement Drift ground support	Concrete lining	Rock bolts	Rock bolts and Bernold-type sheets
Drip Shield materials and thickness	none	20 mm Ti-7	15 mm Ti-7
Waste package capacity	21 PWR	21 PWR	21 PWR
Maximum thermal power	21 kW	11.8 kW	18 kW
Waste Package outer barrier material and thickness	100 mm carbon steel	20 mm Alloy 22	25 mm Alloy 22
Waste Package inner vessel material and thickness	20 mm Alloy 22	50 mm stainless steel	50 mm stainless steel
Commercial SNF canister	none	none	TAD canister

Table II. Evolution of scientific understanding and associated performance assessment differences in the major performance assessment iterations (1998 – 2008).

Performance Assessment Model Component	TSPA-VA (1998)[4]	TSPA-SR TSPA-FEIS (2002)[5, 6, 7]	TSPA-LA TSPA-SEIS (2008)[1 and 2]
Climate and Infiltration	Discrete states Expert elicitation of uncertainty	Discrete states INFIL model	Discrete states NRC specified after 10,000 years MASSIF model
Unsaturated Zone Flow	LBNL model	LBNL model	LBNL model weighted by probability
Engineered Barrier System Environment	Excluded effects of near field chemistry, drift degradation and in-drift condensation	Excluded effects of drift degradation and in-drift condensation. Included effects of near field chemistry.	Included effects of near field chemistry, drift degradation and in-drift condensation
Drip Shield Degradation	Analyzed as design alternative. Excluded effects of seismic and igneous degradation and early failure.	Excluded effects of seismic degradation and early failure. Included effects of igneous degradation.	Excluded effects of seismic degradation. Included effects of igneous degradation and early failure
Waste Package Degradation	Excluded effects of seismic and igneous degradation and localized corrosion and stress corrosion cracking. Expert elicitation of corrosion rate including thermal dependency.	Excluded effects of seismic degradation and localized corrosion and stress corrosion cracking. Included effects of igneous degradation. Corrosion rate based on 2.5 year data without thermal dependency.	Included effects of seismic and igneous degradation and localized corrosion and stress corrosion cracking. Corrosion rate based on 5 year data with thermal dependency.
Waste Form Degradation and Radionuclide Mobilization	Excluded sorption on corrosion products. Expert elicitation of radionuclide solubility. Included credit for CSNF cladding.	Excluded sorption on corrosion products. Laboratory data of radionuclide solubility. Included credit for CSNF cladding.	Included sorption on corrosion products inside waste package. Laboratory data of radionuclide solubility. Excluded credit for CSNF cladding.
Saturated Zone Flow and Transport	Expert elicitation of specific discharge, effective porosity, dispersivity and dilution factor.	Expert elicitation of specific discharge, effective porosity and dispersivity. Dilution determined by annual water demand specified in 10 CFR Part 63.	Expert elicitation of dispersivity. Specific discharge and effective porosity based on field data and models. Dilution determined by annual water demand specified in 10 CFR Part 63.
Biosphere Transport	Three potential receptors. ICRP 30 radiation weighting factors.	RMEI receptor defined in 40 CFR Part 197. ICRP 30 radiation weighting factors.	RMEI receptor defined in 40 CFR Part 197. ICRP 60/72 radiation weighting factors defined in 40 CFR Part 197.

Figure 5 compares the TSPA results for these major programmatic milestones. Two plots are illustrated for the TSPA-SR, to evaluate the effect of alternative conceptual models. The following discussion highlights some of the key aspects of the evolution in design and postclosure performance assessments as well as the evolution of the applicable regulations during this time period.

Design Evolution – Thermal Management

Over the past decade, increased understanding of the complexity of the coupled thermal-hydrologic-chemical response and the associated degradation of the other engineered barrier system features led to an approach of managing the thermal output of the repository by reducing the repository's thermal load. The VA design considered widely-spaced waste packages, closely spaced drifts, a high thermal output from the CSNF waste packages, and limited ventilation; this resulted in a "high" thermal load. The SR-FEIS and LA-SEIS designs utilized a moderate thermal load by placing the drifts further apart, reducing the maximum thermal output of the CSNF waste packages and increasing the ventilation rate. In addition, the SR-FEIS and LA-SEIS designs employed a waste package spacing that is end-to-end in order to reduce the cost and related worker safety issues associated with an increased length of emplacement drifts as well as to reduce the complexity of the analyses of three-dimensional coupled thermal-hydrologic-chemical-mechanical processes.

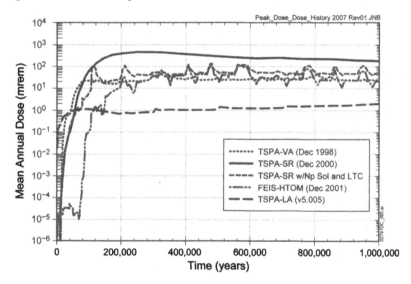

Figure 5. Estimated Mean Annual Dose from DOE TSPAs for Yucca Mountain, 1998-2008. [8] The abbreviation "w/ Np Sol and LTC" indicates "with neptunium solubility limits reduced by sequestration in secondary mineral phases and with long-term climate change." HTOM indicates "high-temperature operating mode."

Design Evolution – Ground Support Materials

The ground support assumed in the VA design was a concrete lining with a concrete invert. To reduce the uncertainty associated with the postclosure models of radionuclide mobilization and transport and to simplify the screening of features, events and processes from the postclosure models, the use of cementitious materials was eliminated in the SR-FEIS and LA-SEIS designs.

Design Evolution – Drip Shield

At the time of the VA, there was no drip shield in the repository design. At the time of the SR-FEIS, a titanium drip shield was added to the design to provide additional defense-in-depth and margin to the postclosure performance and to reduce the uncertainty in the potential effects associated with the range of expected chemical and mechanical environments on the waste package surface. The LA-SEIS design retains the drip shield, but found acceptable results while reducing drip shield material thickness which minimized the costs of fabricating this design feature while maintaining the performance function.

Design Evolution – Waste Package

At the time of the VA, the waste package consisted of two layers, one layer (the outer corrosion allowance material) providing mechanical strength and the other layer (the inner corrosion resistant material) providing long-term corrosion resistance. The SR-FEIS and LA-SEIS designs reversed the configuration of these two layers, placing the Alloy 22 corrosion resistant layer on the outside of the waste package and using stainless steel as the inner vessel for mechanical strength. This difference was affected to enhance the long term performance of the waste package and to reduce uncertainty associated with the potential affects of oxide wedging on the inner vessel.

Design Evolution – Commercial SNF TAD canister

In order to minimize the handling of commercial SNF, reduce the likelihood of preclosure safety events and minimize worker exposure to radiation, DOE simplified the approach to receive and handle commercial SNF at the Yucca Mountain repository in the LA-SEIS design. While the VA and SR-FEIS design utilized an approach where spent fuel assemblies would arrive at the repository in transportation casks to be transferred to the waste package prior to disposal, the LA-SEIS design utilizes a conceptual approach for receiving and handling mostly canisterized commercial SNF waste in transportation, aging and disposal (TAD) canisters that would be loaded at the utilities for transport to the repository. The TAD canisters significantly reduce pre-closure handling and enhance pre-closure safety at the repository. The TAD also has the effect of enhancing the mechanical robustness of the waste package in unlikely post-closure seismic events.

Science Evolution – Drip Shield Degradation

The potential contribution of drip shields to postclosure performance was considered as a sensitivity analysis in TSPA-VA. The rate of degradation was most significantly affected by the general corrosion rate of titanium. In the TSPA-VA and TSPA-SR/TSPA-FEIS, the effects of early failures of the drip shield due to handling or placement defects was not included in the assessment of performance. Because they had the potential to affect performance, these effects were considered and included in the TSPA-LA/TSPA-SEIS, yet were found to contribute only about 0.01% to the total dose in the model case presented in Figure 5.

Science Evolution – Waste Package Degradation

The models of degradation of the waste package have undergone significant evolution over the past decade. In the TSPA-VA, the effects of unlikely seismic and igneous events on the degradation of the waste package were not considered. In addition, the potential effects of localized corrosion and stress corrosion cracking were not included in the model and the uncertainty in the general corrosion rates was based on an expert elicitation given the limited amount of relevant laboratory or analog data. At the time of the TSPA-SR/TSPA-FEIS, the effects of seismic-initiated degradation, localized corrosion and stress corrosion cracking were continued to be excluded from the TSPA and the corrosion rates for the corrosion resistant Alloy 22 were based on 2.5 year laboratory data. The thermal dependency of the corrosion rate was not included in the nominal performance projections but was analyzed as a sensitivity case [6]. In the TSPA-LA/TSPA-SEIS, advances in the understanding of these processes permitted the effects of seismic-initiated degradation, localized corrosion and stress corrosion cracking to be included in the model. The corrosion model utilized 5 year laboratory data and included the thermal dependency of the corrosion rate based on additional analyses of laboratory data developed since the completion of the TSPA-SR/TSPA-FEIS.

Science Evolution – Waste Form Degradation and Radionuclide Mobilization

The models of degradation of the waste form have also undergone significant evolution over the past decade. In the TSPA-VA, radionuclide solubilities were based on an expert elicitation, while in the TSPA-SR/TSPA-FEIS and TSPA-LA/TSPA-SEIS the solubilities were based on a model that used laboratory data relevant to the expected in-package chemical environment. In addition, in the TSPA-VA and TSPA-SR/TSPA-FEIS, no performance credit was taken for the effect of sorption on in-package corrosion products while credit was taken for the expected performance of commercial SNF cladding; while in contrast in the TSPA-LA/TSPA-SEIS performance credit was taken for the effect of sorption on in-package corrosion products while no credit was taken for the potential performance of commercial SNF cladding. The change in cladding credit was the result of the inability to confirm the cladding conditions of commercial SNF that is received at the repository in TAD canisters. Although the commercial SNF cladding is still present and is expected to provide additional protection, the conservative approach of assuming no credit for this postclosure performance capability in the TSPA-LA/TSPA-SEIS model is taken because of the inability to inspect the internals of the TAD canisters upon receipt at the repository. In essence the benefits of the TAD canister in minimizing pre-closure risks outweighed the potential postclosure advantages of cladding in minimizing water contacting the commercial SNF waste form.

Regulatory Evolution

In addition to the design and scientific information evolution over the past decade, the regulatory framework applicable to the Yucca Mountain repository has also evolved. Although generic EPA and NRC regulations existed at the time of the VA (see 40 CFR 191 [9] and 10 CFR 60 [10]), the general environmental standards promulgated by EPA under the Nuclear Waste Policy Act of 1982 [11] had been nullified for application at Yucca Mountain by the Energy Policy Act of 1992 [12] which directed that EPA establish a limit on annual radiation doses to individual members of the public and that the new standard be consistent with findings and recommendations of the National Academy of Sciences (NAS). At the time of the VA, the National Research Council of the NAS had published its findings and recommendations [13] and the DOE had published Yucca Mountain specific siting guidelines in 10 CFR 960 [14], consistent with these findings. At the time of the SR, the EPA and NRC had completed final Yucca Mountain specific rules in 40 CFR 191 [15] and 10 CFR 63 [16], respectively which required a 10,000 year individual protection standard, a separate 10,000 year groundwater protection standard and the inclusion of peak dose analyses in the EIS. In response to a legal challenge, the portions of the standard related to the time period for which compliance must be demonstrated were vacated and remanded to EPA for revision. These remanded provisions were addressed by EPA in the proposed rule in 40 CFR 197 [17] and by NRC in the proposed rule in 10 CFR 63 [18]. These proposed rules, which require analyses of postclosure performance during the period of geologic stability of the Yucca Mountain site (defined as 1,000,000 years), were used as the basis for the TSPA-LA/TSPA-SEIS and the License Application [1 and 2]. Subsequent to the completion of the License Application, EPA has completed the final rule [19].

CONCLUSIONS

The above discussion has illustrated several examples of the integration between design, pre-closure safety, postclosure performance and cost that have resulted in a robust design that is both achievable and relies on passive features rather than active features or controls. In some examples, the postclosure performance is enhanced or the uncertainty reduced by modifying the design (e.g., the TAD canister design, the use of drip shields, the use of a corrosion resistant Alloy 22 for the waste package outer barrier, excluding the use of cementitious materials, or reducing the thermal load). In other examples, additional realism is used in the postclosure performance assessments by gaining additional scientific understanding (e.g., the inclusion of radionuclide sorption on corrosion products inside the waste package, the use of a thermally-dependent corrosion rate of Alloy 22, the inclusion of localized corrosion and stress corrosion cracking processes, the inclusion of drift degradation and near-field chemistry processes, and the use of laboratory and field data and models in lieu of expert elicitations). In yet other examples, the analyzed postclosure performance is insignificantly affected by design solutions that minimize operational and other preclosure safety considerations or costs (e.g., the exclusion of postclosure cladding credit to minimize the operational hazards associated with confirming the characteristics of transported commercial SNF waste and the thinning of the drip shield thickness). These examples illustrate the design, science and performance analyses evolution at

Yucca Mountain that are integrated to provide adequate protection of workers, the public and the environment while minimizing cost impacts.

The iterative nature of the repository design and performance assessment analyses have resulted in a robust safety analysis that has allowed DOE to conclude that it can safely construct and operate a repository for SNF and HLW at Yucca Mountain. The detailed assessments establishing the safety of the repository are presented in the License Application [1].

ACKNOWLEDGMENTS

The authors thank their many colleagues who have contributed to the evolution of the design and scientific understanding of the natural and engineered features of the Yucca Mountain repository over the past more than two decades. This manuscript has been authored by Bechtel SAIC Company, LLC (BSC) under Contract DE-AC28-01RW12101 with the U.S. Department of Energy. The United Sates government retains and the publisher, by accepting the article for publication, acknowledges that the United States government retains a non-exclusive, paid-up, irrevocable, world-wide license to publish or reproduce the published form of this manuscript, or allow others to do so, for United States Government purposes. The statements expressed in this article are those of the authors and do not necessarily reflect the views or policies of the United States Department of Energy, Bechtel-SAIC Company, LLC or the University of Nevada Las Vegas.

REFERENCES

1. DOE, *Yucca Mountain Repository License Application*, DOE/RW-0573, U.S. Department of Energy (2008).
2. DOE, *Final Supplemental Environmental Impact Statement for a Geologic Repository for the Disposal of Spent Nuclear Fuel and High-Level Radioactive Waste at Yucca Mountain, Nye County, Nevada*, DOE/EIS-0250F-S1, U.S. Department of Energy (2008).
3. DOE, *The Safety of a Repository at Yucca Mountain*, U.S. Department of Energy (2008), http://www.ocrwm.doe.gov/ym_repository/license/docs/Safety_of_a_repository.pdf.
4. DOE, *Viability Assessment of a Repository at Yucca Mountain*, DOE/RW-0508, U.S. Department of Energy (1998).
5. DOE, *Yucca Mountain Site Suitability Evaluation*, DOE/RW-0549, U.S. Department of Energy (2002).
6. DOE, *Yucca Mountain Science and Engineering Report*, DOE/RW-0539-1, U.S. Department of Energy (2002).
7. DOE, *Final Environmental Impact Statement for a Geologic Repository for the Disposal of Spent Nuclear Fuel and High-Level Radioactive Waste at Yucca Mountain, Nye County, Nevada*, DOE/EIS-0250F, U.S. Department of Energy (2002).
8. P.N. Swift, K. Knowles, J. McNeish, S.D. Sevougian, M. Tynan, A. Van Luik, "Broader Perspectives on the Yucca Mountain Performance Assessment," *Proceedings of the 2008 International High-Level Radioactive Waste Management Conference,* American Nuclear Society (2008).
9. EPA, "Environmental Standards for the Management and Disposal of Spent Nuclear Fuel, High-Level and Transuranic Radioactive Wastes; Final Rule," U.S. Environmental Protection Agency, Code of Federal Regulations, Title 40, Part 191, 50 FR 38066 (September 19, 1985).

10. NRC, "Disposal of High-Level Radioactive Wastes in a Proposed Geologic Repository at Yucca Mountain, Nevada; Proposed Rule," U.S. Nuclear Regulatory Commission, Code of Federal Regulations, Title 10, Part 60, 48 FR 28194 (June 21, 1983).

11. Nuclear Waste Policy Act of 1982, Public Law 97-425 (1983).

12. Energy Policy Act of 1992, Public Law 102-486 (1992).

13. NRC/NAS, "Technical Bases for Yucca Mountain Standards," National Research Council/National Academy of Sciences, National Academy Press, Washington, D.C. (1995).

14. DOE, "General Guidelines for the Recommendation of Sites for Nuclear Waste Repositories; Proposed Rule and Public Hearing," U.S. Department of Energy , Code of Federal Regulations, Title 10, Part 960, 61 FR 66158 (December 16, 1996).

15. EPA, "Environmental Radiation Protection Standards for Yucca Mountain, Nevada; Final Rule," U.S. Environmental Protection Agency , Code of Federal Regulations, Title 40, Part 197, 66 FR 32074 (June 13, 2001).

16. NRC, "Disposal of High-Level Radioactive Wastes in a Proposed Geologic Repository at Yucca Mountain, Nevada; Final Rule," U.S. Nuclear Regulatory Commission, Code of Federal Regulations, Title 10, Part 63, 66 FR 55732 (November 2, 2001).

17. EPA, "Public Health and Environmental Radiation Protection Standards for Yucca Mountain, Nevada; Proposed Rule," U.S. Environmental Protection Agency, Code of Federal Regulations, Title 40, Part 197, 70 FR 49014 (August 22, 2005).

18. NRC, "Implementation of a Dose Standard After 10,000 Years" (proposed rule), U.S. Nuclear Regulatory Commission, *Federal Register* V. 70, p. 53313. (2005).

19. EPA, "Public Health and Environmental Radiation Protection Standards for Yucca Mountain, Nevada; Final Rule," U.S. Environmental Protection Agency, Code of Federal Regulations, Title 40, Part 197, http://www.epa.gov/radiation/yucca.

Mater. Res. Soc. Symp. Proc. Vol. 1124 © 2009 Materials Research Society   1124-Q01-02

Integration of Postclosure Safety Analysis With Repository Design for the Yucca Mountain Repository Through the Selection of Design Control Parameters

Gerald Nieder-Westermann[1], Robert H. Spencer, Jr. [1], Robert W. Andrews[1], and Neil Brown[2]

[1]Bechtel SAIC Company, LLC 1180 N. Town Center Drive, Las Vegas, NV 89144
[2]Los Alamos National Laboratory, 1180 N. Town Center Drive, Las Vegas, NV 89144

ABSTRACT

The Yucca Mountain repository will use both natural and engineered barriers that work both individually and collectively to limit the movement of water and the potential release and movement of radionuclides to the accessible environment. Engineered structures, systems, and components are designed to function with features of the natural environment in order to meet the postclosure performance objectives established by the Nuclear Regulatory Commission. The features of the natural environment are expected to respond to the presence of the repository through geomechanical, hydrogeologic, and geochemical changes. In order to ensure that conformity between the design basis of the repository and the analyzed postclosure safety basis is maintained, specific features, both engineered and natural, have been identified as requiring design control during repository construction and operations. These features are referred to as postclosure design controls and are intended to maintain the integration of the repository design with the postclosure performance assessment through continued design evolution and changes in scientific understanding.

INTRODUCTION

The U.S. Department of Energy (DOE) has submitted a license application to the U.S. Nuclear Regulatory Commission (NRC) to construct a geologic repository at Yucca Mountain, Nevada to provide permanent storage for spent nuclear fuel and high-level radioactive waste. The repository will be located on Federal property approximately 90 miles northwest of Las Vegas, Nevada and will consist of both surface and subsurface facilities (Figure 1). The 8,600 page license application consists of both the General Information Section and the Safety Analysis Report [1].

The safety analysis presented in the license application evaluates both the preclosure and postclosure performance objectives required at 10 CFR 63.111 and 63.113. The preclosure and postclosure analyses utilize the design features, including both surface and subsurface structures, systems, and components (SSCs) of the repository in the evaluation demonstrating compliance with these performance objectives [2, 3]. For the subsurface facilities of the repository, the anticipated length of the preclosure period (i.e., the time from initial placement until the conditions for closure are satisfied) is expected to be 100 years. The timescales of concern for the postclosure period have been set by the NRC at 10 CFR 63 and include both the first 10,000 years and the period of geologic stability, which is defined to end at 1 million years after disposal.

For the postclosure performance evaluation, the features and associated SSCs of the repository design are constrained by requirements derived from postclosure analyses of the design and the events and processes that will occur in the postclosure geologic setting. These

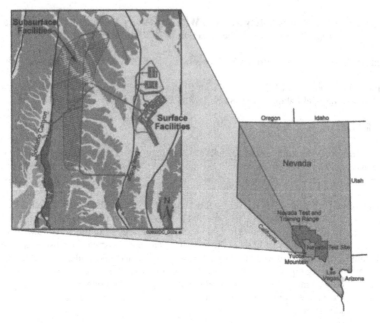

Figure 1. Location Map of the Yucca Mountain Geologic Repository Showing the Two Major Quaternary Faults in the Repository Vicinity; the Bow Ridge and Solitario Canyon Faults [1]

constraints are related to the functions of SSCs in the postclosure analyses. Some constraints are related to the waste isolation functions of the barrier's engineered and natural features, while other constraints are related to maintaining the analyzed postclosure bases.

The constraints are termed control parameters, which specify values or ranges of values as reference bounds for the repository design. Together, the postclosure functions of the SSCs and the postclosure control parameters comprise the postclosure design bases of the repository. The design bases and associated postclosure control parameters are used to define the design criteria.

BACKGROUND

The NRC established at 10 CFR 63.113 that the geologic repository must incorporate multiple barriers, both natural and engineered. A barrier is defined by the NRC as any material, structure, or feature that prevents or substantially reduces the rate of movement of water or radionuclides from the Yucca Mountain repository to the accessible environment, or prevents the release or substantially reduces the release rate of radionuclides from the waste (10 CFR 63.2). DOE is required by 10 CFR 63.21(c)(20) to include in its safety analysis report a description of the quality assurance program to be applied to all structures, systems, and components important to safety, to design and characterization of barriers important to waste isolation, and to related

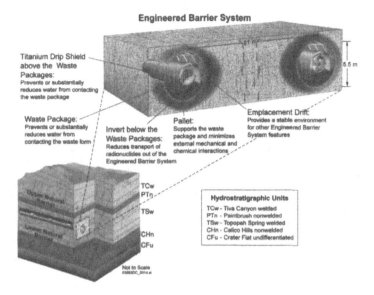

Figure 2. Engineered Barrier System located within the Natural Barriers at Yucca Mountain. [1]

activities. Consistent with this requirement, DOE includes the ability of a feature or SSC in preventing or substantially reducing the probability of criticality as a significant contributor to barrier capability [3].

The Yucca Mountain repository system is composed of natural and engineered features that function together. Two natural barriers, designated as the Upper Natural Barrier and the Lower Natural Barrier, work in conjunction with an engineered barrier, designated as the Engineered Barrier System (EBS). These three barriers are shown in Figure 2. The Upper Natural Barrier consists of the surface topography and the unsaturated zone above the repository, while the Lower Natural Barrier consists of the unsaturated and saturated zones below the repository [1, 3]. The EBS is shown in detail in Figure 2. The principle functions of each of the features and SSCs with respect to barrier capability are identified in Table I.

The postclosure performance assessment analyzes the capability of natural features of the Upper Natural Barrier and Lower Natural Barrier in conjunction with the SSCs of the EBS after closure of the repository. The methodology used to determine which aspects of the natural environment or engineered components are evaluated using the total system performance assessment for the license application (TSPA-LA) model, which involves a series of steps from the collection of data and empirical observations through the identification and screening of features, events, and processes (FEPs) [1, 4].

Initially, a comprehensive, numbered list of FEPs that could potentially affect the postclosure performance of the Yucca Mountain repository was compiled. FEPs identified for consideration in the performance assessment were subsequently screened for either inclusion or exclusion in the TSPA-LA model. Screening criteria for exclusion are based on low probability, low consequence, or by regulation [4, 5].

Table I. Barrier Function of Repository Features or Structures, Systems, and Components [1]

Feature or SSC	Barrier Function
Upper Natural Barrier	
Topography and surficial soils	Prevents or substantially reduces the rate of movement of water
Unsaturated zone	Prevents or substantially reduces the rate of movement of water
Engineered Barrier System	
Subsurface Facilities	None
Ground support and ventilation	Prevents or substantially reduces the rate of movement of water Prevents or substantially reduces the rate of movement of radionuclides
Drip shield	Prevents or substantially reduces the rate of movement of water Prevents or substantially reduces the release rate of radionuclides from the waste Prevents or substantially reduces the rate of movement of radionuclides
Waste package	Prevents or substantially reduces the release rate of radionuclides from the waste Prevents or substantially reduces the rate of movement of radionuclides
Transportation, aging, and disposal canister	Prevents or substantially reduces the release rate of radionuclides from the waste Prevents or substantially reduces the rate of movement of radionuclides
Naval spent nuclear fuel canister	Prevents or substantially reduces the release rate of radionuclides from the waste Prevents or substantially reduces the rate of movement of radionuclides Reduces the probability of criticality
DOE spent nuclear fuel canister and high-level radioactive waste canister	None
Naval spent nuclear fuel canister system components	Reduces the probability of criticality
Codisposal waste package internals	None
Transportation, aging, and disposal canister internals	Reduces the probability of criticality
DOE spent nuclear fuel canister internals	Reduces the probability of criticality
Commercial spent nuclear fuel and high-level radioactive waste form	Prevents or substantially reduces the release rate of radionuclides from the waste Prevents or substantially reduces the rate of movement of radionuclides
Naval spent nuclear fuel	Prevents or substantially reduces the release rate of radionuclides from the waste Prevents or substantially reduces the rate of movement of radionuclides
DOE spent nuclear fuel	None
Commercial spent nuclear fuel/DOE spent nuclear fuel cladding	None
Waste package pallet and invert	None
Lower Natural Barrier	
Unsaturated zone	Prevents or substantially reduces the rate of movement of radionuclides
Saturated zone	Prevents or substantially reduces the rate of movement of radionuclides

For each FEP identified for inclusion in the TSPA-LA model, a disposition was developed that describes how the FEP was to be modeled in the performance assessment. For each excluded FEP, screening analyses were conducted to provide the basis and justification for the exclude decision.

MAINTENANCE OF THE REPOSITORY DESIGN WITH THE ANALYZED POSTCLOSURE SAFETY BASIS

Internal constraints are identified in the screening of a FEP when a specific function is required to be performed by a feature or SSC or when a description or set of descriptions of how specific actions related to the feature or SSC are to be performed. For cases where the performance of a specific feature or SSC is required as the basis for the FEP screening, postclosure control parameters are identified which relate to the postclosure functions of the SSCs to either provide boundary and/or initial conditions for models of FEPs that are included in the TSPA-LA model, or to support the exclusion of FEPs from the TSPA-LA model. These postclosure control parameters ensure that the design is maintained consistent with the analytical basis of the performance assessments [5].

Postclosure control parameters used to support the exclusion of FEPs are important to maintain in order to be able to reasonably exclude the FEP from the TSPA-LA model. Postclosure control parameters are subject to quality assurance controls applied in accordance with 10 CFR 63.142. Through the application of the controls, the analytical basis of the total system performance assessment will be implemented during the preclosure period. Controls include both design configuration controls as well as procedural safety controls.

Design configuration controls are applied to specify and control design parameters in support of the design basis, design criteria, and operational configuration. They include general configuration control as well as fabrication, welding, and other specifications for items that are expected to be procured. Procedural safety controls are documented and controlled specific actions or series of actions taken by the operating staff in preparation for, or during the execution of, waste handling and emplacement operations. Procedural safety controls are applied when there are specific and unique activities requiring operator, inspector, or verification actions by the postclosure control parameter to ensure that operations are within the analyzed conditions of the TSPA-LA model. Procedural safety controls implement human activities that are relied upon to ensure that the control parameters in the FEPs and total system performance assessment are satisfied. Both the procedural safety controls and the design configuration controls will be used to ensure that the postclosure analytical basis conforms to the design basis. Both types of controls are candidates for potential license specifications for the construction or operations of the repository as described at 10 CFR 63.43.

Examples of postclosure control parameters developed to maintain the integration between the repository design basis and the performance assessment are described below. These examples are used to illustrate how the postclosure analytical basis is linked to the design basis.

RELATIONSHIP OF CONTROL PARAMTERES TO POSTCLOSURE FUNCTION

A range of postclosure control parameters has been developed in order to ensure that the design is appropriately maintained with respect to the postclosure performance objectives. The postclosure control parameters that are identified to ensure the postclosure analyzed functions are realized as analyzed include, constraints on the materials or material characteristics or constraints on the handling of materials that could affect the analyzed conditions. Table II provides a representative sampling of control parameters and related FEPs. The discussion presented here will use a subset of the postclosure control parameters to illustrate how these are used to maintain the integration between repository design and postclosure performance assessment [3, 4, 5].

Postclosure Function of Control Parameter—Repository Standoff from Quaternary Fault

In Table II, the Subsurface Facilities SSC includes the control parameter "Repository Standoff from Quaternary Faults." This control parameter states that emplacement drifts shall be located a minimum of 60 m from a Quaternary fault with potential for significant displacement [3, 4, 5] (e.g., the Solitario Canyon Fault). The postclosure function of the 60 m standoff is to preclude block-bounding Quaternary faults with the greatest potential for displacement from damaging EBS components. Therefore in the inclusion in the TSPA-LA of the FEP "Fault Displacement Damages EBS Components" only those intra-block faults with the potential of causing displacement damage to waste packages and resultant releases are evaluated. Because waste packages will not be located within the 60 m stand-off zone, the potential for increased damage to waste packages beyond the bounds of the analyzed basis from seismic events, and the subsequent increase in releases after degradation of the other EBS features, will not occur. The location of the repository with respect to Quaternary faults is shown in Figure 1.

Table II. Relationship of Control Parameters to Postclosure Functions Based on Representative Examples

Structure, System, or Component (SSCs)	Representative Control Parameter	Representative Features, Events, and Processes (FEPs)
Subsurface Facilities	Repository Standoff from Quaternary Faults [1]	Fault displacement damages EBS components
		Flow in the UZ from episodic infiltration
	Emplacement Drift Spacing [1]	Repository resaturation due to waste cooling
		Condensation zone forms around drifts
	Minimum Thickness of the Paintbrush Nonwelded Hydrogeologic Unit above the Repository [1]	Flow in the UZ from episodic infiltration
Emplacement Drift Configuration	Waste Package Spacing [2]	Repository design
		Seismic ground motion damages EBS components
Emplacement Drift Ventilation	Drift-Wall Temperature [1]	Preclosure ventilation
		Drift collapse
Drip Shield	Drip Shield Materials and Thicknesses [1]	Consolidation of EBS
		General corrosion of drip shields
		Localized corrosion of drip shields
Waste Package	Waste Package Surface Marring Prior to Emplacement [2]	Stress corrosion cracking of waste packages
	Waste Package Outer Corrosion Barrier Material Specifications [1]	Thermal sensitization of waste packages
Waste Form and Transportation, Aging, and Disposal Canister	Waste Form Moisture Removal and Inerting [2]	Internal corrosion of waste packages prior to breach
		Radiolysis

[1] Design configuration control; [2] Procedural Safety Control

46

In addition, the 60 m standoff from Quaternary faults in part allows for the exclusion of the FEP "Flow in the UZ from Episodic Infiltration" from the TSPA-LA, because the thickness of Paintbrush nonwelded tuff is too thin near Quaternary faults to dampen flow effectively. Along Quaternary faults (e.g., the Solitario Canyon Fault), relatively small transient pulses of water may penetrate through the fault zones between the ground surface and the repository elevation. Inclusion of the waste packages within the 60 m stand-off zone could result in a higher than analyzed seepage flux contacting the waste packages, potentially resulting in increased releases after degradation of the other EBS features.

Postclosure Function of Control Parameter—Emplacement Drift Spacing

The potential impacts of thermal interactions between adjacent drifts and the effects of thermal mobilization of water are controlled for the Subsurface Facilities SSC by the control parameter Emplacement Drift Spacing. This control establishes the design requirement for the emplacement drifts to be nominally 81 m apart. Because the lateral extent of boiling has been determined to be smaller than the resulting half-spacing between emplacement drifts, thermally mobilized water will continuously drain in the rock pillars between emplacement drifts, as illustrated in Figure 2. As a result, seepage is not expected to occur until later times when the drift-wall temperatures cool to below 100°C and resaturate. This effect is evaluated in the TSPA-LA through the inclusion of the FEP "Repository Resaturation Due to Waste Cooling."

Another aspect of this control parameter for the Subsurface Facilities SSC that is included in the TSPA-LA is the FEP "Condensation Zone Forms Around Drifts." This FEP addresses the impacts from the formation of a condensation zone in the fractured rock above an emplacement drift as a result of evaporation, vapor transport, and condensation during the thermal period, when temperatures above 100°C are achieved. However, the amount of condensation in this zone is mitigated by the continuous gravity-driven drainage of condensation water through cooler regions of rock between drifts. Condensation results in the dilution of pore waters in high-saturation zones above the drifts during the boiling period, but has negligible effect on water chemistry once drift-wall temperatures drop below boiling.

Postclosure Function of Control Parameter—Minimum Thickness of the Paintbrush Nonwelded Hydrogeologic Unit above the Repository

The Subsurface Facilities SSC also includes a design control on it's location within the natural system; specifically, the Subsurface Facilities SSC requires a minimum thickness of 10 m be maintained for the Paintbrush nonwelded hydrogeologic unit above emplacement areas as shown in Figure 2. The minimum thickness of 10 m is needed to maintain the effectiveness of the Paintbrush nonwelded hydrogeologic unit in damping episodic infiltration pulses for the unsaturated zone below this unit. Maintenance of this control in part allows for the exclusion of the FEP "Flow in the Unsaturated Zone from Episodic Infiltration" from the TSPA-LA and the associated potential impacts transient flow would have on waste containment. Without this control to ensure maintenance of the damping effect of the Paintbrush nonwelded hydrogeologic unit, increased shorter duration percolation fluxes could lead to increased shorter duration seepage fluxes, which in the presence of a failed EBS could lead to earlier releases to the accessible environment.

Postclosure Function of Control Parameter—Waste Package Spacing

The Emplacement Drift Configuration SSC includes the control parameter "Waste Package Spacing." This control provides the spacing requirements between adjacent waste packages in emplacement drifts. The waste packages are arranged in a repeating sequence of seven waste packages, as shown in Figure 3, that is used to represent waste package variability in simulations for TSPA-LA. This arrangement corresponds to a thermal line load used in the TSPA-LA of 1.45 kW per meter at fifty years from the initiation of waste emplacement. This parameter provides the interface control between the postclosure performance assessment and design as identified in FEPs "Repository Design" and "Seismic Ground Motion Damages EBS Components." The analyzed basis as implemented in the TSPA-LA for the maximum line-averaged heat load at emplacement of 2.0 kW per meter as and the assessment of the mechanical effects from end to end waste package impacts are maintained through control of waste package spacing.

Postclosure Function of Control Parameter—Drift-Wall Temperature

The Emplacement Drift Ventilation SSC includes the control parameter "Drift-Wall Temperature," which has implications for both preclosure and postclosure analyses. Specifically,

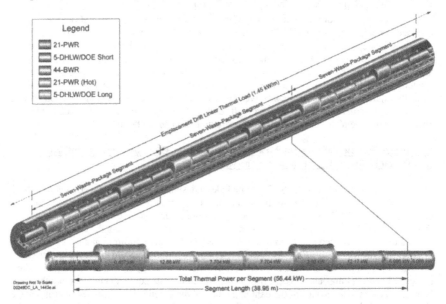

Figure 3. Representation of Contiguous Waste Package Segments Emplaced in a Drift for the TSPA Reference Case Thermal Line Load Based on 10 cm Nominal Waste Package Spacing at Fifty Years from the Initiation of Waste Emplacement [1]

the control requires that the maximum preclosure emplacement drift-wall temperature be held at or below 200°C to avoid possible adverse conditions such as mineralogical transitions or rock weakening through thermal expansion. The end of the preclosure ventilation period provides the initial condition for postclosure analysis. Therefore, the related FEP "Preclosure Ventilation" is included in the TSPA-LA. In addition, maintaining the drift-wall temperature at or below 200°C will confine the extent of permanent changes in rock characteristics that could impact drift opening stability. Thus, adherence to this control is also required by the FEP "Drift Collapse," which has been excluded from the postclosure TSPA-LA, in part on the basis of the initial temperature conditions, which are determined by the preclosure ventilation rates and duration.

Postclosure Function of Control Parameter—Drip Shield Materials and Thicknesses

The Drip Shield SSC will be constructed consistent with the control parameter Drip Shield Materials and Thicknesses. This control requires that the drip shield will be constructed of Titanium Grade 7, with a minimum thickness of 15 mm, while drip shield structural material will be manufactured of Titanium Grade 29. The postclosure function of the control on drip shield materials and thicknesses supports the inclusion in the TSPA of the FEP "General Corrosion of Drip Shields" and the exclusion of FEPs "Consolidation of EBS Components" "Localized Corrosion of Drip Shields" and "Rockfall Damages EBS Components." The typical emplacement drift configuration for the drip shield and waste package is illustrated in Figure 4.

The design controls for drip shields include corrosion resistance as well as structural strength. Over the regulatory life of the repository, the drip shields are expected to be subjected to the effects of general corrosion. These effects are included in the TSPA by FEP "General Corrosion of Drip Shields." The selection of titanium and the associated control parameters are important to the performance of the drip shield because of the inherent corrosion resistance of

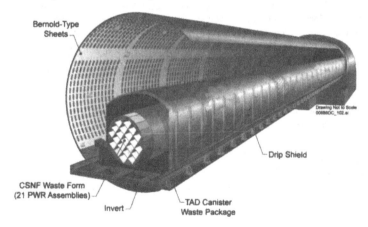

Figure 4. Typical Yucca Mountain Emplacement Drift Configuration with Waste Packages and Installed Drip Shields [1]

the alloy, which tends to form a continuous, highly adherent, and protective oxide-passive film on exposed surfaces, thus providing an important mitigating factor to the processes of general corrosion. The existence of this passive film also precludes the initiation of localized corrosion. Therefore, the control of the drip shield materials also allows for the exclusion of localized corrosion from the TSPA.

The FEP "Consolidation of EBS Components" refers to the collapse and settlement within the repository of the drip shield as a result of physical and chemical degradation. Based on a nominal drip shield thickness of 15 mm and a general corrosion rate of 0.46 mm/10,000 years (aggressive environment), the corrosion lifetime of the drip shield is expected to exceed 100,000 years. Thus maintenance of this control is relevant to the exclusion of this FEP from the TSPA-LA because the corrosion lifetime of the drip shield is expected to exceed 100,000 years.

Postclosure Function of Control Parameter—Waste Package Surface Marring Prior to Emplacement

The Waste Package SSC includes the control parameter "Waste Package Surface Marring prior to Emplacement." The waste package will be required to be certified as suitable for emplacement by process control and/or inspection to ensure that any surface marring is acceptable under the derived internal constraints. The surface marring constraints are: (1) damage to the waste package outer corrosion barrier that displaces material (i.e., scratches) shall be limited to 1/16 in. (1.6 mm) in depth; and (2) modifications to the waste package outer corrosion barrier that deform the surface, but do not remove material (i.e., dents), shall not leave residual tensile stresses greater than 257 MPa. This control parameter directly relates to the inclusion of the postclosure FEP "Stress Corrosion Cracking of Waste Packages" in the TSPA-LA.

Waste package surface marring in excess of the analyzed constraints identified above could result in the formation of incipient cracks on the waste package outer surface, which could act as foci for the formation of stress corrosion cracks. In the analysis of seismic events, it is assumed that 2 mm of the waste packages outer surface have already been removed as a result of general corrosion and therefore any incipient cracks resulting from allowable damage (i.e., scratches not greater than 1.6 mm in depth) are effectively considered in the assumed conditions at closure.

Postclosure Function of Control Parameter—Waste Package Outer Corrosion Barrier Material Specifications

The Waste Package SSC also includes the control parameter "Waste Package Outer Corrosion Barrier Material Specifications." This control is placed on the waste package Alloy 22 material to ensure that it will be manufactured according to ASTM B 575-99a specifications with the following additional, more restrictive, elemental and chemical composition allowable specifications: (a) chromium = 20.0% to 21.4%, (b) molybdenum = 12.5% to 13.5%, (c) tungsten = 2.5% to 3.0%, and (d) iron = 2.0% to 4.5%. Adherence to this control allows for the exclusion from the TSPA-LA analytical basis of the FEP "Thermal sensitization of waste packages." These specifications are the analyzed basis for the selection of Alloy 22 material. Other potential compositions for Alloy 22 may be feasible but have not been analyzed and therefore do not meet the requirements of the control.

Model predictions and extrapolation of higher temperature results to lower temperatures show that formation of tetrahedrally close-packed or ordered phases in Alloy 22 base metal and annealed welds will not occur for repository conditions following closure, based on the standards identified by the control parameter. On this basis, neither the waste package outer corrosion barrier base metal nor weld metal will be subject to enhanced degradation due to the effects of thermal aging.

Postclosure Function of Control Parameter—Waste Form Moisture Removal and Inerting

The Waste Form and Transportation, Aging, and Disposal Canister SSC includes the control parameter Waste Form Moisture Removal and Inerting. This control is designed to ensure that all transportation, aging, and disposal canisters are adequately dried and backfilled with helium. Maintenance of this control allows the exclusion of the FEPs "Internal Corrosion of Waste Packages Prior to Breach" and "Radiolysis" from consideration in the TSPA-LA.

To preclude internal corrosion of the waste package and contained material prior to breach waste packages will be dried and backfilled with helium gas to achieve less than 0.43 mol (7.7 g) of H_2O in a 7 m^3 volume. Drying and inerting the waste form will also preclude the possibility of enhanced corrosion of waste packages and drip shields due to the effects of beta-gamma radiolysis, because beta-gamma radiation is only intense enough to affect corrosion rates while the repository temperature is above boiling when little or no liquid water is present on waste package or drip shield surfaces.

CONCLUSIONS

The postclosure performance assessment analyzes the performance of the natural barriers environment and the engineered components of the EBS after permanent closure of the repository. The analytical basis of the performance assessment relies upon the screening of FEPs for either inclusion in or exclusion from the TSPA-LA. To ensure that the TSPA-LA meets objectives of 10 CFR 63.21 it is important that the identified functions of features and SSCs relied upon in the screening of FEPs are maintained within the analytical basis of the performance assessment and consistent with the design basis. To this end the integration between the performance assessment and repository design is ensured through the maintenance of a range of control parameters.

The above discussion has presented several examples of the use of control parameters in ensuring integration between the post closure performance assessment and the repository design. Postclosure control parameters also are used to define explicit values or ranges of values that provide the boundary or initial conditions of models used as the basis of the performance assessment. Postclosure control parameters used to support the exclusion of FEPs are important to maintain in order to be able to reasonably support the exclusion of the FEP from the TSPA-LA model. The examples presented illustrate the relationship of control parameters to their postclosure functions through the use of representative examples supporting FEP screening justifications.

In accordance with the Yucca Mountain License Application, maintenance of these controls will be ensured through design configuration management and the implementation of procedural safety controls. Configuration management controls will be implemented to ensure that constraints related to material characteristics are within analytical bounds, while those constraints related to the personnel handling of materials are maintained through procedural safety controls. These constraints may form the basis for subsequent license specifications for the construction or operations of the repository in accordance with 10 CFR 63.43.

ACKNOWLEDGMENTS

The authors thank their many colleagues who have contributed to the development of a fully integrated repository design, which adequately addresses the postclosure performance objectives established by the Nuclear Regulatory Commission for the permanent storage of spent nuclear fuel and high-level radioactive waste at Yucca Mountain. This manuscript has been authored by Bechtel SAIC Company, LLC under Contract DE-AC28-01RW12101 with the U.S. Department of Energy. The United Sates government retains and the publisher, by accepting the article for publication, acknowledges that the United States government retains a non-exclusive, paid-up, irrevocable, world-wide license to publish or reproduce the published form of this manuscript, or allow others to do so, for United States Government purposes. The statements expressed in this article are those of the authors and do not necessarily reflect the views or policies of the United States Department of Energy, Bechtel-SAIC Company, LLC, or Los Alamos National Laboratory.

ACRONYMS AND ABBREVIATIONS

DOE	U.S. Department of Energy
EBS	engineered barrier system
FEP	feature, event, or process
kW	kilowatt
MPa	megapascal
NRC	U.S. Nuclear Regulatory Commission
SSC	structure, system, and component
TSPA-LA	total system performance assessment for the license application

REFERENCES

1. DOE (U.S. Department of Energy) 2008. Yucca Mountain Repository License Application. DOE/RW-0573, Washington, D.C.: U.S. Department of Energy, Office of Civilian Radioactive Waste Management.
2. BSC (Bechtel SAIC Company) 2008. Preclosure Nuclear Safety Design Bases. 000-30R-MGR0-03500-000-000. Las Vegas, Nevada: Bechtel SAIC Company.
3. SNL (Sandia National Laboratories) 2008. Postclosure Nuclear Safety Design Bases. ANL-WIS-MD-000024 REV 01. Las Vegas, Nevada: Sandia National Laboratories.
4. SNL 2008. Features, Events, and Processes for the Total System Performance Assessment: Analyses. ANL-WIS-MD-000027 REV 00. Las Vegas, Nevada: Sandia National Laboratories.
5. BSC 2008. Postclosure Modeling and Analyses Design Parameters. TDR-MGR-MD-000037 REV 02. Las Vegas, Nevada: Bechtel SAIC Company.

Mater. Res. Soc. Symp. Proc. Vol. 1124 © 2009 Materials Research Society   1124-Q01-03

A Trade Study for Waste Concepts to Minimize HLW Volume

Dirk Gombert[1], Joe Carter[2], Bill Ebert[3], Steve Piet[1], Tim Trickel[4], and John Vienna[5]

[1]Idaho National Laboratory, Idaho Falls, ID, U.S.A.
[2]Savannah River Nuclear Solutions, Aiken, SC, U.S.A.
[3]Argonne National Laboratory, Argonne, IL, U.S.A.
[4]North Carolina State University, Raleigh, NC, U.S.A.
[5]Pacific Northwest National Laboratory, Richland, WA, U.S.A.

ABSTRACT

Advanced nuclear fuel reprocessing can partition wastes into groups of common chemistry. This enables new waste management strategies not possible with the plutonium, uranium extraction (PUREX) process alone. Combining all of the metallic fission products in an alloy and the balance as oxides in glass minimizes high level waste (HLW) volume. Implementing a waste management strategy using state-of-the-art combined waste forms and storage to allow radioactive decay and heat dissipation prior to placement in a repository makes it possible to place almost 10x the HLW equivalent of spent nuclear fuel (SNF) in the same repository space. However, using generic costs based on preliminary studies for waste stabilization facilities and separations modules, this analysis shows that combining the non-actinide wastes and using only one glass waste form is the most cost-effective.

INTRODUCTION

A new generation of aqueous nuclear fuel reprocessing separates fuel into several fractions, thereby partitioning wastes into groups of common chemistry. Advanced separations make possible recycling of long-lived actinides in nuclear fuel so they can be transmuted into shorter-lived wastes [1]. This also enables development of specialized waste forms to more efficiently immobilize groups of radionuclides to reduce volume, manage decay heat, and enhance overall disposal system performance. Conventional wisdom suggests minimizing HLW volume is desirable, but logical extrapolation of this concept suggests that at some point the cost of reducing volume further may cease to be cost-effective.

Currently, HLW from SNF reprocessing is immobilized in glass containing waste as oxides. Several fission product elements (noble metals Ru, Rh, Pd) have limited solubility, resulting in low waste loading and production of more glass.[a] Matching the waste form to the target waste chemistry allows greater waste loading with comparable or improved performance. The most efficient combination may be an oxidized form like glass for readily oxidized elements and a metallic form for reduced species. This trade-study was designed to juxtapose a combined waste form baseline waste treatment scheme supporting advanced separations with alternate waste combinations to evaluate the relative cost-benefit. The five waste streams considered were:
1. Undissolved solids (UDS): containing Mo, Zr, Ru, Pd, Rh, and Tc in filtered solids

---

a.   An unpublished limit of 3 wt% noble metal oxides is used by Areva at Le Hague, France and it Rokkasho, Japan; a lower limit is used at the Defense Waste Processing Plant at Savannah River, USA.

2. Technetium (Tc): a relatively pure pertechnetate on ion exchange resin
3. Alkali and alkaline earths (Cs/Sr): containing Ba, Cs, Sr, Rb in relatively dilute nitric acid
4. Transition metal fission products (TMFP): containing Fe, Ru, Pd, Rh, Zr, Mo in nitric acid
5. Lanthanide fission products (LNFP): Y, Ln, Ce, Nd, and other lanthanides in lactic acid.

WASTE STRATEGY ALTERNATIVES

The baseline strategy to which alternatives can be compared is depicted in Figure 1 as conceived for UREX+ separations. This waste treatment concept makes use of the potential for tailored waste forms matching waste stream chemistry and the disposal options enabled by advanced separations. It also requires certain specialized process steps to consider in the comparison. Specific points include:
- Four separations processes.
- Three waste immobilization processes producing three different waste forms.
- A dedicated waste form for stabilizing the high-heat generating Cs/Sr waste stream.
- A metal waste form containing reduced Tc and UDS.
- A low-heat generating glass similar to well characterized HLW glasses.
- The potential option for disposal of the Cs/Sr waste form as LLW pending decay.

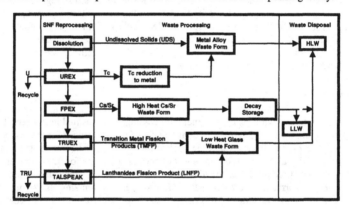

Figure 1. Combined waste form baseline waste treatment concept.

The first alternative (glass option), combining the Cs/Sr, TMFP and LNFP in glass is a simplification of the baseline in which the FPEX separation is eliminated, thereby cancelling the need for creating the dedicated high-heat Cs/Sr waste form. Without FPEX, the alkali/alkaline earth elements are separated with the TMFP in TRUEX. Overall, this change has several positive and potentially negative impacts to consider:
- Simplifies to three separations processes.
- Simplifies to two waste immobilization processes producing two waste forms.
- A larger volume glass waste form including Cs/Sr/LNFP/TMFP is produced.

- Some significant process changes will probably be required for the relatively high-heat glass with greater Cs/Sr waste loading than traditional HLW glasses.
- Same metal waste form proposed in Baseline.
- The option for disposal of the Cs/Sr waste form as LLW pending decay is likely lost, and all wastes will probably be HLW.

The second alternative (alloy option) combines the Cs/Sr and LNFP in glass and the TMFP with the Tc/UDS in a metal alloy. This is actually somewhat more complex in waste preparation than the baseline because the FPEX separation is maintained and a new process is added to reduce the dissolved TMFP from the TRUEX separation to metallic elemental form [2,3]. The metallic TMFP are then added to the Tc/UDS alloy, making a very high density waste form that could be essentially all waste material if the additives needed are supplied by waste fuel cladding and hardware. The major benefit to this concept is that similar to the glass option only two waste forms are made, and in addition, by combining all of the metals in one form the volume is greatly reduced. Overall, this concept has several positive and potentially negative impacts to consider:

- Maintains the need for four separations processes.
- Simplifies to two waste immobilization processes producing two waste forms.
- A glass including Cs/Sr/LNFP requires significant process changes for the relatively high heat generation with greater Cs/Sr waste loading than traditional HLW glasses.
- A new metal waste form is proposed, similar to the metal waste form as above, but requiring a significant reduction step applied to the TMFP TRUEX waste.

A significant benefit to this concept is that the glass is made without the noble metals that restrict waste loading due to their very low solubility. This allows a much greater waste loading in the glass. Combined with the high-density, high waste-loading alloy waste form, this concept minimizes the total HLW volume over all other waste management concepts. However, the option for disposal of the Cs/Sr waste form as LLW pending decay is likely lost, and all wastes will probably be HLW.

TRADE-STUDY APPROACH

The three concepts were compared on a relative basis by using a preliminary conceptual design for the baseline and then adding or subtracting the capital and operating costs for changes made to the baseline. Figure 2 provides a summary of significant cost increments between the baseline, glass option, and alloy option.

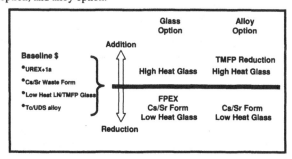

Figure 2. Summary of cost module addition and reductions to make relative comparison.

Cost additions

- High Heat Glass – a process module based on existing HLW glass production with an added relatively small allowance for additional costs to produce glass using high-heat wastes. Costs are incurred to manage solutions that produce appreciable heat which may require cooling and greater radioactivity requiring additional shielding.
- TMFP reduction – chemical or electrochemical reduction of transition metal elements dissolved in nitric acid to form elemental metals, preparation for addition to a metal melter, and modification of the baseline system for the Tc/UDS alloy waste form to include the additional TMFP metals from the TRUEX separation.

Cost reductions

- FPEX extraction – the FPEX separation process is a significant module in the separations facility involving many centrifugal contactors in a hot cell.
- Cs/Sr dedicated waste form – a glass or ceramic process to convert this extremely radioactive, high-heat waste into a solid.
- Low heat glass – a process module based on existing HLW glass production.

Assumptions

- The balance of the facility modules including the UREX, TRUEX, TALSPEAK separations, waste stabilization processes, and supporting facilities including offgas treatment, secondary waste handling, and plant support are all assumed to be fairly comparable and were not considered discriminating costs in the comparison.
- Any waste containing Cs/Sr can be stored sufficiently long (nominally 100 years) to allow it to decay to a heat generation level allowable for placement in a repository, thus waste loading is limited only by solubility of all constituents in the combined waste streams or by the thermal stability of the waste form, whichever is most limiting.
- HLW disposal cost is set on a canister unit cost basis, regardless of the diameter or weight of the canister and includes transportation, placement in the repository and all supporting functions beyond the reprocessing facility, as detailed below:
  - All separations are 100% efficient.
  - Spent fuel composition is based on Origen calculations for 51 gigawatt-days per metric tonne of heavy metal (GWd/MTHM), 20 years cooled.
  - Waste loading and density information are based on current research as shown in Table I. This study used the best-case, most advantageous values updated from the Integrated Waste Management Strategy [4] to evaluate the cost-benefit of advanced waste forms in terms of their maximum expected benefit.

Disposal cost basis

Estimating the unit volume cost of disposal for HLW varies significantly depending, among other things, on what costs are included (e.g., repository development, transportation, placement, closure, etc.) and whether the volume used is the volume of the waste form material, the canister containing the waste, the package containing the container or the space taken up in the repository. In an effort to tie the cost more directly to a waste metric, reference must be made to the Yucca Mountain Science and Engineering Report.[5] The Yucca Mountain design disposal concept uses spent nuclear fuel (SNF) only and SNF/HLW disposal packages for disposal in a repository drift (tunnel). The co-disposal packages, containing five HLW canisters and one SNF canister, (six total) are disposed linearly on a rail system. Using this configuration as a basis, a correlation can be made between units of waste and repository space allocation in rail length (meters) whether the disposal package contains SNF (MTHM) or HLW (MTHM equivalent, or volume m^3). We equate one package to six canisters of waste.

Table I. Waste stream mass, and stabilized waste form data.

Waste Form Matrix	Waste Stream	Mass of Untreated Waste* kg/MTHM	Waste Form Density MT/m3	Waste Loading Wt%	Waste Form Mass** Kg/MTHM	Waste Volume*** m^3/MTHM
Glass	LNFP/TMFP	64	2.8	40%	161	5.69E-02
	CsSr	9	3.4	40%	22	6.46E-03
	CsSr/LNFP	27	3.7	50%	55	1.47E-02
	CsSr/LNFP/TMFP	73	2.7	45%	161	5.99E-02
Metal	UDS/Tc	11	8.0	39%	28	3.46E-03
	UDS/Tc/TMFP	43	8.0	83%	52	6.55E-03

*Mass of untreated waste is the summation of the mass of all elements from the fuel reporting to this stream as well as any chemicals added in the separation process that are included in the final waste form. Wastes are reported as oxides if stabilized in glass, as elements if stabilized in metal.

**Waste Form Mass is calculated as mass of untreated waste divided by waste loading.

***Waste Volume calculated as waste mass (MT=kg/1000) divided by density MT/m^3 = m^3.

However, there are constraints on the capacity of a waste canister in terms of mass and heat. The maximum waste canister size for this disposal configuration is 0.61 m (two feet) in diameter and 4.5 m (15 feet) long, but a solid metal ingot of that size would exceed the weight limits for canister handling. Decay heat can also be limiting, particularly for glasses containing high concentrations of cesium and strontium (Cs/Sr). Canisters containing glass, which has a relatively low thermal conductivity, can devitrify at a centerline above the glass transition temperature 500-700C thereby changing the waste form. To keep the centerline below the glass transition temperature, the canister diameter is reduced as waste loading increases. To account for these variables, canister diameter was reduced as necessary to meet the weight and transition temperature limits but it was assumed a waste package would contain a constant six canisters, at a disposal cost based on the total system life-cycle costs (TSLCC) reported for Yucca Mountain [6].

The results of considering HLW disposal quantity by volume, mass, and heat is that the traditional quantification of HLW by total volume or mass is insufficient to understand the incremental disposal costs and potential savings using advanced separations and tailored waste forms. With metal waste forms limited by mass and glass waste forms limited by heat, the only common basis that can be compared internally and to containerized SNF is the number of packages that must be transported, handled, and placed. The HLW disposal costs estimated for the repository downstream of the reprocessing plant do not correlate with waste mass or volume just as they are independent of what fuel type the commercial SNF packages contain. Only the number of packages affects operations and lineal disposal space in a drift. Therefore, disposal costs can be allocated by waste package. The Yucca Mountain TSLCC in 2007 dollars is $96.2B. This covers repository operations (including transportation) for 17,450 waste packages [6], (commercial SNF and government SNF/HLW), or $5.51M per package ($920K per canister). One could argue that $13.5B has been spent on repository characterization and license development, and therefore only future costs ($82.6B) should be apportioned to SNF/HLW disposal, or $4.7B/package $790K/canister. For the purposes of this analysis a value of $5.1B/package or $850K/canister was used, and sensitivity studies were conducted to determine the effects of varying the value up or down.

COMPARISON

The contrast between the baseline waste management strategy and the two options is stark whether one compares based on total volume or total canisters as shown in Figures 3 and 4. Figure 3 shows the relative total volumes of HLW for each option, broken down by alloy, low-heat glass (no Cs/Sr), and high-heat glass (with Cs/Sr). Note that when evaluated using the traditional metric of total volume the alloy option is clearly the most attractive. The glass option uses only two waste form processes, which is attractive compared to the base case which uses three, but from a volume perspective, there is very little benefit. Also, the base case maintains the potential to dispose of the Cs/Sr glass as LLW after decay, which could offset the small volume difference with the glass option. Based on volume alone the base case and the glass option appear comparable while the alloy option appears to be the best.

In Figure 4 one can see the effects of weight and heat constraints. Starting with the baseline there is a dramatic shift from Figure 3; even though the high-heat glass is only ~10% of the total volume in Figure 3, because of the high heat generation rate, it must be packaged in many more much smaller canisters, accounting for ~60% of the total canister count. The glass option shows a larger benefit over the baseline on a canister number basis. Even though the high-heat glass is volume is much larger (due to the noble metals from the TMFP waste) the container size can now be nearly 3× larger in diameter (9× larger volume) because the decay heat generation is lower for a volumetric advantage of ~9×. While there is roughly a three fold difference in volume between the glass and alloy options, there is only a small difference in canister numbers (~14%). The count of alloy containers has only increased about 2× over the glass option, and the high-heat glass is controlling. The high-heat glass is less dilute, containing only Cs/Sr+LNFP without the TMFP and decay heat generation is intermediate between the baseline and glass options. Container size is still limited by decay heat: less than the baseline, but much more than the glass option. The intermediate size can causes this option to have nearly as many total canisters (86%) as the glass option. Table II summarizes the net affect of waste volume, mass, and decay heat in terms of canister count and potential savings for the high waste loading waste form combinations under consideration.

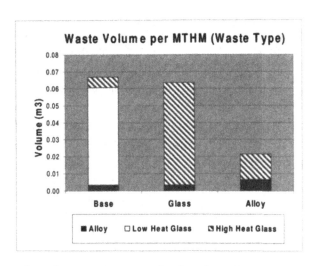

Figure 3. Comparison of waste management options by volume of waste.

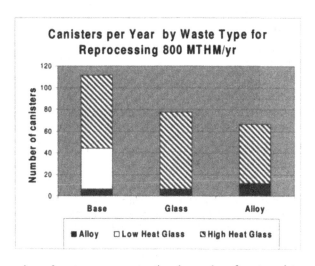

Figure 4. Comparison of waste management options by number of waste canisters.

As shown in Table II, HLW management options with reprocessing could save up to ~$400M/yr in repository costs for an 800 MTHM/yr throughput. Of course, all of the waste processing costs must also be considered. The goal of this trade-study is to evaluate the relative cost benefit among the HLW management strategies following reprocessing. The Table II results show the relative dollar savings of either the glass or alloy option to the three-waste-form base case is ~$30-40M. This sets the bar for the initial differences amongst the strategies; for one strategy to be markedly advantageous to another it will have to be significantly less expensive to implement and operate reliably, or provide some other compelling intangible benefit of greater value than the purely economic differences.

Table II. Waste strategy SNF/HLW disposal comparison for 800 MTHM/yr.

Waste Process/Parameter	SNF Direct Disposal	Baseline	Glass Option	Alloy Option
SNF	●			
Ln/TMFP Glass		●		
Cs/Sr Glass		●		
Cs/Sr/Ln Glass				●
Cs/Sr/Ln/TMFP Glass			●	
UDS/TC Metal		●	●	
UDS/TC/TMFP Metal				●
Total Volume m^3/MTHM	1.97	0.067	0.063	0.021
Canisters/MTHM	-	0.140	0.097	0.083
Canisters/yr 800 MTHM/yr facility	-	112	78	67
Waste Packages yr	89	19	13	11
Package Efficiency vs. SNF	-	79%	85%	88%
Drift Length Used	521	99	69	59
Lineal Drift Efficiency vs. SNF	-	81%	87%	89%
HLW Disposal Cost $M	454	95	66	57
HLW Disposal Savings vs. SNF $M/yr	-	359	388	397
HLW Disposal Savings vs. SNF %	-	79%	85%	88%
HLW Disposal Savings vs. Baseline $M/yr	-	-	29	38

One last observation is that the HLW management strategies also reduce the repository drift length used by up to a factor of ten over direct disposal of SNF. This is ultimately the real differential versus direct disposal of SNF rather than comparison of HLW volume or mass. These waste strategies based on advanced aqueous separations, using combined waste forms and storage to allow radioactive decay and heat dissipation, make it possible to place up to 10x the HLW equivalent of SNF in the same repository space. Conversely, this also means that to accomplish the goal of putting greater than a factor of 10x more SNF equivalent as HLW in the repository, additional measures must be taken.

Process and operating costs

Estimates of process and operating costs based on conceptual designs are highly speculative at early stages of development. While several cost estimates have been made, as research continues the estimates continue to change. Referring back to Figure 2, one can readily see that the differences between the glass and alloy strategies are few. The glass option differs from the baseline by adding one process and subtracting three, whereas the alloy option requires adding two processes as well as subtracting two. On closer examination, it is apparent that both options add high-heat glass production and delete both low-heat glass and the Cs/Sr waste form production, so these cancel from the comparison between options 1 and 2. What is left is the glass option does not require the FPEX separation while the alloy option requires both FPEX and a TMFP reduction process. These processes are totally different, FPEX is a separations module that is based on several years of laboratory development as opposed to the TMFP reduction concept that has had essentially no development, hence the issue with direct comparison of highly speculative costs. Using the data from Table II, the difference in cost savings favors the alloy option over the glass option by only ~$10M/yr. Just the operating costs of the more complex alloy option requiring both FPEX separations and the TMFP reduction processes are likely to exceed this savings. In addition, experience dictates, that for this small difference, one would always choose the simpler system for greater reliability. Therefore, unless there are significant changes to the design constraints, the glass option would be the economic choice.

Comparison to the baseline is somewhat more challenging, and the evaluation becomes more reliant on judgment. Both the glass and alloy options require production of a high heat glass (with Cs/Sr) versus the baseline process that requires both the low heat glass (without Cs/Sr) and the high-heat Cs/Sr glass. At the high waste loadings projected, processing the higher heat glass will be more challenging than the low heat glass. Greater radioactivity will require additional shielding and greater likelihood of failure of non-metallic parts and instrumentation. Greater heat may require cooling or additional engineering to prevent solutions from inadvertently over-heating. However, in both cases the glass melting processes are essentially the same to the HLW glass processes performed worldwide over the last 30 years and both must be conducted in a hot-cell environment, which represents most of the cost. Preliminary design studies suggest that the additional costs for processing the higher activity solutions might add ~10-15% to this single process. This is likely within the accuracy of the conceptual estimates. Offsetting this additional cost, the glass option allows deleting both the Cs/Sr dedicated waste form process and the FPEX separation module. The capital and operating expenses of these two processes are intuitively greater than just a 10-15% increase in a single, well-known process; again the glass option is more economically effective given the current assumptions and design constraints. Put simply the glass option is less costly to build, and less costly to operate than the base case, so the glass option is likely the most cost-effective.

These logical comparisons can be checked for reasonableness based on experience from preliminary design studies over the last decade of fuel reprocessing development [7]. Capital costs are largely controlled by hot-cell space, with second-order impacts due to equipment. Thus similar processes can be compared if they have similar throughputs. Separations processes may use columns, centrifugal contactors, or mixer-settlers, but they all operate in hot-cells and use similar ancillary equipment. Waste vitrification processes all use melters in hot cells equipped to dissipate waste heat, relying on evaporators and material handling systems. Metal alloying processes are more developmental, using much smaller melters, but the support systems and cell

61

design are similar to vitrification. The TMFP reduction system is so conceptual that it is estimated based on a waste immobilization system because it will still rely heavily on material handling and evaporation as described above, with the added cost of a chemical/electrochemical reduction process. While design concepts are in preliminary stages, rough estimates for these processes to support an 800 MTHM/yr reprocessing facility range from $300-500M per separation module and $2-5B per waste immobilization process similar to glass melting. Annual operating costs are generally 1-3% of the capital cost for these facilities. Comparing costs to the baseline as illustrated in Figure 2, and using the minimum values for these costs results in the data summarized in Table III and Figure 5.

Table III. Cost comparison for waste management strategies for 800 MTHM/yr facility.[b]

Process	Glass Option		Alloy Option	
	Capital ($M)	Operating ($M)	Capital ($M)	Operating ($M)
FPEX Separation	-400	-4		
LNFP+TMFP Glass[*]	-2000	-20	-2000	-20
Cs/Sr Glass[**]	-2000	-20	-2000	-20
Cs/Sr+LNFP Glass[**]			2000	20
Cs/Sr+LNFP+TMFP Glass[**]	2000	20		
TMFP Reduction			3000	30
Relative HLW Disposal[***]		-29		-38
Difference vs. Baseline[****]	-2300	-52	1000	-28

*Used in baseline concept only, subtracted from both options

**Using generic data all of the vitrification processes have similar costs, though follow-on studies may more closely evaluate cost vs. throughput

***HLW disposal cost savings at the repository is essentially an annual operating cost to reprocessing, even though it includes capital costs

****Metal melting processes following the chemical reduction of wastes are considered substantially offset by reduced glass production, though follow-on studies may more closely evaluate cost vs. throughput

b. Cost data is approximate and representative of unpublished preliminary design studies.

62

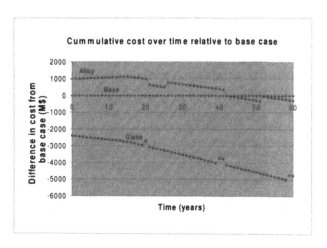

Figure 5. Comparison of waste strategy costs over time normalized to baseline.

This data is too generic to account for 10-15% differences like that described above between high and low heat vitrification, but this was addressed in sensitivity studies. The TMFP reduction process will likely cost more than the Cs/Sr immobilization process, and is assumed here to cost 50% more. Capital costs differences define the y-axis intercepts with the HLW disposal savings combined with process operating costs defining slopes. The periodic irregularities in the curves are the added costs of decay storage modules as they are built and add to operating costs.

Also, HLW disposal cost savings do not start for fifteen years because the reprocessing plant has fifteen year surge storage capacity. The glass option has the lowest capital cost and trends lower because the annual operating costs are less than the base case. The alloy option starts higher because of the added capital cost for TMFP reduction, but eventually crosses the base case due to savings in HLW disposal. The degree of separation of the base case and alloy option are very sensitive to the concept costs for the TMFP reduction, but this projection shows the alloy option would not be competitive with the base case for over 40 years.

Sensitivity studies

Studies evaluating the effects of disposing Cs/Sr as LLW after decay ($2,500/m^3 vs $850,000 per canister) removing the weight limit on alloy canisters, aging SNF up to 120 years or removing the heat limit entirely, reducing waste loading by 50%, and varying HLW canister cost +20% changed the profiles somewhat but did not change the overall ranking of the options. Finally, based on the results with the glass and alloy options, an additional simplification of combining all of the non-actinide waste in one glass waste form was evaluated (using the 3 wt% noble metal oxide limit), and it proved to be the most cost-effective of all. Even though this "all-glass" option produced 65% more glass than the glass option, it produced the same number of canisters per year as the minimum volume alloy option. This option is essentially what is currently used internationally in operating reprocessing plants.

CONCLUSIONS

Advanced separations allow development of waste management strategies that were not conceivable with PUREX reprocessing. For the first time matching the waste form to the waste is possible for HLW; metals can be segregated from oxides and each can be separated and stabilized in forms to minimize the volume of HLW for disposal in the repository. Combining all of the metals in a metallic form and the oxides in glass results in a minimum HLW volume requiring nearly 90% less repository space than SNF, but due to process capital and operating costs, this is not the most cost-effective solution. Using generic costs for typical waste stabilization facilities and separations modules based on preliminary cost studies, it appears that stabilizing all non-actinide wastes together in one HLW glass is the most cost-effective waste management strategy. Sensitivity studies considering variability in weight and heat, waste loading and HLW canister disposal costs indicate that this conclusion is robust over a wide range of these variables. This shows it is very difficult to justify the costs of dedicated HLW treatment facilities purely on reducing HLW disposal costs.

ACKNOWLEDGEMENTS

The submitted manuscript has been authored by a contractor of the U.S. Government under DOE Contract DE-AC07-05ID14517. Accordingly, the U.S. Government retains a nonexclusive, royalty-free license to publish or reproduce the published form of this contribution, or allow others to do so, for U.S. Government purposes.

REFERENCES

1. G. F. Vandegrift, M. C. Regalbuto, S. Aase, A. Bakel, T. J. Battisti, D. Bowers, J. P. Byrnes, M. A. Clark, J. W. Emery, J. R. Falkenberg, A. V. Gelis, C. Pereira, L. Hafenrichter, Y. Tsai, K. J. Quigley, and M. H. Vander Polet, "Designing and Demonstration of the UREX+ Process Using Spent Nuclear Fuel," Atalante 2004, Nimes, France, June 21-25, 2004.
2. S. M. McDeavitt, D. P. Abraham, J. Y. Park, "Evaluation of stainless steel-zirconium alloys as high-level nuclear waste forms," *J. Nucl. Mater.*, **257**(1) 21-34 (1998).
3. T. H. Bauer, S. G. Johnson, and C.T. Snyder, "Modeling the Long-Term Degradation of a Metallic Waste Form," American Nuclear Society Fifth Topical Meeting DOE Spent Nuclear Fuel and Fissile Material Management, Charleston, SC, USA, September 17-20, 2002.
4. D. Gombert II, J. Carter, A. Cozzi, R. Jones, G. Matthern, M. Nutt, S. Priebe, and K. Sorenson, "Global Nuclear Energy Partnership Integrated Waste Management Strategy," GNEP-WAST-WAST-AI-RT-2008-000214, Rev. 1, May 2008.
5. "Yucca Mountain Science and Engineering Report," Rev. 1, U.S. DOE, Office of Civilian Radioactive Waste Management, DOE/RW-0539-1, February 2002.
6. "The Analysis of the Total System Life Cycle Cost of the Civilian Radioactive Waste Management Program," Rev. 0, U.S. DOE, Office of Civilian Radioactive Waste Management, DOE/RW-0591, July 2008.
7. D. E. Shropshire, K. A. Williams, J. D. Smith, E. A. Hoffman, A. Reisman, J. J. Jacobson, T. A. Bjornard, A. M. Phillips, E. A. Schneider, G. S. Rothwell, J. D. Morton, and R. S. Cherry, "GNEP Economic Analysis Report, GNEP-SYSA-ECON-SS-RT-2008-000067, October 17, 2008.

Mater. Res. Soc. Symp. Proc. Vol. 1124 © 2009 Materials Research Society   1124-Q01-04

The National Nuclear Laboratory and Collaborative University Research in the UK

Graham A Fairhall
National Nuclear Laboratory, Sellafield, Seascale, Cumbria, UK CA20 1PG

ABSTRACT

The UK has recently established its first National Nuclear Laboratory (NNL). This has been formed out of the Nexia Solutions organisation, formally the Research and Technology subsidiary of BNFL. Over the next year the NNL will develop so that it can fulfil its mission, which will include the development and maintenance of key skills and undertaking strategic R&D programmes both in the UK and in international collaborations.

A key role of the NNL will be to enhance its interactions with universities to facilitate skills transfer into the nuclear industry as well as support its R&D programmes. Over the past decade the NNL and its predecessor has already established close relationships with leading universities in the UK including Sheffield, Manchester, Leeds and Imperial College London. One of the key objectives of the NNL has been to work in an integrated way with university researchers with programmes spanning fundamental research through to applied R&D.

This paper describes the future plans for working with universities as the NNL develops, building on the success to date.

Joint R&D research programmes already cover many important areas for the nuclear industry. This includes support for the UK Pu disposition programme where R&D at the Immobilisation Science Laboratory, Sheffield has included investigating a range of ceramic and glass wasteforms which has allowed a number of ceramic wasteforms to be down selected for detailed evaluation. Cementation research has included understanding wasteform performance, for example long-term durability. This has involved work on determining free water and the implications for immobilisation of reactive wastes.

Other work involving the NNL and Universities, in particular at Leeds and Manchester, has considered the characteristics and behaviour of intermediate and high level waste sludges. This has included determining the chemical speciation of actinides and fission products, and physical properties of active and simulated sludges using experimental and modelling techniques.

A significant programme on environmental work has been undertaken by the NNL. In research applicable to low level waste disposal and contaminated land reactive transport modelling is utilised to apply experimental and field based research in environmental geochemistry and radiochemistry undertaken by university collaborators. This paper will describe a number of examples of R&D carried out by NNL in association with its university partners.

INTRODUCTION

The UK has recently established its National Nuclear Laboratory (NNL) Figure 1, which has been formed out of the Nexia Solutions organisation, formally the Research and Technology subsidiary of BNFL.

Figure 1: Central Laboratory of the National Nuclear Laboratory

During the next year the NNL will develop so that it can fulfil its mission, which will include the development and maintenance of key skills and undertaking strategic R&D programmes both in the UK and in international collaborations. One role of the NNL will be to enhance its interactions with universities to facilitate skills transfer into the nuclear industry as well as to support its R&D programmes.

Over the past decade the NNL and its predecessor organisations has recognised the importance of close working with universities and has established close relationships with a number leading universities in the UK including Sheffield, Manchester, Leeds and Imperial College London.

This paper describes NNL's plans for working with universities as we develop, building on the success to date. It also describes a number of examples of R&D to illustrate the benefits of close working between the NNL and universities.

UNIVERSITY INTERACTIONS

The NNL view is that universities have a vital role to play, both in carrying out research to underpin the NNL's own R&D programmes and to provide relevant research directly to the nuclear industry. Universities provide the main area in the UK where generic and fundamental research is able to flourish; supported by the UK's research councils. They also provide the main source for the scientists and engineers that will be required, now and in the future, across all the UK's nuclear industry. It follows that the NNL needs to work in an integrated way with university researchers in programmes spanning fundamental research through to applied R&D.

We recognise and support a range of interactions with universities to bring benefit to the nuclear industry, which can be summarised:-

- *To encourage research relevant to nuclear industry needs*

- *To create university networks, focussed on nuclear industry issues*
- *To translate new knowledge into solutions applicable to the nuclear industrial environment*
- *To seek leveraged funding from other industrial groups, research councils and other funding bodies*

To realise the value of universities in providing a source of skilled staff for recruitment into the industry, NNL also needs to engage in activities such as:-

- *Development of undergraduate and Master's courses*
- *Supporting undergraduate students via bursaries, awards and vocational placements at NNL establishments*
- *Attracting students for placement and employment in the NNL*
- *Secondment or placement of personnel between NNL and universities, including visiting roles*

The NNL (and its predecessor organisations) already has a successful track record of working with universities to deliver value to the nuclear industry, notably to the large clean-up programmes that Nuclear Decommissioning Authority (NDA) requires. Over the last 10 years we have developed and implemented a strategy for investing in university research which has seen the creation of 4 major centres of expertise. We designed and implemented contracts with these universities that have created a total of 20 new academic posts and have driven the development of research groups which now comprise a total of more than 160 researchers. An £8M investment has been leveraged and these 4 centres have attracted additional funding of over £50M which has been used to ensure the sustainability of this research capability.

As part of working together, we have developed with some of our university partners a 'seamless team' approach. The idea is to have a way of working which minimises the organisational hurdles to enable staff from each organisation to co-operate fully to the benefit of the project. Each partner obviously focuses on those areas of the programme which most suits their skills and facilities, but everyone is encouraged to perceive themselves as far as possible as part of a single team.

In order to broaden NNL's university links beyond the original 4 centres, we have a strategy to establish a network of key links based on a hub and spoke model, our 3 R's strategy: Research/Recruitment/Reputation.

For research, the NNL will aim to have a strategic link for each of the main research themes that make up the planned R&D portfolio, namely: Fuel and reactor technology; Spent fuel and nuclear materials; Waste and decommissioning; Site end points and disposal; together with key cross cutting activities, such as: Materials; Measurement, analysis and characterisation.

To demonstrate how the NNL is working with universities to deliver value for the UK nuclear industry, the remainder of this paper gives a few examples of recent research where we have either sponsored or provided technical guidance to the research as part of joint programmes.

RESEARCH INTO RADIOACTIVE WASTEFORMS

The long-lived nature of the hazard of radioactive materials and the extended timescales associated with treatment and disposal of radioactive waste mean that it is important to have a thorough understanding of the behaviour of radioactive wasteforms. Valuable insights into the behaviour of matrices for the encapsulation of radioactive wastes have been obtained from research carried in collaboration with the Immobilisation Science Laboratory (ISL) at Sheffield University. Since ISL was established 8 years ago there have been a number of investigations into different aspects immobilisation of highly radioactive materials.

Ceramics for immobilisation of plutonium

Research into high durability wasteforms is important to the nuclear industry and one potential application is for immobilisation of stocks of plutonium dioxide. Joint investigations between the NNL and ISL have been made into a range of ceramic and glass wasteforms which has allowed a number of ceramic wasteforms to be down selected for detailed evaluation [1]. This work has demonstrated the strength of the NNL-ISL alliance to develop technical options which can advise forward NDA strategy. The approach has been to build a database of directly comparable data for the processing, microstructure and performance of candidate materials to allow informed selection of the most promising materials for future elaboration and Pu-active trials. These studies have utilised ISL's capabilities in the materials field to study phase development and the partitioning of plutonium surrogates into those phases, as for example in Figure 2 and this underpinning research is being applied by NNL to the development of processing options for the immobilisation of plutonium.

Figure 2: *HREM image of analogue Pu ceramic*

ISL's work is complementary to additional studies [2] being carried out at our Workington facility where NNL is collaborating with ANSTO in developing glass ceramic based wasteforms for the immobilisation of plutonium containing waste and residues using a hot isostatic pressing process, Figure 3. Development of the process for plutonium dioxide stocks will be undertaken by NNL with support from an ISL EngD student who will work alongside NNL personnel not only developing the process but also ensuring the next generation of engineers are being trained for the implementation of these technologies on nuclear sites.

Figure 3: NNL's Hot Isostatic Pressing Rig

Work to date has established highly durable ceramic forms based on the use of surrogates. Plans are being developed to proceed to the next phase, which will be active demonstration, through the building and operation of a pilot plant in the Central Laboratory. ISL are assisting by developing leaching protocols which will be demonstrated inactively before proceeding to Pu active studies in the Central Laboratory where NNL and ISL personnel will work together in exploring the long term leaching behaviour of these materials, vital in underpinning the case for disposal.

Behaviour of cemented wasteforms

NNL has played a major role in the development of cemented wasteforms for intermediate level radioactive wastes (ILW) and is leading efforts to understand the long term behaviour of existing wasteforms as well as to develop wasteforms suitable for wastes from clean-up and decommissioning programmes. In this context, our alliance with ISL is yielding research into the fundamental chemical and microstructural properties of simulated ILW wasteforms, relevant to both wastes from plants operating on the Sellafield sites and those which are being designed for the treatment of UK legacy wastes both at Sellafield and other UK operational sites.

ISL research projects have provided data on the mechanisms for interaction between the waste and the encapsulant, allowing an improved understanding of the processes affecting wasteform evolution in the longer term. By understanding the waste behaviour at a microstructural level, factors such as the contaminant retention can also be assessed. This in turn can be fed through into applications such as wasteform performance modelling particularly in the repository post closure environment.

One focus has been on the wastes which result from encapsulation of reactive metals, magnesium and aluminium [3,4]. By studying the microstructural behaviour of the metal-cement interface and the reaction products formed from corrosion of the metals, it has been possible to quantify the degree of corrosion which may be accommodated long-term by these wasteforms, Figure 4.

Figure 4: Backscattered electron image of the interface between aluminium and 9:1 BFS:OPC cement

Microstructural analysis of the reaction products of the corrosion of magnesium metal in a composite blast furnace slag: ordinary Portland cement (BFS:OPC) encapsulant has identified brucite (magnesium hydroxide) as the sole product when the magnesium metal corrodes. Products from the corrosion of aluminium in cement are complex with the formation of the calcium aluminosilicate mineral, stratlingite along with bayerite and gibbsite. Such increased understanding of the reaction products and their characteristics and how they evolve with time, is an important step in improving the precision of predicting the degree of corrosion which can be accommodated within the products.

A second focus has been radioactive wastes which are produced by encapsulation of an oxide or hydroxide slurry in cement. One example is ferric floc produced by the precipitation of iron bearing effluents, these act as sorbents to trap radionuclides and are subsequently pre-treated with lime and then cemented using a composite fly ash: ordinary Portland cement matrix. Characterisation of the floc before and after pretreatment and the encapsulated products have been performed as part of a research project at ISL [5,6] to provide data to identify the phases formed and to determine the binding of the waste with the cementitious phases, Figure 5. It has become apparent that a combination of chemical and physical binding of the waste components is occurring in the wasteform and this complicates the understanding of the evolution of the wasteform during storage and disposal.

Figure 5: SEM SEI micrograph of fracture surface of encapsulated ferric hydroxide slurry

A subsequent study [7] has been on high slag replacement composite BFS:OPC cements used to encapsulate barium carbonate slurry, which results from the treatment of gaseous effluents containing carbon-14. This produced interesting data concerning the respeciation of the carbonate as barium sulphate and the effect on the cementitious components. This project also examined the degree of hydration of slag present, in the systems and the results indicate that a substantial proportion of the slag remains unreacted after extended curing of the samples, although this did not affect the physical properties of the wasteforms produced. This research has provided a valuable insight into the degree of reactivity of blast furnace slag in these systems and how the carbonates which were previously understood to be unreactive, behave during the encapsulation process.

Our work with the ISL also illustrates the benefits in terms of increasing the skill-base. Seven PhD students have gained experience in nuclear waste management and of these four are continuing with research directly relevant to treatment of radioactive waste.

BEHAVIOUR OF RADIOACTIVE SLUDGES AND SLURRIES DURING WASTE TREATMENT

Across the nuclear industry, and particularly at Sellafield, there are significant volumes of radioactive waste in the form of slurries. It is important to have effective ways of characterising these slurries so that the most suitable processing routes can be selected. The NNL is working with various customers to develop techniques to retrieve and process sludges and slurries and one aspect which has received attention through our links with the School of Process, Environmental and Materials Engineering at the University of Leeds is in-situ measurement of rheology of slurries. It is important to have techniques which are amenable to deployment in radioactive environments and recently we have been supporting a project which has developed a piezo-electric technique to measure changes in suspension stability. Using an AT-cut quartz silica crystal, the resonating frequency of 5 MHz significantly deviates as the amplitude of oscillation is increased for a coagulated suspension, while remaining unchanged for a dispersed suspension [8]. Studies at Leeds have been undertaken at laboratory scale on model systems in which mono-dispersed silica suspensions were prepared in several different chemical environments and now the NNL is aiming to assist with proving trials at industrial scale.

In a second example the NNL has worked cooperatively with a fundamental study into the effect of particle-particle interaction forces on the flow properties of silica slurries [9]. We recognise its potential to underpin slurry processing operations in the nuclear industry because nuclear waste consists of a complex mixture of particles (different size, shape, density, roughness) surrounded by a complex mixture of ions in solution. We have used our knowledge of various nuclear waste challenges to enable the university to understand how a simplified model of the waste can be constructed and thus be able to explore the underlying mechanisms in a potentially complex system.

Studies have examined the packing nature and reslurrying nature of mono-dispersed colloidal silica spheres in electrolyte solutions of KNO_3 and KCl with the aim of building a fundamental understanding that can be used to guide waste-specific research later, Figure 6.

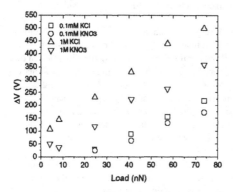

Figure 6: Lateral force interactions between a silica sphere and silica flat

Investigations have covered the effect of adsorbed cations on the stability of a suspension, the packing nature of a sediment and the frictional forces to be overcome during reslurrying. A significant change in the criteria assessed was observed as the electrolyte concentration was increased from 0.1mM to 1M. In relation to industrial processes, such delicate control of the slurry chemistry can greatly influence the optimum operating conditions of non-Newtonian pipe flows.

The work has been extended to explore one area that has still largely gone un-investigated, namely, the reslurrying of compact sediment beds. As mentioned, the two properties that would govern the reslurrying behaviour are the bulk properties of the sediment (yield stress) and the contact forces (lateral and normal) between the individual particles. We have observed that the yield stress for a fully consolidated bed (12 %vol slurry) increases in the following order 1M KCl, 1M KNO_3, 0.1mM KCl, and 0.1mM KNO_3. The strongest adhesion force is observed for the 1M KCl solution, and the weakest force for 0.1mM KNO_3. Therefore, it could be expected that the reslurrying behaviour is characteristic of either the lateral forces, the bulk properties or a balance of the two mechanisms. To investigate this further we have constructed a 25mm horizontal pipe loop which will enable us to monitor the reslurrying behaviour of sediments in different flow regimes.

The NNL continues to work cooperatively with the university in these fundamental studies of studies of slurry behaviour so that the research can appropriately focused onto nuclear industry issues and provide a foundation for waste specific studies within the NNL. We expect to use the mechanistic understanding of model systems to ensure that the correct parameters and range of conditions are studied for the real wastes themselves.

BEHAVIOUR OF ACTINIDES IN UK LEGACY WASTES

The NNL is working with customers at Sellafield on the safe and efficient retrieval and processing of debris from fuel storage ponds. We are using our chemistry knowledge to ensure that the behaviour of radionuclides leached from U metal fuel, particularly key fission products and transuranic elements, is fully understood in the pond environment and also under conditions that will occur during the retrieval operations. The distribution of these elements between solution, colloidal and solid phases under the current pond conditions are obviously key factors, Figure 7.

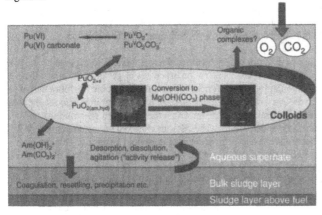

Figure 7: Plutonium chemistry in alkaline systems

Hence, the distribution of radionuclides between solid and liquid/colloidal phases as well as the distribution of radionuclides between liquid and colloidal phases in the liquor and its distribution within the colloidal phase have to be understood.

We have been working collaboratively with the Centre for Radiochemistry Research (CRR), Manchester, where one project has investigated the effectiveness of $Mg(OH)_2$ bulk in trapping radionuclides and of $Mg(OH)_2$ colloids for radionuclide transport in the pond environment. Studies [10] have used a well characterised and controlled model system to provide some significant insights into how trivalent, and possibly tetravalent, f-elements are interacting with corroded magnesium sludges. Detailed study has been made of the chemical and physical behaviour of magnesium hydroxide particles and colloids using a range of analytical and microscopy techniques. [152]Eu has been used to model the behaviour of radionuclides in terms of sorption onto the solids.

The overall outcome is that a good understanding has been developed of the distribution of trivalent, and possibly tetravalent, f-elements between solution, colloidal and solid phases. The results suggest that radionuclides leached from corroding fuel are quickly adsorbed on the bulk brucite phase within the sludge. Solution phase carbonate ions, at concentrations expected to be found in alkaline pond liquors ($10^{-2} - 10^{-3}$ M), however, have been shown to compete with brucite for Eu ions, thus providing a mechanism for the release of activity from the sludge back into solution. Colloids, particularly in the absence of bulk magnesium hydroxide, have been shown to play an important role in the radio-nuclide behaviour.

Using the CRR work as a basis the NNL is conducting studies of the real sludges to quantify their radionuclide behaviour more precisely. We are convinced that the sound mechanistic studies from the CRR together with our experiments on real materials will provide a robust foundation that can guide choices of waste retrieval and treatment options at Sellafield.

ENVIRONMENTAL R&D

The NNL is active in research applicable to low level waste disposal and contaminated land reactive transport modelling and in both areas collaboration with universities is important to access experimental and field based research in environmental geochemistry and radiochemistry.

One important challenge to disposal of low level radioactive waste is to understand the role of cellulose degradation because of its potential effect on processes of gas generation and the chemical speciation and migration of radionuclides. Identifying the anaerobic microorganisms responsible for first stage (hydrolysis) of cellulose degradation is important in predicting the long term biogeochemistry of waste disposal systems, from which the impact on the solubility and sorption of key radionuclides may be determined. The NNL has worked on different aspects of the challenge and had a collaborative research project with Liverpool University. This research was aimed at improving the understanding of the underlying processes that occur in the degradation of disposed cellulosic wastes, in low level radioactive waste and in understanding the microbial processes of cellulose degradation in deep geological disposal. Molecular biological techniques have been utilised on leachate samples from a number of landfill sites in NW England as well as samples from the Low Level Waste Repository (LLWR). A key result was that a group of cellulose degrading bacteria (the fibrobacters), previously found only in the ruminant gut, was found to be active at the Low Level Waste Repository (LLWR) and in landfill sites with areas of high cellulose (paper) disposal. The implications of this are being explored further.

In the context of managing contaminated land at UK reprocessing sites NNL has developed computer modelling methods, Figure 8, that examine the chemical reaction and groundwater transport interactions to assess the mobility of contamination. Such models can be used to evaluate the potential use of remediation technologies and the mobility of fission products such as ^{90}Sr in groundwater [11]. NNL's research in this area is integrated with experimental and field based research at UK universities. Currently, this includes research at Leeds University concerning the properties of Fe(II)/Fe(III) containing green rust minerals to incorporate radionuclides. Also at Leeds a research study into the behaviour of ^{90}Sr in contaminated land is being undertaken. Collaborative research at Sheffield University examines at the laboratory scale how key pH and Eh reactions occurring in zero valent iron permeable reactive barriers can be monitored in space and time using UV fluorescence probes. Such research can be used to test reactive transport models developed by NNL and advance the fundamental understanding of the biogeochemical processes involved.

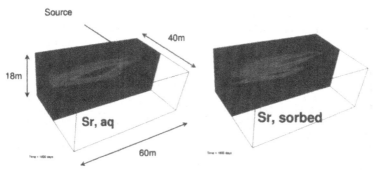

Figure 8: Reactive transport modelling of Sr transport and sorption in ground contamination

A further example relates to knowledge management to improve the industry's ability to manage contaminated land. In this case NNL is collaborating with Southampton University to produce an easily searchable, high level database of up-to-date information on site characterisation methods. The database provides descriptions of all the characterisation techniques identified, thus enabling the user to decide if a proposed technique is likely to provide the information required against a given objective. It also collates the costs, advantages and disadvantages of any investigative or analytical technique that may be used for the identification and quantification of both chemotoxic and radioactive contaminants in soil and water. The NNL's extensive experience in the site characterisation of nuclear licensed sites has been combined with Southampton University's radiochemistry expertise in order to produce a product which will provide useful guidance to industry problem holders looking to assess and manage the land quality on their sites.

CONCLUSIONS

In its first year of operation the National Nuclear Laboratory is set to develop its strategy for close working with universities. Over a number of years we have an established track record of working with a number of UK universities, particularly Sheffield, Leeds and Manchester which were the foundation of our University research alliances. The NNL plan is to build on these existing relationships and to extend our links further by establishing a limited number of additional strategic relationships covering research topics across the nuclear fuel cycle. An example where this is being implemented is with Imperial College, where an agreement is in place to collaborate on research and education in a number of areas. As the NNL develops there will be a greater emphasis placed on international programmes and we intend to work on collaboratively with other international National Laboratories. In addition it is likely that there will be an increased interaction with non UK Universities and some of these may have the potential of developing into a strategic relationship; however at present these will develop on a case by case basis as the NNL strategy evolves. Overall we consider that joint working between universities and the National Nuclear Laboratory has high potential to deliver effective and innovative R&D into all our programmes.

REFERENCES

1. M. C. Stennett, N. C. Hyatt, E. R. Maddrell, C. R. Scales, F. R. Livens and M. Gilbert, in *Scientific Basis for Nuclear Waste Management XXX*, edited by D. Dunn, C. Poinssot and B. Begg (Mater. Res. Soc. Symp. Proc. 985, Warrendale, PA, 2007).
2. M. T. Harrison, C. R. Scales, P. A. Bingham and R. J. Hand, in *Scientific Basis for Nuclear Waste Management XXX*, edited by D. Dunn, C. Poinssot and B. Begg (Mater. Res. Soc. Symp. Proc. 985, Warrendale, PA, 2007).
3. A. Setiadi, PhD. Thesis, University of Sheffield (2006).
4. A. Setiadi , N.B. Milestone, J. Hill and M. Hayes, Cement and Concrete Science Conference 25, Egham, Surrey (2006).
5. N.C. Collier, N.B. Milestone, J. Hill and I.H. Godfrey, Waste Management, 26 (7), 769-775 (2006).
6. N.C. Collier, PhD. Thesis, University of Sheffield (2006).
7. C. Utton, PhD. Thesis, University of Sheffield (2006).
8. D. Harbottle, D. Rhodes, T.F. Jones, and S.R. Biggs, Hydrotransport 17, The Southern African Institute of Mining and Metallurgy and the BHR Group (2007).
9. D. Harbottle, D. Rhodes, M. Fairweather and S.R Biggs, Proceedings of the 11th International Conference on Environmental Remediation and Radioactive Waste Management ICEM07, 7104, Bruges (2007).
10. A. Pitois, P.I. Ivanov, L.G. Abrahamsen, N.D. Bryan, R.J. Taylor and H.E. Sims, J Environ. Monit., 10, 315-324 (2008).
11. S. Kwong and J. Small.. Reactive transport modelling of the interaction of fission product ground contamination with alkaline and cementitious leachates. Proceedings of the 11th International Conference on Environmental Remediation and Radioactive Waste Management ICEM07, Bruges (2007).

Mater. Res. Soc. Symp. Proc. Vol. 1124 © 2009 Materials Research Society 1124-Q01-05

DIAMOND: Academic Innovation in Support of UK Radioactive Waste Management

Neil C. Hyatt[1,] Simon R. Biggs,[1] Francis R. Livens[3] and James C. Young[2]

[1] Department of Engineering Materials, The University of Sheffield, Mappin Street, Sheffield, S1 3JD, UK
[2] School of Process, Environmental and Materials Engineering, The University of Leeds, LS2 9J2. UK
[3] Department of Chemistry, The University of Manchester, Oxford Road, Manchester, M13 9PL. UK

ABSTRACT

The background and planned research activities are outlined for a new UK research consortium focused on Decommissioning, Immobilisation And Management Of Nuclear-wastes for Disposal (DIAMOND). This consortium is the first integrated trans-disciplinary and multi-institution academic research network in the UK, focused on nuclear waste management.

INTRODUCTION

Several reports have defined a new policy framework for management UK radioactive waste legacy, resulting from the 60 years civil nuclear fission activities [1-5]. The overall cost of waste treatment, packaging and storage, together with decommissioning and remediation activities, for which the Nuclear Decommissioning Authority (NDA) are currently responsible, is estimated at £70 Bn [6, 7]. Geological disposal of radioactive wastes, in a mined repository, is now accepted by Government as the "best available approach" for long term management of the UK nuclear waste legacy, supported by a "robust programme of interim storage", as recommended by the Committee for Radioactive Waste Management (CoRWM) [3-5].

It is recognised that academic research and development will play a pivotal role in reducing the cost and timescale of the UK clean up and decommissioning programme [8]. Furthermore, public confidence in nuclear waste management and disposal depends crucially on rigorous peer review of academic understanding, data and models that underpin long term predictions of wasteform behaviour and evolution of the disposal environment [3, 9]. The research challenge is compounded, however, by a critical nuclear skills shortage: it is estimated that the UK nuclear defence, power, and clean up industries require >15,000 new degree level professionals in the next 15 years - excluding potential demand from new nuclear build, repository construction and renewal of the nuclear deterrent [10].

Recognising that "integration of [academic expertise in] biological, environmental, and physical sciences together with engineering" was essential to "stimulate novel ideas/approaches in nuclear waste management", the UK Engineering and Physical Research Council launched a "Call for Research Proposals in the area of Nuclear Waste Management and Decommissioning" in September 2007 [11]. The successful consortium, DIAMOND, was a was awarded £4.3 M (through competitive peer review), to deliver a four year national research and training programme in radioactive waste management across the Universities of Leeds, Loughborough, Manchester and Sheffield, and University and Imperial Colleges, London. This consortium is the first integrated trans-disciplinary and multi-institution academic research network in the UK, focused on nuclear waste management. Broadly, the strategic aims of the consortium are:

- To carry out internationally leading science and engineering in the broad area of decommissioning and nuclear waste management.
- To support research that will underpin the development of innovative technologies for nuclear decommissioning, waste management and disposition.
- To broaden the UK research base in science and engineering that focuses on nuclear waste technologies and thereby help address a developing skills gap.
- To develop and support new links between investigators in universities that have established nuclear science programmes and those universities that are developing such programmes.
- To develop new inter- and intra-university links to facilitate multi-disciplinary collaboration and stimulate new applications of knowledge at the interface between disciplines.
- To train the next generation of UK scientists and engineers with skills and expertise in nuclear waste management and decommissioning issues.

STRUCTURE AND ORGANISATION

The consortium's activities are organised into three Work Packages (WPs) and Cross-Cutting Themes (CCTs), as summarised in the table below. This matrix structure has been designed to maximise opportunities for the sharing and exchange of knowledge between researchers across the consortium. The identifiers within each box of the matrix refer to the individual research projects described in detail on the DIAMOND website (www.diamondconsortium.org). The general aim is to link activities that can take a particular waste type (for example corroded Magnox sludge) from its current state through characterisation and handling to treatment and packing, interim storage and finally into secure disposal. The CCT champions within our consortium are recognised experts who will take responsibility for maximising knowledge and technology transfer opportunities between the WPs. We believe this will provide an environment that optimises opportunities for collaboration, training, and knowledge/technology transfer across the consortium.

			Cross Cutting Themes		
			CCT1 Characterisation	CCT2 Treatment & Packaging	CCT3 Disposal
			Francis Livens (Manchester)	Neil Hyatt (Sheffield)	Howard Wheater (Imperial)
Work Package	WP1 Environment, Migration & Risk	Nick Evans (Loughborough)	1.3.1.1, 1.3.1.2, 1.3.1.4, 1.3.2.1	1.3.1.1, 1.3.1.3, 1.3.1.4, 1.3.2.2, 1.3.3.1, 1.3.3.2	1.3.1.1, 1.3.1.2, 1.3.1.3, 1.3.1.4, 1.3.2.1, 1.3.2.2, 1.3.3.1, 1.3.3.2, 1.3.3.3, 1.3.3.4
	WP2 Decommissioning, Legacy & Site Termination	Mike Fairweather (Leeds)	2.3.1.1, 2.3.2.1, 2.3.2.2, 2.3.3.2, 2.3.3.3	2.3.1.2, 2.3.1.3, 2.3.1.4, 2.3.2.2, 2.3.2.3, 2.3.2.4, 2.3.3.1, 2.3.3.2, 2.3.3.3	2.3.1.4, 2.3.2.4
	WP3 Materials Design, Development & Performance	Bill Lee (Imperial)	3.3.1.3, 3.3.3.2	3.3.1.1, 3.3.1.2, 3.3.1.3, 3.3.2.1, 3.3.2.2, 3.3.2.3, 3.3.2.4, 3.3.3.2, 3.3.3.4, 3.3.4.1, 3.3.4.2, 3.3.4.3	3.3.1.1, 3.3.1.2, 3.3.1.3, 3.3.2.1, 3.3.2.2, 3.3.2.3, 3.3.2.4, 3.3.3.1, 3.3.3.2, 3.3.3.3, 3.3.3.4, 3.3.4.1, 3.3.4.2, 3.3.4.3

Table I: Organisation of research projects in the DIAMOND consortium.

The consortium involves staff from across a wide range of science and engineering disciplines, including radiochemists, radiation chemists, physicists, earth scientists, environmental scientists, process engineers, structural engineers, materials scientists and civil engineers. Each of the research projects includes a minimum of two researchers from different disciplines and/or different universities. We believe this will provide an invaluable opportunity to stimulate new ideas across disciplines where contact rarely occurs. At the same time, we are deliberately interfacing with researchers from disciplines that may not conventionally see a role in the nuclear waste research area.

SCIENTIFIC PLANS

Legacy nuclear waste treatment, storage and disposal, as well as decommissioning and site remediation are conservatively estimated at a cost of £70 Bn for the UK taxpayer [6, 7]. The diversity of issues that must be addressed in tackling this problem is immense and requires a wide range of innovative solutions drawn from an equally wide range of technology discipline areas, as summarised schematically in Figure 1.

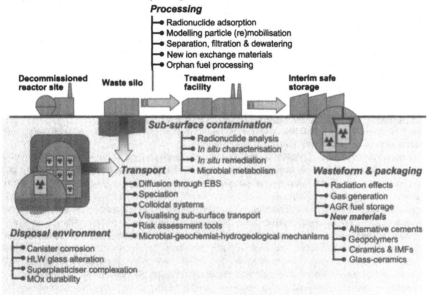

Figure 1: Schematic representation of research questions to be addressed by the DIAMOND consortium.

In considering the balance of the consortium, we examined how we could align with the strategic priorities of the NDA [8], as well as taking into account the key R&D priorities of the CoRWM recommendations [3, 4]. Obviously, it is the primary purpose of this academic research consortium to be adventurous and deliver real innovation, and the majority of our scientific program falls within this description. However, we also believe that there is a role for the consortium to play in the delivery of research that will have a more immediate value to the industry and we will explore how the consortium can have maximum near-term impact. As noted above, a two-way matrix for the management of the work is proposed to ensure the maximum sharing of information. For ease of description, we will concentrate here on the main features of each WP (see also Figure 1), the detailed science plan and objectives for these themes are available on the consortium website.

WP 1: Environment, Migration and Risk.
Aim: to define the processes which control transport of key radioactive contaminants in natural and engineered environments.

Cleaning up contaminated land is a substantial component of site restoration and forms the major focus of this WP. Large volumes of contaminated land are present at many nuclear sites, most notably at Sellafield. The range of radioactive contaminants present, and the potential for mixed contamination, represents a major challenge. In particular, the potential for migration of soluble and colloidal species in the subsurface is a key uncertainty in defining site end points, developing decommissioning and cleanup strategies, and quantifying the associated costs. Currently, most cleanup is expected to be through invasive technologies, so there is an opportunity to innovate through exploitation of more cost-effective *in situ* technologies such as barrier treatment systems. Many of the uncertainties and technical challenges associated with geological disposal of immobilised radioactive wastes are critically dependent on the rate of release from the engineered facility, and on the subsequent behaviour of the radionuclides in the disturbed zone and the far field. These processes have close parallels in site remediation, so it is logical to link these two themes within this work package.

WP 2: Decommissioning, the Historic Legacy and Site Termination.
Aim: To provide new techniques and new technologies in support of legacy waste management, decommissioning and the monitoring of site end points.

The heterogeneous wastes present in fuel storage and fuel handling facilities are a key target of the work described here. These wastes comprise irradiated fuel, contaminated materials and corrosion products. Total volumes and compositions (both chemical and radioactive) are poorly known, although there are believed to be several hundred m^3 in each of the main storage ponds. Improved and accurate characterisation is therefore a priority, and technologies for retrieval and treatment are also required. Furthermore, historical activities have created small volumes of "orphan wastes" for which no clear management route exists. These are present in only small volumes but their diversity requires versatile methods for treatment and immobilisation. Linkages between proposed treatment strategies for all these waste materials will obviously be beneficial and such information exchange is expected to be a key benefit of this consortium.

Progression towards safe site end points is a complex process, requiring the dismantling and removal of plant and buildings, identification, characterisation and removal of contaminated material, passive safe storage of packaged wastes and long term monitoring of the cleared site. However, in contrast to substantial industrial activity in decommissioning, academic activity in the UK is relatively sparse. One approach to this issue is to actively to seek out opportunities for technology transfer into the nuclear industry, particularly from the aerospace and medical sectors. In a more conventional research activity, we focus on quantification of radionuclide inventories, contaminant transport through engineered pathways and waste retrieval as areas where we can make a contribution. For site end points, long-term restriction of contaminant mobility is important, cost effective *in situ* monitoring is a valuable technology and the durability of waste containers in interim safe storage are important research questions.

WP 3: Materials- Design, Development and Performance

Aim: To provide innovations in the processing and immobilisation of problematic wastes, the synthesis of novel wasteform materials and improved understanding of wasteform and container performance in interim storage and disposal environments.

Geological disposal in a mined repository, preceded by a period of interim safe storage, has been accepted by Government as the "best available approach" for the long term management of UK radioactive wastes. However, final agreement on the route and site for disposal of different waste categories must await further, extensive, technical and public consultation. Therefore, in this work package, we focus, primarily, on key challenges in the immobilisation and safe interim storage of nuclear wastes (including irradiated graphite) and fuels, to support clean up of the existing waste legacy and underpin later geological disposal. Decisions regarding the re-use, long term storage and disposal of some strategic nuclear materials, including plutonium and spent fuels, require specific knowledge gaps to be addressed. A key task in this work package is to understand, quantify and predict the evolution during interim storage of spent AGR fuel, separated plutonium and MOX fuel, in order to inform decision making regarding future re-packaging and storage requirements. This is supported by research aimed at improving our understanding of mechanisms of radiation damage in model ceramic materials at a fundamental level. We will also undertake a programme of research, integrated with WP2, to address the immobilisation of high activity sludges, fuel element debris, and defence and pyrochemical wastes, in glass-ceramic wasteforms. Here, we aim to capture the advantage of high temperature processing technologies to deliver high volume reduction, throughput and product durability, while simultaneously achieving acceptable emission thresholds. This will be complemented by research aimed at developing novel encapsulants and improved cementing systems for the encapsulation of reactive metals, graphite and loosely packed wastes. We will also develop novel routes to inert matrix fuels and ceramic wasteforms. A further element of this work package is to address a key knowledge deficit in the corrosion behaviour of UK HLW glasses, with particular reference to actinide and fission product retention, in order to underpin future geological disposal.

INDUSTRIAL ENGAGEMENT

Industrial stakeholders and regulators will be key beneficiaries of the knowledge and understanding generated by the consortium. Engagement with these organisations is therefore an integral part of consortium activities, with aim of:

- Guiding research activity, as appropriate, to ensure delivery against end user requirements, over near and longer term time scales.

- Provision of access to corporate knowledge, technical expertise and laboratories in support of research.

- Provision of training opportunities to research students and associates, through placement / secondment and access to laboratories.

The consortium received strong support from over 20 industrial companies for the successful proposal. Each of the supporting organisations had previously funded one or more of the collaborating partners to undertake research associated with nuclear waste treatment, with a total portfolio of research estimated at £12M over the last five years.

Industrial stakeholders will also contribute to the activities of the consortium as members of the International Advisory Group, advising on the strategic direction of the consortium and providing forward-looking ideas for development of the group.

INTERNATIONAL LINKS

Members of the consortium have a range of important international partnerships in the nuclear research field. Examples of these links include ANSTO, Forschungszentrum Karlsruhe, CEA, Los Alamos, INL, PNNL and Savannah River. We will actively seek routes to allow these links to be further developed for the benefit of the consortium and are especially interested in developing collaborative opportunities for training.

CONCLUSIONS

This paper has summarised the background to, and scope of the first integrated academic research programme in the UK, focused on nuclear waste management and decommissioning. The research programme of this consortium is expected to advance UK nuclear waste management by reducing the cost and timescale of the clean up and decommissioning programme, and delivering new understanding and models to support geological disposal.

REFERENCES

1. Managing Radioactive Waste Safely: Proposals for Developing a Policy for Managing Solid Radioactive Waste in the UK, DEFRA, September 2000.
2. The Nuclear Legacy: A Strategy for Action. Cm 5552 DTI London (2002)
3. Managing our Radioactive Waste Safely: CoRWM's recommendations to Government, July (2006).
4. Moving Forward CoRWM's Proposals for Implementation: CoRWM document 1703, November (2006).
5. Managing Radioactive Waste Safely: A Framework for Implementing Geological Disposal. Cm 7386. DEFRA, London (2008)
6. Taking Forward Decommissioning. National Audit Office Report (2008).
7. Nuclear Decommissioning Authority Business Plan 2008-2011 (2008).
8. NDA Strategy Document, March (2006).
9. Radioactive Waste: Special Eurobarometer 227 / Wave 63.2 – *TNS Opinion and Social*, European Commission, June 2005.
10. Nuclear and Radiological Skills Study- Report of the Nuclear Skills Group, (2002).
11. Engineering and Physical Sciences Research Council, Call for Research Proposals in the area of Nuclear Waste Management and Decommissioning, (2007).

Spent Nuclear Fuel

Mater. Res. Soc. Symp. Proc. Vol. 1124 © 2009 Materials Research Society   1124-Q02-01

Key Scientific Issues Related to the Sustainable Management of the Spent Nuclear Fuel in the Back-End of the Fuel Cycle

Christophe Poinssot[1], Jean-Marie Gras[2]
[1] CEA, Nuclear Energy Division, Department of RadioChemistry and Processes, CEA-Marcoule, F-30200 Bagnols-sur-Cèze Cedex, France.
[2] EDF, R&D division, F-77250 Moret-sur-Loing, France.

ABSTRACT

Direct geological disposal has for a long time being considered in many countries as a reference solution to ultimately manage Spent Nuclear Fuel (SNF). However, the recent concerns about the global climate change as well as the strong increase of energy demands in the world leads to a remarkable renaissance of nuclear energy. In this new context, resources have to be preserved and recycling part of the actinides of SNF is clearly an option of growing interest for many countries instead of SNF direct disposal. The research regarding the spent nuclear fuel evolution in the back-end of the fuel cycle has therefore to be revised to shift from direct disposal and interim storage to new issues related to treatment/recycling. This paper gives a concise state of knowledge of spent fuel key properties in each of these scenarii and focuses on the main scientific issues to be addressed in the next years.

INTRODUCTION

In the past decades, society evolves in a context where energy was abundant and at low price. At that time, direct geological disposal was considered in many countries (USA, Sweden, Canada, more recently Germany ...) as a reference solution to ultimately manage Spent Nuclear Fuel (SNF). Treatment and recycling of spent nuclear fuel was only implemented by some few countries as France, UK and Japan but assessed as non-economically valuable by the others. In this context, studies on SNF long term evolution in direct disposal significantly developed to support the safety analyses of a potential geological repository. They allow understanding the key processes governing the long term alteration processed and therefore deriving reliable and scientifically-sounded radionuclides source term to be used in safety analyses, especially for two main contributions, the Instant Release Fraction and the matrix dissolution.

However, the context of nuclear energy significantly evolves in the past years. The recent concerns about the global climate change leads to renew our relation to energy by aiming to (i) optimize the resources use and (ii) look for new energy sources which does not contribute to the greenhouse gases emission. At the same time, the growing of emerging countries dramatically increases the energy demand and leads to significant international pressure and explosion of fossil fuel prices. This new context completely changes the relative significance of energy sources and perspectives. Occidental societies are thinking to decrease their dependence to fossil fuel whereas non greenhouse gases emitting energies are developing faster than never imagined. Among them, although still disputed by some opponents, nuclear energy undergoes a remarkable renaissance with an increasing interest of many companies and countries, an increasing number of new reactor projects and even some new fuel cycle plants projects. This increasing interest in nuclear energy addresses the crucial issue of fissile material resources (natural ^{235}U and artificial ^{239}Pu) which should obviously be preserved and recycled as far as possible, like other resources. Indeed, 96% of SNF is still composed of U and Pu which are recyclable. The closed cycle option, which allows at least U and Pu recycling, should progressively take more and more interest around the world. On the opposite, directly disposing SNF in deep underground seems not to be the most sustainable solution. Recycling together U and Pu, and potentially also the minor actinides, is therefore an option of growing interest for many countries instead of direct disposal in consistency with the sustainable development.

Surprisingly, this deep strategic and political evolution seems to have little influence on spent nuclear fuel R&D in the last years although new scientific issues are rising: most of the open R&D was up to now only driven by repository issues. This papers aims to depict the current state of knowledge and remaining scientific key issues related to the different options considered for managing spent nuclear fuel, direct disposal and recycling and attempt to draw what should be their respective significance. In particular, it stresses that most of the scientific skills developed for studying fuel storage and disposal should now shift to address some of the new challenges related to close fuel cycle. This paper is widely based on the results acquired in the French research program devoted to the spent fuel long term evolution in storage and disposal entitled the PRECCI program ([1]; funded by Commissariat à l'Energie Atomique CEA, Electricité de France EDF and partially the French waste management agency, ANDRA, and AREVA-NP) and in the relevant European research projects (Framework program of EURATOM): Spent Fuel Stability under Repository Conditions (SFS project;[2]) and Near Field Processes (NFPRO project; [3]). More details can be found in the specific references.

ONCE-THROUGH CYCLE: SPENT FUEL EVOLUTION IN GEOLOGICAL REPOSITORY

In a once-through cycle, spent nuclear fuels aim to be definitely disposed off deep underground. The main function of the SNF is therefore only to restrict the radionuclides release as a function of time, and therefore delay the release towards the biosphere to take benefit from the natural radioactive decay. The major challenge of R&D is gaining a reliable source term for SNF for assessing the long term performance and safety of any SNF repository. The SNF source term is normally described as the combination of two contributions:

(i)   An instantaneous release of radionuclides, often referred to as the Instant Release Fraction (IRF), which corresponds to the radionuclides (RN) which are directly accessible to water [4].

(ii)   A slow long-term matrix contribution which corresponds to the slow dissolution of the uranium oxide matrix.

The relative weight of these two contributions is definitely opposite: IRF significantly dominates the medium term release (up to ~10^5y) due to the presence of highly mobile elements like iodine and chlorine whereas the matrix contribution is rather responsible for the constant and low long term contribution, up to millions years.

Issues related to the Instant Release Fraction

IRF was historically defined and characterized through short-term leaching experiments: it was assumed to be the inventory of radionuclides released in the first hundreds to thousands days of leaching, and for which the release rate significantly decreases with time [4]. Up to the early 2000, most of the studies aim to gather new leaching results and attempt to derive empirical correlation with other parameters like linear power. However, the increase of fuel burnup leads CEA to identify potential mechanisms which could lead to a significant evolution of the fuel microstructure and RN location in the time between irradiation and water ingress in the canister [5]: alpha decay helium ingrowth and fate, potential destabilization of grain boundaries due to pressure increase, potential diffusion of radionuclides and gases due to alpha self-irradiation enhanced diffusion ASIED [6].

In the framework of the SFS European project, the IRF concept was renewed and defined as the RN inventory located within microstructures with low confinement properties: the gap interface, the rim pores, and potentially, the rim grains and grain boundaries, these latter being options in the model [7]. This model allows explicitly accounting for the potential evolution of the fuel microstructure and the potential influence of fuel burnup. Indeed, IRF at a given time is defined as the sum of the initial IRF determined on "freshly-irradiated spent fuel" completed by any RN diffusing later on towards one of the previous low-confinement microstructures. Recent results demonstrate that alpha self-irradiation enhanced diffusion significance is not high enough to lead to a quantitative migration of RN [8]. Furthermore, Ferry et al. [8] developed in the framework of the NFPRO EU Project a micromechanical model which also demonstrates that helium ingrowth within the irradiated UOX pellet should not lead to any grain boundaries cracking, contrarily to what was initially estimated. An updated IRF assessment was recently published which significantly decreases the estimated IRF by decreasing the conservatism [9].

Table I. IRF estimates (% of total inventory) for various radionuclides for PWR UO_2 fuel, Best estimate values, with Pessimistic estimate values in brackets [9]

BURNUP (GWd/tU)	41	48	60	75
RN	IRF	IRF	IRF	IRF
fission gas	1 (2)	2 (4)	4 (8)	8 (16)
^{14}C	10	10	10	10
^{36}Cl	5	10	16	26
^{90}Sr	1 (2)	1 (3)	1 (5)	1 (9)
^{99}Tc, ^{107}Pd	0.1 (1)	0.1 (3)	0.1 (5)	0.1 (9)
^{129}I, ^{135}Cs, ^{137}Cs	1 (3)	2 (4)	4(8)	8 (16)

IRF is currently well defined but remaining uncertainties should be studied.

- First, initial distribution of RN between the various microstructures is not completely understood, in particular for high burnup and MOX fuels. This is particularly the case of highly mobile elements like ^{36}Cl for which recent studies seem to indicate a high mobility in reactor [10]. Since these mobile elements are also highly mobile in geological environment and responsible for most of the repository impact, it is of prime importance to better understand their behavior in the fuel pellet in reactor and their subsequent distribution after irradiation.
- Second, the behavior of rim zone is still poorly understood. Despite the large size of porosity and the small grains, its resistance to alteration seems to be much higher than the central part of the rod [11]. This may be related to the high fission products content of this zone as for oxidation. Anyhow, this result should be confirmed to definitely conclude

that rim does not significantly contribute to the IRF. The significance is potentially high since ..% of the RN inventory is located in the rim zone.

- Finally, the micromechanical model, which demonstrates the long term stability of grain boundary (in bulk and rim), should be confirmed (i) by new independent experimental validation of the most sensitive parameters, among which the critical bubble pressure and the stress intensity factor, (ii) and more micromechanical understanding of SNF.

Basically, the level of confidence of IRF in geological disposal for PWR irradiated fuel has been deeply enhanced by the last decade studies. The figures are below 10% for every radionuclides except ^{36}Cl due to its enhanced mobility in reactor. Although it represents only a fraction of inventory, effort has still to be maintained since its represent the main impact in deep repository in reducing conditions. Furthermore, lack of data on MOX fuels leads to very conservative figures. R&D has to be pursued on this type of fuel to either confirm these figures or decrease the conservatism to reach more realistic figures.

Issues related to the long term matrix alteration

Radiolytic dissolution
Within the reducing conditions encountered in most potential repositories, the main exception being Yucca Mountain, water alpha radiolysis was long assumed to be the governing alteration process by locally producing significant amount of radiolytic oxidants (Figure 1).

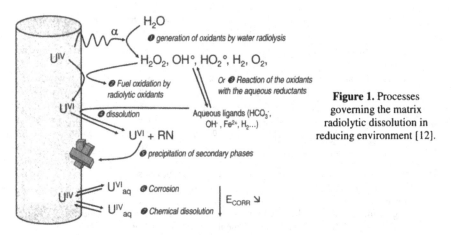

Figure 1. Processes governing the matrix radiolytic dissolution in reducing environment [12].

This process which was referred to as "radiolytic dissolution", has however been demonstrated to be efficient only above a given dose threshold (Fig.2).

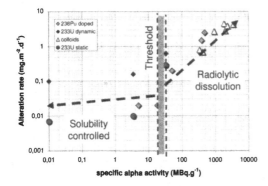

Figure 2. Spent fuel alteration rates as a function of specific alpha activity. Results are selected both from the literature and the SFS project and are obtained both on [233]U and [238]Pu doped samples and [225]Ac doped colloids [12].

This effective dose threshold is a purely empirical measurement of the competition between oxidants production and subsequent matrix alteration on the one side, and the consumption of oxidants by the other aqueous ions. It has been estimated to be in the range 18 – 33 MBq.g^{-1} which corresponds to ages between 3500 and 55000 y. depending on the fuel types: between 4.5 ky and 15 ky for a 60GWd/t UOX fuel, and between 42 ky and 55 ky for a 60 GWd/t MOX fuel [12].

In addition, hydrogen has been demonstrated to be activated in the conditions anticipated at the fuel/water interface and behave as a strongly reducing species [13]. This behavior is very surprising at low temperature and would probably mean that a catalyst is active in the system, the origin of which is not yet fully understood. Several explanations have been proposed, among which the potential reactivity of the epsilon particles. These sub-micrometric particles correspond to the metallic alloy precipitates (Tc,Ro,Pd,Ru ...) which are dispersed within the fuel pellet. These particles are poorly known and attempts have been made to accurately separate them and characterize their reactivity [14]. Other explanation may be the direct reactivity of hydrogen at UO$_2$ surface.

As a conclusion, on the long term in reducing conditions as those in deep repository, the amount of reducing species is expected to be high enough to completely counteract the production of oxidants and therefore inhibit the radiolytic dissolution. However, a more mechanistic understanding of the different processes occurring at the fuel/water interface, in particular the redox balance, is still required to definitely eliminate the potential occurrence of a radiolytic dissolution.

Corrosion and environment couplings
Once the radiolytic dissolution suppressed, other processes will govern the fuel alteration: (i) at the intermediate potential, alteration would proceed by fuel corrosion whereas (ii) at more strongly reducing environment, a purely chemical alteration is expected. Predicting fuel lifetime in such conditions would therefore require development of electro-kinetic model describing the oxidation/reductions reactions at the fuel/water interfaces in couplings with the near-environment. Large developments have been performed by the Canadian team (e.g. [15]) in this direction while accounting for the large influence of container corrosion. These models will have

(i) on the one hand, to be completed at the scientific level to account for the complexity of the natural environment, (ii) on the other hand to be simplified to derive more operational models allowing predicting the fuel alteration rate with a reasonable degree of confidence and small number of parameters. In these conditions, any reaction influencing either the redox potential or the uranium concentration may influence the fuel alteration rate, in particular the potential secondary phases. Secondary phases can also affect the hydrodynamic (pore clogging) and RN behavior (sorption). Among the various potential secondary phases, the reactivity of the U(IV) silicate, $USiO_4.(H_2O)$, coffinite, is still to be understood and its thermodynamic parameters accurately determined. It has extensively been observed in the Oklo reactor and in many uranium ore mining but its solubility is still derived from questionable natural observations and has never been measured [16]. Recent results already allow better understanding the chemical reactivity of this mineral and thermodynamics should in a near-future be achieved [17].

Finally, the effective surface area of the fuel/water interface will also significantly influence the fuel alteration rate. This parameter is poorly understood since it is directly related to the fuel microstructure after irradiation (in particular the first cycles of irradiation when the temperature gradient is still high) and the fuel/cladding/canister interface in repository. A significant R&D effort should therefore be focused on a better characterization of the fuel properties after irradiation since its subsequent evolution has been demonstrated to be negligible (see the previous discussion on helium ingrowth effect).

As a conclusion regarding the evolution of spent fuel in geological repository, very significant improvement bas been made since the late 90's and the current level of knowledge allows to conservatively model the radionuclides release from the fuel in geological repository conditions. However, relevant mechanistic understanding is still lacking to support the expected global fuel performance that is certainly well caught by the current model. Hence, R&D moves in the last year from an "exploration R&D" towards a "confirmation R&D": fuel performance is rather well determined although the understanding of accurate molecular reactions is on a scientific viewpoint still lacking.

Performance of non-classical SNF

Almost all the studies on SNF evolution in repository focused on the main stream, *i.e.* the SNF produced by current power reactors. However, within the perspective of closed fuel cycle where fuels are recycled, the most likely fuels to be disposed off are not the power reactor ones but rather the non-classical ones: material testing reactors, research reactors, fleet reactors, ... Indeed, these fuels are most often very specific and in very small quantity, and their reprocessing may not be economically viable. However, almost no results are available on the performances of these fuels in repository. In the current new perspective, this lack of knowledge should clearly be compensated by adequate R&D program taking into account the specificity of these fuel types and aiming to derive robust model to predict their performances.

FROM OPEN TO CLOSED CYCLE: SPENT FUEL EVOLUTION IN TRANSPORTATION, STORAGE AND RELATED CLADDING ISSUES

Whatever the option chosen for the final fate of spent fuel, interim storage is a likely industrial option to be implemented, either while waiting for the final repository in an open cycle, or as a flexibility tool in the global fissile element management in a closed cycle. Fuel interim storages can be implemented either in cask or vault in dry conditions (and high temperature) or in pool in wet conditions (and low temperature). Some countries are storing their fuels in dry conditions in casks, especially USA, with initial temperature in the range 350 – 450°C. In France, after exploratory studies on dry storage, the wet storage has been chosen as the reference option. Interim storage of irradiated fuels is realized both in the reactor pool on the reactor-sites for the shortest time, and in the centralized storage pools of the La Hague plant for the longest time. The industrial strategy of EDF is not to process all the annual budget of irradiated fuels discharged from reactors, but to limit the quantity recycled in order to strictly balance the plutonium produced by fuel processing and that consumed by Pu recycling in Mixed Oxide Fuel (MOX fuel). Therefore, roughly one third of the annual irradiated fuel budget is not recycled and stored. Before this storage, fuel has also to be transported from reactor-site to centralized storage or reprocessing plant (La Hague in both cases in France) and undergoes relatively high temperature during this operation which is carried out in dry conditions, the effect of which on the fuel assembly have to be assessed. It has to be stressed that in transportation and interim storage, cladding is the first confinement barrier before radionuclides release. Key issues are therefore related to its evolution and its potential risk of rupture.

The fuel pellet evolution in interim storage and transportation

In pool conditions, the very low temperature conditions strongly decrease the significance of most of the evolution driving forces. Regarding the fuel pellet, the only active process is the radioactive decay that modifies the chemical composition of the fuel, increase the irradiation damages contents of the pellet (since no annealing can be observed in such temperature range) and produces significant amount of He (alpha decay). Results obtained in the PRECCI program demonstrate that the significance of these processes is not high enough to have a macroscopic impact on the safety of the fuel storage [18]. Furthermore, large experimental databases have been used to derive robust radionuclides source term [18]. These conservative operational models as well as interpolations of experimental data can easily be used to bracket the potential RN release to support any safety analyses. Finally, any incidental scenario would lead to a potential contact between the fuel matrix and an oxidative atmosphere. Large experimental and theoretical work has also been performed on the irradiated fuel oxidations. These works demonstrate the irrelevance of the current molecular model to describe the two steps fuel oxidation from UO_2 to U_3O_8 through U_3O_7 [19]. New models have been proposed and used to derive operational model to predict the fuel oxidation rate, the subsequent fracture propagation and finally the RN release [20].

Properties of fuel cladding as first confinement barrier in transportation and storage

The main function allocated to cladding is to be the first confinement barrier in both situations.

Transportation
End-of-life internal pressure measured after irradiation in reactor ranges from 40 to 60 bars at room temperature. Thus, the cladding will be subjected to significant hoop stresses (from about 70 to 120 MPa) for the fuel rod temperatures as expected during transport or at the beginning of dry storage (up to 400°C). The main mechanism, activated by temperature, which could modify

the cladding properties in these conditions, is creep. Due to temperature and stress decrease with time, creep is expected to be particularly active during the first stages after irradiation (transport, beginning of dry storage).

Further to the research work already achieved on the Zircaloy-4, creep models have to be developed for the different cladding materials used, taking into account:
- the micro-structural state of the alloys (stress-relieved or re-crystallized),
- the irradiation defects annealing observed at relatively high temperature (this temperature depends on the duration, stress and hydrogen content),

Another challenge is the definition of a realistic strain criterion of rupture of cladding, taking into account the effects of irradiation defects annealing and re-crystallization.

Pool interim storage

Actually, wet storage does not pose any major problem, provided that the fuel assemblies are stored in water, the quality of which is strictly monitored and respect the following specifications: pH > 4.5, conductivity < 1.5 $\mu S.cm^{-1}$, Ca^{2+} < mg.L^{-1}, Na^+, Cl^-, F^- and SO_4^{2-} < 0.05 mg.L^{-1}. Indeed, the zirconium alloys show extremely high resistance to general and localized corrosion under these conditions (mean corrosion rate < 10nm.yr^{-1}). The sole issue likely to be encountered is their potential brittleness due to the significant hydride loading during the reactor life. Indeed, the Zircaloy-hydride precipitates feature low ductility (i.e. brittleness) at low temperatures (< 100°C) and present, due to this very fact, the risk of cracking under particular mechanical loads (flexion, shocks) in the course of fuel assembly or rod handling operations. In case the fuel is exposed to high temperature followed by a cooling under stress (after transportation for instance), hydrides can re-orientate in radial position and significantly decrease the cladding mechanical strength. This process has been well described for Zircaloy-4 alloy: hydrides reorientation takes place for stress higher than 90MPa and involves a significant decrease of the ductility of alloy [21].

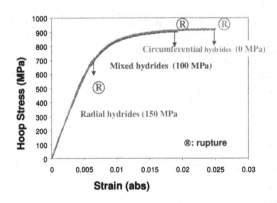

Figure 3 : experimental results on irradiated Zircaloy-4 evidencing the significant decrease of rupture resistance when hydrides become radial [21]

These results have to been extended to any new cladding material, in particular M5^{TM} alloy (Zircaloy with 1% Nb). This process is also to be assessed for structure material that plays an important role for the retrievability of spent fuel even if they do not contribute to confinement.

Dry storage

Mechanical regime will be quite different with creep being the governing process under the influence of the rod internal pressure. Determining the rate and amplitude of creep as a function of rod pressure and applied stress, temperature, and irradiation history is a key issue to assess the potential risk of cladding rupture. Significant effort has been focused on this question mainly for Zircaloy-4 in several research program among which PRECCI in France and EPRI program in the USA. They allow proposing creep model associated to creep rupture criterion [22], which determine together the possible conditions for a safe storage. Application to transportation can be directly valid since relevant timescale is accessible to experiments. For interim storage, long term extrapolation can only be valid if one can demonstrate that the governing elementary mechanisms are similar for short and long term. In particular, microstructure observations demonstrate that the irradiation defects annealing under strain, is one of the main mechanism explaining the observed creep acceleration [23]. R&D has to be supported in this direction. Extension to other cladding material is also to be addressed (e.g. Zirlo®). Finally, these coupled mechanical and microstructural studies have also to be applied to assembly structural material to confirm their retrievability and allows margin to be quantified.

CLOSED CYCLE: SPENT FUEL TREATMENT AND RECYCLING

In previous situations, spent nuclear fuel was kept as it is after irradiation and R&D has to assess their capacity to maintain their allocated functions for the long-term. In this latter scenario, the situation is rather different since spent nuclear fuel is supposed to be ultimately digested in highly acidic conditions, preliminary step for the subsequent extraction of U, Pu and potentially minor actinides. Three main fields of investigations have to be explored: the fuel dissolution (head-end of the process), the actinides extraction cycle and their subsequent conversion, and the confinement of fission products in nuclear glass. The first topic will be more detailed in this paper due to its obvious link with spent fuel, whereas others will only be briefly presented.

The key issue of the head-end of the treatment/recycling process

The head-end of the fuel treatment is of prime importance for the viability of any processing/recycling project. It determines a large part of the actinides loss all over the process as well as impurities contents in the following cycle, it corresponds to the highest active part of the whole plant and it is intimately linked to the characteristics and properties of the fuel itself. Therefore, any modification in the fuel properties (burnup, composition, type of reactor ...) should basically require evaluating the specificity and feasibility of its dissolution, at least for safety demonstration purpose. In the head-end of any treatment plant, the fuel has to be separated from the irradiated cladding (and structural material), made accessible to acidic solutions to be ultimately dissolved as completely as possible. What are the current scientific key issues in relation with the SNF scientific community? Obviously, treatment and recycling of current industrial spent fuels is already well known and more than 24 000 t of UOX fuels have already been successfully reprocessed in La Hague plants. R&D issues are therefore related to the recycling of new fuel types, in particular SFR MOX fuels with high Pu contents (> 15%).

93

Head-end

First of all, the first step of any treatment processing is to access the fuel pellet by cutting or removing the cladding. Although most of the issues are related to technological aspects, the efficiency of the cutting process is closely related to the mechanical properties of the irradiated cladding. Irradiation leads to an embrittlement of the cladding which is beneficial for the cutting step since it allows a clean rupture without any crushing of the extremity which would be detrimental for the subsequent dissolution efficiency. Furthermore, if rupture is not clear enough, it should increase significantly the amount of cutting fines (small cladding scrap). If old fuels are reprocessed after interim storage, being able to ensure that irradiation defects recovery is limited and does not decrease significantly the embrittlement is therefore important.

Another issue related to the head-end is the distribution of fission gases within the pellet. During SNF cutting, free fission gases are released in the atmosphere and have to be treated in the gases treatment process. It is therefore important to assess how much fission gases will be released during that step, and how much will be in the fuel, even after some decades of interim storage. These questions are related to a good understanding and modeling of fuel evolution in reactor but also after irradiation if any evolution could occur.

Dissolution step

The fuel dissolution could appear as a simple acidic digestion process. However, many scientific key issues have still to be better understood to allow a predictive approach of dissolution for any type of fuel. Two main questions are rising concerning fuel dissolution in highly concentrated nitric acid:

First, the dissolution kinetics. It has been experimentally measured that fuel dissolution is slowed by irradiation similarly as fuel oxidation is. This observation is not yet fully understood but could be related to stabilization of the fuel matrix by the increasing FP content and/or Pu content. Any increase in fuel burnup schematically requires to assess the relative fuel kinetics. Moreover, as Pu decreases fuel dissolution rate, moving towards Pu-richer fuel as fast reactors ones will impact the dissolution step design. The knowledge of the effective surface area of spent fuel is also of importance to predict the fuel dissolution rate and design the industrial process. Second, the dissolution process is not complete and yields to significant amount of un-dissolved materials. The un-dissolved material (the so-called dissolution fines) are of different type: (i) metallic epsilon particles composed of platinoids like Ru, Rh, Pd, Tc, Pt ..., (ii) cladding scrap formed during fuel pins cutting during the head-end, (iii) low solubility PuO_2 oxides, (iv) instable fission products in nitric acids as Mo, Zr, which can re-precipitate. Amount of dissolution fines is in the range 0.3 wt% for 33 GWd/t UOX fuel. The impact of these particles is important since they can divert Pu from the main stream (by surface interaction or incorporation for instance) or favor Pu precipitation with the subsequent criticity risk. In addition, they require a specific management in the whole process. Understanding their amount and reactivity as a function of fuel burnup requires a better understanding of the fuel microstructure after irradiation and its potential subsequent evolution. These questions are hence very similar to those addressed by the epsilon particles for fuel disposal.

The actinides partitioning and recycling, the fission products confinement.

Actinides partitioning is also a field of significant research area in order (i) to improve the robustness and proliferation-resistance of the current processes by managing U and Pu together (COEX^{TM} process) and (ii) to allow future recovering of minor actinides (MA). Extraction schemes with specific extracting molecules has already been basically defined and experimentally qualified for the different potential options for the back-end of the fuel cycle: (i) grouped minor actinides recycling (GANEX option), heterogeneous recycling in dedicated reactors or fuels (Enhanced partitioning). Additional works is related to the simplification of the process to move towards industrial application as well the study of the consolidation by considering every industrial steps to be developed around the central extraction scheme: molecules synthesis, solvents radiolysis and purification, actinides conversion ... Most of these studies are dealing with actinides chemistry and process chemistry and will hence not be detailed in that paper.

The fission products confinement in nuclear glass

The increase of burn-up to 60 GWd.t^{-1} for the future UO$_2$ and the anticipated reprocessing of MOX fuels lead to an increase in the quantity of the FP and MA per fuel ton. In order to minimise the number of packages produced per ton of fuel, the waste incorporation rate will have to be increased. The limit rate for the incorporation of FP and actinides in the glass is due to three current limitations to be studied:

- The first limitation is due to the capacity of the glass to "digest" the increased content of FP and MA (Am, Cm). In the glass grades currently produced by Areva NC, the nominal content of FP and MA oxides is approximately 15 wt %, and the maximum value 18.5 wt %. Current studies performed by CEA and AREVA demonstrate that the incorporation limits of the FP and actinides in the glass can be increased by increasing the temperature or reducing the oxygen potential during glass manufacture.
- The second limitation is related to the consequences of self-irradiation, notably alpha radiation produced by actinides. The α self-irradiation is likely to make changes in some properties (vitreous state, mechanical resistance), create stress (swelling, cracking, gas bubbles, etc.), or even change the chemical durability of the glass. The current knowledge shows that the properties of the confinement glass grades are not altered up to to an alpha disintegrations dose of at least about 10^{19} α.g^{-1}. Studies, in particular on Cm-doped glass, are performed to assess whether this threshold can be increased.
- The third limit to the increase in the incorporation rate is due to the capacity of the glass to support the thermal load generated by the disintegration of the radionuclides contained in the glass (in particular fission products and ^{244}Cm). To ensure the stability of the glass, the core temperature of the stored package shall not exceed the vitreous transition temperature. New formulations of borosilicated, rare earth-enriched glass grades are currently under study and should allow an increase in the vitreous transition temperatures and therefore in the thermal load.

Additional work on volatile and gaseous radionuclides confinement is also of high interest since future recycling plant may be located quite away from any potential dilution area.

CONCLUSION: TOWARDS A BETTER UNDERSTANDING OF SNF REACTIVITY

As a conclusion, we currently estimate that there was a significant involvement of the scientific community in the last decades on the issues related to the SNF evolution within a deep repository. This situation was justified in the framework of a once-through cycle for many of the nuclear country. This research was (i) on the one hand, successful in the sense that it allows having reliable source term model to be used in the repository performance assessment, (ii) on the other hand incomplete since most of the molecular mechanisms involved in the long term fuel alteration are not yet fully understood. This leads us to point out an important distinction which has therefore to be made between two complimentary logical that are too often not clearly distinguished: (i) the basic scientific approach which aims to identify the different processes occurring in the fuel alteration and develop the accurate model to describe them in an exhaustive manner, (ii) the more operational approach which uses the previous results to rank the different processes and select the governing ones to be used for developing predictive model. The timeframe of the two approaches is clearly different as well as the objectives. The operational approach is rather a medium term one which has to produce a conservative model to be used to assess the robustness and safety of a facility or a process, whereas the scientific approach is a long term one which aims to increase the scientific knowledge and support the operational one. We do think that this distinction has to be more clearly stated to better balance the R&D effort. Regarding the evolution in repository, successful operational models are available and have been already used to demonstrate the feasibility and safety of a deep underground repository in reducing conditions. We can state that most of the work on the operational approach is probably achieved as the conservative predicted performance is sufficient to ensure the global repository safety. On the other hand, several scientific key issues are still to be tackled to increase the underlying scientific understanding of any long term fuel evolutions, but with a limited investment.

The second important conclusion is that new operational and scientific challenges are rising due to the renaissance of nuclear energy and the concern about sustainability and recycling of fissile resources, among which are fissile element. Closed cycle will probably become progressively the reference industrial scenario in some decades. In that framework, R&D effort has on our mind to shift from SNF long term evolution to new rising issues as those related to the head-end and dissolution steps of the recycling process. Most of the skills and knowledge developed for SNF long term evolution will be useful for these new issues. Furthermore, for most of these questions, both the scientific and operational approaches are lacking and have to be developed.
Finally, in the new demand of research coupling with the main societal issues, the SNF scientific community has to demonstrate its capability to adapt to the new R&D issues derived from the new energy prospective and bring to the society scientific and technical results that will help to meet the sustainability challenge for the XXIst century.

References

[1] C. Ferry, J.P. Piron, C. Poinssot (2006), Mat. Res. Soc. Symp. Proc., vol. 932, 513-520.
[2] C. Poinssot, C. Ferry, B. Grambow, M. Kelm, K. Spahiu, A.Martinez, L. Johnson, E.Cera, J.de Pablo, J.Quinones, D.Wegen, K. Lemmens, T. McMenamin, Mat. Res. Soc. Symp. Proc.Vol.824, 421-432.
[3] K.Lemmens, B.Grambow, K.Spahiu, Y.Minet, C.Poinssot (2008), Euradwaste Conference Proc., Oct.2008, Ghent, Belgium
[4] L.Johnson, N.C. Garisto, S.Stroes-Gascoyne (1985), Proc.Waste Management, 1985, 479.
[5] C. Poinssot, P. Lovera, C. Ferry, J.M. Gras (2003), Mater.Res.Soc.Symp.Proc. Vol.757, 35-42.
[6] C. Ferry, P. Lovera, C. Poinssot, P. Garcia (2005), J. Nucl. Mat. 346, 48-55.
[7] L. Johnson, C. Ferry, C. Poinssot, P. Lovera (2005), J. Nucl. Mat. 346, 56-65
[8] C. Ferry, J.P. Piron, R. Stout, (2006), Mat. Res. Symp. Proc. Vol. 985, 65-70.
[9] C. Ferry, J.P. Piron, A. Poulesquen, C. Poinssot (2007), XXXIst Scientific Basis for Nuclear Waste Management, Material Research Society symposium, Sheffield
[10] Y. Pipon, N. Toulhoat, N. Moncoffre, L. Raimbault, A. Scheidegger, F. Farges, G. Carlot (2007), J. Nucl. Mat. 362, 416-425.
[11] B. Grambow et al. 2008: RTDC-1 Final Synthesis Report, Dissolution and release from the waste matrix. EU NF-PRO Project FI6W-CT-2003-02389
[12] C. Poinssot, C. Ferry, A. Poulesquen (2006), Mater.Res.Soc.Symp.Proc. Vol.985, 111-116
[13] K. Spahiu, L. Werme, U.B. Eklund (2000), Radiochim.Acta, 88, 507-511.
[14] D. Wronkiewicz, C. Watkins, A. Baughman, F. Miller (2001), Mater.Res.Soc.Symp.Proc. Vol.713, p.625-632.
[15] M.E. Brocskowski, J.S. Goldik, B. Santos, J. Noel, D. Shoesmith (2006), Mater.Res.Soc.Symp.Proc.Vol.985, 3-14.
[16] V.Robit-Pointeau (2005), Paris XI University (France), PhD thesis, Dec.05, 208pp.
[17] V.Pointeau, A. Deditius, F. Miserque, D. Reynolds, J. Zhang, N. Clavier, N. Dacheux, C. Poinssot, R. Ewing (subm.), J.Nucl.Mater.
[18] C. Ferry, J.P. Piron, C. Poinssot (2005), Mater.Res.Soc.Symp.Proc. Vol.932, 513-520.
[19] G. Rousseau, L. Desgranges, F. Charlot, N. Millot, J.C. Nièpce, M. Pijolat, F. Valdivieso, G. Baldinozzi and J.F. Bérar (2006) Journal of Nuclear Materials, vol 355, pp10-20
[20] A. Poulesquen, L. Desgranges, C. Ferry (2007) J. Nucl. Mat. 362 (2007), 402-410.
[21] C. Cappelaere, R. Limon, T. Bredel, P. Herter, D. Gilbon, P. Bouffioux, J.P. Mardon (2001), 8th ICEM conference Proc., Brugges, Belgium.
[22] R. Limon, S. Lehmann (2004), J. Nucl. Mat., 335, 322-334
[23] J. Ribis, F. Onimus, J-L. Béchade, S. Doriot, C. Cappelaere, C. Lemaignan, A. Barbu, O. Rabouille (2007), Journal of ASTM International, June 2007

Mater. Res. Soc. Symp. Proc. Vol. 1124 © 2009 Materials Research Society   1124-Q02-02

Influence of the Specific Surface Area on Spent Nuclear Fuel Dissolution Rates

J. Quiñones*, E. Iglesias, N, Rodriguez, J. Nieto
CIEMAT. High Level Waste Unit. Avda. Complutense, 22. 28040. Madrid

ABSTRACT

This paper focuses on the influence of the evolution of the specific surface area during the alteration process of the spent nuclear fuel. We describe results of a spent fuel Matrix Alteration Model (MAM) that allows the alteration rate evolution to be predicted as a function of the host rock considered and evaluation time scale of interest.

The changes produced by the value of the UO_2 specific surface area on the MAM model (presented in a previous MRS conference) are here analyzed. The matrix alteration rates obtained with the MAM model (for granitic environment) are presented and compared to those performed for Spent Fuel Stability project (SFS). Furthermore, a sensitivity analysis study has been performed on the influence of the following variables: influence of the initial power size distribution and the initial oxidation state.

A strong dependence of the alteration rates of the spent fuel on the specific surface area is found. These results are presented for a specific scenario, but they could be extrapolated to different environments depending on the input file.

INTRODUCTION

The Matrix Alteration Model (MAM) developed by CIEMAT, UPC (Universidad Politécnica de Cataluña), ENVIROS, and financially supported by ENRESA (Empresa Nacional de Residuos) [1, 2] works in any scenario, depending on the environment of the final option for high level radioactive waste disposal (Deep Geological Repository: DGR). MAM predicts the matrix altered rate of the spent fuel pellet with a defined geometry under different environmental conditions. Model description and the considered hypothesis are detailed in [1-3]. There are still some processes not considered by the model [2], such as influence of precipitated secondary phases over the pellet; improvement is being developed by the MICADO coordinated action. Obtained values so far using MAM have been used in performance assessment studies for ENRESA [1] and the SFS project [3], in lab experiments for validation [4] and for long term extrapolation [2, 3]. The possibility of working with different host rocks (granite, clay and salt) and the capability for long term repository extrapolation gives the MAM a great versatility.

METHODS

At present, the model assumes that alteration of the spent fuel will start when the groundwater reaches and contact the solid surface and that only the radiolytic species of the groundwater (oxidants generated by α-radiation of spent fuel) produce the surface oxidation process and subsequent matrix dissolution. O_2, H_2O_2 and OH^- are the species that react with $UO_2(s)$ for oxidation of the pellet surface. The alteration process of the U surface sites is modeled in two steps: first, a surface co-ordination of the oxidized layer with aqueous ligands and, second, detachment (dissolution) of the product species. Taking this mechanism into account, the model gives the evolution of the spent fuel matrix alteration rate over periods as long as 1,000,000 years.

The MAM works in sequential steps. For different times in the mean life of the DGR, the dose rate of the radiation fields changes. This fact produces step-functions; the same happens for the chemical environmental. The results presented here are those corresponding to the following conditions.

The scenario is developed for just one pellet surrounded by water, simplifying the geometry. Only α-radiation fields are taken into account (Figure 1 and 2) because the water is assumed to come into the DGR after 10^3 years. Granite is the host rock and bentonite is the buffer material, so the concentration of chlorides and carbonates, and also the pH level, are changing while the calculations are running. The specific surface area for this scenario (base case) is $7 \cdot 10^{-3}$ m^2g^{-1} (best-estimated value: 90 % of the pellet is altered after 1 million y). The U site density value is 165 $sites \cdot nm^{-2}$, obtained by de Pablo et al [5 and their references]. For solving differential equations the Maksima code [6] was used. In this scenario, three cases are taken into account depending on the origin of the specific surface area value: the empirical, theoretical and hybrid cases.

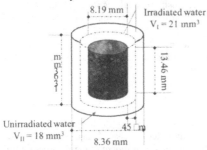

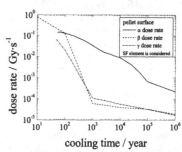

Figure 1. Pellet geometry considered in the exercise [1, 3].

Figure 2. Dose rate evolution in the deep geological repository source [2].

The empirical value for the specific surface area comes from accelerated dissolutions tests carried out at CIEMAT [7]. It has been observed that, when the 10 weight percent of sample mass has been altered, the specific surface area becomes constant. The measured values have been called into question there are several reasons to support calculated values (geometric specific surface area) more than the measured values (BET surface area): different measurement methodologies; technical difficulties of the equipment during the measurement (dead time corrections), or just the fact that BET is not a good approach to the UO_2 specific surface area value. Two adsorbates gases are used to show the influence of the measuring process: N_2 and Kr. None of the measured values is the "real" one: water will be the real adsorbate in the DGR. At the moment, we cannot choose among any of the results obtained to give the "correct" one.

The theoretical approach for calculating specific surface area comes from calculations assuming cylindrical particles for the powder analyzed. This shape is linked to the corresponding value of 3.50 for the roughness factor ($\lambda = 3.50$): λ is a factor needed to make an idealized surface more realistic [8].

The hybrid case works in two steps. It begins with the theoretical results and it follows with the empirical ones (the calculations for the specific surface area are carried on taking into account the aforementioned roughness factor). The experimental values (10 and 0.10 m^2·g^{-1}, for N$_2$ and Kr respectively) substitute the theoretical primitive data when 10 % pellet alteration is reached (around 3.45·10^5 y for this case specific surface area value). It works better following these steps because the empirical values overestimate the alterarion rate. The whole pellet would be altered very early, and this work is devloped from the perspective of performance assessment studies for a DGR [7] .

RESULTS AND DISCUSSION

Empirical approach.

Figure 3 and Figure 5 show the weight percentage of pellet alteration depending on the measurement adsorbate as a function of the cooling time for the spent fuel in DGR conditions. The related matrix alteration rates are shown in Figure 4 and Figure 6. The starting values for the performance assessment study are 0.90 and 0.03 m^2·g^{-1} for N$_2$ and Kr. In the models, 10 % of pellet mass is altered at 2500 y. The values for the specific surface area at this point are 10 and 0.10 m^2·g^{-1} (results from accelerated leaching studies). For the N$_2$ case, after 2·10^4 y, the pellet has been corroded completely. However, for the Kr case, complete corrosion takes over 10^5 y. The values obtained are always higher than in the base case (SFS [2, 3]): the best estimated value for the SFS (7·10^{-3} m^{2}g^{-1}) is very far from this empirical reality.

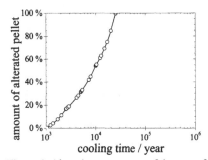

Figure 3. Alteration percentage of the spent fuel pellet in granite case, N$_2$ surface area.

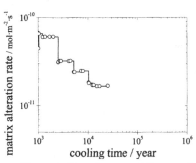

Figure 4. Alteration rate of the spent fuel pellet in granite case, N$_2$ surface area.

Theoretical approach: geometrical contribution.

Model values for the specific surface area are now calculated. The cylindrical shape of the particles and the correspondent value of 3.50 for the roughness factor [8] give a value (2.1 ·10^{-4} m^2·g^{-1}) that delays 10 % of alteration to 3.5·10^5 y. Results are presented in Figure 7 and Figure 8. After a million years the alteration of the pellet is about 30 % for the theoretical calculations versus a 90 % for the base case (SFS).

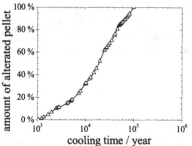

Figure 5. Alteration percentage of the spent fuel pellet in granite case, Kr surface area.

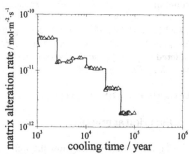

Figure 6. Alteration rate of the spent fuel pellet in granite case, Kr surface area.

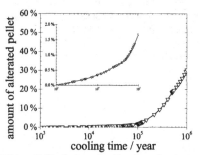

Figure 7. Evolution of pellet alteration in granite case with geometrical estimations.

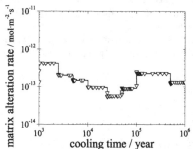

Figure 8. Alteration rate of the spent fuel pellet in granite case with geometrical estimations.

Hybrid approach. Coupled alteration rates: calculated and measured values.

Figure 9 to Figure 12 illustrate the hybrid model results. For the N_2 case the pellet is completely corroded by $9 \cdot 10^5$ y, but more gradually than the case that only takes the empirical values into account. However, for the Kr case, after a million of years, the pellet alteration will be around 90 % (similar to the base case).

Figure 13 and Figure 14 show all the models studied. The hybrid model is a very useful way of approximation to a very difficult unsolved problem. The result for the base case shown in Figure 13 runs between the results for the hybrid case. After that, the obtained values for the hybrid case lead to a similar final result in the case of Kr, despite considering even more realistic scenarios than the base case (updating the specific surface area value). Taking a look at the results here presented, from a performance assessment studies point of view, the empirical case is the most conservative, the theoretical one is the most optimistic and the hybrid approximation runs between them. It will depend on the scientific criteria of the researcher to choose the most suitable option for his work.

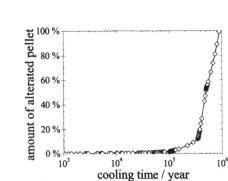

Figure 9. Evolution of pellet alteration in granite case, hybrid approach with N_2.

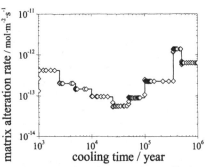

Figure 10. Alteration rate of the pellet in granite case, hybrid approach with N_2.

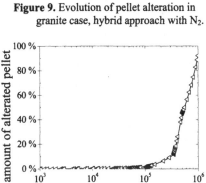

Figure 11. Evolution of pellet alteration in granite case, hybrid approach with Kr.

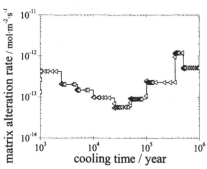

Figure 12. Alteration rate of the pellet in granite case, hybrid approach with Kr.

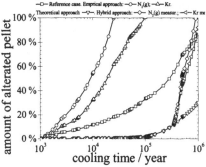

Figure 13. Evolution of the alteration in granite case. Summary of cases.

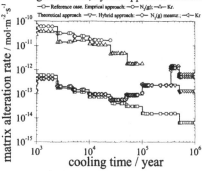

Figure 14. Alteration rate evolution in granite case. Summary of presented cases.

CONCLUSIONS

Different proposed approaches for MAM gave results capable of providing alternative pellet alteration rates for spent fuel pellet storage in a DGR. Results confirm that the specific surface area model, a controlling factor of the corroding process, can affect drastically the predicted rate of spent fuel pellet alteration.

The base case for Spain ($7 \cdot 10^{-3}$ m^{2}g^{-1}) shows that the pellet is altered 90 % after the mean life of the DGR (one million years): the alteration rate diminishes 2 orders of magnitude, achieving values under 10^{-14} mol m^{-2}s^{-1}. For the empirical approach here presented, in the N$_2$ (0.90 m^{2}g^{-1}) case, after $2 \cdot 10^4$ y the pellet has been completely corroded. For Kr (0.03 m^{2}g^{-1}), however, complete corrosion requires 10^5. For the theoretical approach ($2.1 \cdot 10^{-4}$ m$^2 \cdot$g^{-1}), calculations assume cylindrical shaped particles combined with a roughness factor of 3.50: after a million years, 30 % of the pellet is altered. The hybrid model for N$_2$ (10 m^{2}g^{-1}) case shows that, with a cooling time of $9 \cdot 10^5$ y, the pellet is completely altered, but for the Kr (0.01 m^{2}g^{-1}), the alteration is about 90 % after one million years of cooling time. From the perspective of performance assessment studies of a DGR point of view, the hybrid approximation is the more realistic, while the empirical one is the most conservative. All the results show that the specific surface area is a controlling factor of the corrosion process.

ACKNOWLEDGEMENTS

This work has been financially supported under the agreement of ACACIAS project (ENRESA-CIEMAT framework).

REFERENCES

1. A. Martínez Esparza, et al., *Development of a Matrix Alteration Model (MAM)*, in *Publicaciones técnicas*. ENRESA. (2005).
2. J. Quiñones, et al. in *Scientific Basis for Nuclear Waste Management XXIX,* edited by P. Van Iseghem, (Mater. Res. Soc. Proc. **932**, Warrendale, PA, 2006) pp. 433-440.
3. C. Poinssot et al. Ed. CEA. Vol. CEA-R-6093. Saclay. (2005).
4. K. Spahiu et al., *D10. Contract N° FIKW-CT-2001-20192 SFS. European Commission. 5th Euratom Framework Programme 1998-2002. Key Action: Nuclear Fission.* K. Spahiu (Editor) p. 63. (2004).
5. J. de Pablo et al., *Contribución experimental y modelización de procesos básicos para el desarrollo del modelo de alteración de la matriz del combustible irradiado*, in *Publicaciones técnicas*. ENRESA. Madrid, SPAIN. (2003).
6. M.B Carver, D.V. Hanley, and K.R. Chaplin, *Maksima Chemist. A program for mass action kinetics simulation by automatic chemical equation manipulation and integration by using Stiff techniques.* AECL: Chalk River, Ontario. (1979).
7. E. Iglesias, J. Quiñones, and N. Rodriguez Villagra in *Actinides 2008 - Basic Science, Aplications and Technology*, edited by D.K. Shuh, et al. (Mater. Res. Soc. Proc. **1104**, Warrendale, PA, 2008). pp. 101-105.
8. E. Iglesias and J. Quiñones. *Applied Surface Science*. **254** (21), 6890 (2008).

Mater. Res. Soc. Symp. Proc. Vol. 1124 © 2009 Materials Research Society

UO_2 Corrosion in an Iron Waste Package

Elizabeth D.A. Ferriss[1], Katheryn B. Helean[2], Charles R. Bryan[2], Patrick V. Brady[2] and Rodney C. Ewing[1]

[1] Department of Geological Sciences, University of Michigan, Ann Arbor, MI, 48109-1005 USA
[2] Sandia National Laboratories, P.O. Box 5800, MS 0779, Albuquerque, NM, 87185-0779 USA

ABSTRACT

In order to investigate the interactions among spent nuclear fuel, corroding iron waste packages, and water under conditions likely to be relevant at the proposed repository at Yucca Mountain, six small-scale waste packages were constructed. Each package differed with respect to water input, exposure to the atmosphere, and temperature. Two of the packages contained 0.1 g UO_2. Simulated Yucca Mountain process water (YMPW) was injected into five of the packages at a rate of 200 µL per day for up to two years, at which point the solids were characterized by X-ray powder diffraction, scanning electron microscopy, and electron microprobe analysis. In all cases, the dominant corrosion product was identified by X-ray diffraction to be magnetite or the structurally similar maghemite; wet chemical analysis using the ferrozine method showed that a significant fraction of the iron present in the corrosion products was ferrous. Under the conditions tested, UO_2 is expected to alter to the uranyl silicate uranophane $(Ca[(UO_2)SiO_3(OH)]_2 \cdot 5H_2O)$. Neither oxidation of the UO_2 nor any oxidized (uranyl) solid was observed, suggesting that conditions were sufficiently reducing to kinetically hinder U(IV) oxidation.

INTRODUCTION

The focus of this paper is on the nature of corrosion products of steel waste packages planned for use at the proposed nuclear waste repository at Yucca Mountain (YMR), the local redox and pH conditions inside these waste packages, and the resulting uranium mineralogy and mobility. Steel and steel corrosion products may play an important role in limiting radionuclide release [1-4]. Corroding steel inside of these waste packages may also influence the redox potential (Eh or pe) and pH, two variables that will strongly influence the degradation behavior of the spent nuclear fuel and the subsequent mobility of radionuclides. The availability of electrons from corroding steel may be an important factor in establishing and maintaining overall reducing conditions, possibly leading to the stabilization of spent nuclear fuel (SNF), kinetic hindrance of SNF oxidation, or a change in the nature of any SNF alteration products.

EXPERIMENT

Six small-scale (~1:40 by length) miniature waste packages were constructed using 316 stainless steel (nominal composition: $Fe_{62.0}Cr_{18.0}Ni_{14.0}Mo_{3.0}Mn_{2.0}N_{0.08}Si_{0.75}P_{0.045}S_{0.03}C_{0.02}$ [5]), the same material as the proposed Yucca Mountain waste packages [6], for the body, end-caps, and fittings (Figure 1). This steel corrodes less rapidly under most conditions than the A-516 carbon steel (nominal composition: $Fe_{97.87}C_{0.31}Mn_{1.3}P_{0.035}S_{0.035}Si_{0.45}$ [7]) proposed for use as guides for spent nuclear fuel inside the waste packages. The A-516 steel and 316 stainless steels were obtained from Laboratory Testing, Inc., a DOE-approved supplier, and are certified to meet the ASTM standards for those materials. Twenty-five 1 x 10 x 0.1 cm strips of A-516 carbon steel (mass 7.9g each) were inserted into each waste package, and inert polytetrafluoroethylene (PTFE) balls with a diameter of 9.53 mm were used to separate the steel strips and fill some of the excess void space. The six packages differed with respect to water input, exposure to the atmosphere, temperature, and the presence of uranium (Table I). The uranium in packages E and F was present as 0.1 g of crushed synthetic UO_2. Package D is significantly different from the other three packages because it was initiated as a scoping study, a short test to determine if the steel would corrode quickly under the chosen conditions. A simulated Yucca Mountain process water (YMPW) was injected into packages A, B, E, and F at a rate of 200 μL per day five days a week using a calibrated needle syringe. YMPW and YMPW-2, which was injected into mockup D, are similar to water currently found at Yucca Mountain (Table II).

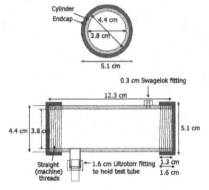

Figure 1. Schematic of miniature waste package viewed from two angles. The Swagelok fitting could be left open to the atmosphere or closed off with a Swagelok snubber.

Table I. Test matrix for waste package experiments, measured pH, and calculated pe after 1.5-2 years.

Package	Exposure to atmosphere	YMPW Volume	Temp. (°C)	Relative Humidity	UO_2 present	pH	pe
A	Snubbered	200 μl/day (~ 50 mL total)	25	100%	none	8.9	-1.4
B	Open	200 μl/day (~ 50 mL total)	25	100%	none	(6.6)[a]	(5.21)[a]
C	Open	none	25	100%	none	8.7	0.63
D	Snubbered	1 mL/week YMPW-2	60	varied	none	-	-
E	Snubbered	200 μl/day – (~ 25 mL total)	25	100%	0.1 g	10.2	-4.39
F	Open	200 μl/day – (~30 mL total)	25	100%	0.1 g	9.5	-2.16

[a] Values determined after 6 months

Table II. Saturated zone (J-13) and selected pore water compositions compared with initial concentrations in YMPW and YMPW-2.

Component	unit	J-13	Ca pore water	Na pore water	YMPW [a]	YMPW-2
pH [b]	pH	7.4 (7.8)	7.6 (8.1)	7.4 (8.3)	7.8	7.9
$SiO_{2(aq)}$	mmolar	0.95	0.66	0.66	0.83	0.90
HCO_3^-	mmolar	2.34	6.51	5.93	0.45	1.95
Cl^-	mmolar	0.20	0.59	0.68	1.00	0.02
Na^+	mmolar	1.99	1.70	5.22	1.66	2.00
Ca^{2+}	mmolar	0.32	2.35	2.02	5.02	0.87

[a] All but pH are calculated values given known Na and Si inputs, atmospheric carbon dioxide levels and calcite equilibria. Actual values produced may vary slightly.
[b] Values given in parentheses for J-13 and pore waters assume the solution is equilibrated to log fCO_2 = -3.0

The surfaces of the A-516 steel and UO_2 were examined using scanning electron microscopy with energy dispersive spectroscopy (SEM/EDS; Hitachi S3200N) prior to being placed in the packages. Packages A, B, E, and F were allowed to corrode at room temperature and 100% relative humidity until the test tubes in the lower port were nearly full of effluent. Package C was opened at the same time as packages A and B, and package D was sampled at 30 and 90 days. Characterization of corrosion products included powder X-ray powder diffraction (XRD, Scintag X1, Cu Kα radiation), SEM/EDS, and electron microprobe analysis (Cameca SX100) with wavelength dispersive spectroscopy (EMPA/WDS). SEM analyses were conducted at an accelerating voltage of 20 kV, and EMPA/WDS analyses used 25 kV and a beam current of 40 nA and typical counting times of 30 s. Total U and Fe were measured in the effluent of packages E and F using inductively-coupled mass spectrometry (ICP-MS, Thermo Fisher Finnigan Mat Element). Effluent pH was analyzed using a Ross electrode with a Symphony SB70P meter, and Fe(II)/Fe(III) values were measured using the ferrozine method [8]. These values, along with initial YMPW water chemistry, were used as input in the software package EQ3NR [9] to calculate speciation, solution-mineral equilibria, pe, and oxygen fugacities using a thermodynamic database developed specifically for application to the Yucca Mountain Project [10]. These calculations assume that all aqueous and gas phases have reached thermodynamic equilibrium. This assumption is almost certainly incorrect, and the calculated oxygen fugacities and pe's determined and reported here based on the Fe(II)/Fe(III) redox couple should be considered only estimates of a theoretical system pe that may bear little resemblance to the actual potential of the U(IV)/U(VI) couple. Although the pe values calculated using the Fe system are unlikely to be meaningful for the U redox couple, the trends that they show are useful for understanding the overall system.

RESULTS

Table I shows the pe-pH conditions inside of the packages determined from the measured pH and Fe(II)/Fe(III) ratio in the effluent. Over time, the internal chemistry becomes both more basic and more reducing as the steel corrodes, releasing more electrons from the iron. The open packages are more oxidizing than similar closed packages.

The extent of steel corrosion varied from virtually nothing, most notably in package C, which was not injected with water, to up to 50% on some strips. Typically 5% or less of an individual strip's surface appeared to be corroded. EDS analysis showed a major chemistry of Fe or Fe-O in all areas, but Cu, S, Si, and Mn were also noted in corrosion products. The major corrosion product identified by XRD in packages A, B, D, E, and F was either magnetite, Fe_3O_4, or the structurally similar maghemite, Fe_2O_3 (Figure 2).

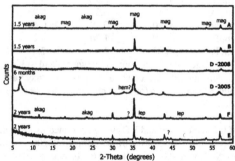

Figure 2. X-ray diffraction spectra (Cu Kα radiation) of corrosion products in packages A, B, D, F, and E showing magnetite/maghemite (mag), hematite (hem), akaganeite (akag), lepidocrocite (lep), and an unidentified mineral (?). All patterns were obtained within six hours of opening the packages except D-2008, which was obtained 3.5 years after the initial analysis. The original data for the high temperature study (D-2005) has been lost, and all peak positions for that pattern should be treated as approximate values.

Several particles containing U and O were found with SEM/EDS associated with the corrosion products. Most of these were only a few μm wide, although one 20 μm-wide grain was located and analyzed using EMPA/WDS. Measured wt % UO_2 for points on the grain were close to 100%, and very little U was noted associated with the surrounding steel corrosion products. These measurements, along with the generally sharp appearance of the boundary between the grain and surrounding corrosion products and low concentration of dissolved U in the water (0.868 ppb in package E and 4.536 ppb in package F) strongly suggest that overall UO_2 corrosion was minimal.

DISCUSSION

Lowered redox conditions resulted in kinetic hindrance of UO_2 corrosion

The high Fe(II)/Fe(III) ratios in the package effluent (assuming Fe redox chemistry dominates the system) and negligible UO_2 corrosion suggest that the internal chemistry of the packages is much more reducing than the outside air. Although the packages were filled with air initially and allowed some contact with the surrounding environment, the rate of oxygen transport into the packages was a controlling factor. The higher value of dissolved U in package F as compared with package E indicates a greater degree of oxidative dissolution of the uraninite, which follows directly from the greater access to oxygen through the open port (assuming no uranyl phases precipitated in either case).

The pe-pH diagram for U system depends largely on the water chemistry. In YMPW, which contains high levels of both Ca^{2+} and silica, geochemical modeling using the Yucca Mountain Project thermodynamic database [10] predicts that uranophane $(Ca[(UO_2)SiO_3(OH)]_2 \cdot 5H_2O)$ is the thermodynamically stable phase. The UO_2 grains in these experiments are not expected to be thermodynamically stable with the oxygenated fluids of the experiments, and their persistence indicates that oxidation was kinetically hindered. The decrease in measured redox conditions reflects a decrease in availability of the chemical reactant O_2, and limited reactant availability leads to slower reaction times. A layer of corrosion products surrounding the UO_2 grains may also have contributed by forming a protective barrier against oxygen and water.

Implications for radionuclide release from waste packages

Standard conceptual models used in performance assessments of the long-term behavior of the Yucca Mountain repository consider only two possible scenarios during the lifetime of the waste package: prior to breach, when no water is available and the corrosion rate is negligible, and after breach, when water infiltrates into the package and the spent nuclear fuel (SNF) is exposed to the open air and oxidized relatively quickly [6]. The results of this study suggest that a more accurate description of the rate of oxidative corrosion of spent nuclear fuel at the proposed repository site at Yucca Mountain should include at least one intermediate stage: when the steel begins to corrode, conditions inside of the canister are more reducing, and SNF does not corrode significantly.

Given the many tons of carbon and stainless steel contained within the actual waste packages, reducing conditions are certain to exist inside of the waste package for some time, but any estimate of that time that is based only on these short-term experiments is unlikely to be meaningful. The relative rates of oxygen ingress and steel oxidation will be largely determined by the form and extent of the breach and the rate of water flow into the package, both of which may change over time. Even assuming a constant water flow and breach size, these experiments are too short in duration to determine any relation between the complete oxidation of carbon steel (expected in tens to hundreds of thousands of years [11]) and the final redox chemistry. Oxidizing conditions may, for instance, become established well before all of the carbon steel has corroded if the breached area is large, or reducing conditions may persist long afterwards due to the effect of corrosion of the stainless steel container and/or the formation of a protective layer of corrosion products. More details and discussion of this study can be found in [12].

CONCLUSIONS

This study has examined the corrosion products of A-516 steel and synthetic UO_2 over a two-year period under conditions likely to prevail at the proposed nuclear waste repository at Yucca Mountain shortly after a waste package is breached. Over this time, the redox potential for the Fe(II)/Fe(III) couple decreased steadily, and the UO_2 did not experience significant alteration. These data suggest that for several years after breach, spent nuclear fuel corrosion and radionuclide release will be minimal.

ACKNOWLEDGMENTS

E.D.A. Ferriss is thankful for fellowships from the Office of Civilian Radioactive Waste Management and the National Science Foundation. This work was supported by the Office of Science and Technology and International (OST&I) of the Office of Civilian Radioactive Waste Management (DE-FE28-04RW12254) and NSF EAR 99-11352. The views, opinions, findings and conclusions or recommendations of the authors expressed herein do not necessarily state or reflect those of DOE/OCRWM/OSTI. Sandia is a multiprogram laboratory operated by Sandia Corporation, a Lockheed Martin Company, for US DOE's NNSA under contract DE-AC04-94AL85000.

REFERENCES

[1] C. W. Eng, G. P. Halada, A. J. Francis, C. J. Dodge, and J. B. Gillow, "Uranium association with corroding carbon steel surfaces," *Surface and Interface Analysis*, vol. **35**, pp. 525-535, 2003.

[2] C. J. Dodge, A. J. Francis, J. B. Gillow, G. P. Halada, C. Eng, and C. R. Clayton, "Association of uranium with iron oxides typically formed on corroding steel surfaces," *Environmental Science & Technology*, vol. **36**, pp. 3504-3511, 2002.

[3] B. Gu, L. Liang, M. J. Dickey, X. Yin, and S. Dai, "Reductive precipitation of uranium(VI) by zero-valent iron," *Environmental Science & Technology*, vol. **32**, pp. 3366-3373, 1998.

[4] D. Q. Cui and K. Spahiu, "The reduction of U(VI) on corroded iron under anoxic conditions," *Radiochimica Acta*, vol. **90**, pp. 623-628, 2002.

[5] ASTM, "Standard Specification for Chromium and Chromium-Nickel Stainless Steel Plate, Sheet, and Strip for Pressure Vessels and for General Applications," American Society for Testing and Materials, West Conshohocken, PN A 240/A 240M-02a, 2002.

[6] SNL, "Total System Performance Assessment Data Input Package for Requirements Analysis for Transportation Aging and Disposal Canister and Related Waste Package Physical Attributes Basis for Performance Assessment," Sandia National Laboratories, Las Vegas, NV TDR-TDIP-ES-000006 Rev 00, 2007.

[7] ASTM, "Standard Specification for High-Strength, Low-Alloy Structural Steel, up to 50ksi (345Mpa) Minimum Yield Point, with Atmospheric Corrosion Resistance," American Society for Testing and Materials, West Conshohocken, PN A 588/A588M-05, 2005.

[8] L. L. Stookey, "Ferrozine - a new spectrophotometric reagent for iron," *Analytical Chemistry*, vol. **42**, pp. 779, 1970.

[9] T. W. Wolery and R. L. Jarek, "EQ3NR Speciation-Solubility Code (EQ3/6-V8-EQ3NR-EXE-R43-PC)," The Regents of the University of California, Lawrence Livermore National Laboratory, 2002.

[10] SNL, "Qualification of Thermodynamic Data for Geochemical Modeling of Mineral-Water Interations in Dilute Systems," Sandia National Laboratories, Las Vegas, NV ANL-WIS-GS-0000003 Rev 01, 2007.

[11] SNL, "EBS Radionuclide Transport Abstraction," Sandia National Laboratories, Las Vegas, NV ANL-WIS-PA_000001 Rev 03, 2007.

[12] E. D. A. Ferriss, K. B. Helean, C. R. Bryan, P. V. Brady, and R. C. Ewing, "UO_2 corrosion in an iron waste package," *Journal of Nuclear Materials*, pp. in press, 2009.

Mater. Res. Soc. Symp. Proc. Vol. 1124 © 2009 Materials Research Society   1124-Q02-04

Immobilization of Radionuclides on Iron Canister Material at Simulated Near-Field Conditions

D. Cui [1], Y: Ranebo [2,3], J. Low [1], V.V. Rondinella [2], J. Pan [4] and K. Spahiu [5]
[1] Studsvik Nuclear AB, SE-71182 Nyköping, Sweden
[2] Inst. for Transuranium Elements (ITU), 76125 Karlsruhe, Germany
[3] Dept. of Medical Radiation Physics, Lund University, SE-221 85 Lund, Sweden
[4] Div. of Corrosion Science, Royal Inst. of Tech., 10044 Stockholm, Sweden
[5] SKB, SE-10240 Stockholm, Sweden

ABSTRACT

This work is a continuation of a long-term spent fuel leaching and radionuclides immobilization (by iron canister) experiment under simulated near-field conditions, in deoxygenated 2 mM $NaHCO_3$ solution with 1 Gy/h γ irradiation. The corrosion of iron canister material was investigated by electrochemical and microanalytical methods. Significant amounts of radionuclides (U, Np, Tc, Sr) were found to be immobilized on the corrosion layer of iron canister material by using SEM_WDS and SIMS methods. The observation is useful for bettering our understanding of near-field chemical processes at earlier canister failure conditions.

INTRODUCTION

In Sweden and Finland, spent nuclear fuel (SNF) will be placed in canisters with a corrosion resistant copper shell and a massive cast iron insert, embedded in compacted bentonite and disposed in a deep hard rock repository [1, 2]. In the safety assessment of SNF repository, it is often conservatively assumed that SNF will contact groundwater at early canister failure conditions soon after closing the repository. At this time, airborne oxygen has been consumed by ferrous minerals and radiation at the outer surface of the SNF container is about 1 Gy/h [3]. The near-field behaviors of canisters, spent fuel, and radionuclides might be influenced by O_2, H_2O_2 and oxidizing radicals generated by water radiolysis. The amounts of O_2 and H_2O_2 are relatively small (reaching a stable value ~ 10^{-7} M after several days) [4] and orders of magnitude lower than the concentration imposed by atmospheric oxygen. Therefore, in experiments on near-field redox chemistry, air contamination of the leaching system should be avoided. Corrosion behaviors of canister material at simulated near field conditions during the initial disposal period are important for the safety assessment but has not been reported previously. Metallic iron has been used as a reductant to remove U(VI) from contaminated groundwaters [5]. The reducing effect of iron canisters is expected to dominate over the oxidizing effect of radiolysis in a damaged canister [6]. Under anaerobic conditions, iron reacts with water by producing hydrogen and forming magnetite (Fe_3O_4) or green rust, ($Fe^{II}_4Fe^{III}_2(OH)_{12}CO_3$) as corrosion products [7, 8]. Recent results have shown that under anoxic conditions, metallic iron can reduce U(VI), Se(IV) and Tc(VII)to $UO_2(s)$ [8], $FeSe_2$ and TcO_2 [9], respectively. It is important to investigate the how fast radionuclides can be dissolved from SNF and be immobilized on iron canister surfaces. Earlier results from our experiments on SNF leaching have been reported [10]. It was found that, during the first period (287days), the leaching rates (inventory fractions leached per day) of all radionuclides (except Mo) are constants, higher for fission products (10^{-6}) than for actinides ($<10^{-7}$). After adding a canister material probe into the leaching vessel, the concentrations of redox sensitive radionuclides U, Tc and Np rapidly dropped. As a continuation of this long-term experiment, the near field behaviors of the radionuclides dissolved from SNF and iron canister material under anaerobic conditions with 1Gy / h radiation have been investigated by electrochemical and microanalytical methods.

EXPERIMENTAL

In the previous experiment, a 2 cm-long SNF segment with 17.7 g UO_2 and cladding (47 MW/kg burn-up, BWR) was leached in a glass vessel with 350 mL solution containing 10mM NaCl and 2 mM $NaHCO_3$ and flushed with gas mixtures (99.97%Ar + 0.03%CO_2). The pre-oxidized layer on SNF was carefully rinsed away. Iron coupons (0.5 mm) with 99.998 % pure cast iron (1 mm, 1002, SS0717-00) were cut into 4mm × 8 mm pieces and polished with Silicon Carbide paper (#4000). Four 30 mm^2 sized iron coupons were fixed on a PEEK probe (Poly-Ether-Ether-Keton) by using epoxy glue. An iron coupon and a cast iron coupon were connected by cables for electrochemical measurements (open circle potential and polarization analysis) and the other two iron coupons were used as references. The detailed procedures and results of the SNF leaching experiment were reported previously [10].

Electrochemical measurement of corrosion behaviors of canister materials

Because of the radiation field in the hot cell and the long leaching experiment period, a newly deposited Ag/AgCl electrode inserted directly in the leaching solution was used as the reference electrode for pH, Eh and corrosion potential measurements. The glass electrode and the Pt electrode were pre-calibrated by titration with a hydrogen electrode as described in [11,12]. Eh of leaching solution, corrosion potentials and impedance spectra for canister materials were measured.

Characterizing the corroded canister materials and immobilized radionuclides on iron surface

After about two years interaction in the SNF leaching vessel, the canister material probe was removed, rinsed immediately by alcohol and dried in a vacuum chamber. The metal coupons were embedded with epoxy. The cross section was polished by sand papers in the hot cell and analyzed by SEM-EDS and Secondary Ion Mass Spectroscopy (SIMS). SIMS was performed using a Cameca IMS-6f instrument. A primary O_2^+ ion was produced from a duoplasmatron ion source and accelerated by a voltage of 10 kV. Ion beam currents of 1-20 nA were used. Secondary positive ions were accelerated by a voltage of 5 kV and analysed by an electron multiplier detector. A contrast aperture of 150 um and a field aperture of 1800 um were used.

RESULTS AND DISCUSSION

Redox potential (Eh)), corrosion potentials (E_{corr}) and polarization resistances

During the experiments, the pH value was buffered around 8.60 by 2 mM HCO_3^- in solution and 0.03% CO_2 in the flushing gas. Eh values of the leaching solution measured are shown in Figure 1. The slight increase of Eh before adding SNF may reflect the effect of water radiolysis due to the radiation background inside the hot cell. Measurements of corrosion potentials, E $_{corr}$ of iron coupons provide information on electrochemical processes. The difference between corrosion potentials for iron and cast iron is a small and constant value at 15 mV, The initial corrosion potentials of iron and cast iron were around -440 mV, then sharply increased (to -320 mV) and dropped, and finally stablizing to a plateau ($E_{corr.}$ = -500 to -600 mV). At the pH-Eh-[HCO_3^-] conditions during this plateau period, carbonate green rust, $Fe(II)_4Fe(III)_2(OH)_{12}CO_3$, is expected to be a stable iron corrosion product [8]. It is indicated that iron and cast iron display similar reductive behaviors for a long period of time. The higher initial E_{corr} of iron samples may reflect that magnetite (Fe_3O_4) was the iron corrosion product at the time. The polarization resistance R_p was obtained by fitting the impedance spectra, and then corrected by the surface area of the sample. The corrosion current density I_{corr} was calculated according to the Stern-Gary relationship, and using the same Tafel constants (26 mV) for all samples, i.e., $I_{corr} = 0.026/R_p$. The

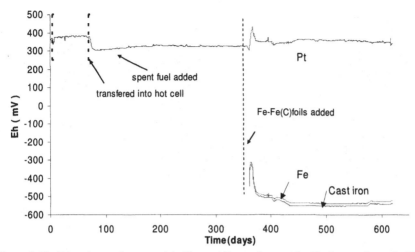

Figure 1. Eh (Pt) and corrosion potentials (E $_{Fe}$ and E $_{cast\ iron}$) vs standard hydrogen electrode (SHE).

corrosion rates were calculated based on Faraday's law, using a density value 7.8 g/cm³ and an equivalent weight 27.9 g for both pure iron and cast iron. The momentary corrosion rates of iron and cast iron calculated at two different times are listed in Table 1. It is important to keep in mind that the electrochemical corrosion rate is an instantaneous rate at the time of the measurement, and this instantaneous rate often decreases with exposure time due to formation and thickening of corrosion of products on the metal surface. During this period of time, Tafel constants were decreased (several tens percents) and the rates of further iron oxidation decreased several times [13].

Corrosion rates of canister materials measured by microscope and by electrochemical method

The polished cross sections of two reacted canister material coupons are shown in Figure 2. For corroded pure iron (Figure 2a), the white, grey and black regions represent iron metal, iron oxide and epoxy, respectively. The dark coloured grains in cast iron (Figure 2b) are carbon grains. Similar thicknesses of corrosion product layers (non-uniform, 20 - 50 μm thick after two years corrosion) observed on pure iron and cast-iron confirm that the two materials exhibit similar corrosion behaviour under these experimental conditions. This observation corresponds to the very similar E$_{corr}$ values for pure iron and cast iron (Figure 1). Corrosion of cast iron is more localized, as shown by some corroded "caves" around connected carbon grains (see Figure 2b).

Table 1. Corrosion rates (μm/year) of iron and cast iron measured by electrochemical methods and from the thickness of corrosion layers.

Corrosion rate, μm/year	Iron	Cast iron
Electrochemically measured, 2nd day	480	500
Electrochemically measured, 6th month	50	110
Estimated from corrosion layer, Figure 2	10-30	15-50

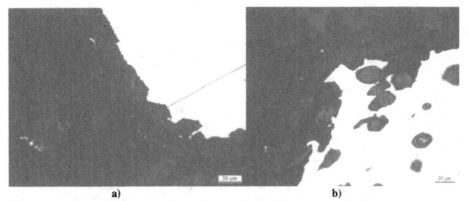

a)   b)

Figure 2. Cross sections of pure iron a) and cast iron b) canister materials (not connected for electrochemical analysis) reacted for two years. Reference bars represent 20 μm.

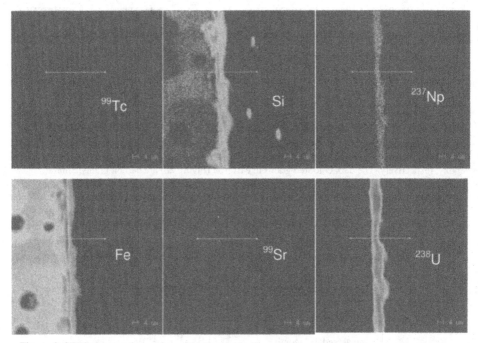

Figure 3. SIMS elemental mapping of Si, Fe, U, Np, Tc and Sr on a corrosion layer on the cast iron.

This may explain why the E_{corr} for cast iron is slightly lower while the corrosion rate of cast iron is slightly higher than that for pure iron. Furthermore, part of the corrosion product re-dissolves into the solution, which implies that the corrosion layer remaining on the metal surface represents only part of the total corrosion product. Taking into account all these sources of uncertainty, the rates derived from measuring the thickness of the corrosion product layer are in a reasonable agreement with those determined by the electrochemical method. The corrosion rates of iron coupons observed in this work are 3 to 10 times faster than that observed in synthetic groundwater under 11 Gy.hr^{-1}[14] The oxidative species such as U(VI) and H_2O_2 dissolved from SNF apparently enhanced the iron corrosion processes.

Immobilized radionuclides and Si on the iron corrosion layer

The metal coupons that reacted in the leaching vessel were embedded by epoxy and the cross section was polished. The corrosion layers and immobilized radionuclides on the polished cross sections were analyzed by SEM-WDS and SIMS. The sensitivity of SEM-WDS analysis is limited to elements present in amounts>1% and the radionuclides immobilized on iron samples are much lower than this value. The left and right parts of the images in Figure 3 represent the cast iron metal matrix and epoxy, respectively. The voids observed on the Fe image in Figure 3 correspond to carbon inclusions (carbide or graphite). As shown in Figures 3 and 4, more Si exists with the iron corrosion product layer than in the cast iron matrix as an impurity. Si in iron corrosion product was mainly from the dissolution of glass vessel wall [8,15]. ^{238}U, ^{237}Np, ^{99}Tc, ^{90}Sr and Si are observed with peaks at the middle of the corrosion product layer near the metal surface (Figure 4), where Fe(II) rich corrosion products dominate. In some Fe(III)-Si rich particles separated from iron metal (see Fe-Si mappings in Figure 3), no radionuclides were detected. Apparently, U(VI), Np(V) and Tc(VII) are reduced and immobilized in the Fe(II) rich corrosion layer on the iron surface. ^{99}Sr^{2+} may be sorbed or co-precipitated on iron corrosion products. Because primary O_2^+ was used as the ion beam source during SIMS analysis, no conclusions about element stoichiometry can be made. Actinide dioxide signals were preferred over the elemental only because of their higher count-rates (C/S). Due to the overlapping of ^{239}PuO$_2$ and ^{238}UHO$_2$ peaks in SIMS analysis, no

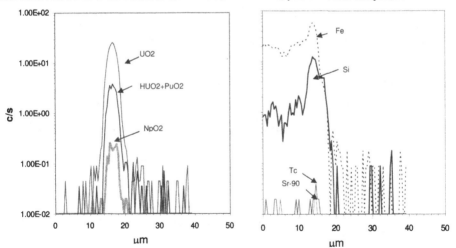

Figure 4. SIMS of radionuclide distribution in the corrosion layer of cast iron, along the line in Figure 3.

information about ^{239}Pu distribution can be given in this work. The chemical forms of iron corrosion products and the immobilized radionuclides (including ^{240}Pu) on the metal surfaces, and the effect on SNF leaching of precipitated ferric oxides in fuel-cladding gap and fuel cracks, are under investigation and will be reported separately.

SUMMARY

• The momentary corrosion rates of pure iron and cast iron determined by electrochemical measurements decrease with exposure time due to the formation and thickening of the corrosion product layers.
• Under the simulated early canister failure conditions, the corrosion rates of iron samples are about 3 to10 times faster than that in anaerobic groundwater without oxidative species dissolved from SNF, and 20 to 100 times slower than that at air saturated conditions (1 mm. yr^{-1}) [16].
• Si dissolved from the vessel wall was found in the iron corrosion layer; U, Np, Tc and Sr dissolved from SNF are immobilized in the iron corrosion layer.

ACKNOWLEDGMENTS

This work has been supported by Swedish nuclear fuel and waste management Co, (SKB), and was conducted mainly at Studsvik Nuclear AB. SIMS analysis was carried out at ITU-JRC-EC.

REFERENCES

1. L. Werme, P. Sellin, and N. Kjellbert: *Copper Canisters for Nuclear High Level Waste Disposal. Corrosion Aspects*, SKB TR 92-26 (SKB, Stockholm,1992).
2. SKBF/KBS, *Final Storage of Spent Nuclear Fuel*, KBS-3 (Swedish Nuclear Fuel Supply Co., Stockholm, 1983).
3. SKB, *Deep Repository for spent nuclear fuel, SR97 - Post closure safety*, Main Report. Vol. I, E - Main Report summary (SKB, Stockholm,1999).
4. T. Eriksen, U-B Eklund, L. Werme, J. Bruno, J. Nucl. Mater. **227**, 76 (1995).
5. B. Gu, L. Liang, J. Dickey and S. Dai,: J. Environ. Sci. Technol. **32**, 3366 (1998).
6. L. Loida, B. Grambow, H. Geckeis, and P. Dressler, Mat. Res. Soc. Symp. Proc **353**, 577 (1995).
7. N. R. Smart, D. J. Blackwood and L. Werme, Corrosion, **58**, 547 (2002).
8. D. Cui and K. Spahiu, Radiochimica Acta **90**, 623 (2002).
9. D. Cui, A. Puranen, A. Scheidegger, D, Grolimund, P. Wersin, O. Leupin and K. Spahiu, *On the interactions between iron canister material and ^{79}Se and ^{99}Tc*, NF-PRO –WP2.5 final report 2007, contract FI6W-CT-2003-02389, European Commission.
10. D. Cui, J. Low, M. Lungdsström and K. Spahiu, Mat. Res. Soc. Symp. Proc. **807**, 89 (2004).
11. G. Gran, Acta Chem. Scand. **4**, 559 (1950).
12. K. Spahiu, L. Werme, J. Low and U-B Eklund, Mat. Res. Soc. Symp. Proc. **608**, 55 (1999).
13. B. Rosborg, J. Pan, C. Leygraf, Corros. Sci., **47**, 3267 (2005).
14. N. R. Smart and A. P. Rance, *Effect of radiation on anaerobic corrosion of iron*. SKB TR-05-05, (SKB, Stockholm, 2005).
15. K. Lemmens and M. Aertsens, Mat. Res. Soc. Symp. Proc. **932,** 329 (2006).
16. G.P. Marsh, K. J. Taylor and S.H. Harker, *The kinetics of pitting corrosion of carbon steel*. SKB TR 91-62, (SKB, Stockholm, 1991).

Mater. Res. Soc. Symp. Proc. Vol. 1124 © 2009 Materials Research Society   1124-Q02-06

A new criterion for the degradation of a defective spent fuel rod under dry storage conditions based on nuclear ceramic cracking

L. Desgranges[1], F. Charollais[1], I. Felines[1], C. Ferry[2], J. Radwan[2]

[1] CEA, DEN, DEC F-13108 Saint-Paul lez Durance, France
[2] CEA, DEN, DPC F-91191 Gif sur Yvette, France

ABSTRACT

Experimental results using environmental SEM on intentionally defected fuel particles showed that oxidation induced cracking could lead to the degradation of HTR coated particles. The interpretation proposed for the swelling resulting from cracking can be extended to irradiated nuclear fuels. That is why a new criterion was proposed to defined safe handling of defective fuel in dry storage condition. This criterion defines the time needed to create an oxidized layer thickness leading to significant cracking.

INTRODUCTION

An accident scenario for nuclear spent fuel dry storage consists in cask and fuel rod simultaneous failures that will put nuclear ceramic, mostly made of uranium dioxide, in contact with air. As the temperature expected during the first 100 years of dry storage lies in the range 100-300°C, the nuclear ceramic will be oxidised, leading to the transformation of UO_2 into U_3O_8. Experimental simulation of this accidental scenario showed evidence that the swelling of the ceramic induced by oxidation could lead to the formation of cracks in the cladding and to the ruin of the spent fuel rod. Up to now the swelling of UO_2 was assigned to the formation of U_3O_8, which has a molar volume 36% higher than the one of UO_2. That is why previous criteria for safe behaviour of defected fuel rod in contact with air took into account U_3O_8 molar fraction. The fraction of U_3O_8 was deduced from analysis of weight gain curves measured during the isothermal oxidation of UO_2 un-irradiated powders [1]. The oxidation of UO_2 powders is usually partitioned in two stages: the first one is associated to a pseudo parabolic weight gain curve while the second one is associated to a sigmoid weight gain curve. The pseudo-parabolic curve, attributed to the formation of U_4O_9 and U_3O_7 on the surface of UO_2 powders, indicates a diffusion-controlled mechanism [2], for the modelling of which a finite difference algorithm [3] was recently developed. The sigmoid curve is generally interpreted as the oxidation of U_3O_7 into U_3O_8 with a nucleation and growth mechanism [4].

In the case of an un-irradiated fuel pellet, the oxidation process is also characterised by the formation of cracks. Bae showed that two types of cracking are observed [5]. First macro-cracks, associated to intermediate oxide (U_4O_9, U_3O_7) formation, occurred at the grain boundaries. It is followed by micro-cracking corresponding to U_3O_8 formation starting at cracked surfaces, which only enhances spallation.

Used nuclear fuel is made of pellets and oxidation induced cracking also has to be taken into account in the safety assessment of an accident scenario in a dry storage facility. Because macrocracking occurs before U_3O_8 formation as shown by Bae and confirmed by [6], could it lead to ceramic swelling that would be detrimental to the cladding?

This aim of this paper is to answer to this question. For that purpose new experiments were performed with environmental scanning electron microscope (ESEM) in which oxidized UO_2 sample mechanically interacted with another material.

EXPERIMENT

A coated fuel particle, used for the fabrication of fuel element for high temperature gas-cooled reactors [7], was used for an in-situ ESEM experiment. The coated fuel particle design consists of a UO_2 kernel surrounded by a low density pyrocarbon layer (buffer), a high density and isotropic inner pyrocarbon layer (IPyC), a dense silicon carbide layer (SiC) and a high density and isotropic outer pyrocarbon layer (OPyC). Each layer fulfills particular functions [8]. For our purpose it has to be noted that the SiC layer constitutes the pressure vessel of the coated fuel particle and provides structural rigidity and dimensional stability to the particle. Such a particle was intentionally damaged so that the UO_2 kernel was not protected from the atmosphere by the different layers, but was still in mechanical contact with the coating layers.

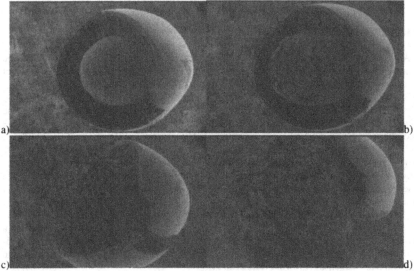

Figure 1. evolution of a damaged coated particle during isothermal oxidation at 350°C

In Figure 1, ESEM images of the defective particle are shown at different times during its isothermal annealing at 350°C. The first image (Figure 1a) was taken at the beginning of the experiment; the UO_2 kernel appears as an intact sphere inside the broken layers which surround it. The scale of the images is given by the initial diameter of the UO_2 kernel equal to 0.5 mm, all four images have the same scale. After 1 hour incubation time, cracks were formed at the surface of the UO$_2$ kernel and significant strain was visible in the layers (Figure 1b). After around 1 hour and a half of oxidation the coated layers were broken into two pieces (Figure 1c) and split the one from the other as the UO$_2$ kernel continued swelling (Figure 1d). After 5 hours of oxidation (not shown here), the apparent volume of the UO$_2$ kernel was multiplied at least by 3.

These results clearly indicated that the swelling occurring during the UO$_2$ oxidation are sufficient to break down the SiC layer in our experimental conditions. Our results also evidenced that the coated layer was broken down at the beginning of macrocracking appearance. Moreover the observed swelling (more than 300%) is at least one order of magnitude larger than the swelling expected from the crystalline transformation of UO$_2$ into U$_3$O$_8$ (36%).

Because macrocracking occurs before U$_3$O$_8$ formation [5], these results make it obvious that macrocracking is the driving force for the degradation of the coating layers.

DISCUSSION

Our ESEM results proved that UO$_2$ macrocracking can be detrimental for coated layer in SiC. In this discussion three points will be addressed related to safety criterion for an incidental scenario in dry storage:
- what is the mechanism leading to the mechanical degradation due to macrocracks formation ?
- are there any proofs that this mechanism could occur in irradiated nuclear fuel?
- if it was so, how is it possible to define a criterion warranting no degradation of the cladding?

Mechanical stresses induced by macrocracking

Macrocracking is interpreted as the result of the formation of intermediate oxides (U$_4$O$_9$, U$_3$O$_7$) at the surface of UO$_2$. Because these intermediate oxides have a smaller unit cell parameter than UO$_2$ they induce tensile stresses parallel to the surface and cracks formation as soon as these tensile stresses reach a threshold. Because these intermediate oxides have a smaller molar volume than UO$_2$, they are expected to produce some shrinkage of the sample rather than its swelling as observed. This apparent contradiction can be resolved taking into account 3D effects.

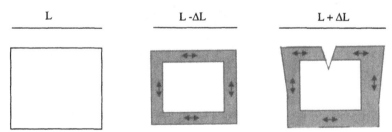

| L | L -ΔL | L + ΔL |

Figure 2: schematic representation of tensile stress (black arrow) and strain (line at the top) in an oxidized layer on a UO$_2$ sample.

On figure 2, a schema of a UO_2 sample at different oxidation times is represented. At the beginning a uniform oxidized layer covers the sample surface; tensile stresses in this layer induce some shrinkage of the sample. When a crack is formed, the tensile stresses are relaxed at the crack position but still remain on the oxidized layer far enough from the crack. These remaining stresses tend to open the crack tips and consequently, when the sample does not interact mechanically with other elements, to move the edges of the sample leading to an apparent increase of the sample length. This description is consistent with Figure 1 where crack opening is clearly observed on the free surface of the UO_2 kernel. At the contact between the coated layer and the UO_2 kernel, a strong mechanical interaction occurred which prevented the swelling of the UO_2 kernel and generated high stresses in the coated layer responsible of its breakdown. It is also important to notice that the sample was not transformed into a powder, as it would have been expected if the transformation of UO_2 in U_3O_8 occurred without Macro-cracking.

Evidence of macrocracking in irradiated fuel

Several studies have been already performed in which macrocracks were clearly evidenced. However the observation was performed on irradiated fuel fragments only, thus the relationship between cracking and the mechanical degradation of the cladding is not easy to establish. The morphology of those cracks is nevertheless very compatible with the swelling mechanism presented in Figure 2. Two types of cracks have to be considered: intra and inter-granular ones. An example of each type of crack as observed on oxidized irradiated fuel is presented on figure 3. On Figure 3a [9], an intra-granular crack was observed on an irradiated UO_2 fuel after approximately 800 hours of oxidation at 200°C; within a grain, the oxidized layer is brighter than the un-oxidized core and the crack morphology is very similar to the schema of Figure 2 where the sample would be a UO_2 grain. On figure 3b [10], several inter-granular cracks are visible on an irradiated UO_2 fuel after approximately 4000 hours of oxidation at 200°C; here the inter-granular cracks joined the one to the other forming long lines. These lines are very similar to the cracks observed on the UO_2 kernel on Figure 1.

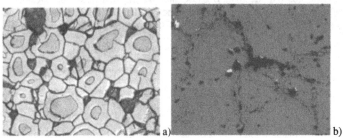

Figure 3. Optical observation of a cross section through a UOX fragment (a) on a 40x30μm² after 800 hours of oxidation, and (b) on a 400x300 μm² area after more than 4000 hours of oxidation

These results bring into prominence the similarity between the cracks formed on un-irradiated UO_2 pellet or kernel with the ones observed on oxidised irradiated nuclear fuel. This implies that macrocracking also has to be taken into account for the degradation of a defective fuel cladding in the same manner it was taken into account to interpret coated particle degradation.

A new safety criterion

The previous criterion, proposed for the safety of a dry storage facility [11], defined the time during which no detrimental effect can occur during fuel oxidation based on the amount of U_3O_8 formed within the irradiated fuel. We have seen that macrocraking occurring before U_3O_8 formation can lead to the degradation of the cladding by the induced swelling. This led us to propose a new criterion depending on macrocrack formation rather than on U_3O_8 formation.

As said in the interpretation of Figure 2, the oxidized layer of intermediate oxide has to reach a critical depth before a crack could be formed. So we proposed to use the time needed to reach a critical depth of the oxidized layer as a new criterion. Taking the time needed to form the first crack is too conservative, because significant cracking is needed to get macroscopic stress field. In order to define a critical depth representative of significant swelling, we referred to the oxidation weight gain curves on irradiated fragment. This curve has a similar shape than for un-irradiated powder with a pseudo parabolic curve followed by a sigmoid one. After a comparative study of UO_2 pellet morphology and their weight gain curves, Bae et al. [5] stated that, for pellet oxidised at 400°C, "after incubation time, crack propagation rate could control the constant oxidation rate". Considering the weight gain curve of irradiated fragment, the pseudo-parabolic curve would correspond to the incubation time and the sigmoid curve would correspond to kinetic controlled by crack propagation. Thus the time at which crack propagation has a significant influence on the irradiated fragment would be the onset of the sigmoid curve.

Thus the new criterion we propose is related to the time needed to form 1µm thickness of $U_4O_{9\gamma}$ phase with a stoichiometry higher than 2.399 (if it is assumed that the O/M value on plateau is at maximum 2.4), in irradiated grains. This $U_4O_{9\gamma}$ layer thickness was chosen to correspond to the onset of the sigmoid curve. An evaluation of this safety duration was performed thanks to the model predicting the irradiated fuel oxidation kinetic [12], which derives from a previously published modeling of oxidation weight gain of un-irradiated powder [3]. The calculated safety duration is consistent with the $t_{2.4}$ criterion proposed by Einziger et al. [11]. In fact both criterions led to similar values because they both take into account the onset of the sigmoid curve in the weight gain curves. Yet, our new criterion is mechanistic since it is based on oxidation mechanisms as observed on UO_2. Thus it is robust and can safely be extrapolated over periods up to a few hundred years for which no experimental data are available. It must also be noted that previous criterions, which considered U_3O_8 formation as the driving force for fuel rod degradation, could have be considered very conservative, because some time was still needed after the onset of the sigmoid weight gain curve before a significant amount of U_3O_8 was formed. We show that there is less conservatism than previously thought because significant macro-cracking occurs near the beginning of the sigmoid curve.

CONCLUSIONS

Experimental results with ESEM on intentionally defected fuel particles showed that oxidation induced cracking has to be considered as the driving force for the degradation of defective fuel rod in dry storage conditions.

Based on this result a new safety criterion was defined The apparition of significant cracking and consecutive bulking is associated to the formation of a critical depth of $U_4O_{9\gamma}$; the time, at which the $U_4O_{9\gamma}$ critical depth is formed, is calculated thanks to a finite difference model derived from a previously published model dedicated to un-irradiated UO_2 oxidation. Basically, this criterion leads to similar numerical values compared to previous ones, but it is now physically based and can thus be extrapolated with confidence over time periods for which no experimental data are available. Moreover the criterion is not over-conservative, because it defines the time at which macrocracking becomes significant and from which macrocraking could induce significant damage.

ACKNOWLEDGMENTS

The authors are grateful to Electricité de France (EDF) for its financial support within the Research Program on the long term Evolution of Spent Fuel Waste Packages of the CEA (PRECCI).

REFERENCES

1. R. J. McEachern and P. Taylor, J. Nucl. Mater. **254**, , 87-121(1998).
2. R. J. McEachern, J. Nucl. Mater. **245**, 238-247 (1997).
3. A. Poulesquen, L. Desgranges, C. Ferry., J. Nucl. Mater. **367**, 402-410 (2007).
4. R. J. McEachern, J.W. Choi, M. Kolar, W. Long, P. Taylor. and D. D.Wood, J. Nucl. Mater. **249** , 58-69 (1997).
5. K.K.Bae, B.G.Kim, Y.W.Lee, Yang, H.S.Park, J. Nucl. Mat. **209**, 274-279 (1994).
6. L. Quémard [1], L. Desgranges [1], V. Bouineau [1], M. Pijolat [2], G. Baldinozzi [3], N. Millot [4], J.C. Nièpce [4] and A. Poulesquen [5] submitted to the Journal of Nuclear Materials
7. F.Charollais et al. Nuclear Engineering and Design **236**, 534–542 (2006).
8. Verfondern, K. (Ed.), 1997. Fuel Performance and Fission Product Behaviour in Gas Cooled Reactors. IAEA Report IAEA-TECDOC-978
9. L.Desgranges et al., "Study of the evolution of a defective fuel rod in contact with the atmosphere within the framework of the PRECCI Program", ICEM'01, Proc. of the 8th International Conference on Radioactive Waste Management and Environmental Remediation (TABOAS, A., VANBRABRANT, R., BENDA,G., Eds), Bruges, Belgium, Sept.30- Oct. 4, 2001.
10. L. Desgranges, G. Rousseau, M-P. Ferroud-Plattet, C. Ferry, H. Giaccalone, I. Aubrun, P. Delion, J-M. Untrau, Mater. Res. Soc.Symp. Proc., 932, paper 18 (2005)
11. R.E.Einziger, L.E.Thomas, H.C.Buchanan, R.B.Stout J. Nucl. Materials **190**, 53 (1992)
12. Poulesquen A. to be published

Mater. Res. Soc. Symp. Proc. Vol. 1124 © 2009 Materials Research Society 1124-Q02-07

Corrosion Studies with High Burnup LWR Fuel in Simulated Groundwater

Ella Ekeroth[1], Jeanett Low[1], Hans-Urs Zwicky[2], Kastriot Spahiu[3]
[1]Studsvik Nuclear AB, Hot Cell Laboratory, SE-611 82 Nyköping, Sweden
[2]Zwicky Consulting GmbH, Mönthalerstr. 44, CH-5236 Remigen, Switzerland
[3]SKB, Box 250, SE-101 24, Stockholm, Sweden

ABSTRACT

The burnup of future spent fuel to be disposed will be higher than the burnup of today's fuel. Actinides accumulate in the rim zone and the content of lanthanides and other fission products will also increase in spent fuel as a consequence of higher burnup. The increased actinide and fission product content leads to higher α-dose rate and higher β- and γ-dose rates initially in the surrounding water. The dissolution rate is expected to increase with higher burnup due to higher dose rates and due to an increased surface area in the rim zone. On the other hand, the presence of fission products like lanthanides in the UO_2 matrix has been shown to have an inhibiting effect on UO_2 dissolution.

Previous static corrosion tests on spent fuel with a burnup range of 27 to 49 MWd/kg U showed that the cumulative release fractions increase with burnup to reach a maximum at approximately 40-45 MWd/kg U. At higher burnup (up to 49 MWd/kg U) the release rates decreased. This study has been extended to comprise PWR spent fuel between 55 and 78 MWd/kg U burnup. Preliminary results from four contact periods, for a cumulative contact time of 182 days, show no systematic increase in cumulative release fractions with increasing burnup. The $^{236}U/^{235}U$ ratio in the leaching solutions is also discussed.

INTRODUCTION

In the Swedish concept for disposal of high level waste [1], the spent nuclear fuel will be encapsulated in copper canisters with an inner cast-iron insert. The canisters will then be placed in a deep repository built at a depth of about 500 m, surrounded by compacted bentonite clay. This arrangement constitutes a multiple barrier deep repository system.

Spent fuel is largely $UO_2(s)$ with only a small fraction of other actinides and fission products. The majority of these radionuclides are dispersed or in solid solution in the $UO_2(s)$ matrix [2]. It is important to evaluate the rate of dissolution of the spent fuel matrix and the release rate of the various radionuclides, the so-called source term. The rate of dissolution of spent fuel depends on a variety of factors such as the composition of the spent fuel itself and of the groundwater, as well as the redox conditions under which the dissolution takes place. Fuel radiation causes radiolysis of water and reactive species are formed of which H_2O_2 was shown to be of highest importance [3]. In this work we focus on the influence of material properties of the fuel related to fuel burnup on fuel leaching under oxidizing conditions.

The majority of LWR fuels are largely $UO_2(s)$, enriched in ^{235}U (typically in the range 3 to 5 %). In the reactor, ^{235}U is consumed ('burned') by nuclear fission. At the same time, higher actinides are produced through neutron capture and decay reactions. At present, the majority of the fuel to be disposed in Sweden has an average burnup of around 40 MWd/kg U. Older fuels have in general lower burnup, but the current trend is towards higher burnups [4, 5]. Fuel burnup leads to complex and significant changes in the composition and properties of the fuel. The transformed microstructure, which is referred to as the high burnup structure (HBS) or rim structure in the outer region of the fuel, consists of small grains of submicron size and a high concentration of pores of typical diameter 1 to 2 μm. This structure forms in UO_2 fuel at a local burnup above 50 MWd/kg U, as long as the temperature is below 1000-1100°C. The high burnup at the pellet periphery is the consequence of plutonium build-up by neutron capture in ^{238}U followed by fission of the formed plutonium.

The chemical composition and microstructure of nuclear fuel have been studied extensively [2, 5, 6] and only a short summary is given below. Fission products which are stable in metallic form (Mo, Ru, Pd, Tc, Rh) tend to form metallic alloy particles, often referred to as 4d-alloy particles or ε-particles. Fission products which are stable as oxides but incompatible with the UO_2 matrix (Rb, Cs, Ba, Zr, Nb, Mo, Te, Sr) separate into precipitates sometimes referred to as grey phases. Elements that form stable oxides in solid solution with UO_2 matrix include actinides (Np, Pu, Am, Cm), lanthanides (La, Ce, Pr, Nd, Pm, Sm, Eu, Gd) and Y as well as Sr, Zr, Ba, Te and Nb within the limits of their solubility in UO_2 and to the extent that they have not precipitated in perovskite-type oxides.

Usually when discussing high burnup fuel dissolution, the effect of the increased radiation field with burnup, as well as of the influence of the smaller grain size and increased porosity at the rim are mentioned as factors which contribute to increased dissolution rates. A third factor, which is the increase of fission product and actinide doping in high burnup fuel, has been discussed extensively in connection with increased resistance to air oxidation of the fuel [7-9], but only recently in connection with fuel dissolution [10-13].

In an experimental series (further on referred to as *series 11*) performed using fuel material from a segment of a stringer rod with burnup varying between 21 and 49 MWd/kg U along the fuel column [14], the cumulative fractional release increased slightly, almost linearly, with burnup up to values of 40-45 MWd/kg U, but afterwards decreased. Therefore, a series of experiments has been started at Studsvik, aiming at extending the data base acquired in the series 11 corrosion tests to higher burnup fuel.

EXPERIMENTAL

Leaching experiments were performed with four spent fuel segments from fuel rods irradiated in different PWRs, see Table I.

Table I. Fuel rod data

Specimen lot	PWR UO$_2$ 57.9 MWd/kg U	PWR UO$_2$ 62.7 MWd/kg U	PWR UO$_2$ 65 MWd/kg U	PWR UO$_2$ 78 MWd/kg U
Reactor	Ringhals 3	Ringhals 4	Ringhals 3	North Anna 1/2
Irradiation period	1989 – 1994	1998-09-13 – 2003-07-31	2000-07-12 – 2005-05-25	1987-07-01 – 2001-03-12
Initial ^{235}U enr.	3.6 wt%	3.8 wt%	3.7 wt%	4.0 wt%
Rod average burnup	52.2 MWd/kg U	60.0 MWd/kg U	62.8 MWd/kg U	70.2 MWd/kg U
Exp. det. Kr release	0.9 %	2.7 %	2.3 %	5.0 %

From each spent fuel rod, a fuel pin segment, containing one complete and two half pellets, is leached under oxidizing conditions in synthetic groundwater at ambient temperature. The present paper covers the first four contact periods up to a cumulative contact time of 182 days.

Previous to the experiment, the samples were washed by exposing them to a 10 mM NaCl/2 mM NaHCO$_3$ solution for about two hours. A certain part of the instant release fraction was thereby washed away. Then they were rinsed with pure water and air-dried. The samples, kept in position by a platinum wire spiral, were exposed to 200 ml of synthetic groundwater; see Table II, in a Pyrex flask. After each contact period, samples are collected for Inductively Coupled Plasma - Mass Spectrometry (ICP-MS) isotopic and γ-spectrometric analyses and for pH and carbonate determination (γ-spectrometric, pH and carbonate results are not presented here due to limited space). The contact periods were 7, 21, 63 and 92 days and data collection is ongoing according to the same experimental scheme as in [14]. Only an insignificant amount of soluble elements were found in the vessel strip solutions; therefore corrosion performance was assessed on the basis of the results from the centrifugates.

The release fractions are based on generic CASMO [15] calculations of nuclide number densities at the end of irradiation as a function of burnup for a typical PWR power history and an initial ^{235}U enrichment of 4%. The chemical determination of the inventory is in progress and will be presented in future work. Release fractions (Fraction of the Inventory in the Aqueous Phase, FIAP) are calculated by dividing the total amount of a nuclide of concern in the analyzed solution by the total amount in the corroding fuel sample. Cumulative release fractions are the sum of release fractions up to a certain cumulative contact time.

Table II. Composition of synthetic groundwater

Element	Ca	Mg	K	Na	Si	HCO$_3$	SO$_4$	Cl	F	Phosphate
Conc. (mM)	0.45	0.18	0.10	2.84	0.21	2.01	0.10	1.97	0.20	0.001

RESULTS AND DISCUSSION

Analytical data for all nuclides discussed in [14] were collected and the results for a selected number of nuclides allow a first comparison to previous data (series 11) [14]. It should be kept in mind that all fuel segments of series 11 were from the same BWR fuel rod while the fuel segments used in this study are from PWR fuels of different origin, irradiated under different conditions.

The concentrations of uranium for the fourth contact period (92 days) are relatively low (less than 4×10^{-6} M) and quite similar for the four fuel samples investigated. They are much lower than the solubility of the kinetically favored phase shoepite under such conditions and in good agreement with the concentrations measured with a similar groundwater by Jegou *et al.* [16, 17] up to 313 days. Another observation is that they are several (2-10) times lower than the uranium concentrations measured under the same time period for the 10 fuel samples of series 11. These observations indicate that very probably no secondary uranium minerals affect uranium releases in our data.

Figure 1 shows present and series 11 [14] data on cumulative release fractions for a cumulative time of 182 days plotted as a function of sample burnup. As seen from the figure, the Cs fractional release is slightly higher than for the previous series 11. This may be due to our short washing time to eliminate completely the instant release fraction (IRF), which usually increases at higher burnups [18]. However, the trend points at lower release fractions with increasing burnup. The uranium releases are lower than the previous series 11 results and no clear trend with burnup can be observed. The cumulative release fractions for Sr and Ba are quite similar to the corresponding releases from the lower burnup series 11 samples. On the other hand Rb releases are lower than those of series 11 and no increase with burnup is observed. Mo and Tc cumulative release fractions are lower than those for series 11 and seem to decrease with increasing burnup. This behavior may be due to the fact that a large part of their inventory is in the form of metallic particles.

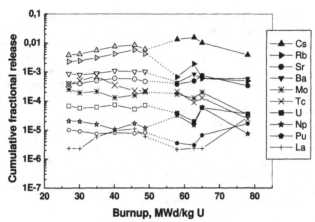

Figure 1. Cumulative release fractions for a cumulative contact time of 182 days (four contact periods) in comparison with a selection of corresponding data from [14] (open symbols).

The absence of any marked influence of increased burnup in fuel dissolution rates has been observed in other work carried out under quite similar conditions [16, 17]. The same trend is observed in flow-through fuel leaching tests with fuel powder in carbonate solutions [13], where the cumulative uranium releases are highest from a fuel in the intermediate burnup interval (44 MWd/kg U) even as compared to high burnup fuels (up to 70 MWd/kg U). Finally, tests carried out at ITU during the EU-Project NF-PRO [19] with fuel samples taken from the drilled central part of a high burnup pellet and the outer part which contains also the rim, showed slightly lower releases from the outer part (i.e. higher burnup) for almost all radionuclides.

The $^{236}U/^{235}U$ ratio in the fuel increases as the irradiation proceeds. The ^{236}U content increases with burnup at the expense of ^{235}U, which decreases due to both fission and neutron capture. The ratio varies radially in the pellet with a marked increase at the rim where the burnup is up to 2.5 times higher than the average pellet burnup [20]. The $^{236}U/^{235}U$ ratio can serve as an indicator of corrosion site by comparing the $^{236}U/^{235}U$ ratio in the leaching solution with the inventory average $^{236}U/^{235}U$ ratio for fuel being leached.

In Figure 2, the calculated average $^{236}U/^{235}U$ ratio for a PWR fuel with initial 4 wt% ^{235}U enrichment is plotted as solid line, together with the values of the ratio measured for the four fuel samples during the four first contact periods. In general the values measured for each of the four fuel samples coincide with the average pellet ratio. For each sample, the ratio increases with the number of contact periods. It is interesting to note that the increase of the $^{236}U/^{235}U$ ratio in progressive contact periods was observed also in series 11 [14]. However, data for centrifugates in series 11 seem to be below the pellet average $^{236}U/^{235}U$ ratio during the first contact periods, indicating that the corrosion occurs to a lesser extent in the rim zone despite a higher surface area and dose rate in this region. In our case, it is not possible to draw any conclusion if the measured $^{236}U/^{235}U$ ratios for each centrifugate are below or above the average $^{236}U/^{235}U$ ratio for each pellet until the chemical inventory analyses are concluded.

In the discussion of high burnup fuel dissolution, the influence of the smaller grain size and increased porosity in the rim region are mentioned as factors which cause an increase of the surface area. The increased actinide content in spent fuel at higher burnups will lead to a higher α-dose rate in the surrounding water and the higher content of fission products, see Figure 3, will also contribute to a higher β- and γ-dose rate initially. Both these factors should contribute to higher cumulative release fractions for uranium and for other "matrix bound nuclides" with increasing burnup. Only in recent work, the influence of a third factor, namely the increase of the dopant content, i.e. non uranium fission product and actinide atoms in the UO_2 matrix of the spent fuel with fuel burnup has been discussed in connection with fuel dissolution studies [10-13]. The most plausible explanation to our results seems to be that the increased amount of doping at higher burnup causes an increased resistance to oxidative dissolution of the UO_2 matrix, which counteracts effectively the increased surface area and radiation dose.

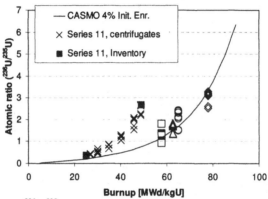

Figure 2. Atomic ratio $^{236}U/^{235}U$ in spent fuel as a function of burnup. Open symbols correspond to centrifugate data for four contact periods. Crosses refer to corresponding data for series 11. The two filled squares are measured inventory values in series 11 [14].

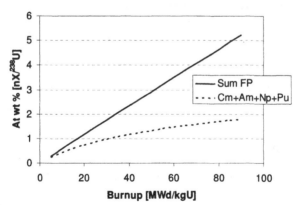

Figure 3. Evolution of fission product and actinide content with burnup based on CASMO calculations (initial 235U enrichment of 4 %).

SUMMARY

A selection of data acquired in corrosion tests with high burnup PWR fuel under oxidizing conditions in simulated groundwater during a cumulative contact time of 182 days has been evaluated and compared to corresponding series 11 results.

The cumulative release fraction is suggested as a reliable way to evaluate spent fuel dissolution [11]. With the limited data presently available we can only note that the releases from the high burnup samples tested in this work are not systematically higher than those from series 11, and in some cases a decrease is observed. It can thus be concluded that the corrosion performance of high burnup PWR fuel under oxidizing conditions is not completely different from the performance of lower burnup BWR fuel. With further leaching contact periods, the influence of the rapid release fraction will decrease and a better discussion of the matrix dissolution trends is expected.

REFERENCES

1. *Final Storage of Spent Nuclear Fuel - KBS-3*. SKB, Stockholm, Sweden (1983)
2. H. Kleykamp, *J. Nucl. Materials*, **131** 221-246 (1985)
3. E. Ekeroth, O. Roth, M. Jonsson, *J Nucl Mater* **255** 38-46 (2006)
4. L. Werme, L.H. Johnson, V.M. Oversby, F. King, K. Spahiu, B. Grambow, D.W. Shoesmith, Spent fuel performance under repository conditions, SKB TR-04-19 (2004)
5. L.H. Johnson and D.W. Shoesmith, *Spent fuel. In: Radioactive Waste Forms for the Future, Eds.:*W. Lutze and R.C. Ewing, North-Holland Physics Publishing, The Netherlands (1988)
6. H. Kleykamp, *Nuclear Technology*, **80** 412-421 (1988)
7. L.E. Thomas, R.E. Einziger and H.C. Buchanan, *J. Nucl. Mater.*, **201** 310-319 (1993)
8. B.D. Hanson, The burnup dependence of light water reactor spent fuel oxidation. Pacific Northwest National Laboratory Report, PNNL-11929 (1998).
9. R.E. Einziger, L.E. Thomas, H.C. Buchanan, R.B. Stout., *J. Nucl. Mater*, **190** 53-60 (1992)
10. H. He, P.G. Keech, M.E. Brockowski, J.J. Noel, D.W. Shoesmith, *Can. J. Chem*, **85** 1-12 (2007)
11. B.D. Hanson, R.B. Stout, *Mat. Res. Symp. Proc.* **824** CC2.4.1-2.4.6 (2004)
12. B.D. Hanson, J.I. Friese, C.Z. Soderquist, *Mat.Res.Symp. Proc.*, **824** CC8.6.1-8.6.6 (2004)
13. B. Hanson, Flow through dissolution of CSNF under oxidizing conditions, DOE/CEA Technical Meeting, 2/09/2005, www.ocrwm.doe.org.
14. R. Forsyth, The SKB Corrosion Programme. An evaluation of results from the experimental programme performed in the Studsvik Hot Cell Laboratory, SKB TR-97-25 (1997)
15. D. Knott, B. Forsen, M. Edenius, SR SOA-95/2, Studsvik Core Analysis AB (1995) and J. Rhodes, M. Edenius, Studsvik Scanpower SSP-01/400 Rev.4.
16. C. Jegou, S. Peuget, J.F. Lucchini, C. Corbel, V. Broudic, J.M. Bart, *Mat. Res. Soc. Symp. Proc.* **663** 399-407 (2001)
17. C. Jegou, S. Peuget, V. Broudic, D. Roudil, X. Deschanels, J.M. Bart, *J. Nucl. Mater.* **326** 144-155 (2004)
18. L. Johnson, C. Ferry, C. Poinssot, P. Lovera, *J. Nucl. Mater.* **346** 56-65 (2005)
19. F. Clarens, D. Serrano-Purroy, A. Martínez-Esparza, D. Wegan, E. Gonzalez-Robles, J. de Pablo, I. Casas, J. Giménez, B. Christiansen, and J.P. Glatz et, *Mat. Res. Soc. Symp. Proc.* **1107** 439-446 (2008)
20. H. Matzke, J. Spino, *J. Nucl. Mater.* **248** 170-179 (1997)

Nuclear Waste Glasses
and Vitrification

Mater. Res. Soc. Symp. Proc. Vol. 1124 © 2009 Materials Research Society   1124-Q03-02

Iron Redox Reactions in Model Nuclear Waste Glasses and Melts

Benjamin Cochain[1,2], Daniel R. Neuville[2], Jacques Roux[2], Dominique de Ligny[3], Denis Testemale[4], Olivier Pinet[1] and Pascal Richet[2]

[1]CEA, DEN, DTCD, SECM, LDMC, Bagnols-sur-Cèze, France
[2]CNRS-IPGP, Physique des Minéraux et des Magmas, Paris, France
[3]LPCML, UCBL, Lyon, France.
[4]Institut Néel, MCMF, Grenoble, France

ABSTRACT

The influence of boron on the kinetics of oxidation of iron in silicate melts relevant to nuclear waste storage has been investigated by XANES experiments. The measurements have been performed isothermally as a function of time at the iron K-edge. The redox kinetics become slower with increasing B_2O_3 content either close to the glass transition range, where the redox kinetics are controlled by diffusion of network-modifying cations, or at superliquidus temperatures where oxygen diffusion is the rate-limiting factor. In both ranges the kinetics can be interpreted in terms of boron speciation and interaction with alkali cations. Below the liquidus, however, the long times needed to reach redox equilibrium allow sintering of the powders investigated to take place so that the resulting changes in sample geometry prevent determinations of oxidation kinetic parameters from being made.

INTRODUCTION

Borosilicate glasses are privileged materials for nuclear waste storage because they allow a wide variety of elements to be incorporated in a stable and compact matrix. These glasses include various multivalent elements (Ce^{4+}/Ce^{3+}, Cr^{4+}/Cr^{3+}, Fe^{3+}/Fe^{2+}) whose redox state influences the vitrification process as well as the properties of the final material. For iron, the multivalent element which is most readily amenable to experimental studies, this influence is complex because not only the abundance but also the structural role of Fe^{2+} and Fe^{3+} ions depend markedly on temperature, chemical composition and oxygen fugacity [1].
Knowledge of the iron redox ratio is thus important to control processes such as elaboration and crystallization, to optimize the physical properties of glasses and to ensure in this way their long-term durability. The dependence of the iron redox ratio on intensive thermodynamic variables can be estimated accurately with empirical models for chemically simple or complex silicate melts [2,3]. But these models do not deal with borosilicate melts, because of the lack of relevant experimental redox data and, in view of their thermodynamic nature, they do not address at all the kinetics and the mechanisms of the redox reactions. Hence, our goal is to determine how the presence of boron in melts affects the equilibrium redox state and the kinetics of iron redox reactions to gain information on the mechanisms of these reactions through determination of rate limiting parameters.
As a matter of fact, it has long been assumed that redox reactions are rate limited by diffusion of either molecular or ionic oxygen [4,5]. However, at supercooled liquid temperatures (i.e. near the glass transition temperature range), extensive work has shown that oxygen diffusion is a too slow process. The kinetics of redox reactions are then limited instead by diffusion of

network-modifying cations toward the melt interface where crystallization of metastable phases takes place [6,7,8]. But this mechanism cannot operate at superliquidus temperatures (i.e. viscosities ranging from 10^1 to 10^3 Pa.s) where crystallization of metastable phases, which provides the driving force for cation diffusion, becomes impossible and oxygen diffusion thus represents the rate-limiting parameter [9].

These results have been mainly obtained for iron-bearing alkaline earth silicate and aluminosilicate mets. To investigate the influence of boron, we have used as a starting material an iron-bearing sodium disilicate composition so as to make structural interpretation of the results easier thanks to the wealth of available information for sodium borosilicate glasses and melts[10,11,12]. As described by Magnien et al [9,13,14], we have measured the time evolution of the redox state of iron as a function of temperature by XANES experiments at the iron K-edge. This technique presents the important advantage of yielding not only the $Fe^{3+}/\Sigma Fe$ ratio at any temperature, but also information on the local environment of iron [15,16]. The redox kinetics were measured for three compositions with B_2O_3 contents ranging from 0 to 18 mol % (Table I) either in the stable liquid domain or on supercooled liquids down to glass transition range. For comparison purposes, the temperatures of the experiments were chosen in such a way that the viscosities of the various melts were similar in the investigated 10^1 - 10^{14} Pa.s range.

EXPERIMENTAL METHODS

Samples

The samples were prepared from reagent grade SiO_2, Na_2CO_3, Fe_2O_3 and $Na_2B_4O_7$ powders dried for 24 h at 1270, 520, 770 and 420 K, respectively. To limit iron loss, the samples were prepared in platinum crucibles that had been previously used for synthesizing other iron-bearing glasses. The starting mixtures were slowly heated in Pt-Rh 10% crucibles to decompose the carbonate components and then to melt the products. After heating at 1500 K for 1 h, the melts were poured on a copper plate for quenching. Each sample was ground and remelted three times to ensure chemical homogeneity. The densities included in Table I were measured at room temperature with an Archimedean method, toluene being used as the immersion fluid. To achieve reduced iron redox states, the samples were melted for several minutes in graphite crucibles at 1500 K.

Table I. Chemical composition (wt %)a, iron redox ratio and room-temperature density of the synthetic samples investigated

	SiO_2	B_2O_3	Na_2O	FeO^b	$Fe^{3+}/\Sigma Fe^c$	$XANES^d$	d
NBF67.0	62.11 (4)		32.03 (4)	5.86 (6)	0.93	0.95	2.625 (3)
NBF67.0 -R12	62.87 (5)		31.51 (4)	5.62 (7)	0.28	0.30	2.583 (3)
NBF67.10	62.05 (5)	11.04 (30)	21.14 (4)	5.79 (3)	0.97	1.00	2.558 (3)
NBF67.10-R12	61.48 (6)	11.82 (30)	20.84 (4)	5.94 (3)	0.28	0.28	2.488 (3)
NBF67.18	61.78 (4)	19.57 (20)	12.99 (2)	5.74 (5)	0.92	0.95	2.434 (3)
NBF67.18-R15	62.81 (5)	19.07 (30)	12.83 (4)	5.60 (4)	0.22	0.22	2.367 (3)

[a]Average of 9-20 analyses made with a CAMECA SX50 electron microprobe, (x) correspond on the uncertainty on the last number; [b]Total iron reported as FeO; [c]Iron redox ratio r determined by wet chemical analyses; [d]Iron redox ratio r determined by XANES spectroscopy

The experimental conditions and the resulting iron redox ratios are summarized in Table I where the label-Rxx indicates the time (in min) during which the sample was melted in a graphite crucible. As expressed as $Fe^{3+}/\Sigma Fe$, the redox ratio r of the starting glasses was determined from the Fe^{2+} content measured with the wet chemical method of Wilson [17] and the total iron contents as determined by electron microprobe analyses. The reproducibility of the reported redox ratios is ± 0.05 [13].

Viscometry

We first measured the viscosity (η) of the three oxidized melts at superliquidus temperature and near the glass transtion range to determine the temperatures at which the kinetic measurements were to be done (Figure 1). For viscosities higher than 10^8 Pa.s, the measurements were made under uniaxial stress to within 0.02 log units with the creep apparatus described by Neuville and Richet [18]. Low viscosities were measured in air to within 0.04 log units with the concentric viscosimeter described by Sipp et al. [19]. The measurements were made at different rotating speeds for the low viscosities or different stresses for the high viscosities to ascertain the Newtonian nature of rheology. The redox ratio of the viscometry samples was checked before and after the experiments. In agreement with the results of Bouhifd et al. [20], no significant change in the iron redox ratios was observed during the high-viscosity measurements. For the high-temperature measurements, the final redox ratio was equal to 0.96 ± 0.04. For the NBF67.0 sample this result is consistent with the values predicted from the model of Kress and Carmichael [2] which ranges from 1.00 at 1180 K and 0.92 at 1600 K and, as such, did not vary enough to have affected significantly the viscosity of the samples.

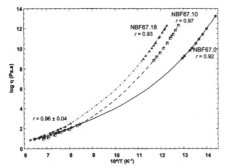

Figure 1. Viscosity of the investigated melts. For the high-viscosity samples, the $Fe^{3+}/\Sigma Fe$ ratios (r) are indicated at room temperature.

XANES spectroscopy experiments

For high-temperature XANES experiments, the samples were loaded as 6 μm-powders in the 1-mm hole of the Pt-Ir10 % heating wire of the microfurnace developed by Richet et al. [21] and previously used for in situ XANES high temperature study [9,13,14,22].The XANES spectra at the Fe K-edge were recorded at the BM30B FAME beamline of ESRF (Grenoble, France). The X-ray energy was monochromatized with a double Si(220) crystal. We scanned in 75 s the photon energy between 7050 and 7300 eV by changing the Bragg angle of the monochromator using a QuickExafs mode that allowed the monochromator to be bent dynamically in a continuous fashion during the scan [23]. After normalization with the Xafs© software [24], we analyzed the XANES spectra with the procedure described by Magnien et al. [9]. A spline function was used to interpolate the background over several eV intervals below and above the pre-edge. The intensity and centroid position of the pre-edge were calculated from fits to the data made with two pseudo-Voigt functions (see Figure 2). The $Fe^{3+}/\Sigma Fe$ ratios (r) listed in Table I were finally determined to within 0.05 from the calibration established for a variety of minerals [15]. Following Magnien et al. [9,13,14] we also checked that the iron redox ratios obtained at room temperature by XANES experiments were similar to those obtain by wet chemical analyses.

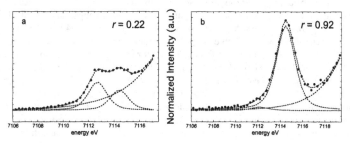

Figure 2. Pre-edge of the XANES spectra at the iron K-edge of NBF67.18-R15 (a) and of NBF67.18 (b) glasses at room temperature. The baseline, individual peaks and fitted spectra are shown as dashed lines.

RESULTS

Iron redox ratio and structural environment of iron

We determined the kinetics of the oxidation reaction between 693 and 1573 K for initially reduced samples with $Fe^{3+}/\Sigma Fe = 0.28$ for both NBF67.0 and NBF67.10 and with $Fe^{3+}/\Sigma Fe = 0.22$ for NBF67.18. We focused our attention on the pre-edge feature of the XANES spectra, which is sensitive to the iron redox state and oxygen coordination [15]. This is shown in Figure 2 for the room-temperature pre-edge of NBF67.18 glass. Whereas the pre-edge of the reduced glass is made of two contributions at 7112.5 and 7114 eV, that of the oxidized glass shows an intense contribution at 7114 eV along with a very weak shoulder at lower energy. This contrast indicates that the sample NBF67.18-R15 was essentially reduced and NBF67.18 largely oxidized, a conclusion confirmed by the quantitative treatment made whose results agree with the redox ratios determined by wet chemical analyses (Table I).

Wilke et al. [15] showed that pre-edge features with characteristic noncentrosymmetric site and strong intensity point to iron in tetrahedral coordination. We thus conclude that Fe^{3+} is in tetrahedral coordination in NBF67.18 glass. This statement is consistent with stabilization of Fe^{3+} in tetrahedral coordination by a charge compensating alkali-element [25]. Similar observations are made for the two others compositions NBF67.0 and NBF67.10 at room temperature. As for Fe^{2+}, the spectral features are consistent with octahedral coordination as indicated by the very small pre-edge area [15].

Temperature and time dependences of the XANES spectra

All kinetic XANES experiments were made with initially reduced samples. The difference between the spectrum of the starting glass and that of the melt at the end of the experiments is shown in Figure 3 for the NBF67.18-R15 sample heated at 1073 K. Likewise, the pre-edge is shown in the insert for the starting glass and the melt at three different times. Temperature-induced oxidation is clearly signaled by the shift to higher energies of both the pre-edge region and the main EXAFS resonance [9,13,14] due to the opposite variations of Fe^{2+} and Fe^{3+} concentrations. In addition, the increasing intensity of the pre-edge indicates a change from Fe^{2+} in octahedral coordination to Fe^{3+} in tetrahedral coordination [15].

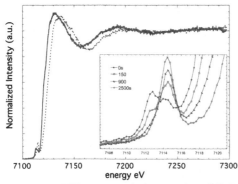

Figure 3. XANES spectra at the iron K-edge of NB67.18 -R15 glass at room temperature and of NBF67.18-R15 melt in equilibrium with air at 1073 K (dashed line); in the insert, time evolution of the pre-edge feature of NBF67.18-R15 melt at 1073 K.

KINETICS OF IRON REDOX REACTIONS

Superliquidus temperatures

Redox reactions are rate limited by diffusion because redox ratios vary linearly with the square root of time [6-9]. The XANES experiments made at 1273 and 1573 K on NB67.18-R15 melt (see insert of Figure 6) thus conform to this feature. But comparisons between the kinetics of redox reactions can be made only if the influence of the sample size is properly taken into account. This has been made by Magnien et al [9] through the so-called redox diffusivity, D, viz.

$$D = r^2/4t_{eq}, \qquad (1)$$

where t_{eq} is arbitrarily defined as the time needed to reach 99 % of the equilibrium redox ratio and r is the sample thickness determined from XANES experiment as described by Munoz et al.[26].

Redox diffusivities thus characterize the kinetics of the redox reaction regardless of both the sample size and the kind of the diffusive mechanism implied. The values determined from our experiments are shown in Figure 4 for the three compositions investigated. Since they decrease by more than one order of magnitude at constant temperature from NBF67.0 to NBF67.18 melts, it appears that the kinetics of the oxidation reaction of iron become markedly slower upon addition of B_2O_3 to Fe-bearing sodium disilicate.

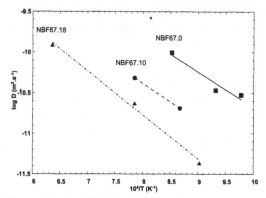

Figure 4. Redox diffusivity against reciprocal temperature for the three compositions investigated

Supercooled liquid range

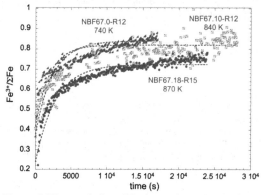

Figure 5. Time evolution of the iron redox ratio of NB67.0-R12, NB67.10-R12 and NBF67.18-R15 melts at 740, 840 and 870 K, respectively.

Even near the glass transition range, characteristic times of the order of hundreds of seconds were determined for iron oxidation in experiments made on alkaline earth silicate melts [9,13,14]. For the present samples, in contrast, the kinetics appeared to be much slower so that constant redox values typical of equilibrium could not be obtained (Figure 5). That equilibrium was far from being reached is in addition indicated by the fact that thermodynamic models [2,3] predict a $Fe^{3+}/\Sigma Fe$ ratio of 1 for the NBF67.0 sample below 900 K.

Such slow kinetics of iron oxidation could be spurious, however, and result from the fact that sintering of the initially powdered samples took place during the experiments. The sample geometry would have then changed and the surface/volume ratio be reduced so much as to make attainment of equilibrium considerably longer and determinations of characteristic times impossible.

It nonetheless appears that addition of boron again causes the oxidation kinetics of iron to slow down. This is shown in Figure 5 by the differences between the kinetics of NBF67.0-R12 at 740 K and of NBF67.10-R12 at 840 K: the cross-over observed near 10^4 s indicates that the former sample underwent slower oxidation than the latter despite a higher initial redox state (0.40 vs. 0.36), a higher temperature and a lower viscosity ($10^{9.3}$ vs. $10^{10.1}$ Pa.s). But we did not attempt quantitative treatment of this effect in view of the impossibility to know how the sample geometry changed with time during the XANES experiments.

These problems are also observed in the experiments made on the NBF67.18-R15 sample at 938, 1059 and 1108 K (Figure 6). Again, the redox state of the sample does not tend to an equilibrium value and does not vary linearly against the square root of time. In particular, the kink apparent in the 1059 K results indicate that discontinuous changes, likely due to partial crystallization, can take place and make interpretation of such results still more problematic. Further work will nonetheless be done with high-temperature X-ray diffraction and Scanning Electron Microscopy to investigate in some detail these transformations.

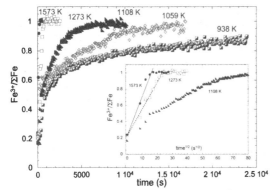

Figure 6. Time evolution of the iron redox ratio of NBF67.18-R15 glass at different temperatures; in the insert evolution of the iron redox ratio against the square root of time at 1108 K, 1273 K, 1573 K

Discussion

At supercooled liquid temperatures Cooper and coworkers have shown that redox reactions are controlled by diffusion of modifier cations toward the melt interface [6,7,8]. In our previous XANES studies on iron-bearing alkaline earth silicate melts, this conclusion has been borne out and quantitative information has been derived on the kinetics of the process [9,13,14]. Introduction of a few mol % alkali oxides in these melts has in particular been shown to increase the kinetics by several orders of magnitude [9].

In view of the previously mentioned sintering problems, the present results for supercooled B-free, Fe-bearing disilicate melt cannot be compared to our previous data. We may nonetheless assume that iron ions and Na^+ are the cations that carry the charges during the internal oxidation process. The slower oxidation kinetics that result from addition of boron can then be explained simply in terms of two effects. The first is the decrease in the concentration of Na^+ ions when the Na_2O content decreases from 32 to 13 mol %. The second results from addition of boron since transformation of BO_3 triangles into BO_4 tetrahedra takes place as described by the reaction

$$BO_3 + NBO \Leftrightarrow BO_4, \tag{2}$$

where NBO designates non-bridging oxygens [11]. Indeed, BO_3 triangles, that constitute B_2O_3, and the NBOs from the melt network are transformed in BO_4 tetrahedra due to the polymerization reaction (2). Because this transformation requires charge compensation of B in tetrahedral coordination by Na^+, the mobility of Na^+ should decrease and thus reduce the rate of oxidation.

At superliquidus temperatures oxygen diffusion is the rate-limiting factor of the iron oxidation reaction. The slower kinetics observed in the present work thus indicate that oxygen diffusion also becomes slower upon addition of boron. Such an effect is expected from the polymerization reaction (2) that is induced by introduction of boron. Although reaction (2) shifts to the left-hand side with increasing temperature, such a change is minor with respect to its net polymerizing effect [27] since recent Nuclear Magnetic Resonance data for borosilicate melts indicate decreases of the BO_4 concentration of less than 30 % between 800 and 1570 K [27,28]. In summary, the slow down of the iron oxidation reaction upon addition of boron can be interpreted simply in both the supercooled liquid and superliquidus temperature domains as a consequence of melt polymerization. Work is in progress to put this conclusion on a quantitative basis as made in our previous papers on Fe-bearing alkaline earth silicate melts [9,14].

ACKNOWLEDGMENTS

Grateful thanks are due to E. Strukelj for help in XANES spectra acquisition and to an anonymous reviewer for thoughtful comments.

REFERENCES

1. B.O. Mysen, P. Richet, 2005. Silicate Glasses and Melts: Properties and Structure. Elsevier. 550pp
2. V.C Kress., I.S.E. Carmichael, Amer. Min., 73 (1988), 1267-1274.

3. G. Ottonello, R. Moretti, L. Marini, M.V. Zuccholini, Chem. Geol. 174 (1985), 157-179.
4. H.D. Schreiber, S.J. Kozak, A.L. Fritchman, D.S. Goldman, H.A. Schaeffer, Phys. Chem. Glasses, 27 (1986), 152-177.
5. D.S. Goldman, P.K. Gupta, J. Amer. Ceram. Soc. (1983), 66, N°3, 188-190.
6. G.B. Cook, R.F. Cooper, T. Wu, J. Non-Cryst. Solids, 120 (1990), 207-222.
7. R.F. Cooper, J.B. Fanselow, D.B. Poker, Geochim. Cosmochim. Acta., 60 (1996), 3253-3265.
8. G.B. Cook, R.F. Cooper, Amer. Min., 85 (2000), 397-406.
9. V. Magnien, D.R. Neuville, L. Cormier, J. Roux, J-L Hazemann, D. de Ligny, S. Pascarelli, I. Vickridge, O. Pinet, P. Richet, Geochim. Cosmochim. Acta., 72 (2008), 2157-2168.
10. Y.H. Yun, P.J. Bray, J. Non-Cryst. Solids (1978), 27, 363
11. S. Sen, Z. Xu, J.F. Stebbins, J. Non-Cryst. Solids (1998), 226, 29-40.
12. R. Martens, W. Müller-Warmuth, J. Non-Cryst.Solids (2000), 265, 167-175.
13. V. Magnien, D.R. Neuville, L. Cormier, B.O. Mysen, V. Briois, S. Belin, O. Pinet, P. Richet, Chem. Geol., 213 (2004), 253-263.
14. V. Magnien, D.R. Neuville, L. Cormier, J. Roux, J-L. Hazemann, O. Pinet, P. Richet, J. Nucl. Mat., 352 (2006), 190-195.
15. M. Wilke, F. Farges, P-E. Petit, G.E.Brown, Jr, F. Martin, Amer. Mineral., 86 (2001), 714-730.
16. A.J. Berry, H.St.C. O'Neill, K.D. Jayasuriya, S.J. Campbell, G.J. Foran, Amer. Mineral., 88 (2003), 967-977.
17. A.D. Wilson. Analyst, 85 (1960), 823-827.
18. D.R. Neuville, P. Richet, Geochim. Cosmochim. Acta, 55 (1991), 1011-1021.
19. A. Sipp, D.R. Neuville, P. Richet,. J. Non-Cryst. Solids 211/3 (1997), 281-293.
20. M.A. Bouhifd, P. Richet, P. Besson, M. Roskosz, J. Ingrin, Earth and Planetary Science Letters, 218 (2004), 31-44.
21. P. Richet, Ph. Gillet, A. Pierre, A. Bouhifd, I. Daniel, G. Fiquet, J. Appl. Phys. 74 (1993), 5451-5456.
22. D.R. Neuville, L. Cormier, D. de Ligny, J. Roux, A-M. Flank, P. Lagarde, Amer. Min., 93 (2008), 228-234.
23. O. Proux, X. Biquard X. E. Lahera, J.-J Menthonnex, A. Prat, O. Ulrich, Y. Soldo, P. Trévisson, G. Kapoujvan, G. Perroux, P. Taunier, D. Grand, P. Jeantet, M. Deléglise, J.-P. Roux, J.-L. Hazemann, Phys. Scripta, 115 (2005), 970-973.
24. M. Winterer, J. Phys., IV, C2, (1997), 243-244.
25. B.O. Mysen, F.A. Seifert, D. Virgo, Amer. Mineral., 65 (1980), 867– 884.
26. M. Munoz, Ph D thesis, Université de Marne-la-Vallée (2003).
27. S. Sen, J. Non-Cryst. Solids (1999), 253, 84-94.
28. J.F. Stebbins, Chem. Geol. (2008), 256, 80-91

Mater. Res. Soc. Symp. Proc. Vol. 1124 © 2009 Materials Research Society   1124-Q03-03

Widening the Envelope of UK HLW Vitrification–Experimental Studies With High Waste Loading Formulations Containing Platinoids

Carl. J. Steele[1], Charlie Scales[2] and Barbara Dunnett[2]
[1]Sellafield Ltd., Seascale, Cumbria, CA20 1PG, U.K.
[2]National Nuclear Laboratory, Sellafield, Seascale, Cumbria, CA20 1PG, U.K.

ABSTRACT

Platinoid species containing ruthenium, rhodium and palladium will increase with increasing waste oxide loading to an extent that settling of the platinoids in molten glass may occur, leading to heel enrichment, poor melter performance and difficulty in draining the melter. The viscosity will also increase, which may require higher melter temperatures to mix and pour the molten glass and could result in enhanced corrosion of the melter. Inactive laboratory scale experiments with different glass frit formulations have been performed to determine how physical properties change with higher platinoid concentrations. Scanning electron microscopy (SEM) examination has shown that ruthenium, palladium or rhodium were insoluble in the melt and were not evenly distributed throughout the glass but clustered together. At higher ruthenium concentrations there was a larger increase in viscosity of the molten glass exhibiting non-Newtonian behaviour. These results will be used as a basis for further development work.

INTRODUCTION

As part of its operation of the site license for the Sellafield site, Sellafield Ltd operates the Waste Vitrification Plant which immobilises high level wastes resulting from the reprocessing of spent nuclear fuel from power reactors. As part of the drive to accelerate immobilisation and thus reduce hazard, Sellafield Ltd are exploring ways of enhancing the vitrification operation through increased throughput and higher waste loading. In order to facilitate this the National Nuclear Laboratory (NNL) are supporting Sellafield Ltd through the operation of a full scale inactive development rig, Vitrification Test Rig (VTR) and associated development programmes.

The presence of platinoid fission product species containing ruthenium, rhodium and palladium present in highly radioactive liquor form insoluble phases in the glass product. Increased waste incorporation will thus inevitably lead to higher levels of platinoids in the melt. Viscosity is increased requiring higher temperatures to mix and pour the molten glass. This can result in enhanced corrosion of the melter and possible melter failure. The design of the AVH (Atelier de Vitrification La Hague) melter employs a heel and settling of the platinoids in the glass may occur, leading to heel enrichment.

In order to explore the limits of operation for noble metals levels within the melt, an inactive experimental study has been carried out. Using glass compositions with enhanced (38 wt%) waste simulant loading, the effects of varying levels of noble metal contents on viscosity and settling have been studied at laboratory scale and compared to levels of noble metal build up in the melter heels obtained following full scale inactive operation.

EXPERIMENT

Existing inactive simulant formulations containing platinoids were expanded to widen the operational envelopes to observe inactive laboratory trial performance and determine any changes to resulting glass physical properties. Also, glasses with high waste incorporations have been produced to test process capability and to ascertain any potential phase separation or devitrification issues that could affect either the process or production performance. Physical properties of the different glass formulations were performed to measure changes in viscosity to examine how readily miscible the glasses were. Glass melts containing 38 ± 0.5 wt% Magnox or 75:25 wt% Oxide: Magnox Blend simulant waste with a targeted total platinoid (RuO_2, Rh_2O_3 and PdO) concentration as indicated in Tables 1 and 2 were prepared in the laboratory.

Table I: Nominal wt.% of platinoids in Magnox glasses prepared

Wt%	1% RuO_2 Magnox	1.5% RuO_2 Magnox	2% RuO_2 Magnox	2.5% RuO_2 Magnox	3% RuO_2 Magnox	3.5% RuO_2 Magnox
RuO_2	1.00	1.50	2.00	2.50	3.00	3.50
PdO	0.48	0.71	0.95	1.19	1.43	1.67
Rh_2O_3	0.27	0.40	0.53	0.67	0.80	0.93

Table II: Nominal wt.% of platinoids in 75:25 wt.% Oxide:Magnox Blend glasses prepared

Wt%	1% RuO_2 Blend	1.5% RuO_2 Blend	2% RuO_2 Blend	2.5% RuO_2 Blend	3% RuO_2 Blend	3.5% RuO_2 Blend
RuO_2	1.00	1.50	2.00	2.50	3.00	3.50
PdO	0.74	1.11	1.48	1.85	2.22	2.59
Rh_2O_3	0.26	0.39	0.52	0.64	0.77	0.90

The 38 wt.% waste simulant loading included 0.95 wt% RuO_2 in the Magnox glasses and 0.61 wt.% RuO_2 in the blend glasses. The wt.% of MgO, SiO_2, B_2O_3, Na_2O, Al_2O_3, Fe_2O_3, Li_2O, Cr_2O_3 and NiO was kept constant as these elements may affect the viscosity or glass forming properties. TeO_2 was also kept constant as it is known that Pd tends to form solid solutions with Te.

SEM examination of glasses

The glasses from 1 wt.% RuO_2 Magnox, 2.5 wt.% RuO_2 Magnox, 3.5 wt.% RuO_2 Magnox, 1 wt.% RuO_2 Blend, 2 wt.% RuO_2 Blend and 3.5 wt.% RuO_2 Blend were mounted in resin and polished to 1 µm finish. The samples were cleaned in methanol and gold coated for examination under a JEOL JSM-5600 SEM and Energy Dispersive Spectroscopy (EDS) was carried out using a Princeton Gamma-Tech Prism Digital Spectroscopy to determine the chemical elements present in the undissolved particles. The density concentration of the platinoids, which are not soluble in the glass, was found to be variable throughout the glass (Figure 1). In all of the samples, the majority of the insoluble particles present in the glass were RuO_2 as shown in Figure 2.

Figure 1: Two areas of 2 wt% RuO_2 Blend glasses showing differing amounts of insoluble material

Figure 2: X-ray mapping of blend glass frit containing nominally 3.5 wt.% RuO_2

EDS analysis of a selection of the RuO_2 particles indicated that in the Blend glasses no other elements were associated with them whereas in the Magnox glasses in some instances ruthenium was found to be associated with Cr, Fe, Ni Spinels. Larger palladium particles alloyed with tellurium was also present (Figure 2) in all glasses as spherical droplets indicating that they were liquid at the temperature of glass formation, which was 1100 °C for these glasses. In one instance, ruthenium was also detected to be associated with the Pd/Te alloy[1]. Rhodium was only observed alloyed with palladium in the samples containing the highest concentration of platinoids. Although rhodium particles were not seen in any of the glass frits examined, they may have been present as metallic particles dispersed amongst the ruthenium particles.

Viscosity

Viscosity measurements of each glass were carried out to ISO standard 7884-2:1987(E) using a rotating viscometer. Viscosities were measured at temperatures of 1000, 1100, 1200 and 1250 °C. All the glasses exhibited non-Newtonian behaviour. This effect has also been observed by other researchers for nuclear glasses containing platinoids [2, 3] whom also noted that the rheological properties of the platinum group metal loaded glass samples were affected by the particle inclusions rendering their behaviours non-Newtonian. The viscosity measurements are shown graphically in Figures 3 and 4 for the Magnox and Blend glasses respectively.

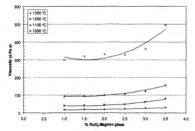

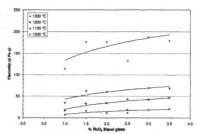

Figure 3. Viscosity of Magnox glasses with increasing platinoid concentration

Figure 4. Viscosity of Blended glasses with increasing platinoid concentration

For the Magnox glasses the results show an increase in viscosity with increasing platinoid concentration, with the largest increase being between the 3 and 3.5 wt.% RuO_2 Magnox glasses. The viscosity results for the glasses containing Blended waste is lower than those containing Magnox glass due to the lower aluminium content in the glass. The measured viscosities of the 1.5, 2, 3 and 3.5 wt.% RuO_2 Blend glasses were similar with 1 wt.% RuO_2 and 2.5 wt.% RuO_2 being significantly lower especially at the lower temperature.

Residue heel left in a VTR melter

Pours of masses between 190 ~212 Kg and of 25-31 wt.% Blend waste incorporation were made using this melter which was used for Campaigns 2 and 3 in the VTR [4]. A drain pour was carried out prior to it being taken out of service and sectioned. The drain pour was of similar composition to the last product pour. Although the waste did not contain palladium or rhodium for full scale inactive trials, ruthenium was present in the simulants. The residue glass heel, which was never poured, was found to be situated around the pour nozzle as shown in Figure 5.

Figure 5: Residual glass heels remaining in the VTR melter.

Samples were taken for chemical, SEM and EDS analysis from the top surface and then progressively through the glass towards the base with the final sample being that which was in contact with the inconel melter. The residue heel which was in-homogenous was found to have increased concentration of Ru, Mo and Fe/Cr/Ni. Chemical analysis indicated a significant increase in the ruthenium present in all the residue heel samples (Table 3) with up to 13 wt.% RuO$_2$ being analysed compared to 0.76 wt.% in the drain pour.

Table III: RuO$_2$ (wt.%) of samples taken from the residue heel

	Base (next to melter surface)	0.5 - 2 cm from base	~2-4 cm from base	~5 cm below surface	Top surface
RuO$_2$	2.7	3.3	5.4	12.9	5.6

SEM and EDS analysis revealed that in addition to ruthenium dispersed throughout the glass, as generally observed in product glass, ruthenium particles had aggregated (up to 0.5 mm), and were visually present on polished sections of the samples (Figure 6).

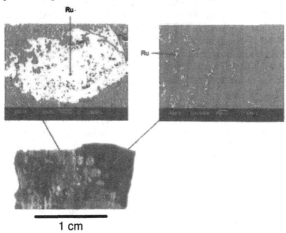

1 cm

Figure 6: Ruthenium present in the residue melter heel.

DISCUSSION

The viscosity measurements taken in this study have demonstrated that ruthenium containing glass exhibits non-Newtonian behaviour, which has also been observed previously [2,3]. It has been possible to establish the relationship between viscosity, temperature and platinoid concentration. Bart et al [2] measured the viscosity of glass samples containing 4-6 wt.% platinoids and reported glass viscosity to be in the range of 15-20 Pa s at a temperature of 1100 °C, which is comparable with the Magnox simulant glasses prepared in this study. They also noted that ruthenium was present as RuO$_2$ needles with palladium metal being observed in combination with tellurium. Durability studies were also performed [2] which suggested that even though ruthenium formed insoluble and separate regions in the bulk glass the bulk leach rate was unaffected. Further work would be required to study the durability of UK HLW glasses prepared in this study.

A disproportionate increase in viscosity with increasing ruthenium concentration was measured. This is in agreement with the work previously performed by Igarashi et al [3]. A greater concentration range of ruthenium was studied ranging from 1 to 10 wt. %. At high ruthenium concentration the viscosity was found to be between 3 to 7 times higher than that for ruthenium free glass. This may support the theory for failures to pour the residual heel during VTR melter operation since the glass temperature would not be sufficient to allow molten glass to flow at high viscosity caused by the presence of high concentrations of ruthenium ~ 13 wt. %.

CONCLUSIONS

Laboratory glasses have been prepared using inactive simulants and a range of platinoid concentrations to measure the change in physical properties of glass. A relationship has been established between viscosity, temperature and platinoid concentration for a given simulant and non-Newtonian behaviour was displayed. Ruthenium is present as RuO_2 needles found as an insoluble phase in the bulk glass. In Magnox glass simulant it was also found to be present associated with other insoluble phases, e.g. Ni/Cr/Fe spinels. Palladium was found to be present with tellurium and only at the highest concentrations was rhodium detected alloyed with palladium. Analysis of the VTR melter heel measured high levels of platinoids and other insoluble phases, e.g. molybdates and spinels. This provides an explanation for failures to pour observed in melter operation.

Further research is required to study the durability of the glasses with high platinoid concentration to assess the suitability for long term storage and disposal. Information from this study may also be useful to support further computational fluid dynamic modeling to predict glass physical properties aimed at preventing accumulation in melter operations and investigating potential recovery strategies if this occurred.

REFERENCES

1. 'On the issue of Noble Metals in the DWPF Melter', S. K Sundaram and J. M. Perez, PNNL-13347, September 2000.
2. Influence of platinium-group metals on nuclear glass properties:- viscosity, thermal stability and alterability', F. Bart, J. L. Dussossoy and C. Fillett, Mat. Res. Soc. Symp. Proc, Vol. 663, p161 (2001).
3. 'Effect of noble metal elements on viscosity and electrical resistivity of simulated vitrified products for high-level liquid waste', H. Ingarashi, K. Kawamura and T. Takahashi, Mat. Res. Soc. Symp. Proc., vol. 257, p169-176 (1992).
4. 'Progress in the vitrification of HA waste in the UK', N. R. Gribble, K. Bradshaw, D. O. Hughes, A. D. Riley, C. J. Steele, International conference on Glass, ICG Strasbourg, 1-6 July 2007.

ACKNOWLEDGEMENTS

I would like to acknowledge the support of Andrew Riley, Sellafield Ltd and the Nuclear Decommissioning Authority for funding this research.

Mater. Res. Soc. Symp. Proc. Vol. 1124 © 2009 Materials Research Society   1124-Q03-05

Microwave processing of glasses for waste immobilisation

Martin C. Stennett and Neil C. Hyatt
Immobilization Science Laboratory, Department of Engineering Materials, University of
Sheffield, Sir Robert Hadfield Building, Mappin Street, Sheffield, S1 3JD, UK.

ABSTRACT

Microwave (or dielectric) heating is recognised as a fast, clean and economical
preparation route for a wide range of inorganic solids. The microwave spectrum lies in the
frequency range 0.3 to 300 GHz which is mainly used for communication purposes, although a
narrow frequency window centred at 2.45 GHz is allowed for microwave heating purposes.
Materials in general fall into three categories: microwave reflectors, transmitters, and absorbers.
A key requirement for direct microwave heating is that one or more of the major constituents of
the batch must be a microwave absorber and couple strongly to the microwave field at room
temperature. Microwave heating has a number of key advantages over conventional heating.
Microwave heating is a rapid process and, unlike conventional heating, the batch is heated from
the inside out. This minimises undesirable decomposition, oxidation / reduction, loss of volatile
materials, and other kinetically slow processes which can occur during conventional melting.

In this work, microwave heating has been applied to processing of iron-phosphate based
glass compositions suitable for the immobilisation of radioactive wastes. Chemically
homogeneous glasses have been successfully prepared by microwave heating over timescales of
the order of several minutes. The resulting glasses have been characterised by X-ray diffraction
(XRD), optical and scanning electron microscopy (SEM).

INTRODUCTION

The use of microwave heating as an alternative to conventional heating was first
developed over 50 years ago although its application to materials processing has only happened
relatively recently. Microwave processing has been investigated for a variety of applications but
of particular interest from the waste immobilisation perspective is the use of microwaves for the
synthesis of inorganic solids and glass [1-3]. Vitrification is the current route for the
immobilisation of high level nuclear waste in a number of countries and thousands of metric tons
of high level waste (HLW) glass have been processed ready for geological disposal worldwide
[4, 5]. Although batch melting is a well established and mature technology it is energy intensive
and economic justification for the use of microwave processing is based on the comparatively
inefficient nature of conventional melting processes [6]. In addition to saving energy,
microwave heating has many additional advantages over conventional heating. These include:
very rapid heating rates (> 400 °C min $^{-1}$), reduced processing time and temperature, reduction in
loss of volatile batch components, and prevention of undesirable decomposition or redox
processes.

A fundamental constraint of microwave heating is that a major constituent of the batch
must be a microwave absorbing species. Whilst a detailed description of the interactions of
materials with microwaves is outside the remit of this work it should be noted that in general

materials with a high dielectric loss make good susceptors. Figure 1 shows a schematic representation of the heating profiles of three oxides when subjected to microwave irradiation. SiO_2 has a low dielectric loss and shows only a small increase in temperature during irradiation whilst NiO has a high room temperature dielectric loss and rapidly heats up to > 800 °C. Cr_2O_3 exhibits higher dielectric loss at elevated temperature and shows an abrupt rise in temperature after several minutes due to increased coupling to the microwave field.

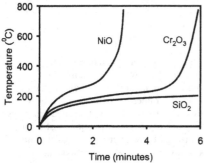

Figure 1. Schematic representation of temperature response of several oxides to microwave irradiation (adapted from [1]).

Table 1 below lists several microwave active compounds that contain suitable glass forming elements. The temperature values given indicate the maximum reported temperatures achieved by heating in a conventional microwave oven. Most glass compositions used for HLW immobilisation are silicate based glasses [5]. They readily dissolve a wide range of waste components and are widely used in the commercial glass industry so their properties are well known and characterised. Durable high silica containing glasses do, however, require high processing temperatures. Phosphate glasses have also been studied for HLW immobilisation and they exhibit lower melting temperatures, high solubility for transition metal oxides and some systems demonstrate equivalent or superior chemical durability to borosilicate systems [8, 9]. This combination of characteristics makes phosphate-based systems attractive for microwave processing.

Table I. Summary of microwave active elements of interest for glass production [1, 7].

Compound	T (°C)	Compound	T (°C)
Fe_3O_4	1258	NiO	1305
CuO	1012	V_2O_5	714
MnO_2	1287	WO_3	1270
Co_2O_3	1014	$NaH_2PO_4 \cdot 2H_2O$	678

A survey of the literature on phosphate glasses revealed several systems containing elements for which suitable microwave suscepting precursor species exist. For the initial feasibility study a number of potential options were highlighted including glasses in the Na_2O-Fe_2O_3-P_2O_5 system. After consulting the ternary phase diagram the $10Na_2O$-$30Fe_2O_3$-$60P_2O_5$ (mol %) composition was chosen [10].

EXPERIMENTAL

Microwave heating was carried out in a domestic microwave oven (DMO) with a nominal power output of 800 W. The glass batch constituents were weighed out under flowing nitrogen in a sealed glovebox to prevent hygroscopic oxide components absorbing moisture. The batches were then homogenised in a mortar and pestle and transferred into silica crucibles before being transferring from the glovebox into the microwave for irradiation. Figure 2 shows schematically the experimental setup inside the DMO. The silica crucibles were placed inside a recessed alumina block mounted on alumina spaces to thermally isolate it from the microwave turntable. The alumina block was machined to leave the top of the silica crucibles exposed and allow them to be removed easily. The exposed region of the crucibles was surrounded by silica wool and finally a silica lid was used to cover the crucibles and reduce thermal loss. Care was taken to ensure the block was positioned in the centre of the microwave cavity. Samples were irradiated at maximum nominal power (800 W) for the required time and then the alumina block was removed from the microwave cavity. The crucibles were removed from the block to maximise the cooling rate.

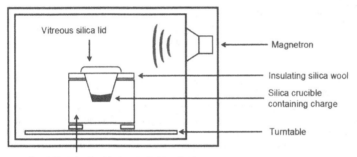

Figure 2. Schematic representation of the experimental setup inside the DMO.

Once cool the silica crucibles were cut in cross section and one half was resin mounted and polished to a 1 μm optical finish. The samples were coated with conducting carbon and analysed using a JEOL JSM 6400 scanning electron microscope (SEM) equipped with a LINK energy dispersive spectrometer (EDS) system for elemental analysis. The remaining half of the crucible was broken to allow the glass to be removed for further characterisation. Room temperature X-ray diffraction (XRD) patterns were collected to check for evidence of crystalline phases.

RESULTS AND DISCUSSION

Glass samples with the $10Na_2O-30Fe_2O_3-60P_2O_5$ (mol %) composition were irradiated at 0.75, 1.5, 3 and 12 minutes and examined. After approximately 20 seconds of irradiation a dull red glow emanating from the crucible was observed through the silica lid. The intensity of the glow increased until, after approximately 1 minute, an incandescent red glow could be clearly observed. This continued until the DMO was switched off. Figure 3 shows the $10Na_2O-$

$30Fe_2O_3-60P_2O_5$ sample irradiated for 12 minutes at 800 W in cross section. A homogeneous bubble free black glass had formed at the bottom of the crucible and there was no indication of any residual batch around the crucible. Inspection of the samples irradiated for 1.5 and 0.75 minutes also revealed a bubble free black glass at the bottom of the crucible but re-crystallised batch components were observed on the sides of the crucible.

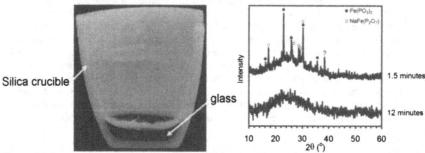

Figure 3. Cross section of $10Na_2O-30Fe_2O_3-60P_2O_5$ sample irradiated for 12 minutes at 800 W. The crucible was filled with resin to prevent sample fracturing during cutting. XRD pattern obtained from glass samples irradiated for 1.5 and 12 minutes shown on right. Question mark indicates reflection due to unknown phase.

XRD diffraction patterns for the glass irradiated for 12 minutes and residual batch removed from the side of the crucible of the glass irradiated for 1.5 minutes are shown in Figure 3. For the 12 minute and 3 minute samples broad humps in the diffraction pattern was observed typical of amorphous materials. There were no indication of any sharp reflections characteristic of crystalline phases in the patterns. Samples of glass removed from the bottom of the crucibles for the 1.5 and 0.75 minute samples also showed broad diffuse amorphous signals but XRD patterns from the residual batch removed from the sides of the crucibles showed reflections which could be attributed to the presence of $Fe(PO_3)_3$, $NaFe(P_2O_7)$ and an unknown phase. Polished cross sections of all the samples were examined by electron microscopy.

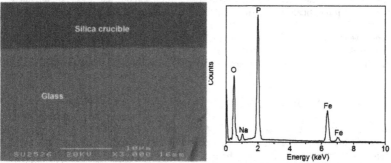

Figure 4. Electron micrograph (left) of the 12 minute sample. Typical EDS spectra obtained from glass region shown on right.

Figure 4 shows an electron micrograph obtained from the 12 minute sample. No crystalline inclusions were observed in the glass and elemental analysis by energy dispersive spectroscopy (EDS) indicated the composition of the glass formed was close to the target stoichiometry. Examination of the interface between the glass and the silica crucible indicated negligible reaction had occurred and no silica was observed in the glass close to the interface by EDS. Figure 5 shows a crystalline inclusion observed in the 1.5 minute sample. Elemental mapping indicated that the inclusion was rich in iron and deficient in P with respect to the surrounding glass matrix.

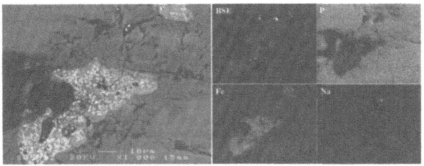

Figure 5. Electron micrograph (left) showing crystalline inclusion observed in the 1.5 minute sample. Backscattered electron imaging and elemental maps shown on right.

Preliminary studies are underway to synthesise homogeneous samples by microwave processing in a number of other transition metal phosphate systems including $CuO-P_2O_5$, $MnO_2-P_2O_5$, $NiO-P_2O_5$ and $WO_3-P_2O_5$. In addition, it has been shown that with pro-rata additions of up to 10 mol % cerium (as a plutonium surrogate), to the $10Na_2O-30Fe_2O_3-60P_2O_5$ base glass, a homogeneous sample can still be fabricated using the same processing conditions.

CONCLUSIONS

Chemically homogeneous glass samples in the $Na_2O-Fe_2O_3-P_2O_5$ system have been prepared by microwave irradiation over very short timescales. For irradiation times ≥ 3 minutes no crystalline inclusions were observed, by XRD or SEM, in the glasses formed. For shorter irradiation times re-crystallised batch components were observed around the sides of the crucible and crystalline inclusions were observed in the glasses. This suggests that although the temperature required for melting the batch components was reached very rapidly, longer irradiation times were required to produce homogeneous glass samples. Microwave heating has been studied for 30 years as a potential vitrification route although the radioactive waste related work has been conducted either on liquid wastes [11, 12], or on systems where the microwave absorbing species is present as an element in the waste stream or as a batch addition rather than as a major glass forming element [13, 14]. Work is underway to study the effect of radionuclide additions (using suitable non-radioactive surrogate species) to glass forming ability under microwave processing conditions.

ACKNOWLEDGMENTS

This work was carried out as part of the TSEC programme KNOO and as such we are grateful to the EPSRC for funding under grant EP/C549465/1.

REFERENCES

1. K. J. Rao, B. Vaidhyanathan, M. Ganguli and P. A. Ramakrishnan. Chem. Mater. 11, 882 (1999).
2. D. Agrawal. Curr. Opin. Solid State Mater. Sci. 3, 480 (1998).
3. D. Clark and W. H. Sutton. Annu. Rev. Mater. Sci. 26, 299 (1996).
4. I. Muller and W. J. Weber. Mater. Res. Soc. Bull. 26, 698 (2001).
5. I. W. Donald, B. L. Metcalf, and R. N. J. Taylor. J. Mater. Sci. 32, 5851 (1997).
6. A. C. Metaxas and R. J. Meredith. Industrial Microwave Heating. (Institute of Engineering and Technology, London, 1983).
7. D. M. P. Mingos and D. R. Baghurst. Chem. Soc. Rev. 20, 1 (1991).
8. C. Mercier, G. Palavit, L. Montagne, and C. Follet-Houtternane. Compt. Rend. Chim. 5, 693 (2002).
9. D. E. Day, Z. Wu, C. S. Ray, and P. Hrma. J. Non-Cryst. Solids. 241, 1 (1998).
10. Y. Kin, Y. Zhang, W. Huang, K. Lu, and Y Zhao. J. of Non-Cryst. Solids. 112, 136 (1989).
11. M. S. Morrell and M. Smith, Report AERE-M 3373, 1986.
12. A. A. Kurkumeli, M. I. Molokhov, O. D. Sadkovskaya, V. I. Kononov, G. B. Borisov, V. F. Savel'ev, and A. I. Kachurin. Atom. Energy. 73, 210 (1992).
13. C. M. Jantzen and J. R. Cadieux in *Microwaves: Theory and Applications in Materials Processing*, edited by D. E. Clark, F. D. Gac, and W. H. Sutton, (Ceram. Trans. 21, Westerville, OH, 1991) pp. 441-449.
14. R. L. Schulz, Z. Fathi, D. E. Clark, and G. G. Wicks in *Microwaves: Theory and Applications in Materials Processing*, edited by D. E. Clark, F. D. Gac, and W. H. Sutton, (Ceram. Trans. 21, Westerville, OH, 1991) pp. 451-458.

152

Mater. Res. Soc. Symp. Proc. Vol. 1124 © 2009 Materials Research Society   1124-Q03-06

^{95}Mo NMR Study of Crystallization in Model Nuclear Waste Glasses

Scott Kroeker,[1] Ian Farnan,[2] Sophie Schuller[3] and Thierry Advocat[4]

[1]Department of Chemistry, University of Manitoba, Winnipeg, Manitoba, R3T 2N2, Canada
[2]Department of Earth Sciences, University of Cambridge, Cambridge, CB2 3EQ, U.K.
[3]CEA Valrhô Marcoule, DEN/DTCD/SECM/LDMC, BP 17171, 30207 Bagnols/Céze, France
[4]CEA Saclay, DEN/DANS/DPC/SECR, 91191 Gif-sur-Yvette, France

ABSTRACT

^{95}Mo magic-angle spinning nuclear magnetic resonance (MAS NMR) spectroscopy is surprisingly sensitive to the local environment of tetrahedral molybdate species. A series of compounds related to expected crystallization products in nuclear waste glasses are probed to calibrate their spectral characteristics. Glasses formed with fast and slow quenching show a glassy peak corresponding to tetrahedral molybdate species. With slow quenching, a prominent sharp peak is observed, representing crystallinity. In sodium-borosilicate glasses with 2.5 mol% MoO_3, the sharp peak corresponds to pure crystalline sodium molybdate. Cesium-sodium and lithium-sodium borosilicate glasses with Mo show crystalline peaks as well, and suggest that NMR may potentially be used to characterize mixed-cation molydates and more complex phase assemblages. While precise quantification of Mo in different phases is likely to be time-consuming, reasonable estimates can be obtained routinely, making ^{95}Mo MAS NMR a useful tool for investigating phase separation and crystallization in model nuclear waste materials.

INTRODUCTION

The requirement of nuclear wasteforms to incorporate a large number of chemically diverse ions without compromising long-term durability is exacting, and has motivated extensive research into suitable materials. Borosilicate glasses have been selected in many countries as wasteforms for high-level liquid waste from fuel reprocessing, due to their excellent solvent properties, high durability and favourable processing procedures. However, certain fission products can induce macroscopic liquid-liquid phase separation during formation, and crystallization during cooling. Depending of the nature (alkali or alkaline earth) of the crystalline phases, the durability of the bulk wasteform can suffer, and melter corrosion can be accelerated. Molybdenum, for example, is known to form a complex crystalline molybdate referred to as "yellow phase" in English and French nuclear glasses, [1, 2] which possesses a soluble fraction. To prevent yellow phase formation, waste loading is currently limited to keep Mo concentrations below a certain threshold. While this solution may be *effective* in eliminating yellow phase formation, it is not particularly *efficient*, as it increases the waste glass volume. To increase the level of molybdenum oxide that can be accommodated, new wasteforms prepared at higher temperatures (1200°C) have been designed to meet both microstructural and long-term behaviour standards. [3] Studying the roles of boron, sodium and calcium oxides to determine the chemical and structural effects on crystallization phases will be aided by methodology which can conveniently measure the extent of crystallization and the identity of the precipitate(s).

^{95}Mo nuclear magnetic resonance (NMR) spectroscopy is generally unpopular due to its low resonance frequency, sizeable quadrupole moment, low natural abundance and typically long relaxation times. These inherent obstacles are exacerbated in the present case by structural disorder – which broadens the peaks – and low Mo concentrations. Despite these sensitivity challenges, magic-angle spinning (MAS) NMR of relatively large samples turns out to reveal good resolution between Mo in crystalline and amorphous fractions. In this contribution, we show that ^{95}Mo MAS NMR is an appropriate tool for studying phase separation in Mo-bearing borosilicate glasses.

EXPERIMENTAL DETAILS

Crystalline alkali molybdates (Na_2MoO_4, Li_2MoO_4, Cs_2MoO_4) were obtained from Aldrich and used without further purification. A single-crystal specimen of mixed-alkali $CsLiMoO_4$ was received from Dr. Petra Becker (Institute of Crystallography, University of Cologne). Three different compositions of borosilicate glasses containing Na or mixed-alkali Na-Cs and Na-Li were made using conventional melt-quench methodology. Glass samples (100 g) were heated for 3 hours at 1350°C. Samples (~ 5 g) quenched between copper plates had a quench rate of approximately 10000°C/minute and resulted in optically homogeneous glasses; samples cooled slowly in Pt-Au crucibles at about 1°C/minute resulted in visually opaque materials, indicative of a significant fraction of crystallinity.

All glass samples were polished, carbon-coated and studied by scanning electron microscopy (SEM) and energy-dispersive x-ray analysis (EDX). Glass samples were also analysed by x-ray diffraction (XRD) with a Philips X'Pert Pro Theta/Theta powder x-ray diffractometer using monochromatized Cu Kα radiation ($\lambda = 0.15406$ nm). The crystalline phases were identified by comparison with XRD powder patterns in the International Center of Diffraction Data crystallographic database.

^{95}Mo MAS NMR was carried out on a Varian CMX 500 tuned to 32.57 MHz. Powdered samples weighing about 1 g were spun at 5-6 kHz in a 7.5 mm MAS probe. A simple one-pulse experiment was used, with a radiofrequency field of 16 kHz and short pulses corresponding to a 30° tip angle. Recycle delays ranged from 10-20 s. Chemical shifts are externally referenced using 2 M Na_2MoO_4 (aq).

RESULTS AND DISCUSSION

Crystalline reference compounds

Figure 1 compares the ^{95}Mo MAS NMR spectra of several crystalline molybdate compounds expected to be relevant for understanding Mo-related crystallization. The sharp peaks of the sodium, cesium and cesium-lithium molybdates reflect highly symmetrical MoO_4^{2-} anions with diagnostic chemical shifts. The lithium salt exhibits a non-zero quadrupolar interaction due to its distorted tetrahedral environment, [4] and is in fact more representative of the alkali and alkaline-earth molybdates. [5] The chemical shift of $CsLiMoO_4$ is much more shielded than the normal range of alkali molybdates (-17 to -34 ppm). [5] The origin of this large shift is unknown, but suggests that ^{95}Mo NMR chemical shifts are sufficiently sensitive to aid in structural determinations. This would be particularly advantageous for studying partially crystallized model waste glasses, where the resulting precipitates are not pure phases, but contain a variety of cationic substituents. [6]

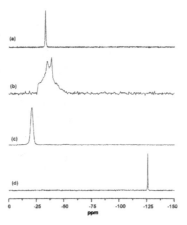

Figure 1. [95]Mo MAS NMR of selected model compounds: (a) Na_2MoO_4, (b) Li_2MoO_4, (c) Cs_2MoO_4, (d) $CsLiMoO_4$.

Model nuclear waste glasses

Typical levels of MoO_3 in nuclear waste glasses are restricted to a few percent. In the model glasses studied here, 2.5 mol% is found to be the upper incorporation limit for producing an homogeneous glass at 1350°C with rapid quenching (10000°C/min). Figure 2b illustrates [95]Mo MAS NMR spectra of a sodium borosilicate glass composition quenched rapidly between copper plates, and more slowly, by programmed cooling. The fast-quenched glass yields a single broad peak indicating a fully amorphous material, located in the region of the spectrum corresponding to four-coordinate, oxygen-bound Mo. [7] The slow-quenched sample exhibits this peak as a minor component, accompanied by a sharp peak at the expected position of Na_2MoO_4. X-ray diffraction of this material verifies the presence of crystalline cubic Na_2MoO_4, while SEM shows separated phases ranging in size from 50 nm to 20 μm, enriched in molybdenum and sodium (Figure 2c).

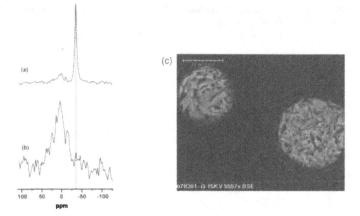

Figure 2. Sodium borosilicate glass of nominal composition 62 mol% SiO_2 : 16.5 mol% B_2O_3 : 19 mol% Na_2O : 2.5 mol% MoO_3. ^{95}Mo MAS NMR of glasses formed by (a) slow quenching and (b) fast quenching. Spectra are the result of 8516 and 10344 co-added transients, respectively, and were acquired with recycle delays of 10 seconds. Dotted line gives the position of Na_2MoO_4. (c) SEM image of separated phases in slow-cooled material.

A similar glass containing a small, but industrially typical, amount of Cs_2O was studied in the same way. Figure 3 (a and b) show that the fast-quenched glass is free of crystalline phases and contains four-coordinate MoO_4^{2-}, and that the slow-quenched glass contains a substantial fraction of crystallized material. Considering the very small amount of added Cs_2O (0.3 mol%), it is no surprise that the resulting crystalline phase is predominantly Na_2MoO_4. However, the x-ray diffraction data reveal the presence of unidentified minor peaks as well, and the scanning electron microscope image shows evidence of complex crystal growth (Figure 3c). Separated phases ranging in size from 50 nm to 30 μm contain a main phase high in Mo and Na, (grey) and a white phase enriched in Mo and Cs.

Careful comparison of the sodium molybdate peak position with the crystalline peak present here hints that the crystalline phase may not be pure, but may contain some level of Cs incorporation. Given the sensitivity of the ^{95}Mo chemical shifts to alkali metals in crystalline compounds, NMR may prove to be more useful for studying substituted molybdate phases than x-ray diffraction.

With care, the relative intensities of the NMR peaks can be used to quantify the fraction of Mo in the crystalline component with respect to that in the glass, provided the whole sample is crushed and thoroughly mixed to ensure statistical sampling of the specimen. Accurate quantification also demands that ^{95}Mo nuclei in all different environments are uniformly excited by the radiofrequency field, and that pulse repetition rates are sufficiently slow to ensure complete relaxation of all species. Due to the sensitivity challenges faced in these experiments, neither of these conditions were stringently verified, although both were probably met within reasonable uncertainties. Reliable quantification is possible, but would require more attention to these experimental aspects.

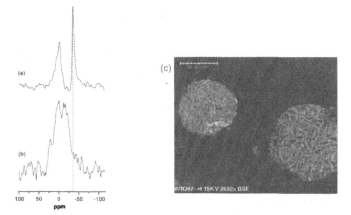

Figure 3. Sodium borosilicate glass with 2.5 mol% MoO_3 and 0.3 mol% Cs_2O. ^{95}Mo MAS NMR of glass formed by (a) slow quenching and (b) fast quenching. Spectra are the result of 23582 and 32768 co-added transients, respectively, and were acquired with recycle delays of 10 and 2 seconds. Dotted line gives the position of Na_2MoO_4. (c) SEM image of slow-quenched sample.

A lithium-containing mixed-alkali borosilicate glass with Mo was also studied by ^{95}Mo MAS NMR (Figure 4). The results are similar to the foregoing, but less clean due to the formation of hexagonal Li_2MoO_4 in the slow-quenched material, as indicated by x-ray diffraction. SEM detects only separated phases enriched in Mo and Na, as Li cannot be detected by EDX. However, the substantially different morphology of the separated phase (Figure 4c) suggests that the precipitate differs chemically from those in above.

With reference to Figure 1, it can be seen that the Li-containing phase possesses significant second-order quadrupolar broadening, which reduces the resolution obtained above. While clearly exhibiting spectral evidence of Na_2MoO_4, the slow-quenched sample suffers from peak overlap due to multiple crystalline phases and a somewhat broadened glass peak. This observation suggests that the ^{95}Mo chemical shift range for molybdates may be too small to provide optimal resolution for complex phase assemblages. The use of higher magnetic fields should improve this situation, as the chemical shift difference scales directly with applied field, while the quadrupolar broadening decreases. Preliminary ^{95}Mo MAS NMR experiments (data not shown) on the slow-quenched $Na_2O-B_2O_3-SiO_2-MoO_3$ sample run at 18.8 T in a 4 mm probe (52.2 MHz) confirm that the linewidth is primarily due to a distribution of chemical shifts, and that sample volume (i.e., large rotors for high-field magnets) will be an important parameter for collection of quantitative spectra.

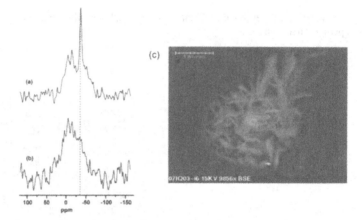

Figure 4. A mixed-alkali borosilicate glass of nominal composition 62 mol% SiO_2 : 16.5 mol% B_2O_3 : 13.6 mol% Na_2O : 5.4 mol% Li_2O : 2.5 mol% MoO_3. [95]Mo MAS NMR of (a) slow-quenched, (b) fast-quenched samples. Spectra are the result of 25124 and17284 co-added transients, respectively, and were acquired with recycle delays of 20 and 5 seconds. Dotted line gives the position of Na_2MoO_4. (c) SEM image of separated phases in slow-quenched material.

CONCLUSIONS

These data show that [95]Mo NMR is sufficiently sensitive to detect molybdenum in a variety of environments common to phase-separated borosilicate glasses, even at very low concentrations. Chemical shifts are reasonably sensitive to the local molybdate environment and can possibly be used for "fingerprinting" the phases present. This sensitivity may prove capable to determine more detailed structural information not easily obtained by x-ray diffraction, such as the nature and degree of substitution in non-stoichiometric mixed-cation compounds. However, further work will be required to define the spectral characteristics of such systems. The measures necessary to ensure that data are quantitatively accurate are simple, but time-consuming. Extensions to more complex glasses are expected to be straightforward, except those containing more than trace levels of paramagnetic ions, the unpaired electrons of which can cause rapid relaxation and significant peak broadening. Nevertheless, this work shows that even at moderate magnetic fields, valuable information about phase separation can be obtained from routine MAS NMR experiments on model nuclear waste glasses.

ACKNOWLEDGMENTS

We are grateful to Dr. Petra Becker (University of Cologne) for the CsLiMoO₄ crystal. The UK EPSRC is acknowledged for Grant number EF/F011008/1 to IF, which also provided a visiting fellowship for SK. High-field NMR experiments were performed using EMSL, a national scientific user facility sponsored by the Department of Energy's Office of Biological and Environmental Research located at Pacific Northwest National Laboratory.

REFERENCES

1. R.J. Short, R.J. Hand, N.C. Hyatt, and G. Möbus, *J. Nucl. Mater.* **340**, 179 (2005).
2. G. Calas, M. Le Grand, L. Galoisy, and D. Ghaleb, *J. Nucl. Mater.* **322**, 15 (2003).
3. M. Magnin, S. Schuller, D. Caurant, O; Majerus, D. Deligny, and C. Mercier, *Ceramic Trans.*, submitted.
4. A.V. Barinova, R.K. Rastsvetaeva, Y.V. Nekrasov, and D.Y. Pushcharovskii, *Doklady Chem.* **376**, 16 (2001).
5. J.-B. d'Espinose de Lacaillerie, F. Barberon, K.V. Romanenko, O.B. Lapina, L. Le Pollès, R. Gautier, and Z. Gan, *J. Phys. Chem. B* **109**, 14033 (2005).
6. R.J. Hand, R.J. Short, S. Morgan, N.C. Hyatt, G. Möbus, and W.E. Lee, *Glass Technol.* **46**, 121 (2005).
7. S.H. Santagneli, C.C. de Araujo, W. Strojek, H. Eckert, G. Poirier, S.J.L. Ribeiro, and Y. Messaddeq, *J. Phys. Chem. B* **111**, 10109 (2007).

Mater. Res. Soc. Symp. Proc. Vol. 1124 © 2009 Materials Research Society 1124-Q03-07

Glass Development for Vitrification of Wet Intermediate Level Waste (WILW) From Decommissioning of the Hinkley Point 'A' Site

Paul A. Bingham [1], Neil C. Hyatt [1], Russell J. Hand [1] and Christopher R. Wilding [2]
[1] Immobilisation Science Laboratory, Dept. of Engineering Materials, University of Sheffield, Sir Robert Hadfield Building, Mappin Street, Sheffield S1 3JD, UK
[2] Magnox South Ltd., Hinkley Point A Site, Nr Bridgwater, Somerset TA5 1YA, UK

ABSTRACT

The Immobilisation Science Laboratory, University of Sheffield, is working with Magnox South Ltd to develop a range of glass formulations that are suitable for vitrification of the Wet Intermediate Level Waste (WILW) envelope arising from decommissioning of the Hinkley Point 'A' (HPA) power station. Four waste mixtures or permutations are under consideration for volume reduction and immobilisation by vitrification. The inorganic fractions of several of the wastes are suitable for vitrification as they largely consist of SiO_2, MgO, Fe_2O_3, Na_2O, Al_2O_3 and CaO. However, difficulties may arise from the high organic and sulphurous contents of certain waste streams, particularly spent ion exchange (IEX) resins. IEX resin wastes may be the key factor in limiting waste loading, and possible thermal pre–treatments of IEX resin to decrease C and S contents prior to vitrification have been investigated. Our results suggest that low–temperature (90 °C) pre–treatment is more favourable than high–temperature (250, 450, 1000 °C) pre–treatment. A thorough desktop study has provided initial candidate glass compositions which have been down–selected on the basis of glass forming ability, melting temperature, viscosity, liquidus temperature, chemical durability and potential sulphate capacity. Early results for two of the candidate glass formulations indicate that formation of an amorphous product with at least 35 wt % (dry waste) loading is achievable for HPA IEX resin wastes.

INTRODUCTION

Hinkley Point 'A' Power Station (HPA) was a twin reactor Magnox nuclear power station commissioned in 1965. The power plant was permanently shut down in 2000 and the nuclear fuel was removed from the site by the end of 2004. The site is now in the process of being decommissioned. Wet Intermediate Level Waste (WILW) was generated during operation of HPA and essentially occurs in three forms: ion exchange resins, sludges and sand. It is a requirement that any WILW waste stream that can not be shown to be in a passively safe state be retrieved, processed and packaged in approved product forms. Following a review of process options, treatment of WILW by vitrification has been identified as a potential process for HPA. Subsequently HPA has been granted by the Nuclear Decommissioning Authority (NDA) Radioactive Waste Management Directorate (RWMD) a Conceptual Stage Letter of Compliance based on volume reduction by a vitrification process of the WILW wastes at HPA. As part of future submissions to the RWMD for interim and final LoC's there is a requirement to produce a

Waste Product Specification (WPrS) which describes the waste package that is to be manufactured by defining the key limits and controls on the waste package production process. This key document will eventually be used, together with other records, to demonstrate that waste packages meet the repository Waste Acceptance Criteria (WAC) and are compatible with all future stages of waste management including ultimate disposal to the planned UK National ILW Repository.

The objectives and criteria requirements of the current project are as follows:

- Develop glass formulations to enable the HPA WILW to be successfully immobilised within a vitreous matrix
- Define the formulation envelope and associated product properties to enable the immobilised product to meet the waste product specifications for safe interim storage
- Provide data for incorporation in the WPrS
- Maximise ^{137}Cs and ^{106}Ru retention in the waste product (Cs >80 %)
- Produce a product with a homogeneous distribution of phases
- Viscosity < 100,000 cP at the melting system operating temperature
- Use of minimal additives
- Target upper melt temperature of 1200 °C
- Quality of glass products formed must meet wasteform product qualities of the NDA "specification of wasteform for 500 litre drum waste packages" [1]
- Leach resistance of better than 10^{-3} g cm^2 d^{-1} for Cs, Sr and Na at pH = 10.5, T = 50 °C over a 28 day test period.
- Utilise data to develop glass formulations for the processing of ILW at other sites

This paper gives an overview of the background to and preliminary results from a research programme aimed at supporting the ongoing HPA WILW vitrification project. Four HPA WILW streams are currently under consideration for vitrification. These are Spent Ion Exchange (IEX) resin; Active Effluent Treatment Plant (AETP) sludge; Pond Water Treatment Plant (PWTP) sludge; and Sand Pressure Filter (SPF) waste. Four waste Permutations, which consist of different combinations of these four wastes, are under investigation:

Permutation A. All wastes (IEX resin + AETP sludge + PWTP sludge + SPF sand)
Permutation B. All wastes (Pre–treated IEX resin + AETP sludge + PWTP sludge + SPF sand)
Permutation C. Sludge and sand wastes (AETP sludge + PWTP sludge + SPF sand)
Permutation D. IEX resin only

The inorganic fractions of the HPA wastes are suitable for vitrification as they largely consist of oxides and hydroxides of Si, Mg, Fe, Na, Al and Ca. However, it is possible that difficulties may arise due to the high organic and sulphurous contents of spent ion exchange (IEX) resin wastes (see Table I). As a result the waste loading limit for Permutation D waste may be lower in order to safely produce an amorphous product with the required properties. This IEX resin waste therefore represents one potential "worst–case" scenario in terms of upper waste loading limits. Thermal pre–treatment of IEX resin may provide a means of increasing the waste loading limit of the vitrified product by reducing the C and S contents of the waste prior to vitrification. In this paper we discuss our investigations to date on thermal pre–treatment of IEX resin and early–stage vitrification of pre–treated Permutation D IEX resin waste.

EXPERIMENT

An inactive waste Permutation D simulant has been considered here, as described in Table I. Differential scanning calorimetry has been conducted on this waste in air and N_2 at a heating rate of 1 °C / min from 20–1000 °C. Bulk resin samples have subsequently been held isothermally at 90, 250, 450 and 1000 °C in air and N_2 for 24 h, and their products analysed for mass loss and by Leco combustion furnace analysis of C, O and S contents. Six glass compositions have been selected for vitrification trials: two are discussed here. Both are $SiO_2–R_2O_3–R'_2O–R''O$ glasses for which R = B, Al, Fe; R' = Li, Na; and R'' = Ca, Zn, Ba. Glass frits were melted using reagent–grade raw materials. Pre–treated Permutation D waste was mixed thoroughly with (< 1000 μm) glass frit to provide an equivalent (dry) waste loading of 35 weight %. Mixtures were then melted on a 50g scale in open silica crucibles in an electric furnace. Crucibles were heated from room temperature at 2 °C / minute to 1200 °C and held for 8 hours before being poured onto a steel plate and allowed to cool.

Table I. Simulated inactive Permutation D (IEX resin) waste composition reported to 1 d.p.

Component	Abundance / wt %
Al(OH)$_3$)	0.3
CaCO$_3$	0.3
Fe$_2$O$_3$	0
Mg(OH)$_2$	0.4
K$_2$SO$_4$	0.2
SiO$_2$	0.1
NaOH	2.7
Free Water	65.9
IEX resins	29.6
Hydrocarbons	0.5

RESULTS AND DISCUSSION

IEX Resin Wastes

The HPA IEX resin wastes primarily consist of a mixture of (a) phenol formaldehyde resin with sulphonated active groups, ~$C_7H_6O_4S$ and (b) sulphonated styrene/dibenzene copolymer, ~$C_9H_9O_3S$. Their high S and C contents present unique challenges for vitrification [2, 3]. Sulphur solubility in alkali borosilicate glasses is typically < ~1 wt % SO_3 (equiv.) [4–8]. If this limit is exceeded an alkali sulphate "gall" layer forms on the melt surface. Cesium may preferentially enter this water–soluble sulphate layer. The S content of the vitrified waste must therefore be maintained below its solubility limit, and in the case of wastes such as IEX resin this provides the waste loading limit. Organic material also causes reducing melting conditions which, if not controlled, can lead to gas accumulation and formation of metallic or metal sulphide inclusions which are electrically conductive and can damage melting vessels [9].

Direct vitrification of IEX resin wastes has met with some success [2, 3 and refs therein]. However, literature suggests that thermal pre–treatment of IEX resin wastes could decrease their S and C contents [4, 10–15]. If this were carried out prior to vitrification, and if volatilisation of ^{137}Cs and [106]Ru can indeed be minimised during pre–treatment [10], it could provide an increase in the upper waste loading limit for IEX resin wastes. Our TGA analyses indicate that key thermal events occur in the HPA waste IEX resins at 50–160 °C, 370–405 °C and 850–940 °C (see Figure 1). On this basis, pre–treatment has been carried out at plateau temperatures of 90, 250, 450 and 1000 °C, as described in Table II.

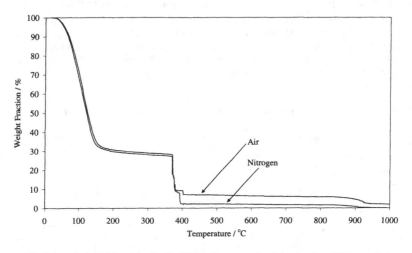

Figure 1. TGA analyses of Permutation D IEX resin waste simulants in air and N_2 atmospheres

Simply drying the waste at 90 °C offers an effective treatment (~ 80 % weight reduction) and further heating provides additional weight losses. Sulphur and carbon in the Permutation D IEX resin wastes are not evolved in significant quantities at temperatures ≤ 450 °C. Our results for sulphur differ from the results of studies on pure styrene / divinylbenzene IEX resins [10–13] which indicate that strong sulphur evolution occurs ~350–400 °C. Increasing temperature has a concentrating effect wherein water, oxygen and other components are evolved but S and C remain present until higher temperature (450 < T ≤ 1000 °C) . The differences between treatment at 90, 250 and 450 °C are small; however, ^{137}Cs and [106]Ru volatilisation increases with temperature [10]. Only by treating at 1000 °C were the S and C contents of the HPA IEX resin wastes substantially lower, and it is more likely that unacceptable levels of ^{137}Cs and [106]Ru volatilisation will occur at these high temperatures [4, 10]. Therefore our results, in conjunction with other published data, suggest that low temperature pre–treatment (~100 °C) may be favourable to higher temperature (450 – 1000 °C) methods for pre–treatment of the HPA IEX resin wastes prior to their vitrification. The potential advantages of high–temperature treatment are outweighed by its apparent disadvantages.

Table II. Weight loss and elemental analyses of heat–treated IEX resin wastes (Permutation D)

Temperature / °C	Initial Wt / g	Final Wt / g	Residual Wt %	Residual Wt (S) / g	Residual Wt (C) / g	Residual Wt (O) / g	Residual Wt (other) / g
Air 90	98.98	19.19	19.39	0.62	7.69	4.41	6.47
Air 250	99.76	17.09	17.13	0.58	7.61	3.54	5.35
Air 450	99.26	14.91	15.02	0.58	7.14	2.48	4.71
Air 1000	99.75	8.48	8.50	0.57	1.55	2.94	3.41
N_2 90	99.61	19.84	19.92	0.37	4.90	4.68	9.89
N_2 250	99.66	16.27	16.33	0.53	7.05	2.91	5.78
N_2 450	99.7	13.63	13.67	0.49	6.63	2.38	4.13
N_2 1000	99.95	9.19	9.19	0.66	3.97	2.23	2.32

Glass Formulation Development and Testing

All vitrified Permutation D waste samples shown in Figure 2 exhibit good glass forming ability. All glasses were poured, after removal from the furnace at 1200 °C, with ease indicating low viscosities, and all vitrified products are confirmed to be XRD amorphous (diffraction patterns not shown). Jantzen *et al.* [2] summarised the key considerations for vitrification of IEX resin wastes: chemical durability, homogeneity, thermal stability, viscosity, liquidus temperature, volatility, melting temperature, melt corrosivity and waste solubility. We have applied a similar approach to our candidate glass selection process, using published data where available and modelled data elsewhere. In addition we have considered the sulphate capacity of the glass using a recently–developed methodology [5]. Early–stage vitrification studies therefore indicate that Permutation D IEX resin waste can be successfully vitrified at 35 weight % (dry) waste loading. Further work is currently underway to establish the upper waste loading limits for untreated and pre–treated IEX resin wastes and to determine the physical properties of the vitrified products.

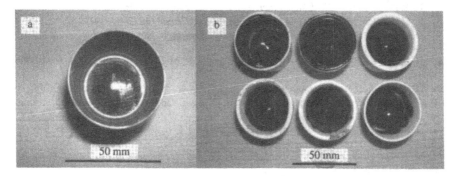

Figure 2. (a) Untreated Permutation D (IEX resin waste) simulant; (b) Vitrified products (Glass A top row, Glass B bottom row) with 35 wt% (dry) waste loading of IEX resin waste pre–treated at (left to right) 90, 250 and 450 °C in air for 24 hours

CONCLUSIONS

The high C and S contents of IEX resin waste streams arising from decommissioning of the Hinkley Point 'A' (HPA) site may limit waste loading during vitrification of HPA Wet ILW, and potential methods of mitigating such an effect are under investigation. Results of thermal pre–treatment trials of IEX resin waste suggest that low–temperature (90 °C) drying is more favourable in this case than high–temperature (250, 450, 1000 °C) treatment. Vitrification studies of simulated HPA IEX resin wastes indicate that formation of an amorphous product with at least 35 wt % (dry waste) loading is achievable at all pre–treatment temperatures studied. This in turn suggests that untreated resin could also be successfully vitrified at this level of waste loading.

ACKNOWLEDGMENTS

The authors acknowledge the contributions of N. Hamodi and O. M. Hannant (ISL) and L. J. Rust, A. Burnett, S. Kenney, A. Peppin, C. Fox, R. Streatfield and M. Pick (MSL).

REFERENCES

1. Generic Waste Package Specification, Vol. 1: Specification, Nirex Report N/104, June 2005; http://www.nda.gov.uk/documents/upload/Generic–waste–package–specification–2005.pdf
2. C. M. Jantzen, D. K. Peeler and C. A. Cicero, Report WSRC–MS–95–0518, 1995; http://www.osti.gov/bridge/servlets/purl/237361-VKs8Cg/webviewable/
3. T. N. Sargent, T. J. Overcamp, D. F. Bickford and C. A. Cicero–Herman, *Nucl. Technol.* **123** (1998) 60–66.
4. M. I. Ojovan and W. E. Lee, An introduction to nuclear waste immobilisation (2005) Elsevier, Amsterdam.
5. P. A. Bingham and R. J. Hand, *Mater. Res. Bull.* **43** (2008) 1679–1693.
6. C. M. Jantzen, M. E. Smith and D. K. Peeler, *Ceram. Trans.* **168** (2005) 141–152.
7. D. Manara, A. Grandjean, O. Pint, J. L. Dussossoy and D. R. Neuville, *J. Non–Cryst. Solids* **353** (2007) 12–23.
8. H. D. Schreiber, S. J. Kozak, P. G. Leonhard and K. K. McManus, *Glastech. Ber.* **60** (1987) 389–398.
9. D. M. Bickford, R. B. Diemer and D. C. Iverson, *J. Non–Cryst. Solids* **84** (1986) 285–291.
10. P. Antonetti, Y. Claire, H. Massit, P. Lessart, C. Pham Van Cang and A. Perichaud, *J. Anal. Appl. Pyrol.* **55** (2000) 81–92.
11. U. K. Chun, K. Choi, K. H. Yang, J. K. Park and M. J. Song, *Waste Man.* **18** (1998) 183–196.
12. M. A. Dubois, J. F. Dozol, C. Nicotra, J. Scrose and C. Massiani, *J. Anal. Appl. Pyrol.* **31** (1995) 129–140.
13. R. S. Juang and T. S. Lee, *J. Hazard. Mater.* **B92** (2002) 301–314.
14. P. Nassoy, F. P. Scanlan and J. F. Muller, *J. Anal. Appl. Pyrol.* **16** (1989) 255–268.
15. F. P. Scanlan, J. F. Muller and J. M. Fiquet, *J. Anal. Appl. Pyrol.* **16** (1989) 269–289.

Mater. Res. Soc. Symp. Proc. Vol. 1124 © 2009 Materials Research Society   1124-Q03-08

Infrared Spectroscopy Structural Analysis of Corroded Nuclear Waste Glass K-26

A.S. Barinov[1], G.A. Varlakova[1], I.V. Startceva[1], S.V. Stefanovsky[1] and M.I. Ojovan[2]

[1] Scientific and Industrial Association Radon, Moscow, Russia.
[2] Immobilisation Science Laboratory, University of Sheffield, UK.

ABSTRACT

The dissolution of silicate glasses in near-neutral water solutions is relatively slow and long term tests are required to fully understand glass dissolution behaviour. Long-term field test on K-26 high alkali-borosilicate nuclear waste glass were conducted in an experimental shallow ground repository. Infrared (IR)-spectroscopy of pristine bulk glass and near surface altered glass showed that the structure consisted of $[SiO_4]$ tetrahedra and boron $[BO_4]$ tetrahedra. The IR-spectra indicated that the glass surface had a highly ordered structure in comparison to the bulk. It is suggested that this is due to corrosion of the glass in the disposal environment and formation of crystalline phases on the surface.

INTRODUCTION

Vitrification is currently attracting significant interest for the immobilisation of HLW and operational radioactive wastes from nuclear power plants as well as radioactive and toxic legacy wastes [1]. Large scale vitrification programmes are underway in a number of countries. The selection of durable wasteforms minimises the environmental impact of radionuclides and enables simpler and less expensive waste disposal facilities, such as near-surface repositories, to be used for storage. Silicate glasses in near-neutral water solutions, such as natural groundwater from clayey soils, have low corrosion rates and therefore require long term tests to understand their dissolution behaviour. Many field tests on simulated nuclear waste glasses have been conducted [2,3] and it has been shown that waste glasses exhibit a much higher durability in the natural environment than in accelerated laboratory test conditions. To date there are have been three documented burial experiments on actual radioactive waste glasses. These were conducted at Chalk River Nuclear Laboratories (CRNL) in Ontario in Canada [4], Savannah River National Laboratory (SRNL) at Savannah River Site in the USA [5], and at the Moscow Scientific and Industrial Association "Radon" in Russia [6]. This work presents the results on the structure of radioactive K-26 borosilicate glass after long term testing in a wet disposal environment.

EXPERIMENTAL

The borosilicate glass K-26 has been designed to vitrify radioactive operational NPP waste from channel type reactors [7]. Its composition in oxide mass % is: $23.93+0.71$ Na_2O; $13.72+0.14$ CaO; $3.06+0.30$ Al_2O_3; $43.01+1.93$ SiO_2; $6.55+0.60$ B_2O_3; $0.90+0.17$ K_2O; $1.90+0.14$ Fe_2O_3; and $0.50+0.05$ MgO. The glass has a density of 2.46 g/cm^{-3} [6]. Operational NPP waste is an evaporator concentrate with salt content ~340 g/L and sodium nitrate is the main non-radioactive component (86 % on a dry basis). Caesium isotopes account for 97 - 98 % of the total waste radioactivity. Raw materials used to manufacture the were quartz sand, loam and datolite concentrate, which is a natural mineral containing 17.8 wt.% of B_2O_3. The specific radioactivity of the glass was 3.74MBq/ kg (beta emitters) and 13.0 kBq/kg (alpha emitters). The molten borosilicate waste glass was cast into carbon steel containers and allowed to cool naturally to ambient temperature. The glass was not annealed for this pilot campaign. The field tests of K-26 to evaluate the glass behaviour under near surface disposal conditions were started on the 17th August 1987 [6].

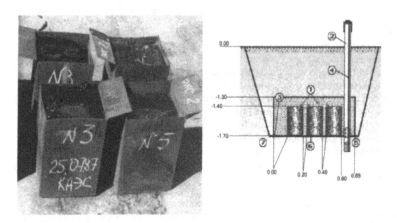

Figure 1. Carbon steel containers with K-26 waste glass (left). Schematic of experimental burial (right). 1 – waste glass containers, 2 – water trap, 3 – tray, 4 – sampling pipe, 5 – perforation, 6 – backfill sand, 7 – host rock (loamy soil).

The containers were left open to the atmosphere (Fig. 1, left image) and six blocks of waste glass (each weighing about 30 kg and 30 cm in height) were placed in a single 40 cm high stainless steel tray supplied with a water trap and a tube for water extraction by pumping (Fig. 1, right image). The experimental repository was 1.7 m deep, well below the freezing depth of the soil (0.7 m) and the average annual temperature at this depth was 4.5 °C. Pure coarse sand was used to backfill the glass blocks in the tray to facilitate infiltration of water but isolate them from direct contact with the host rock. The space outside the containers was backfilled with host loamy soil until flush with the surrounding land surface.

The experimental repository was opened periodically for inspection and to conduct sample analysis. X-ray diffraction (XRD) analysis of glass samples was conducted using a DRON-4 diffractometer (FeK$_\alpha$ radiation). The microstructure of glass was analysed using a JSM-5300 (JEOL) scanning electron microscope (SEM). Elemental analysis of groundwater was performed by inductive-coupled plasma mass spectrometry using an ELAN DRC II PerkinElmer-Sciex unit. Infra red (IR) spectra were recorded using a IKS-29 spectrophotometer (compaction of powdered glass in pellets with KBr).

RESULTS AND DISCUSSION

The XRD pattern of glass K-26 is shown in Fig. 2 (left image). The broad diffuse hump indicated the glass K-26 was X-ray amorphous with a negligibly small fraction of crystalline inclusions and a homogeneous internal structure. A SEM micrograph from the glass K-26 is shown in Fig. 2 (right image). SEM analysis of glass samples showed no signs of phase separation [6]. The effects of waste glass leaching on groundwater chemistry are summarised in Table I.

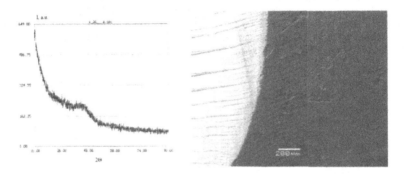

Figure 2. XRD pattern of glass bulk (left) and SEM image of altered glass surface (right).

Table I. Ionic concentrations (mg/L) and pH values of groundwater in contact with glass K-26.

Parameters	After 1st year	After 8th year	After 12th year	After 16th year	Background average
pH	7.5	7.8	7.9	7.9	7.6-7.7
Na^+	34.64	54.55	63.40	64.64	13.11
K^+	3.70	3.82	3.72	4.24	1.26
Ca^{2+}	78.93	54.26	46.30	44.74	48.74
Mg^{2+}	32.30	30.26	27.75	28.57	20.91
Fe_{total}	9.33	2.64	2.02	1.72	1.16
Cl^-	33.11	26.49	24.08	21.22	13.17
NO_3^-	2.22	4.18	4.23	3.89	2.30
HCO_3^-	435	390	373	362.45	258.03
SO_4^{2-}	13.93	12.45	10.62	16.50	4.73
B	Not measured	24.75	27.72	3.59	Not measured
Total mineralisation	710	830	810	810	600

Over time the mineralisation of the water that has been in contact with the glass has altered from (HCO_3^- - Ca^{2+})- type to (HCO_3^- - Na^+ - K^+ - Ca^{2+} - NO_3^- - Cl^-)- type (higher than background by a factor of about 1.35). A progressive decrease in concentration with time was observed for Ca^{2+}, $Mg2^+$, Fe, Cl^-, HCO_3^- and SO_4^{2-} whilst the opposite trend was observed for B, Na^+ and NO_3^-, In addition a small but progressive pH change, from 7.5 to 7.9, was measured. Decrease in the radioactivity due to radionuclide leaching and surface glass dissolution were very low. Long-term monitoring of the repository showed that the average leaching rates of radionuclides from the glass continuously decrease from 0.94 µg cm-2 day in the first year to 0.22 µg/cm^{-2} day by the 16th year [6]. The effective interdiffusion coefficient, D_{Cs} = 4.5 10^{-12} cm^2/d and the rate of glass hydrolysis, r_h = 0.1 µm/yr, were calculated [8]. Based on these data it was concluded that the fractional inventory of radioactive ^{137}Cs leached out of waste glass for all the storage period will not exceed the maximal value φ_{Cs} = 27.5 ppm [9]. The fractional inventory of radionuclides leached by groundwater between1987-2005, was 14.1 ppm.

Sample taken from the bulk and from the surface of the glass were analysed by infrared (IR) spectroscopy. The IR spectra of glasses in the range 4000-400 cm^{-1} (Fig. 3, left) showed the following absorption bands: 3200-3500 cm^{-1} (strong), 2800-3000 cm^{-1} (very weak), 1500-1700 cm^{-1} (shoulder), 1300-1500 cm^{-1} (strong), 850-1300 cm^{-1} (very strong), 700-800 cm^{-1} (weak) and 400-600 cm^{-1} (strong). Fig. 3, right, shows typical IR spectra in the wavelength range 1600-400 cm^{-1}.

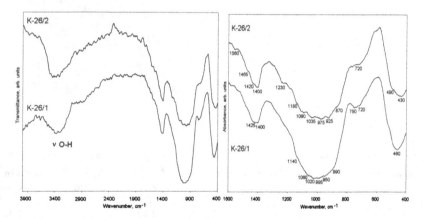

Figure 3. The IR-spectra of K-26 glass. K26/1 – bulk glass sample, K-26/2 surface glass sample.

The IR spectra seemed to suggest that the glass from the surface region K-26/2 had higher content of secondary crystalline phases then the bulk [6].
Absorption in the wavelength range 3200-3500 cm^{-1} can be assigned to vibrations of hydroxide bonds (OH$^-$), absorption in the range 2800-3000 cm^{-1} to hydrogen bonds ≡Si—O···H—O—Si≡ or ≡Si—O···H—O—B≡, and absorption near 1600 cm^{-1} to deformation vibrations of water molecules [10-14]. In the IR spectra of the glass sample the absorption band in this range had a truncated shape (shoulder), which indicated very low content of water molecules in the glass structure. The hydroxide groups available are probably concentrated on the glass surface. The absorption observed in the range 1600-400 cm^{-1} can be assigned to valence and deformational vibrations of silicon-oxygen and boron-oxygen polyhedra in the glass structure [10-14]. In this range of the spectrum there were three main absorption bands (Fig. 3): the strongest broad absorption band is between 800-1300 cm^{-1}, two strong but narrower bands were observed at 1350- 1600 cm^{-1} and 400-550 cm^{-1}, and a weak band was seen at 700-770 cm^{-1}. All these bands

are multicomponent, which indicated the presence of different Si—O and B—O bonds in the glass structure.

The absorption band at 1300-1500 cm^{-1}, with the maximum at ~1400 cm^{-1}, is typically present in the spectra of alkali silicate glasses and can be attributed either to stretching modes (valence vibrations) of C—O bonds in the carbonate $(CO_3)^{2-}$ groups, or to deformational (bending) vibrations of bonded hydroxide groups. In boron-containing glasses the absorption bands, corresponding to vibrations of bridging B^{III}—O—B^{IV} bonds which link triangles and tetrahedra made of boron and oxygen, and are also located in this wavelength range. As one can see from Fig. 3 the band at 1300-1500 cm^{-1} has a multicomponent structure and is most probably composed of several narrower bands characteristic of the presence of a number of different structures in the glass.

The absorption band at 800-1300 cm^{-1} can be assigned to bridging oxygen bonds Si—O—Si, which connect [SiO$_4$] tetrahedra, as the content of SiO$_2$ in the glass K-26 is rather high (~43.0 mass %) whereas the content of B$_2$O$_3$ is relatively low (~6.6 mass %). This band was shifted to lower wavenumbers in the IR spectrum of the sample taken from the near-surface region, possible due to increased contributions from non-bridging Si—O$^-$ bonds. The structure of glass K-26 is made up of [SiO$_4$] tetrahedra with various degrees of binding. The disordered glass structure with lower content of silica is assumed to be composed of [SiO$_4$] tetrahedra connected in one-dimensional metasilicate chains (with two non-bridging oxygen atoms). Boron enters into the glass network as [BO$_4$] tetrahedra with the formation of [BO$_{4/2}$]Na$^+$ structural units, which increases the degree of polymerisation of glass [15]. The absorption bands in the range 1000-1100 cm^{-1} are characteristic of asymmetric stretching modes (valence vibrations) of the B—O bonds in the [BO$_4$] tetrahedral units. These fall in the same region of the spectra as the Si—O—Si bonds and it is not possible to separate the individual B—O and Si—O contributions. Vibrations of bridging B^{III}—O—B^{III} bonds are characterised by absorption in the range 1150-1300 cm^{-1}. The IR-spectra detected show only a shoulder (Fig. 3). This means that bridging bonds of this type and [BO$_3$] groups are present in the glass at relative low concentrations. This is consistent with available models of boron speciation in silicate glasses [15].

The parameter that controls the state of boron:

$$\Psi_B = \frac{\gamma_{Na_2O}}{\gamma_{B_2O_3}} = \frac{W_{Na_2O} m_{B_2O_3}}{W_{B_2O_3} m_{Na_2O}},$$

where γ_{Na2O} and γ_{B2O3} are the molar fractions, W_{Na2O} and W_{B2O3} are the weight fractions, and m_{Na2O} and m_{B2O3} are the molar masses of sodium and boron oxides. Boron remains in a three-coordinated position in a sodium-borosilicate glass when the concentration of alkali metal is low $1 << \Psi_B$ [15].

The glass K-26 has a value of $\Psi_B = 4.08$ which means that most of boron should be in the four coordinated state. The band at 400-550 cm-1 can be assigned to deformation vibrations of [SiO$_4$] tetrahedra. Weak absorption in this range can be due to overlapping of symmetric valence vibrations of Si—O bonds in [SiO$_4$] tetrahedra (with one or two non-bridging oxygens) and Al—O bonds. However the contribution from Al—O bonds in this glass should be small as the content of Al$_2$O$_3$ is low (~3 mass %).

Overall we can characterise the K-26 glass structure as mainly made of [SiO$_4$] terahedra, the majority of which contain two bridging oxygen atoms. Boron is present mainly as a network modifier, [BO$_4$] tetrahedra are present in the network chains of [SiO$_4$] tetrahedra. A small amount of [BO₃] triangles are present and these can form tetraborate and/or diborate groups, connected with [BO₄] tetrahedra.

An important issue is the effect of self-irradiation on the durability of nuclear waste glasses [16]. Self irradiation causes the formation of point defects such as non-bridging oxygen hole centres which contribute to mobilisation of alkalis in glasses [17]. The four coordinated state of boron in the glass means that a significant number of alkali atoms (Na, 137Cs) are bound to negatively charged [BO₄] – units. Recent investigations [18] have revealed that in borosilicate glasses tetravalent BO_4 groups convert to trivalent BO_3 groups during electron irradiation in a transmission electron microscope (TEM). These investigations also showed that the boron was converted from four to three coordination after gamma irradiation characteristic of nuclear waste glasses (1 MGy ^{60}Co-source γ-irradiation) [18]. The conversion reaction:

$$[BO_4]^- A^+ + \text{irradiation } (\gamma, e^-) \rightarrow [BO_3] + A + O^-,$$

releases alkalis A such as radioactive caesium or non-radioactive sodium from the negatively charged [BO4]– tetrahedra. These could provide explanation as to the higher sensitivity of K-26 glasses to high gamma irradiation doses when compared with other glasses such as British Magnox glasses [9,16]. Indeed the glass K-26 has more four-coordinated boron available for radiation induced destruction ($\Psi_B = 4.08$) compared to British Magnox glass (17.17 wt. % B_2O_3, 3.82 wt. % Li_2O and 8.51 wt. % Na_2O [9] ($\Psi_B = 1.07$).

CONCLUSIONS

Long-term field test data on the behaviour of K-26 high-alkali-borosilicate nuclear waste glass have shown that of the glass is durable in an experimental shallow ground repository. Infrared spectroscopy has shown the structure of samples taken from both the bulk and from the near surface altered glass layer are made up of [SiO₄] tetrahedra with boron present as a network modifier in the form of [BO₄] tetrahedra. Formation of [BO₄]₄/₂Na structural units increased the degree of polymerisation of the glass. The IR spectra show that the [BO₃] groups were present in low concentrations. The IR-spectra suggest that the bulk of glass had a lower ordered structure when compared with the altered glass surface which was possibly due to the formation of secondary crystalline phases on the surface as a result of corrosion.

ACKNOWLEDGMENTS

Authors acknowledge A.V. Timofeeva, N.P. Penionjkevich, A.V. Mokhov for their assistance in the research and B.P. McGrail for methodological help and consultation.

REFERENCES

1. M.I. Ojovan and W.E. Lee, *New Developments in Glassy Nuclear Wasteforms*. ISBN: 1-60021-783-4, Nova Science Publishers, New York, 131p. (2007).
2. G.G. Wicks, *J. Nucl. Mater.*, **298**, 78-85 (2001).
3. P. Van Iseghem, *Mat. Res. Soc. Symp. Proc.*, **333**, 133 (1994).

4. J.C. Tait, W.H. Hocking, J.S. Betteridge, and G. Bart, *Advances in Ceramics, 20: Nuclear Waste Management II*, Eds. D.E. Clark, W.B. White, and A.J. Machiels. American Ceramic Society, Columbus, OH, 1986. P. 559-565.
5. C.M. Jantzen, D.I. Kaplan, N.E. Bibler, D.K. Peeler, and M.J. Plodinec, *J. Nucl. Mater.*, **378**, 244 (2008).
6. M.I. Ojovan, W.E. Lee, A.S. Barinov, I.V. Startceva, D.H. Bacon, B.P. McGrail, and J.D. Vienna, *Glass Technology*, **47**, 48 (2006).
7. I.A. Sobolev, F.A. Lifanov, S.V. Stefanovsky, S.A. Dmitriev, N.D. Musatov, A.P. Kobelev, and V.N. Zakharenko, *Atomic Energy*, **69**, 233 (1990).
8. M.I. Ojovan, R.J. Hand, N.V. Ojovan, and W.E. Lee, *J. Nucl. Mater.*, **340**, 12 (2005).
9. M.I. Ojovan, A.S. Pankov, and W.E. Lee, *J. Nucl. Mater.*, **358**, 57-68 (2006).
10. I.I. Plyusnina, *Infra-Red Spectra of Silicates*, Moscow University Press (1967).
11. W.L. Konijnendijk, *Phil. Res. Rep. Suppl.*, 1, (1975).
12. J. Wong and C.A. Angell, *Glass Structure by Spectroscopy*, Marcel Dekker, Inc. (1976).
13. V.G. Tchekhovsky, *Phys. Chem Glass*, **11**, 24 (1985).
14. S.V. Stefanovsky, I.A. Ivanov and A.N. Gulin, *J. Appl. Spectr.*, **54**, 648 (1991).
15. A.K. Varshneya. *Fundamentals of Inorganic Glasses*. Society of Glass Technology, Sheffield, (2006).
16. M.I. Ojovan and W.E. Lee, *Proc. WM'06 Conference*, Feb. 27 – Mar. 3, 2006, Tucson, AZ, WM-6239 (2006).
17. M.I. Ojovan and W.E. Lee. *J. Nucl. Mat.*, **335**, 425-432 (2004).
18. G. Möbus, G. Yang, Z. Saghi, X. Xu, R.J. Hand, A. Pankov and M.I. Ojovan, *Mat. Res. Soc. Symp. Proc.*, **1107** (2008).

Mater. Res. Soc. Symp. Proc. Vol. 1124 © 2009 Materials Research Society   1124-Q03-09

Electron Energy-Loss Near-Edge Structure of Zircon

Nan Jiang
Department of Physics, Arizona State University, Tempe AZ 85287-1504, USA

ABSTRACT

This work demonstrates that absorption near-edge structures of the O K-edge in zircon are directly related to atom structure, including bond length, bond angle and coordination. Small variations in the structure can induce measurable changes in absorption spectrum.

INTRODUCTION

Zircon is an extremely durable and resistant material with the added capability of accommodating a large quantity of actinides. It has been proposed as a candidate waste form for the geologic disposal of excess plutonium from dismantled nuclear weapons [1, 2]. Its structure and behavior under radiation have been intensively investigated in many experimental studies [3]. Radiation damage mainly involves a crystalline-to-amorphous transition, which results in a pronounced decrease in intensity of diffraction maxima [4] and a gradual loss of long-range periodicity in high-resolution electron microscopy (HREM) images [5]. Understanding the mechanisms for this crystalline-to-amorphous transition requires support from experimental observations towards structural and compositional changes at nanometer scale or even at atomic scale. Various spectroscopic techniques have been used to observe radiation damage at short-range order, which includes the changes in the nearest neighbor coordination geometries [6] and the polymerizations of Si and Zr polyhedra [7]. However, the currently used techniques have not provided a unique description of the fully-damaged state of zircon [3].

Electron energy-loss spectroscopy (EELS), combining with transmission electron microscopy (TEM), has a great advantage in probing local structure and chemistry around specific species at very high spatial resolution. However, such a powerful technique has not been used in study of the radiation damage in zircon in literature. Lack of direct relation between the observed electron energy-loss near-edge structure (ELNES) and atom structure information could be one among various reasons. In this work, we demonstrate that the O K-edge ELNES of zircon can be directly interpreted in terms of local structure, including distortion of Si tetrahedron, bond lengths of Si – O and Zr – O, and structure beyond the nearest neighbor distance. Previously, very few studies have been carried out on simulations of O K-edge EELS of zircon [8, 9], and no attempt has been made to identify the structural information from the O K-edge ELNES.

EXPERIMENT and CALCULATION

Synthetic zircon ($ZrSiO_4$) (Aldrich Chemical Company, Inc.) crystals were used in this study. The transmission electron microscope (TEM) specimens were prepared by grinding the samples into powders in acetone, and picking them up using a Cu grid covered with a lacy carbon thin film. The sample was then immediately transferred into and observed in a JEOL 2010 TEM with a field-emission gun operating at 200 keV and a

Gatan's Enfina parallel electron spectrometer. The EELS spectra were acquired in diffraction mode. The collection semi-angle was about 5 mrad. The full width at half maximum (FWHM) of the zero-loss peak without specimen was about 1.0 eV. In order to reduce the noise from channel-to-channel gain variations of the photodiode, a series of spectra was acquired with each spectrum shifted prior to acquisition by 0.5 eV relative to the previous spectrum, which is not equal to the inter-diode spacing. The resulting spectra were realigned before being added together. The background intensities were fitted by the intensities prior to the absorption edges using power-law functions and then subtracted from the original data.

The simulations of O K-edge ELNES of zircon were carried out using a full-multiple-scattering code, FEFF8 [10]. In brief, the method takes into account multiple scatterings of the excited core electron by the surrounding atoms, and the scattering is calculated by including a large number of atoms within a cluster. Self-consistent muffin-tin (MT) potentials were used in the calculations. The MT radii are 1.478, 0.997, and 0.942Å for Zr, Si and O, respectively, with 10% overlap to roughly correct for nonspherical potentials. The core-hole effect was included in the calculations using the "final state" approximation, i.e. an electron was removed form the O 1s orbital. The cluster size was 0.7 nm in radius. The original cluster was constructed from the experimental lattice parameters and atom positions of $ZrSiO_4$ [11].

RESULTS and DISCUSSION

Experimental O K-edge ELNES are compared with the calculation in Figure 1. The calculated spectrum is arbitrarily aligned to its peak b to the experiment at peak B position. However, if both spectra are aligned to the threshold position, the calculated spectrum shifts about 1.0 eV to lower energy. Nevertheless, although the discrepancies between the calculation and experiment exist, their overall appearances are remarkably similar. Approximately, the near-edge structures of O K-edge can be divided into two regions: region I and region II, as shown in Figure 1. The fine structures in these two regions reflect the different bonding characteristics of O associated with Zr and Si respectively.

The origins of absorption fine structures can be generally interpreted with the aid of electronic structure calculations. In Figure 2, the calculated O p-DOS, including a core hole in the O 1s state, is compared with the total ground state DOS projected on Zr and Si, respectively. In the energy window between 0 and 5eV, which corresponds to region I in absorption spectrum, the intensity is overwhelmingly dominated by Zr DOS, which is mostly from the localized Zr d-DOS. Therefore, the fine structures in region I are likely caused by Zr – O interaction. In other words, a change in Zr – O bonding geometry may result in a change in the O K ELENS in region I. However, between 5 and 15eV, which corresponds to region II, both Si and Zr have comparable DOS, with which Si has stronger DOS between 5 and 10eV, while Zr has slightly stronger DOS between 10 and 15eV. Evidently, the association of the absorption fine structures in this region with either Si – O or Zr – O geometry is complex.

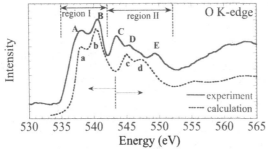

Figure 1 Comparison of experimental and calculated O K-edge ELENS in synthetic zircon.

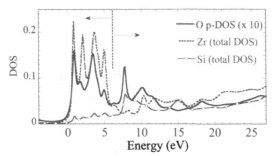

Figure 2 Comparison of O p-DOS (including core-hole effect) with total ground state DOS projected on Zr and Si.

In order to use ELNES to study atom migration, the absorption near-edge fine structures need to be interpreted directly in terms of atom structure in real space, rather than the density of states in reciprocal space. Here we introduce a structural variation method. In this method, the absorption spectra are calculated in structural models, in which the structure parameters such as bond lengths and angles are varied slightly from the experimental values. By comparing with calculated spectrum from the original structure, we can deduce structural information from changes in the absorption fine structures.

In zircon, the isolated SiO_4 tetrahedra share corners and edges with ZrO_8 dodecahedra. Due to repulsion between the Zr^{4+} and Si^{4+} cations, the SiO_4 tetrahedra in zircon are distorted (elongated along z-axis) from the regular SiO_4 tetrahedron [12]. Two O – Si – O angles are about 97° and 116°, as shown in Figure 3(a). The Si – O bond lengths are the same (~1.62Å) in this distorted tetrahedron. The Zr atom is coordinated by eight O atoms that define a triangular dodecahedron with symmetry $\overline{4}2m$. Four Zr – O bond lengths are 2.13Å, while the other four are 2.27Å. Each O coordinates with one Si and two Zr atoms.

Here we construct five modified zircon structural models by slightly varying the O atom's coordination, which are summarized in Table I. The only differences between the modified structure models and original zircon structure are O coordination, while other parameters are the same. In Model I (Figure 3(b)), the Si tetrahedra are further elongated along z-axis but the Si – O bond length is kept the same with the original zircon structure. Meanwhile, all the Zi – O bond lengths are the same, which is approximately the average of Zr – O bond length in the original zircon. In Model II (Figure 3(c)), the Si tetrahedra are in regular shape: all O – Si – O angles are the same and the Si – O bond length is kept the same with that of the original zircon structure. The operation results in a further distortion in Zr dodecahedra, in which the difference in Zr – O bond lengths is further apart. In Model III, the Si tetrahedra have very long Si – O bond length, ~1.66Å, while other parameters are very similar to those in the original zircon structure. In Model IV, on contrast, the Si tetrahedra have very short Si – O bond length, ~1.57Å. In Model V, Si – O is also small, but the Si tetrahedra are in regular shape.

Table I: A list of O coordination in original zircon and five modified structure models. In original zircon, space group is I4$_1$/amd and lattice parameters a and c are 6.607 and 5.982Å, respectively. Si and Zr coordinations are (0, 3/4, 5/8) and (0, 3/4, 1/8), respectively [11].

Zircon	Model I	Model II
(0, 0.0661, 0.1953)	(0, 0.0761, 0.1843)	(0, 0.0495, 0.2185)
Model III	Model IV	Model V
(0, 0.0661, 0.1853)	(0, 0.0761, 0.1953)	(0. 0.0565, 0.2239)

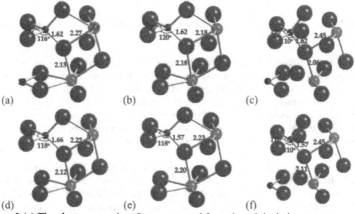

Figure 3 (a) The cluster around an O atom created from the original zircon structure. (b) – (f) The clusters created from the modified zircon structures of Model I – Model V, respectively. The bond lengths of Si – O and two Zr – O, and two Si – O – Si angles are given in graphs.

178

The calculated O K ELNES in these models are compared in Figure 4. The energy scales in these calculations were not adjusted. The spectrum "zircon" is from the original zircon structure. Between original zircon and Model I, the differences in spectra are very small, although peak a and peak b are slightly further apart and peak d is slightly higher than peak c in the later. This indicates that further distortion of Si tetrahedra and equal Zr – O bond lengths in Zr dodecahedra may not have significant effect on the O K-edge ELENS. On the contrary, if the Si tetrahedra become regular, peak c becomes much stronger than peak d as shown in Model II. It is also noticed that the relative intensities of peak a and peak b are inverse in Model II. This may relate to the very short Zr – O bond length (2.06Å) in this model. In Model III, peak a is still weaker than peak b, but it becomes stronger. This is probably due to the slightly shorter Zr – O bond length. In this model, Si tetrahedra have very long Si – O bond length. As a result, peak c moves to lower energy dramatically. On the other hand, when Si – O becomes shorter in Si tetrahedra as shown in Model IV and Model V, the fine structures in region II are very different from those in the original zircon: the multiple peaks are replaced by a single broad peak, no matter Si tetrahedra are distorted (Model IV) or in regular shape (Model V). It is also worth noting that the relative intensities of peak a and peak b in Model IV are similar to the original zircon, although they split further. In this model, shortest Zr – O length is longer than that in the original zircon. In Model V, peak a and peak b have the same intensity. This may be explained by a slightly shorter Zr – O length in this model.

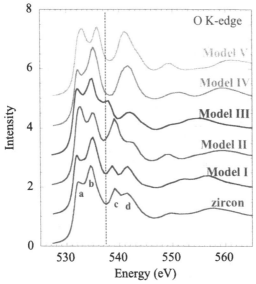

Figure 4 Comparison of calculated O K-edge in various modified zircon structure models. The vertical broken line is a guide for eyes.

In summary, the O K-edge ELNES in region I strongly relate to Zr polyhedron structure. The relative intensities of peak a and peak b and their splitting energy are associated with the shortest Zr – O bonding length. The shorter the Zr – O, the stronger the intensity of peak a and the narrower the splitting energy is. The geometry changes of Zr dodecahedra, on the other hand, have small impact on the O K-edge ELNES. This phenomenon can be explained by considering Zr – O bonding character. Due to the high coordination around Zr, Zr – O bond can be characterized as ionic, which is isotropic. Therefore, the bond angles may not have strong effect on the orbital overlap, but a single short Zr – O distance certainly does. On contrast, Si – O has directional covalent character, and thus the geometry of Si tetrahedra do have impact on the O K-edge ELNES in region II. The distortion of tetrahedron tends to move the absorption peak toward higher energy. Meanwhile, Si – O distance also has strong effect on orbital overlap and thus on the absorption structure.

CONCLUSION

In conclusion, we have demonstrated that the O K-edge ELNES directly relate to structure information, including bond length, bond angle and coordination. Small variations in structure can induce measurable changes in the O K-edge ELENS. This suggests that ELNES technique should be able to probe atom migration rearrangement induced by radiation and other external fields and the dynamic process of crystalline-to-amorphous transition could also be observed at atomic scale.

ACKNOWLEDGEMENTS

This work is supported by NSF DMR0603993. The use of facilities with the center for solid state science of ASU is also acknowledged.

REFERENCES

1. R. C. Ewing and W. Lutze, Science 275 (1997) 737.
2. B. E. Burakov, J. M. Hanchar, M. V. Zamoryanskaya, V. M. Garbuzov, and V. A. Zirlin, Radiochim Acta 89 (2002) 1.
3. R. C. Ewing, A. Meldrum, L. M. Wang, W. J. Weber, and L. R. Corrales, Reviews in Mineralogy and Geochemistry 53 (2003) 387; and references therein.
4. H. D. Holland and D. Gottfried, Acta Crystal. 8 (1955) 291.
5. L. M. Wang and R. C. Ewing, Nucl. Instru. Method Phys. Res. B65 (1992) 324.
6. F. Farges and G. Calas, Am. Mineral 76 (1991) 60.
7. I. Farnan and E. K. H. Salje, J. App. Phys. 89 (2001) 2084.
8. D. W. McComb, R. Brydson, P. L. Hansen, and R. S. Payne, J. Phys.: Condens. Matter 4 (1992) 8363.
9. Z. Wu and F. Seifert, Solid State Communications 99 (1996) 773.
10. A. L. Ankudinov, B. Ravel, J. J. Rehr, and S. D. Conradson, Phy. Rev. B 58, 7565 (1998).
11. K. Robinson, G. V. Gibbs, and P. H. Ribbe, American Mineralogist 56 (1971) 782.
12. J. A. Speer, Rev. Mineral. 5 (1982) 67.

Mater. Res. Soc. Symp. Proc. Vol. 1124 © 2009 Materials Research Society 1124-Q03-10

Effects of Silica Particle Size and pH on Rheological Properties of Simulated Melter Feed Slurries for Nuclear Waste Vitrification

Hong Zhao and Ian L. Pegg

Vitreous State Laboratory, The Catholic University of America, Washington, DC, 20064

ABSTRACT

In the process of nuclear waste vitrification, nuclear waste is usually mixed with glass-forming additives to form an intermediate aqueous slurry, followed by vitrification through a melter at high temperature (typically 1100-1200°C) to form a durable glass product with the desired composition and properties. The rheological behavior of such an intermediate melter feed slurry is important in the slurry mixing stage and in the subsequent slurry transport to the melter. A high solid content is desirable in terms of glass production rate but this must be offset against the requirements for favorable apparent viscosity and yield stress for slurry mixing and transport. In the present work, a model slurry system, prepared by adding commonly used glass-formers into a simulated waste, was studied to investigate the effects of pH and the particle size of silica (as the major glass forming additive) on the rheological properties of simulated melter feed slurries. The results show that whereas silica particle size significantly affects both the apparent viscosity and the yield stress, pH changes modify the yield stress while only slightly altering the apparent viscosity. Such relationships between rheological properties and controllable variables are useful for engineering melter feed slurries with desired rheological properties and potentially for recovery from upset conditions that lead to unfavorable slurry rheologies.

INTRODUCTION

Vitrification is a proven technology for safe and permanent disposal of highly radioactive wastes. In a typical vitrification process, the nuclear waste is usually mixed with glass-forming additives to form an intermediate aqueous feed slurry, followed by vitrification through a melter at high temperature (typically 1100-1200°C) to form a durable glass product with the desired composition and properties. The rheological behavior of such an intermediate melter feed slurry is important in the slurry mixing stage and in the subsequent slurry transport to the melter. Previous studies have examined the rheological and physical properties of aqueous feed slurries over large ranges of slurry total solid content and waste loading in the resulting glass products [1, 2]. The dependence of the rheological properties on various processing variables, such as total solid content in the feed slurry, waste loading in the glass, sodium concentration in the waste, and suspended solid content, as well as the effects of aging, temperature, and dispersing agents, were examined in previous work [1-5].

Relatively high solid content in the melter feed slurry is desirable in terms of glass production rate, but this must be offset against the requirements for favorable rheological properties (such as apparent viscosity and yield stress) that are required for slurry mixing and transport [1, 2]. In general, the challenge is to maintain sufficiently low yield stress and viscosity as the solid content is increased. In typical feed slurries that are developed for high-sodium low-activity waste streams at the Hanford Tank Waste Treatment and Immobilization Plant (WTP), the added glass-forming chemicals represent the principal suspended solids components that dominate the rheological properties of the melter feed slurry at a given solid content. Once the types and quantities of glass formers are determined based on the desired product glass composition, a number of variables remain that do not affect the glass composition but that can influence slurry rheology, such as the particle size distribution of the glass-formers and pH level of the feed slurry. Understanding the effects of these variables is therefore useful for optimization of the overall process.

In the present work, a high-solid-content (70 wt%) model slurry system, prepared by adding commonly used glass-formers into a simulated waste, was used to investigate the effects of particle size and the pH level on the rheological properties of simulated melter feed slurries. Since silica is used as a major component, accounting for more than half of all glass-formers by mass, the effect of variations of the silica particle size were investigated. In addition, the rheological effects of pH were investigated over the range from 10 to 13, reflecting the high-sodium nature of the LAW streams at the Hanford site.

EXPERIMENT

Feed slurries were prepared by adding a set of glass-formers into a simulated high-sodium low-activity waste stream [6, 7]. The resultant base feed slurry had a relatively low solid content of 38 wt%. The pH level of the base feed slurry was adjusted to predetermined values by adding either sodium hydroxide granules or boric acid powder. Finally, additional quantities of the same set of glass-formers (excluding boric acid) were added to the pH-adjusted slurries to increase the solid content to 70 wt%. For this purpose, two types of silica were used with mean volume diameters of 33 μm (Sil-Co-Sil 75, US Silica) and 10 μm (Min-U-Sil 15, US Silica). All feed slurries were homogenized by constant stirring during and after all additions.

Densities were measured by weighing a known volume (typically 1000 ml) of the material. The pH measurements were made using a calibrated pH meter. A Haake RS75 rheometer was employed to determine the apparent viscosity of feed slurries by measuring the shear stress on the slurry at controlled shear rates that are increased stepwise from 0.01 to 1000 s^{-1} with an equilibration delay of up to 30 seconds between steps. All measurements were performed at 25°C after the samples had been aged for one day.

RESULTS AND DISCUSSION

The physical properties of feed slurries with coarse SiO_2, blend (50:50) SiO_2, and fine SiO_2 are listed in Table 1. Note that for all three slurries, the pH is preset at 13 prior to adding glass-formers to formulate the high solid content slurries. Figure 1 shows the shear stress vs. shear rate curves for these three feed slurries; the corresponding yield stresses are given in Table 1.

Table 1. Physical properties and yield stress for feed slurries with various silica particle sizes.

SiO₂ Particle Size	Density (g/cm³)	pH	Total Solid Content (wt%)	Yield Stress (Pa)
Coarse	1.89	12.7	69.8	7.3
Blend (50:50)	1.90	12.6	70.3	15.5
Fine	1.90	12.7	70.6	30.9

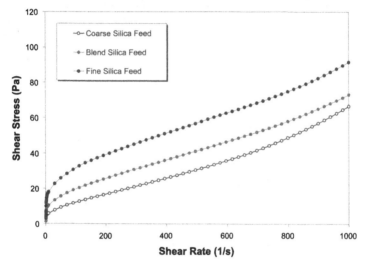

Figure 1. Shear stress vs. shear rate curves for feed slurries with various silica sizes.

The above results show the significant increase in the apparent viscosity and yield stress as more coarse silica is replaced by fine silica for three slurries having almost the same solid content, density, and pH. The results suggest that selection and adjustment of the silica particle size provides a significant measure of control over both the effective viscosity and yield stress of the resulting slurry, which is advantageous for formulating feed slurries with high solid contents and acceptable slurry rheologies. It should be noted, however, that for very coarse glass forming chemicals settling can become an issue, especially for slurries with low effective viscosities. Consequently, settling rates should also be considered when selecting the viable range of particle sizes for glass forming chemicals [4, 5].

The 50:50 blend silica feed slurry was used to further examine the effect of pH level on feed rheology. In this series of tests, the pH value of the base slurry was adjusted to a predetermined value prior to addition of glass-formers to increase the solid content to 70 wt%. Table 2 shows the physical properties of the resulting feed slurries at various pH levels. Figure 2 shows shear stress vs. shear rate curves at various pH levels; the corresponding yield stresses are plotted in Figure 3.

Table 2. Physical properties and yield stress for feed slurries
using the 50:50 blend silica with various pH values.

Adjusted pH Prior To Adding Glass-Formers	Solid Content (%)	Density (g/cm³)	pH After Adding Glass-Formers
10.0	70.0	1.88	10.0
11.0	70.1	1.88	11.0
12.0	70.5	1.89	11.7
13.0	70.3	1.90	12.6

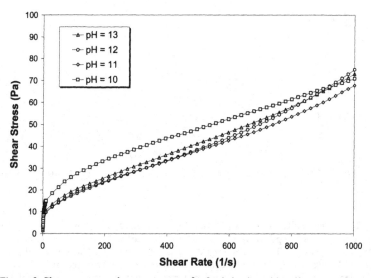

Figure 2. Shear stress vs. shear rate curves for feed slurries with various pH values.

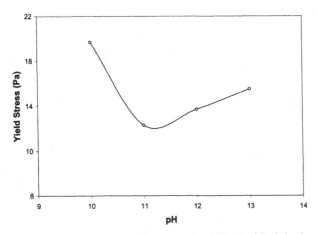

Figure 3. Dependence of yield stress on the pH level of feed slurries.

The above results indicate that the yield stress shows a non-monotonic variation with pH over the pH range studies in this work, showing a minimum yield stress at about pH = 11. Around this minimum, the yield stress increases rather steeply as the pH is reduced from 11 to 10 and more gradually as the pH rises from 11 to 13. In particular, the yield stress increased by nearly 70% when the pH was reduced by one pH unit from the value at the minimum yield stress. In contrast, there is relatively little change in the apparent viscosity with pH level. Again, the largest effect occurs when the pH is reduced from 11 to 10 but the relative change in effective viscosity of about 30% (at about 200 s^{-1}) is much less than the effect on the yield stress; the relative change in the effective viscosity when the pH is increased from 11 to 13 is within ~10%. Thus, unlike the silica particle size, which affects both viscosity and yield stress, changes in pH have a much more pronounced impact on yield stress. The ability to adjust the effective viscosity and yield stress, and to decouple their variations to a large extent, is useful for engineering specific rheological properties for melter feed slurries of this type. In particular, this provides a means for optimization by effecting adjustments to maintain acceptable effective viscosity and yield stress while maximizing solids content. In addition, understanding of these relationships could be useful for recovery from upset conditions that lead to unfavorable slurry rheologies in a large-scale glass production.

CONCLUSIONS

The results of this study on a model slurry system at high solids content (70 wt%) indicate that whereas silica particle size significantly affects both the apparent viscosity and the yield stress, pH changes modify the yield stress while only slightly altering the apparent viscosity. Such relationships between rheological properties and controllable variables are useful for engineering feed slurries with desired rheological properties and potentially for recovery from upset conditions that lead to unfavorable slurry rheologies.

REFERENCES

1. H. Zhao, I.S. Muller, and I.L. Pegg, "Characterization of Simulated WTP LAW Melter Feeds", Final Report, VSL-04R4500-1, Rev. 0, Vitreous State Laboratory, The Catholic University of America, Washington, D.C., May, 2004.
2. I.S. Muller, H. Gan, and I.L. Pegg, "Physical and Rheological Properties of Waste Simulants and Melter Feeds for RPP-WTP LAW Vitrification", Final Report, VSL-00R3520-1, Rev. 0, Vitreous State Laboratory, The Catholic University of America, Washington, D.C., January, 2001.
3. H. Zhao, I.S. Muller, and I.L. Pegg, "Effects of Aging and Temperature on the Rheological Properties of Simulated Melter Feed Slurries for Nuclear Waste Vitrification," Ceram. Trans., Vol.176, pp. 223-229, (2006).
4. H. Zhao, I.S. Muller, and I.L. Pegg, "Rheological Properties of Aqueous Melter Feed Slurries for Nuclear Waste Vitrification," Ceram. Trans., Vol.168, pp.149-58, (2004).
5. H. Zhao, I.S. Muller, I.L. Pegg, "Effects of Poly(Acrylic Acid) on the Rheological Properties of Aqueous Melter Feed Slurries for Nuclear Waste Vitrification," Ceram. Trans., Vol.155, pp.149-158, (2003).
6. R.E. Eibling, R.F. Schumacher, E.K. Hansen, "Development of Simulants to Support Mixing Tests for High Level Waste and Low Activity Waste," WSRC-TR-2003-00220, Rev. 0, Westinghouse Savannah River Company, December 2003.
7. R.R. Russell, S.K. Fiskum, L.K. Jagoda, A.P. Polaski, "AP-101 Diluted Feed (Envelope A) Simulant Development Report," WTP-RPT-057, Battelle Pacific Northwest Division, Richland, WA, February, 2003.

Mater. Res. Soc. Symp. Proc. Vol. 1124 © 2009 Materials Research Society   1124-Q10-01

Infra-Red and Electron Paramagnetic Resonance (EPR) Spectroscopy Studies of Simulated Vitrified Wastes Produced in Cold Crucible

L.D. Bogomolova,[1] S.V. Stefanovsky,[2] and J.C. Marra[3]
[1] Lomonosov Moscow State University, Vorobyovygory, Moscow
[2] SIA Radon, 7th Rostovskii per. 2/14, Moscow 119121, Russia
[3] Materials Science and Technology Directorate, Savannah River National Laboratory, Building 773-42A, Aiken, SC 29808, U.S.A.

ABSTRACT

Two borosilicate glasses simulating vitrified Idaho National Laboratory (INL) sodium bearing waste (SBW) and Hanford high alkaline wastes (HAW) were produced at the Radon cold crucible based bench-scale plant. The SBW glass was predominantly amorphous and composed of a borosilicate matrix with minor inclusions of unreacted quartz, baddeleyite and zircon. An infra-red spectroscopic study showed the structure consisted of silicon-oxygen and boron-oxygen tetrahedra. Electron paramagnetic resonance (EPR) spectra showed a superposition of lines due to Fe^{3+} and V^{4+}. The line with g = 4.3 was due to Fe^{3+} in a strong electric field and the broad g = 2 line to clusters of Fe^{3+} ions. The spectrum obtained for V^{4+} was typical of vanadium-doped borosilicate glasses and hyperfine structure was observed due to the interaction of unpaired electron with I = 7/2 nuclear spins. The vitrified HAW glass was composed of a vitreous matrix, minor aluminosilicate and spinel (high manganese) phases and rare silver inclusions. Mn^{2+} ions gave the major contribution to the EPR spectra due to high manganese content in the glassy products.

INTRODUCTION

An inductive cold crucible melting (ICCM) technology is being successfully applied in Russia at SIA "Radon" for vitrification of liquid institutional and Russian nuclear power plant intermediate-level wastes [1, 2]. These wastes are both sodium nitrate based and form sodium borosilicate glasses. They have 30-40 wt.% waste loading, low leachability of both radionuclides ($^{134,137}Cs$, ^{90}Sr, activated corrosion products, actinides) and macrocomponents (Na, K, B, Si), strong mechanical integrity, and high stability with respect to devitrification. It is of interest to evaluate the feasibility of using the ICCM technology for vitrification of DOE wastes, including high alkaline waste which is currently stored in tanks at the Hanford, Idaho and Savannah River Sites. During the first stage of the contract between US DOE and SIA "Radon", tests on vitrification of INL SBW and Hanford HAW were performed, the conclusions of which are presented elsewhere [3, 4]. In our previous work [5, 6] we investigated in detail phase composition, element partitioning among co-existing vitreous and crystalline phases, and leach resistance of the vitrified sodium bearing and high-alumina wastes. The goal of the present work was to study the structure of the vitrified products to provide insight into the effect of glass structure on the suitability for a long-term storage, specifically the effect on leach resistance.

EXPERIMENTAL

The glassy materials were produced from waste surrogates and glass-forming additives (Table I) in the Radon bench-scale cold crucible facility. The equipment and process variables were described in details elsewhere [3, 4, 6]. Sampled from canisters were analyzed by X-ray fluorescent spectroscopy using a PW-2400 spectrometer (Philips Analytical B.V., The Netherlands) equipped with quantitative analytical Philips SuperQuantitative & IQ-2001 Software, atomic absorption spectroscopy using a Perkin-Elmer 403 spectrometer, and emission flame photometry (EFP) to determine Na and K using a PFM-U 4.2 flame photometer (Russian design). The amounts of Fe^{2+} and Fe^{3+} were determined by chemical analysis. Glass samples were also studied by infra-red (IR) spectroscopy using a Perkin-Elmer 983G spectrophotometer (using KBr and powdered glass compacts), and electron paramagnetic resonance (EPR) using a RE-1306 radiospectrometer (Russian model) operated within the X-range at room temperature.

Table I. Chemical compositions of SBW and HAW glasses.

Oxides	SBW-0 (calc)	SBW-2	SBW-5	SBW-9	SBW-12	HAW-0 (calc)	HAW-1	HAW-2	HAW-4	HAW-6	HAW-8	HAW-10	HAW-14	HAW-16	HAW-18	HAW-20	HAW-21	HAW-23	HAW-DV
Li_2O	2.04	2.34	2.97	3.16	3.34	-	-	-	-	-	-	-	-	-	-	-	-	-	-
B_2O_3	5.10	2.40	1.80	1.37	1.17	3.00	2.50	3.30	2.90	3.60	3.10	3.00	2.30	2.40	2.70	2.40	2.90	2.50	2.30
Na_2O	12.8	13.9	12.5	13.4	12.8	11.2	9.7	10.0	9.1	8.7	9.1	8.6	8.8	8.7	8.7	9.2	9.8	8.8	9.9
MgO	1.54	1.40	1.35	1.65	1.62	0.30	0.27	0.27	0.27	0.40	0.27	0.26	0.30	0.27	0.29	0.28	0.28	0.27	0.21
Al_2O_3	5.91	6.60	7.85	10.52	10.78	14.9	12.5	12.9	14.3	14.8	14.9	14.2	14.4	14.4	14.2	13.9	14.4	13.5	13.1
SiO_2	56.1	55.9	53.1	51.9	52.6	34.2	37.5	38.7	35.7	33.7	35.1	34.3	35.8	34.0	35.1	34.7	36.2	34.2	34.0
P_2O_5	0.16	0.35	0.29	0.19	0.16	0.18	0.20	0.24	0.27	0.28	0.21	0.27	0.25	0.28	0.18	0.24	0.22	0.26	0.17
SO_3	0.95	0.69	0.77	0.85	0.64	0.03	0.18	0.24	0.18	0.15	0.12	0.12	0.15	0.15	0.12	0.15	0.12	0.15	0.15
Cl	0.18	0.03	0.03	<0.01	0.03	-	-	-	-	-	-	-	-	-	-	-	-	-	-
K_2O	1.58	1.34	1.54	1.44	1.47	0.51	0.49	0.52	0.50	0.49	0.38	0.47	0.46	0.48	0.33	0.45	0.41	0.46	0.34
CaO	4.64	4.96	4.31	4.85	4.24	0.99	1.53	1.40	1.28	1.21	1.24	1.23	1.24	1.21	1.24	1.24	1.25	1.24	1.68
TiO_2	0.01	0.10	0.06	0.05	0.05	0.06	0.12	0.12	0.14	0.13	0.13	0.13	0.13	0.13	0.12	0.13	0.13	0.13	0.10
V_2O_5	4.11	4.67	5.23	6.21	6.53	-	-	-	-	-	-	-	-	-	-	-	-	-	-
Cr_2O_3	0.04	-	<0.01	<0.01	-	0.22	0.23	0.25	0.23	0.24	0.24	0.24	0.23	0.23	0.22	0.25	0.26	0.25	0.21
MnO	0.17	0.23	0.27	0.27	0.30	16.2	15.3	15.6	17.3	18.0	17.6	17.8	17.7	17.7	17.5	17.4	17.6	17.7	14.6
FeO_x^*	1.57	2.84	2.25	1.83	1.84	6.10	3.14	2.11	1.58	1.52	1.64	1.53	1.36	1.34	1.36	1.69	1.57	1.41	4.73
NiO	0.03	0.03	0.04	0.05	0.05	0.21	0.24	0.25	0.26	0.26	0.26	0.26	0.26	0.26	0.26	0.24	0.24	0.24	0.17
CuO	0.01	<0.01	<0.01	<0.01	0.02	-	-	-	-	-	-	-	-	-	-	-	-	-	-
ZnO	0.02	<0.01	0.02	0.02	0.03	-	-	-	-	-	-	-	-	-	-	-	-	-	-
SrO	<0.01	<0.01	<0.01	<0.01	0.00	10.10	11.7	14.0	15.9	15.7	15.5	15.4	14.6	16.2	15.2	15.9	15.1	16.2	14.9
ZrO_2	2.04	0.96	1.61	1.88	2.05	0.16	0.27	0.26	0.25	0.25	0.25	0.24	0.24	0.25	0.24	0.24	0.25	0.24	0.27
Ag_2O	-	-	-	-	-	0.32	0.26	0.28	0.35	0.37	0.31	0.25	0.31	0.34	0.30	0.30	0.30	0.32	0.26
CdO	-	-	-	-	-	0.06	0.05	0.04	0.05	0.03	0.03	0.03	0.04	0.04	0.03	0.04	0.06	0.04	0.01
Cs_2O	-	-	-	-	-	0.14	0.12	0.16	0.15	0.13	0.11	0.11	0.13	0.13	0.11	0.11	0.11	0.12	0.09
BaO	0.002	0.08	0.15	0.17	0.21	0.09	0.40	0.24	0.21	0.18	0.20	0.33	0.16	0.17	0.15	0.16	0.16	0.16	0.59
La_2O_3	-	-	-	-	-	0.15	0.13	0.14	0.15	0.14	0.14	0.16	0.14	0.15	0.10	0.09	0.15	0.08	0.06
CeO_2	-	-	-	-	-	0.05	0.03	0.03	0.04	0.04	0.05	0.03	0.04	0.05	0.04	0.05	0.04	0.05	0.03
PbO	0.05	0.04	0.05	0.05	0.05	0.31	0.26	0.27	0.27	0.27	0.29	0.33	0.33	0.33	0.33	0.35	0.35	0.25
Total	99.05	98.86	96.19	99.86	99.98	99.48	97.12	101.32	101.38	100.59	101.17	99.29	99.37	99.27	98.82	99.49	101.9	98.67	98.12
FeO_x	1.57	2.84	2.25	1.83	1.84	-	-	-	-	-	-	-	-	-	-	-	-	-	-
FeO	-	0.42	0.42	0.64	0.83	-	-	-	-	-	-	-	-	-	-	-	-	-	-
Fe_2O_3	-	2.42	1.83	1.19	1.01	-	-	-	-	-	-	-	-	-	-	-	-	-	-

* recalculated to Fe_2O_3.

RESULTS AND DISCUSSION

Twelve containers of SBW glass and twenty four containers of HAW glass were selected, and glass samples from the containers and from the "dead volume" of the cold crucible were examined. X-ray diffraction (XRD) and scanning electron microscopy (SEM) with energy dispersive spectroscopy (EDS) indicated that the bulk of the SBW glasses were amorphous with homogeneous elemental distribution. Minor amounts of unreacted quartz, baddeleyite, and zircon were however observed [5, 6]. The HAW glasses contained spinel-type, nepheline and carnegieite phases. The highest content of the nepheline and carnegieite was observed in the "dead volume" of the cold crucible where cooling rate of the glassy product was the slowest. Very minor inclusions of metallic silver were also observed.

IR spectra obtained from vitrified SBW samples within the range of 1800-400 cm^{-1} were typical of alkali-silicate and borosilicate glasses with low boron content [7-9]. They consisted of four strong absorption bands at 1350-1450 cm^{-1}, 850-1200 cm^{-1}, 700-800 cm^{-1} and 400-550 cm^{-1}, as well as a weak band centered at 1635 cm^{-1} visible only in the SBW-2 glass spectrum (Figure 1). The strongest band at 850-1200 cm^{-1} in the SBW-2 glass spectrum had a peak at 1022 cm^{-1} and shoulders at 1160, 1075, and 883 cm^{-1}. These bands were due to asymmetric v_{as} valence vibrations of bridging Si–O–Si bonds connecting SiO$_4$ tetrahedra and non-bridging Si–O$^-$ bonds. The weak band with a peak at 1430 cm^{-1} was attributed to vibrations of the bridging B–O–B bonds connecting BO$_3$ triangles and BO$_4$ tetrahedra. The weak band with components at 780 and 715 cm^{-1} were due to symmetric v_s valence vibrations of the bridging Si–O–Si bonds and Al–O bonds in the AlO$_4$ tetrahedra. Vibrations within the spectral range of 400-500 cm^{-1} were attributed to deformation vibrations in the SiO$_4$ tetrahedra. The bands observed at 850-1200 cm^{-1} and 700-800 cm^{-1} were more symmetric in the IR spectra of glasses SBW-5, SBW-9, and SBW-12 and their peaks shifted to 1000-1007 cm^{-1} and 729 cm^{-1}, respectively. This indicated increased symmetry of the SiO$_4$ tetrahedra and a minor decrease in the degree of connectedness of the SBW glass structural network. The peak of the band near 1350-1450 cm^{-1} was shifted to 1383 cm^{-1} in the SBW-5 glass spectrum, indicating an increase in the fraction of tetrahedrally-coordinated boron, and slightly reduced in intensity due to a decrease in boron content. In the spectrum of the glass SBW-12, sampled at the end of the test, the peak at 1403 cm-1, and the band were broadened and stronger. The boron content in this glass was low (1.17 wt.%) and, therefore, the band cannot be due only to vibrations of B–O bonds. The band at ~1400 cm^{-1} is normally present in IR spectra of alkali-silicate glasses [9,10], and it is reported to be due to the presence of CO$_3^{2-}$ ions in the glass structure.

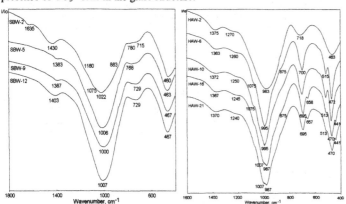

Figure 1. IR spectra of the SBW glasses (left) and HAW glass-ceramics (right).

IR spectra of the HAW glasses within the range of 1600-400 cm^{-1} (Figure 1) consisted of a weak band centered at 1363-1375 cm^{-1}, a shoulder at 1200-1300 cm^{-1}, an intense band at 850-1150 cm^{-1}, and strong bands at 650-750 cm^{-1} and 400-550 cm^{-1}. In the HAW-2 glass spectrum, these bands were centered at 1375, 1270, 983, 718, and 463 cm^{-1}. The bands at 1375, 1270 and 718 cm^{-1} were attributed to vibrations of B–O–B bonds coupling boron-oxygen triangles. The strongest band with a peak at 983 cm^{-1} was due to vibrations in the SiO$_4$ tetrahedra. The low wavenumber pointed to a low degree of connectedness for the glass network. The basis of the glass structure is probably silicon-oxygen chains rather than framework. The HAW glasses have a relatively low silica content (38.7 wt.% SiO$_2$ in HAW-2) and a low number of Si–O–Si bonds, but formation of the Si–O–Al and Si–O–B bonds connecting SiO$_4$ and AlO$_4$ or BO$_4$ tetrahedra compensate for the destruction effect and increases the total degree of connectivity. Vibrations of the Al–O and Fe–O bonds in network modifying AlO$_4$ and FeO$_4$ tetrahedra contribute to the band at 718 cm^{-1} in the HAW-2 glass spectrum. The band at 400-550 cm^{-1} was attributed to deformational vibrations in the network former tetrahedra. It should be noted that the HAW-2 glass was almost completely amorphous with only traces of spinel structured crystalline inclusions, and the spectrum obtained from this glass represented the "true" structure of HAW glasses. The HAW-6 glass contained both nepheline, carnegieite and spinel structured phases [6]. The major bands were narrower and appeared to have multiple components consistent with the presence of lower symmetry SiO$_4$ structure forming units in the crystalline phases. An extra band at 515 cm^{-1} was also observed probably due to the presence of the spinel structured phase. Incorporation of some silica and alumina in the aluminosilicate phases increased the boron fraction in the glass which was characterized by a shift in the band at 1375 cm^{-1} in HAW-2 to 1363 cm^{-1} in HAW-6. This also corresponded to an increase the amount of B–O–B bonds connecting BO$_4$ tetrahedra and BO$_3$ triangles. The spectra of the HAW-10, HAW-16, and HAW 21 glasses with high crystalline phase content showed splitting of the bands due to valence vibrations of the Si–O bonds in the SiO$_4$ tetrahedra (between 987 and 1007 cm^{-1}) due to de degeneration. Also extra bands at 657-658 cm^{-1}, 513-515 cm^{-1}, and ~441 cm^{-1} were observed assigned to vibrations in the MeO$_4$ tetrahedra and MeO$_6$ octahedra (Me = Fe, Mn, Cr, Ni) forming the spinel structure. Occurrence of two different maxima at 987 and 1007 cm^{-1} pointed to the formation of the SiO$_4$ tetrahedra with various degrees of connectedness. The location of the bands due to B-O bonds remained approximately the same in all the glasses studied (HAW-10, HAW-16, and HAW-21). This indicated that the amount of three-coordinated boron did not vary significantly.

The glasses contained various transition metal oxides and the appearance of additional EPR signals from some of them was expected. In all the glasses studied strong spectrum for Fe^{3+} and weak spectrum for V^{4+} were observed. The fraction of Fe^{3+} in the SBW glasses reduced from ~ 85% at the start, to ~ 55% at the end of the cold crucible run (Table I). The EPR spectrum of Fe^{3+} consisted of two lines (Figure 2). One of them was an asymmetric line with g ~ 4.3 and a peak-to- peak width (ΔH_{pp}) between 200 to 300 G, depending on the glass composition. Correlation between the integrated intensity of this line and the concentration of Fe$_2$O$_3$ was observed. The second line observed was a broad line with g ~ 2 and ΔH_{pp} that was almost independent of glass composition. Both these lines have been observed in numerous oxide glasses [9]. According to theory, the g ~ 4.3 line is due to tetrahedrally coordinated Fe^{3+} ions that are located in a strong electric field. The broad g ~ 2 line can be attributed to clusters of Fe^{3+} ions, predominantly in octahedral coordination, coupled by strong dipole-dipole interactions. The broad line observed for the SBW-1 sample was asymmetric and contained weak responses associated with the superposition of the vanadium EPR spectrum. Although the intensity and width of this line varied slightly in the different samples it is indicative the formation of stable clusters of Fe^{3+} ions. With increasing vanadium concentration, new lines were superimposed on the Fe^{3+} signal. The spectrum due to V^{4+} (shown in the inset in Figure 2) was obtained by subtracting the spectrum of the SBW-1 sample from the spectrum of the SBW-12 sample. This differential spectrum has a complex shape and is similar to the spectra of V^{4+} observed for numerous oxide glasses. EPR of silicate glasses containing vanadium has been studied in numerous works [9, 10] and it has been shown that vanadium is present mainly as V^{5+}, which does not give an EPR signal. Minor amounts of V^{4+} present gave a complex hyperfine structure caused by the interaction of unpaired electrons with nuclear spins $I = 7/2$ due to ^{51}V. The approximate parameters determined from the differential spectrum, $A_{||} = (170 \pm 5) \times 10^{-4}$ cm-1 and $g_{||} = 1.93 \pm 0.05$, were consistent with reference data for V^{4+} in silicate glasses with SiO_2 concentration of ~ 50%. The spectra of other transition metal ions were not observed and they were presumably obscured by the intense Fe^{3+} spectrum.

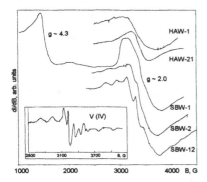

Figure 2. EPR spectra of vitrified SBW and HAW.

Table II. Linewidths and relative intensities of the EPR responses in the HAW glasses.

Glass sample	Linewidth, G	Relative intensity
HAW-1	640	1
HAW-4	310	0.75
HAW-8	300	1.24
HAW-10	340	1.20
HAW-14	320	0.74
HAW-18	310	1.16
HAW-20	310	1.14
HAW-23	460	1.27
HAW-DV	470	1.98

Figure 2 also shows EPR spectra of the vitrified HAW. Both spectra consisted of a single line with g ~ 2 although the linewidths were different (Table 2). Broader linewidths were observed for the glass samples with low crystalline content, taken at the beginning of the test. The high MnO content in all of the samples suggests that the g ~ 2 was due to the presence of Mn^{2+}. Quantitative evaluation of the EPR spectra show that concentration of the Mn^{2+} ions in the HAW-4 to HAW-21 samples to be ~ 1.3×10^{24} kg^{-1}, i.e. almost all the manganese was present as Mn^{2+}. Hyperfine splitting (HFS) due to ^{55}Mn is observed at low concentrations of Mn^{2+} ions (normally $\leq 1\%$). At higher concentrations, the distance between the ions decreases leading to exchange interactions. As a result the HFS disappears and a single line is observed. The presence of different ions (Fe^{3+}, Fe^{2+}, Mn^{3+}, etc.) can result in anisotropic exchange broadening of the observed spectral lines. This was seen in the spectra of the HAW-1 and HAW-DV samples which containing high concentrations of iron. The high integral intensity of the line in the spectrum of the HAW-DV sample may have been due to the contribution of Fe^{3+} ions. Moreover, broadening of the EPR line for the HAW-23 sample could have been due to the relatively high content of transition metal ions in the spinel structured crystalline phase in this sample.

CONCLUSION

The structure of the INL SBW glass consisted of chains of SiO_4 tetrahedra modified by BO_4 and AlO_4 tetrahedra. The structural network of the vitrified Hanford HAW glass had a lower degree of connectedness (polymerization) and a higher content of non-bridging oxygen ions. These glasses contained spinel and nepheline structured crystalline phases distributed throughout the vitreous matrix. Fe^{3+} and Mn^{2+} ions were found to be major contributors to the EPR spectra of the SBW and HAW glasses, respectively. In the EPR spectra of the vanadium-containing SBW glasses a response due to V^{4+} was observed. High concentrations of paramagnetic ions in their $^6S_{5/2}$ ground state (Fe^{3+} and Mn^{2+}) yield strong dipole-broadened unresolved lines which can camouflaging weak lines due to the other paramagnetic ions present.

REFERENCES

1. I.A. Sobolev, S.A. Dmitriev, F.A. Lifanov, A.P. Kobelev, S.V. Stefanovsky and M.I. Ojovan, *Waste Management '04 Conf.* February 29 – March 4, 2004, Tucson, AZ. 2004. CD-ROM. WM-4300.
2. I.A. Sobolev, S.A. Dmitriev, F.A. Lifanov, A.P. Kobelev, S.V. Stefanovsky and M.I. Ojovan, *Glass Technol.*, **46**, 28 (2005).
3. A.P. Kobelev, S.V. Stefanovsky, V.N. Zakharenko, M.A. Polkanov, O.A. Knyazev, T.N. Lashchenova, V.I. Vlasov, C.C. Herman, D.F. Bickford, E.W. Holtzscheiter, R.W. Goles and D. Gombert, *Waste Management' 05 Conf.* February 27-March 3, 2005, Tucson, AZ. CD-ROM.
4. A.P. Kobelev, S.V. Stefanovsky, O.A. Knyazev, T.N. Lashchenova, C.C. Herman, D.F. Bickford, E.W. Holtzscheiter and R.W. Goles, *Proc. 107th Amer. Ceram. Soc. Annual Meeting*. Baltimore, MD, April 10-13, 2005; p159-169.
5. S.V. Stefanovsky, T.N. Lashchenova, L.D. Bogomolova, C.C. Herman, D.F. Bickford, E.W. Holzscheiter, R.W. Goles and D. Gombert, *European Mater. Res. Soc. Spring Meeting*, May 31-June 3, 2005. Strasbourg, France. Abstracts. 2005. N – 12/28.
6. A.P. Kobelev, S.V. Stefanovsky, T.N. Lashchenova, V.N. Zakharenko, M.A. Polkanov, O.A. Knyazev, C.C. Herman, D.F. Bickford, E.W. Holzscheiter, R.W. Goles and D. Gombert, *ICEM'05 The 10th Int. Conf. on Environmental Remediation and Radioactive Waste Management*. September 4-8, 2005, Glasgow, Scotland. ASME, 2005. CD-ROM. ICEM05-1222.
7. J. Wong and C.A. Angell, *Glass Structure by Spectroscopy*, Marcel Dekker, Inc. (New York, 1976).
8. W.L. Konijnendijk, *The Structure of Borosilicate Glasses, Phil. Res. Rep. Suppl.* [1] 1975.
9. S.V. Stefanovsky, I.A. Ivanov and A.N. Gulin, *J. Appl. Spectr.*, **54**, 648 (1991).
10. *Infra-Red Spectra of Alkali Silicates (Russ.)*, Khimiya (Leningrad, 1970).
11. D.L. Griscom, *J. Non-crystal. Solids*, **40**, 211 (1980).
12. L.D. Bogomolova, V.A. Jachkin, V.N. Lazukin and V.A. Shmucler, *J. Non-Crystal. Solids*, **27**, 427 (1978).

Ceramic Wasteforms

Mater. Res. Soc. Symp. Proc. Vol. 1124 © 2009 Materials Research Society   1124-Q04-01

HIPed Tailored Pyrochlore-Rich Glass-Ceramic Waste Forms for the Immobilization of Nuclear Waste

Melody L. Carter, Huijun Li, Yingjie Zhang, Andrew L. Gillen and Eric. R. Vance
Ansto, New Illawarra Rd, Lucas Heights, NSW 2234, Australia

ABSTRACT

Hot isostatically pressed (HIPed) glass-ceramics for the immobilization of uranium-rich intermediate-level wastes and Hanford K-basin sludges were designed. These were based on pyrochlore-structured $Ca_{(1-x)}U_{(1+y)}Ti_2O_7$ in glass, together with minor crystalline phases. Detailed microstructural, diffraction and spectroscopic characterization of selected glass-ceramic samples has been performed, and chemical durability is adequate, as measured by both MCC-1 and PCT-B leach tests.

INTRODUCTION

The use of glass-ceramics to combine the process and chemical flexibility of glasses with the excellent chemical durability of ceramics is a well-known method to immobilize nuclear waste. This can be achieved by exploiting the glass forming components present in the waste, along with appropriate additives to modify the glass and produce the desired crystalline phases. Hot-isostatic pressing (HIPing) technology offers zero off-gas emissions during the high-temperature consolidation, which mitigates volatility concerns, which in turn reduce the footprint of the system. Secondly the process places minimal constraints on the waste form chemistry insofar as the redox conditions can be adjusted from near-neutral to quite reducing by the addition of metal powders to the waste form precursors. Also the HIP can material can be selected to be reasonably compatible with the waste form and these factors permit significantly higher waste loadings to be employed. In addition, the HIP process readily produces a dense monolithic waste form, which both minimizes disposal volume and reduces surface area available for aqueous attack once emplaced in a repository.

The development of glass-ceramics for nuclear waste has been studied by many workers [1-7]. Many of the glass-ceramics developed were targeted at zirconolite-glass ceramics and the resultant waste form were very durable, with aqueous leach rates 10-100 times better than glass [7].

Earlier work [8-9] on pyrochlore-glass ceramics was effected by mixing glass with crystalline pyrochlore and making a composite by sintering at <700°C. The dense and durable crystalline pyrochlore were in any case prepared by simple but well controlled ceramic processing methods. This was not of interest in the present work because the actinide would first need to separated out of the waste to fabricate the pyrochlore and then the pyrochlore would have to be powdered and reacted with the glass powder, raising questions of glass/pyrochlore compatibility.

For U-rich wastes, Vance et al [10-11] developed sphene glass ceramics to be melted in air and these contained pyrochlore and brannerite; the resulting waste form was more durable than boro- and aluminosilicate glasses.

In this paper we designed HIPed glass-ceramics waste forms for the immobilization of a hypothetical high-U waste and two Hanford K-Basin sludges. The hypothetical waste stream was utilized to try to understand the pyrochlore- glass interaction in an all-melted glass-ceramic. The information so obtained was used to develop waste forms for the K-Basin sludges. The Hanford K-East and K-West Basins were used to store irradiated fuel prior to Spent Nuclear Fuel (SNF) processing. In 1980, irradiated N-Reactor fuel was placed under water in the pools previously used for temporary storage of irradiated fuel from the K-East/K-West Reactor production complex. Over the lifetimes of these K-West and K-East Basins, debris, silt, sand, and material from operations resulted in the formation of sludge that accumulated in the bottoms of these basins. In addition, the extended storage of the irradiated fuel resulted in corrosion of the fuel cladding and the storage canisters.

P. D. Rittmann [12] reported five K- Basin general sludge compositions. For the present study a simplified waste composition was used for two K-Basin wastes with the largest compositional differences (see Table I).

EXPERIMENT

Table II lists the composition on an oxide basis of the U-rich waste containing samples. Table III lists the additives on an oxide basis used for the K-Basin wastes in table I.

Sample 1 was prepared by shaking together the SiO_2 Na_2CO_3, Al_2O_3, H_3BO_3, TiO_2, $CaCO_3$ and UO_2 (40 g on an oxide basis) and placing the mixture in a Pt crucible. This was melted in argon at 1250° C for 0.1 h with heating and cooling rates of 3° C /min. (the slow heating rate was required because the tube furnace used had a 150 mm diameter Al_2O_3 tube and the tube was prone to cracking at faster heating and cooling rates).

Samples 2 and 3 (40 g and 300g on an oxide basis respectively) were prepared by mixing all precursors in water (5 ml water per g of precursor), allowing the nitrates to dissolve and the titanium (IV) isopropoxide (TiPT) to hydrolyze. The mixture was then heated to ~110° C to drive off the alcohol and water. The dry product was then calcined in air for 1 hour at 750° C. The sample was hand ground in a mortar and pestle. 30 g aliquots of sample 2 and 3 were put in Pt crucibles and then melted in argon at 1250° C as above.

HIPing was carried out on three aliquots (60 g) of sample 3. The samples had Ti, Ni, or Fe metal -325 mesh powders (2wt%) added to them to control the redox during HIPing. The samples were placed in stainless steel cans, evacuated and sealed, and HIPing was carried out at 1200°C/30MPa/1h under argon.

The K-basin sludge waste form samples (Main Basin floor and Weasel Pit), weighing 100g on an oxide basis, were prepared in the same manner as samples 2 and 3 above. HIPing was carried out on 60 g samples of each Main Basin floor and Weasel Pit sample. The samples had 2 wt% Ni metal added to them to control the redox during HIPing. The samples were placed in stainless steel cans, evacuated and sealed, and HIPing was carried out at 1200°C/30MPa/1h under argon.

Table I. K-Basin waste compositions.

Oxide (added as)	Weasel pit g*/100cc	Main Basin Floor g*/100cc
Al_2O_3 (Al_2O_3)	6.3	4.1
CaO ($CaCO_3$)	1.6	0.3
Fe_2O_3 (Fe nitrate)	40.3	12.9
SiO_2 (SiO_2)	24.0	7.4
UO_2 (U nitrate)	6.3	3.8
Cs_2O (Cs nitrate)	1.0	1.0
SrO (Sr nitrate)	1.0	1.0
MoO_3 (MoO_3)	1.0	1.0
La_2O_3 (La nitrate)	1.0	1.0

g* of oxide

Table II. The composition on an oxide basis of the additives used for the U-rich waste containing samples.

Sample No	SiO_2 wt % (pre*)	Na_2O wt % (pre*)	B_2O_3 wt % (pre*)	Al_2O_3 wt % (pre*)	TiO_2 wt % (pre*)	CaO wt % (pre*)	UO_2 wt % (pre*)
1	44.33 (SiO_2)	7.62 (Na_2CO_3)	4.28 (H_3BO_3)	6.27 (Al_2O_3)	12.33 (TiO_2)	4.33 ($CaCO_3$)	20.84 (UO_2)
2	44.33 (SiO_2)	7.62 (Na_2CO_3)	4.28 (H_3BO_3)	6.27 (Al_2O_3)	12.33 (TiPT)	4.33 (Ca nitrate)	20.84 (U nitrate)
3	44.33 (SiO_2)	7.62 (Na_2CO_3)	4.28 (H_3BO_3)	6.27 (Al_2O_3)	13.73 (TiPT)	6.02 (Ca nitrate)	17.40 (U nitrate)

pre* = precursor

Table III. The additives on an oxide basis that were added to the K-Basin wastes.

Oxide	SiO_2	Na_2O	B_2O_3	Al_2O_3	TiO_2	CaO
wt %	34.28	9.39	5.27	6.00	38.36	6.70
Precursor	SiO_2	Na_2CO_3	H_3BO_3	Al_2O_3	TiO_2	$CaCO_3$

Samples were mounted in epoxy resin and polished to a 1 µm diamond finish for analysis by scanning electron microscope (SEM) and diffuse reflectance (DR) spectroscopy.

A JEOL JSM6400 SEM equipped with a Noran Voyager energy-dispersive spectroscopy system (EDS) was operated at 15 keV for microstructural work.

The X-ray diffraction (XRD) was carried out powdered samples using a Philips 1050 diffractometer and CuKa radiation; data were collected over the angular range $5° \leq 2\theta \leq 80°$ with a step size of $0.05°$ and an acquisition time of 5 seconds.

DR spectra were collected at ambient temperature using a Cary 500 spectrophotometer equipped with a Labsphere Biconical Accessory. Spectra are referenced to that of a Labsphere certified standard (Spectralon), and transformed into Kubelka–Munk optical absorbance units, $F(R) = (1-R)^2/2R$ [13].

Leach testing was carried out using the both the Product Consistency Test (PCT-B)[14] and a modified MCC-1 [15] leach tests. The PCT-B protocol involved crushing the samples and sieving them to obtain particles 75-150 μm in diameter (100-200 mesh). The particles were washed in non-polar cyclohexane to remove the fines, rather than water, to prevent pre-leaching of Cs, and 1g samples were leached in 10 ml of deionised water at 90°C for 1 and 7 days. Specimens for the MCC-1 test were polished cuboids (~ 0.25 μm finish, 8 x 8 x 2mm) and were leached using a modified version of the ASTM C – 1998 standard [15], in which the leachates were completely replaced with fresh water at the end of the one day time interval, to avoid steady-state conditions being reached. Aliquots of leachate solutions were acidified with nitric acid and analyzed for elemental concentrations of the ions using a Perkin Elmer PE-SCIEX Elan 6000 Inductively Coupled Plasma-Mass Spectrometer (ICP-MS).

RESULTS AND DISCUSSION

U-rich waste

The waste form for the U-rich waste was designed, in the first instance, to consist of 62.5 wt% of a simple glass ($NaAl_{0.5}B_{0.5}Si_3O_8$) and the rest pyrochlore ($CaUTi_2O_7$), the pyrochlore composition has an idealized target composition. Given that these waste forms were designed to be ultimately HIPed the pouring properties of the glass were unimportant. Thus the glass chosen was one we had previously found in unpublished work to have PCT-B normalized release rates of <1 g/L on all elements.

The composition, on an oxide basis, is identical between samples 1 and 2 but different precursors were used to make the samples. This was to ascertain if a glass-ceramic pyrochlore-rich waste form could be prepared from a waste stream containing UO_2 without an oxidation step in the processing of the waste.

XRD showed sample 1 after melting to contain UO_2, sphene ($CaTiSiO_5$), brannerite (UTi_2O_6) and there was a diffuse signal in the pattern characteristic of glass. The pattern from sample 2 showed a glass diffuse signal, pyrochlore, rutile and a small amount of UO_2. As there was no pyrochlore found in sample 1 it is clear that if a waste stream contained UO_2 an oxidative step would have to be added to the process for pyrochlore to be produced.

Both samples were examined by SEM. As expected from the XRD, sample 1 (Figure 1a) contained brannerite, sphene, UO_2 and glass. The presence of sphene is not unexpected in this sample since the glass would have become saturated with TiO_2 and CaO due to the absence of pyrochlore. EDX analysis showed the glass to contain ~4 wt% CaO and ~10 wt% TiO_2. The glass in sample 1 also contained ~10 wt% UO_2. The SEM analysis (Figure 1b) showed sample 2 to contain pyrochlore, rutile and glass; a small amount of UO_2 was found (not shown) and estimated at < 0.5 wt% abundance. The EDS analysis of the pyrochlore in sample 2 was found to have a Ca content of greater than 1 formula unit, as reported previously[16-17] .

James et al.[18] showed that it was possible to produce pyrochlore with the composition $Ca_{1.25}U_{0.75}Ti_2O_7$ by solid state sintering methods in an Ar atmosphere and an air-sintered pyrochlore with the composition $Ca_{1.4}U_{0.6}Ti_{1.9}O_7$. Both pyrochlores had mixed U oxidation states (U^{4+}, U^{5+} and U^{6+}), as expected from charge balance considerations and experimental diffuse reflectance studies.

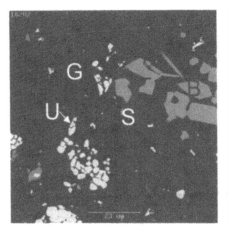
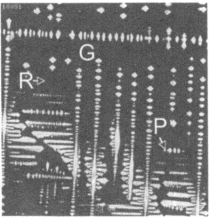

(a)   (b)

Figure 1. Backscattered SEM images of sample 1 (a) and sample 2 (b). G = glass, U = UO_2, B = brannerite, P = pyrochlore, S = sphene and R = rutile. Bar = 23 µm

Following the results obtained from samples 1 and 2, sample 3 (same composition as sample 1 and 2) a pyrochlore composition of $Ca_{1.25}U_{0.75}Ti_2O_7$ was targeted, as the sample was to be prepared in an argon atmosphere.

XRD (Figure 2) showed the sample to contain pyrochlore, rutile and glass (diffuse signal in pattern centered around ~ 20° 2θ). The [111] peak in the pyrochlore pattern however appears to be missing (~ 14 ° 2θ) indicating that the pyrochlore formed in this glass-ceramic was not $Ca_{1.25}U_{0.75}Ti_2O_7$ but $Ca_{1.4}U_{0.6}Ti_{1.9}O_7$ [18]. SEM analysis (Figure 3) showed the sample to contain pyrochlore, rutile and glass and the EDS analysis confirmed that the pyrochlore present in the sample had a composition close to $Ca_{1.4}U_{0.6}Ti_{1.9}O_7$, although the small grain size of the pyrochlore made it difficult to obtain an accurate composition.

DR spectra of U ions in polycrystalline pyrochlore have been reported previously [19] and it has been demonstrated that they can qualitatively be used to identify U^{4+} and U^{5+} in pyrochlore. To confirm the oxidation state of the U in the pyrochlore determined by SEM-EDS, the DR spectrum of sample 3 was collected . As discussed above, the presence of U^{5+} in the pyrochlore would confirm that the pyrochlore composition has higher Ca content than originally designed. Figure 4 shows the DR spectra of sample 3 together with a reference U^{4+} containing pyrochlore ($CaU_{0.6}Zr_{0.4}Ti_2O_7$). It clearly shows sample 3 not to contain the strong absorption bands between 4000 and 5000, and at ~6650 cm^{-1}, associated with U^{4+}. Instead it shows characteristic of U^{5+} with strong absorption bands close to 6000 cm^{-1}, consistent with the SEM-EDS results.

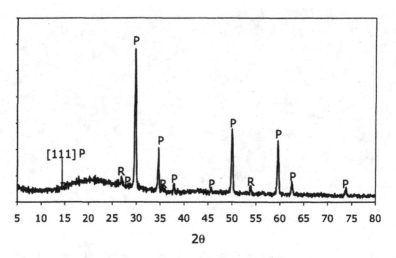

Figure 2. XRD pattern of sample 3. R = rutile and P = pyrochlore

Figure 3. Backscattered SEM images of sample 3 melted in argon 1250 °C. G = glass, P = pyrochlore and R = rutile. Bar = 50 μm

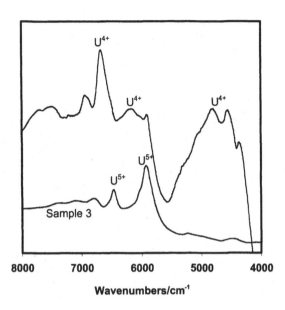

Figure 4. DR spectra of a reference U^{4+} pyrochlore and sample 3.

Leach testing was carried out on sample 3 (see Table IV). The normalized concentrations were < 1 g/L for all elements, well below the normalized PCT-B leachate concentrations of Na, Li and B are 13-16 g/L for the reference EA glass.

Table IV. PCT-B results from Sample 3 melted in argon 1250 °C.

Element	Normalized Concentration g/L
Al	0.19
B	0.64
Ca	0.007
Na	0.94
Si	0.18
Ti	0.0005
U	0.0009

Note: Error on normalized concentration < 10%.

Precursors of sample 3 were HIPed at 1200°C/30MPa/1h with either 2 wt% Ti, Ni or Fe metal powder added. The experiment was carried out to determine the redox conditions required in the HIP can to optimize the pyrochlore formation in the glass-ceramic.

After HIPing the samples were examined by XRD. The pattern of sample 3 (Ti) showed peaks corresponding to UO_2, sphene, and TiO_2. The pattern of sample 3 (Fe) showed peaks corresponding to UO_2, brannerite and sphene. The pattern of sample 3 (Ni) showed peaks corresponding to pyrochlore.

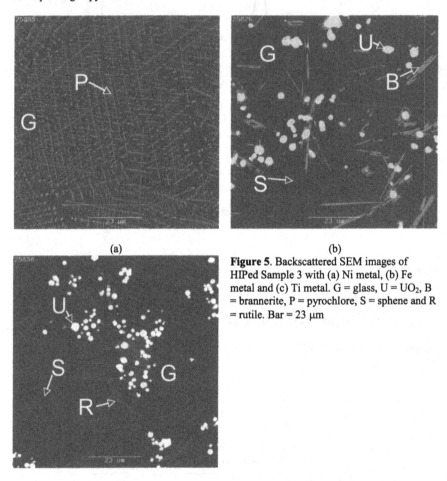

(a)

(b)

(c)

Figure 5. Backscattered SEM images of HIPed Sample 3 with (a) Ni metal, (b) Fe metal and (c) Ti metal. G = glass, U = UO_2, B = brannerite, P = pyrochlore, S = sphene and R = rutile. Bar = 23 μm

The SEM analysis confirmed the XRD results (Figure 5). The SEM analysis of sample 3 (Ni) showed the pyrochlore to have a composition similar to sample 3 (~$Ca_{1.4}U_{0.6}Ti_{1.9}O_7$) and the

202

glass to contain ~10 wt % TiO_2, ~4 wt % CaO and ~2 wt % UO_2. Thus the required redox conditions for HIPing would be governed by Ni metal.

K-Basin sludge waste

The waste form for the K-Basin sludges were designed so as to have a single composition of the precursor additives to different amounts of the two wastes. From the results of the above experiments the following conditions were used in designing the waste form. The pyrochlore would nominally have the composition $Ca_{1.4}U_{0.6}Ti_{1.9}O_7$. Fe_2O_3 would be targeted towards $FeTi_2O_5$ and the glass would be in the compositional range $NaAl_{(0.75-0.5)}B_{(0.25-0.5)}Si_{(3.0-3.5)}O_x$. An excess of TiO_2 and CaO would be available to saturate the glass and to form a small amount of sphene. The small excess of sphene was to allow for the different amounts of U in the various K-basin sludges. The sample would have Ni metal as the redox control agent and be HIPed at 1200°C/30MPa/1h. On this basis the additives listed in table II were used. The additives were combined with 50 wt % Weasel K-basin sludge and 45 wt % Main Basin Floor sludge (on an oxide basis).

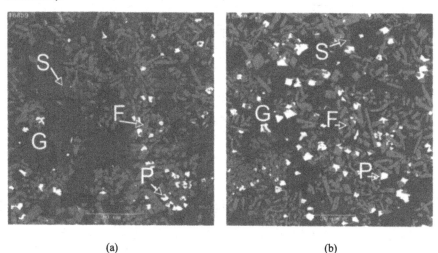

(a)   (b)

Figure 6. Backscattered SEM images of HIPed K-basin sludge waste forms (a) Weasel and (b) Main Basin Floor G = glass, F = Fe_2TiO_5, P = pyrochlore and S = sphene. Bar = 30 μm

After HIPing the sample were X-rayed. XRD showed both samples had developed the designed phases, pyrochlore, $FeTi_2O_5$ and sphene. The difference between the samples was the quantities of the phases.

SEM analysis (Figure 6) confirmed this phase assemblage. The EDS analysis of the glass showed ~ 4 wt % of TiO_2, ~ 8 wt % of CaO , ~ 3 wt% of UO_2 and ~ 1 wt% of Fe_2O_3 in both the Weasel and Main Basin Floor samples. The Cs, Sr, Mo and La had all partitioned into the glass. The pyrochlore composition was~ $Ca_{1.4}U_{0.6}Ti_{1.9}O_7$ as designed.

Leach testing was carried out on both HIPed waste forms. The two samples had very similar normalized concentrations for all elements (Table V). This would be expected because the samples have the same phase assemblage, just different amounts of each phase. The MCC-1 results are in Table VI. As with the PCT-B procedure both samples behave in a similar manner. These leach rates are similar to those obtained for a glass-ceramic designed for immobilization of the high-level nuclear waste generated at the Idaho chemical processing plant (ICPP) HIPing conditions [20]. For the 7-day MCC-1 test, it is easily calculated that the leach rate in $g/m^2/d$ should be around 6 times the g/L extraction values in the PCT test if surface area/volume effects do not influence the results. It can be seen that this is approximately obeyed by the most soluble elements such as Cs and B.

Table V. PCT-B results from HIPed K-Basin sludge waste forms.

Element	Weasel Normalized Concentration g/L	Main Basin Floor Normalized Concentration g/L
Al	0.15	0.13
B	0.17	0.18
Ca	0.035	0.029
Cs	0.31	0.30
Fe	0.0012	0.0015
La	0.00046	0.0026
Mo	0.15	0.16
Na	0.56	0.38
Si	0.25	0.15
Sr	0.027	0.037
Ti	0.00008	0.00004
U	0.0009	0.0004

Note: Error on normalized concentration < 15%.

Table VI. MCC-1 results from HIPed K-Basin sludge waste forms.

Element	Weasel Normalized Release Rate 0-1 day g/m²/day	1-7 day g/m²/day	Main Basin Floor Normalized Release Rate 0-1 day g/m²/day	1-7 day g/m²/day
Al	1.1	0.82	0.9	0.7
B	1.6	1.2	1.3	1.2
Ca	1.0	<0.6	0.84	0.68
Cs	2.3	1.7	2.0	1.5
Fe	<0.06	<0.01	<0.07	<0.01
La	<0.02	<0.004	<0.03	<0.004
Mo	1.7	1.0	1.3	0.81
Na	1.8	1.0	1.5	0.89
Si	1.3	0.89	1.1	0.78
Sr	1.5	0.83	1.2	0.80
Ti	<0.0009	<0.0002	<0.0008	<0.004
U	0.022	0.0037	0.017	0.012

Note: Error on normalized concentration < 5%.

CONCLUSIONS

Dense pyrochlore-rich glass ceramics were developed to immobilize a U-rich waste and K-Basin sludge by HIPing using Ni metal as the redox control. The leach resistance in short-term regulatory tests was very satisfactory. The pyrochlore that formed in the glass-ceramic had a composition of $Ca_{1.4}U_{0.6}Ti_{1.9}O_7$ and the uranium existed mainly as U^{5+}. In waste streams containing UO_2 an oxidative step would have to be carried out for a pyrochlore-rich glass-ceramic to be produced. In immobilizing the K-Basin sludges it is possible to use one set of additives and adjust the waste loadings to form a constant phase assemblage.

The resultant glass-ceramic waste forms all had PCT-B release rates of less than 1 g/L, an order of magnitude better than EA glass. The MCC-1 release rates were below 2 g/m^2/day for all elements at 7 days and 90°C.

ACKNOWLEDGMENTS

We wish to thank K. Olufson and P. Yee for carrying out the leach testing. I. Watson for preparing the samples, and T. Eddowes and N. Webb for carrying out the HIPing of the samples.

REFERENCES

1. P. Loiseau, D. Caurant, N. Baffier and C. Fillet, in *Scientific Basis for Nuclear Waste Management XXIV* edited by K. P. Hart and G. R. Lumpkin (Mater. Res. Soc. Symp. Proc. **663,** Pittsburgh, PA, 2002) pp. 169-177.

2. D. Zhao, L. Li, L. L. Davis, W. J. Weber and R. C. Ewing, in *Scientific Basis for Nuclear Waste Management XXIV* edited by K. P. Hart and G. R. Lumpkin (Mater. Res. Soc. Symp. Proc. **663,** Pittsburgh, PA, 2002) pp. 199-206.

3. P. Loiseau, D. Caurant, O. Majerus, N. Baffier and C. Fillet, *J. Mater. Sci.* **38** (4), 843-852 (2003).

4. P. Loiseau, D. Caurant, O. Majerus, N. Baffier and C. Fillet, in *Scientific Basis for Nuclear Waste Management XXVII* edited by V. M. Oversby and L. O. Werme, (Mater. Res. Soc. Symp. Proc. **807,** Pittsburgh, PA, 2004) pp 333-338.

5. P. Loiseau, D. Caurant, L. Mazerolles, N. Baffier and C. Fillet, *J. Nucl. Mater.* **335,** 14-32 (2004).

6. D. Caurant, O. Majerus, P.Loiseau, I. Bardez, N. Baffier and J. L. Dussossoy, *J. Nucl. Mater.* **354** (1-3), 143-162 (2006).

7. R.A. Day, J. Ferenzy, E. Drabarek T. Advocat, C. Fillet, J. Lacombe, C. Ladirat, C. Veyer, R. Do Quang and J. Thomasson, WM'03 Conference, February 23-27, 2003, Tucson, AZ on CD.

8. A.A. Digeos, J.A. Valdez, K.E. Sickafus, S. Atiq, R.W. Grimes and A.R. Boccaccini, *J. Mater. Sci.* **38** (2003) 1597-1604.

9. A.R. Boccaccini, E. Bernardo, I. Blain and N.D. Boccaccini , *J. Nucl. Mater.* **327,** 148-158 (2004).

10. E. R.Vance, P. J.Hayward, C. D.Cann, S. L.Mitchell, M. A. T.Stanchell and D. J.Wren, *Glass Technology* 25 (5), 232-239 (1984).

11. E.R. Vance, S. Urquhart, D. Anderson and I.M. George, in Advances in Ceramics, Vol. 20 edited by D.E. Clark, W.B. White and A-J. Machiels, (American Ceramics Society,

Westerville, OH, U.S.A. 1987), pp. 249-258.

12. P. D. Rittmann, Report No. SNF-5066, Rev. 0, 1999.
13. W. W. Wendlandt and H. G. Hecht, Reflectance Spectroscopy, Wiley Interscience, New York, 1966.
14. PCT is based on the ASTM Designation: C 1285-02 Standard Test Methods for Determining Chemical Durability of Nuclear, Hazardous, and Mixed Waste Glasses and Multiphase Glass Ceramics: The Product Consistency Test (PCT). (2002)
15. ASTM C 1220 - 98. "Standard Test Method for Static Leaching of Monolithic Waste Forms for Disposal of Radioactive Waste". ASTM International. (1998).
16. M. W. A. Stewart, E. R. Vance, A. Jostsons, and B. B. Ebbinghaus, *Journal of the Australasian Ceramic Society,* **39** (2) 130-148 (2003).
17. Y. Zhang, K. P. Hart, M. G. Blackford, B. S. Thomas, Z. Aly, G. R. Lumpkin, M. W. Stewart, P. J. McGlinn, and A. Brownscombe, in *Scientific Basis for Nuclear Waste Management XXIV* edited by K. P. Hart and G. R. Lumpkin (Mater. Res. Soc. Symp. Proc. **663**, Pittsburgh, PA, 2002) pp. 325-332.
18. M. James, M.L. Carter Y. Zhang, Z. Zhang and ER. Vance To be published.
19. Y. Zhang, E. R. Vance, K. S. Finnie, B. D. Begg and M. L. Carter, *Royal Society of Chemistry,* (2006) 343-345.
20. Y. Zhang, H. Li, P.J. McGlinn, B. Yang and B.D. Begg, *J. Nucl. Mater.* 375 (3), 315-322 (2008).

Calcium Phosphate: A Potential Host for Halide Contaminated Plutonium Wastes

Brian L. Metcalfe[1], Ian W. Donald[1], Shirley K. Fong[1], Lee A. Gerrard[1], Denis M. Strachan[2] and Randall D Scheele[2]

[1] Atomic Weapons Establishment, Aldermaston, Reading, RG7 4PR, UK
[2] Pacific Northwest National Laboratories, Richland, WA, USA

ABSTRACT

The presence of significant quantities of fluoride and chloride in four types of legacy wastes from plutonium pyrochemical reprocessing required the development of a new wasteform which could adequately immobilize the halides in addition to the Pu and Am. Using a simulant chloride-based waste (Type I waste) and Sm as the surrogate for the Pu^{3+} and Am^{3+} present in the waste, AWE developed a process which utilised $Ca_3(PO_4)_2$ as the host material. The waste was successfully incorporated into two crystalline phases, chlorapatite, $[Ca_5(PO_4)_3Cl]$, and spodiosite, $[Ca_2(PO_4)Cl]$. Radioactive studies performed at PNNL with ^{239}Pu and ^{241}Am confirmed the process. A slightly modified version of the process in which $CaHPO_4$ was used as the host was successful in immobilizing a more complex multi-cation oxide–based waste (Type II) which contained significant concentrations of Cl and F in addition to ^{239}Pu and ^{241}Am. This waste resulted in the formation of cation-doped whitlockite, $Ca_{3-x}Mg_x(PO_4)_2$, β-calcium phosphate, $β-Ca_2P_2O_7$ and chlor-fluorapatite rather than the chlorapatite and spodiosite formed with Type I waste.

INTRODUCTION

The presence of significant concentrations of chloride and fluoride ions in the intermediate level legacy wastes derived from pyrochemical reprocessing of Pu prevents the waste from being successfully vitrified in borosilicate glass. The approach taken at AWE was to react the waste with a host material to produce a durable ceramic which would immobilize both the actinides and the halides. Of the ceramic phases, which can contain significant proportions of chloride and fluoride, apatite and spodiosite were chosen for further investigation. Apatite is a naturally occurring mineral of general formula, $A_5(BO_4)_3(OH, F, Cl)$, where A can be a variety of 1 to 3 valent cations and B is commonly P, V or As [1]. The ability to incorporate a variety of elements into the apatite structure and its good aqueous durability at pH >5 [2] offered potential as a waste-form. Calcium phosphate $[Ca_3(PO_4)_2]$ was chosen as the host leading to the formation of chlorapatite $[Ca_5(PO_4)_3Cl]$ and spodiosite $[Ca_2(PO_4)Cl]$. The initial waste investigated (Type I) was chloride based with the Pu being present as Pu(III) whereas the second waste (Types II) was oxide based but containing varying quantities of Cl⁻ and F⁻. Pu is present as Pu(IV) in this waste which necessitated different surrogates for the Pu when performing the initial non-radioactive studies. For Type I waste Sm was used as the surrogate for both Pu and Am but Hf was used as the surrogate for Pu in the other waste-stream. A simple process was developed for Type I waste whereby waste was mixed with the host and calcined at 750 °C for 2h during which time the waste ions were incorporated into the host [3]. Leach trials on the waste-form confirmed the expected good durability and the process was transferred to PNNL where it was found that a more consistent product was obtained if the process time was increased to

4hrs. Samples were then manufactured with actinide-doped Type I waste in which the Sm was replaced by [239]Pu and only a small amount of [241]Am in order to minimise the dose received by the operators. Leach trials demonstrated the durability of this material was comparable to that obtained with the Sm-surrogate material. Having established the viability of the process with actinides, radiation-induced damage trials were carried out on the powder by substituting [238]Pu (half life 87.7 y) for [239]Pu (half life 2.4 x 10[4] y). Thus the accumulation of radiation-induced damage could be accelerated by a factor of 280.

When applying the process to Type II waste we replaced $Ca_3(PO_4)_2$ with $CaHPO_4$ as the host, but all other processing conditions were the same. The change permitted an increase in the waste loading whilst simultaneously retaining a 5% excess over the stoichiometric phosphate requirement. Another change was to the surrogate used. Of the non-radioactive surrogates Ce is the most common option. Its variable valency can mimic that of Pu but differences can be introduced if redox conditions change. Hafnium has been used [4] but does not have variable valency. Although the atomic radius of Ce(IV) is much closer to Pu(IV) than Hf (IV) is [5], the size differential was considered to be less significant than the variable valency under the processing conditions both as currently employed and potential future modifications. Experiments were performed in order to compare the difference in phase assemblage between Ce and Hf at different temperatures.

The waste-form generated is a powder and as such is not considered inherently safe. Sintering to convert powder into a monolithic form failed to produce an acceptable dense product and so a second stage sinter following the addition of a sodium aluminophosphate glass powder was introduced [6].

In this paper we summarise the results from radioactive studies on Type I and II wastes and compare the effectiveness of Ce and Hf as surrogates for Pu.

EXPERIMENTAL

The composition of non-radioactive Type II simulant waste and its Ce equivalent is given in Table I. A batch of each waste was made by ball-milling the appropriate constituents listed in Table I together for 20 hrs. Waste-forms were made by blending 7.5g of the simulant waste with 22.5g of $CaHPO_4$ and calcining in alumina crucibles in air for 4 hrs at the required temperature.

The compositions of the Type I and II simulant wastes used at PNNL to prepare samples are given in Table II. For accelerated ageing trials the PuO_2 used was 80.48% [238]Pu and 19.52% [239]Pu. Simulant wastes were mixed with the host, 20% loading in $Ca_3(PO_4)_2$ for Type I and 23.1% in $CaHPO_4$ for Type II.

At PNNL the powders were mixed in polythene jar overnight using zirconia milling media and were calcined in alumina crucibles under flowing argon at 750°C for 4 h. Type II waste was further treated by adding 25% of finely ground (<45μm) sodium aluminophosphate glass and sintering at 750°C for 4 h to produce monolithic samples.

Durability of the actinide-doped powder in aqueous solution was assessed using the following test protocol. Samples (approximately 1gm) were washed with 40ml of deionised water to remove any unreacted chloride which would mask the leached chloride. Triplicate 1g samples of the washed ceramic powder were leached with 100ml of deionised water in either Teflon® ([239]Pu samples) or stainless steel ([238]Pu samples) vessels at 40°C. At intervals

Table I. Composition of non-radioactive simulant wastes

Component	Waste Composition, mass%	
	Hf waste	Ce waste
HfO_2	20.8	
CeO_2		17.7
Ga_2O_3	28.0	29.0
Al_2O_3	9.8	10.2
Sm_2O_3	4.5	4.7
MgO	6.3	6.5
Fe_2O_3	1.3	1.4
Ta_2O_5	1.3	1.4
NiO	1.3	1.4
CaF_2	10.4	10.8
KCl	16.3	16.9

Table II. Composition of radioactive simulant wastes

Component	Waste Composition, mass%	
	Type I	Type II
PuO_2		20.2
Am_2O_3		0.2
Ga_2O_3		28.2
Al_2O_3		9.9
Sm_2O_3		4.3
MgO		6.35
Fe_2O_3		1.31
Ta_2O_5		1.31
NiO		1.31
$PuCl_3$	19.0	
$AmCl_3$	1.0	
$CaCl_2$	80.0	
CaF_2		10.5
KCl		16.4

over the 28 day test period 10ml leachate aliquots were removed for chemical analysis and replaced with an equal quantity of fresh deionised water. Each aliquot was analysed by ICP-MS and ICP/OES, the chloride concentration measured with a chloride specific electrode. Monolithic material was tested according to a protocol whereby an individual specimen was placed on a stainless steel support grid and the deionised water added. The stainless steel container was then sealed and held at 40 °C for the required test period, 1, 3, 7, 14 or 28 days after which both the specimen and deionised water were recovered and the container rinsed with more deionised water. A 1 vol% HNO_3 strip solution was then added to the container which was resealed and heated to 90 °C for 16 h to remove any material from the container surface. Analyses of the leach and strip solutions were by ICP-MS and ICP/OES. Duplicate specimens were used for this test. In order to compare data the elemental mass losses for each element were normalized and calculated according to Equation 1:

$$N_i = m_i/(f_i \cdot S) \tag{1}$$

where m_i is the mass loss for element i, f_i is the mass fraction of element i in the ceramic, and S is the specific surface area. For the powder specimens, the BET method was used and, for monolithic specimens, the geometric surface area.

Powder X-ray diffraction analysis was performed on samples using a Scintag PAD V (PNNL) and either a Philips PW-1700 or a Bruker D8 Advance powder diffractometer a position sensitive VÅNTEC-1 detector and Bragg-Brentano flat plane geometry (AWE). All diffractometers employed monochromatic Cu Kα_1 radiation ($\lambda = 1.54056$ Å).

RESULTS AND DISCUSSION

Type I waste

Diffraction patterns obtained by PNNL from Type I non-radioactive and actinide-loaded waste-forms show that the major phases in both waste-forms are chlorapatite and spodiosite although the proportions of the two phases differ in the two waste-forms.

The results of the 28 day durability trials on the un-aged ^{239}Pu and ^{238}Pu and aged ^{238}Pu loaded waste-form samples are given in Table III. The ^{238}Pu-bearing specimens had accumulated 4×10^{18} α/g of damage over 1820 days at ambient conditions.

Table III Normalized mass losses and pH values for leachates from Type I ^{239}Pu and ^{238}Pu loaded waste-forms after 28 days at 40 °C

Specimen	pH (25 °C)	Normalized Elemental Mass Loss (g m^{-2})				
		Ca	P	Cl	Pu	Am
^{239}Pu /^{241}Am	6.64	1.6×10^{-3}	2.3×10^{-3}	2.7×10^{-3}	1.2×10^{-5}	2.4×10^{-7}
^{238}Pu /^{241}Am	5.97	1.6×10^{-4}	5.1×10^{-5}	1.6×10^{-4}	1.6×10^{-5}	8.0×10^{-7}
^{238}Pu /^{241}Am aged	6.83	1.9×10^{-2}	1.7×10^{-2}	8.8×10^{-2}	1.4×10^{-5}	4.7×10^{-5}

The similarity in the mass loss for Ca, P and Cl suggests that for ^{239}Pu loaded material the phosphate matrix is dissolving congruently and the lower mass loss for Pu and Am is probably due to the very low solubility of their phosphates. This behaviour was observed previously on both Sm loaded material prepared at AWE and Pu (no Am) loaded material prepared at PNNL [3]. This explanation however does not apply to the ^{238}Pu loaded material where the mass losses for P and Cl vary by a factor of 300. After ageing the durability of the powder decreased.

Type II waste

Preliminary leach test data from the dissolution tests are given in Table IV and show an increase in leach rate above that observed at AWE for non-radioactive monolithic samples.

Table IV Normalized mass losses and pH values for monolithic wasteform after 28 days at 40 ℃

Normalized Elemental Mass Loss (g m^{-2})			
^{238}Pu	0.91	Ni	3.8
^{239}Pu	0.96	Ta	0.11
Am	0.33	P	3.3
Ga	0.12	Fe	20.0
Sm	0.048	K	8.5
Ca	3.1	Na	7.2
pH (25 ℃)	6.3		

In the non-radioactive material the major phases have been identified as a chloride-substituted $Ca_5(PO_4)_3F$ and β-$Ca_2P_2O_7$ (PDF No 04-009-0346). Other peaks are difficult to positively identify but have been tentatively assigned to $Ca_9Ga(PO_4)_7$ and a cation-doped whitlockite, ideally $Ca_{18.53}Mg_{1.6}Fe_{0.4}(PO_4)_{14}$ (PDF No 1-83-2082). Again the proportions of the phases differed between samples prepared at AWE and PNNL. In the Pu/Am loaded waste-form, diffraction peaks corresponding to unreacted PuO_2 were observed which cast doubt on the suitability of Hf as a surrogate because HfO_2 had not initially been observed in the non-radioactive material. However on re-examining the latter material with a higher resolution diffractometer a small peak corresponding to HfO_2 was observed. A similar effect occurred in the Ce doped material where the presence of CeO_2 was observed. Increasing the calcination temperature to 850 ℃ and 950 ℃ led to a reduction in the un-reacted surrogate content. Rietveld refinements of the data were carried out with the GSAS code [7, 8] and a quantitative assessment made of the phases present. As the calcination temperature increased from 750 ℃ to 950 ℃ the mass of un-reacted HfO_2 decreased slightly from 5.4 to 4.0% and that of CeO_2 from 4.5 to 1.5%. This is perhaps not unexpected because Ce(III) is more thermodynamically favourable than Ce(IV) at the higher temperatures and incorporating trivalent cations into whitlockite is easier than tetravalent cations. A major difference between the two surrogates was the presence of monazite in the Ce loaded samples.

Sintering the powder after the addition of the glass leads to a change in the phase assemblage. The glass does not merely act as a binder for the powder but reacts with β-$Ca_2P_2O_7$ to produce more whitlockite. During that process further immobilization of the surrogates occurs with the remaining Ce but only a little of the Hf being immobilized. It is also observed that the quantity of unreacted PuO_2 is reduced.

CONCLUSIONS

Both radioactive and non-radioactive Type I simulant waste have been found to perform in a similar manner thereby confirming Sm as a suitable surrogate for both Pu and Am in that system. Accelerated radiation damage trials with ^{238}Pu have shown that after an estimated radiation dose of 4 x $10^{18}\alpha$/g the durability of the waste-form decreased.

In Type II powder waste-form the presence of unreacted PuO_2 mirrors the presence of HfO_2 and CeO_2 in the surrogate waste-forms calcined at 750 ℃. During the subsequent sintering of the wasteform/glass mixture, reaction occurred between the glass and β-$Ca_2P_2O_7$ to produce more whitlockite whilst the unreacted PuO_2 and surrogate oxides reacted. At the calcination temperature currently used it would appear that Ce and Hf are both acceptable as surrogates in this system. Calcination of non-radioactive material at higher temperatures suggests that CeO_2 is incorporated into the whitlockite structure more readily than HfO_2. One significant difference in the phase assemblage of the non-radioactive waste-forms formed at higher temperatures is the presence of monazite in the Ce doped material. The durability of the Type II monolithic waste-form is less than that of the powder wasteform.

ACKNOWLEDGEMENTS

The authors are grateful to R. Elovich, R. Sell and A. Kozelisky, PNNL for carrying out the radioactive studies.
© British Crown Copyright 2009/MoD. Published with the permission of the Controller of Her Britannic Majesty's Stationery Office.
Pacific Northwest National Laboratory is operated by Battelle for the United States Department of Energy under Contract DE-AC06-76RLO1830.

REFERENCES

1   T. Kanazawa, editor, Inorganic Phosphate Materials, Materials Science Monograph 52, Kodancha (Tokyo), and Elsevier (Amsterdam), 1989, pp. 55-77.
2   T. S. B. Narasaraju, U. S. Rai and K. K. Rao, *Indian Journal of Chemistry*, **16A**, 952, (1978).
3   B. L. Metcalfe, I. W. Donald, R. D. Scheele and D. M. Strachan in *Scientific Basis For Nuclear Waste Management XXVI*, edited by. R. J. Finch and D. B. Bullen. (Mat. Res. Soc. Symp. Proc., **757**, Warrendale PA 2003) pp. 265-271.
4   X. Deschanels, C. Lopez, C. Denauwer and J. M. Bart, in *Plutonium Futures- The Science,* edited by G. D. Jorvinen, Amer. Inst. Physics, **59**, (2003).
5   R. D. Shannon, *Acta Crystallographica Section A* 32:751, (1976).
6   B. L. Metcalfe, S. K. Fong and I. W. Donald, in *Scientific Basis For Nuclear Waste Management XXIX*, edited by P. Van Iseghem, (Mat. Res. Soc. Symp. Proc. **932**, Warrendale PA, 2006) pp. 727-734.
7   A.C. Larson and R.B. Von Dreele, *Los Alamos National Laboratory Report LAUR*, 2000, 86-748.
8   B.H. Toby, *J. Appl. Cryst.*, **34**, 210, 2001.

Mater. Res. Soc. Symp. Proc. Vol. 1124 © 2009 Materials Research Society   1124-Q04-09

Experimental and atomistic modelling study of ion irradiation damage in thin crystals of the TiO2 polymorphs

G.R. Lumpkin[1], K.L. Smith[1], M.G. Blackford[1], B.S. Thomas[1], K.R. Whittle[1], D.J. Attard[1], N.J. Zaluzec[2], and N.A. Marks[3]

[1] ANSTO, Private Mail Bag 1, Menai 2234, NSW, Australia
[2] Argonne National Laboratory, 9700 South Cass Avenue, Argonne, IL 60439, USA
[3] Curtin University of Technology, Perth, WA 6845, Australia

ABSTRACT

Thin crystals of rutile, brookite, and anatase were irradiated in-situ with 1.0 MeV Kr using the IVEM-TANDEM facility. Synthetic rutile and cassisterite (SnO_2, rutile structure) remained crystalline up to 5×10^{15} ion cm^{-2} at 50 K. Natural brookite and anatase with low impurity levels became amorphous at $8.1 \pm 1.8 \times 10^{14}$ and $2.3 \pm 0.2 \times 10^{14}$ ions cm^{-2}, respectively, at 50 K. Irradiation at higher temperature revealed $T_c = 170$ K for brookite and 242 K for anatase. Natural rutile with about 2 wt% impurities became amorphous at $9.4 \pm 1.8 \times 10^{14}$ ions cm^{-2} at 50 K and has a $T_c = 207$ K. The available data reveal both a structural effect in the polymorphs with low levels of chemical impurities and a chemical effect in natural rutile specimens containing up to about 1.7 wt% impurities.

INTRODUCTION

TiO_2 compounds have a number of important industrial and technological applications, including use in pigments, photocatalysis, oxygen sensors, and thin film devices (e.g., antireflective coatings, waveguides, and optical amplifiers) [1]. TiO_2 is also an important chemical component of polyphase and single-phase ceramic nuclear waste forms [2]. We recently tested the structure type criterion and other potential factors affecting susceptibility to amorphization using samples of the low-pressure TiO_2 polymorphs rutile, brookite, and anatase [3,4]. In this paper, we summarize our initial studies of ion irradiation damage and atomistic modelling of synthetic rutile and natural brookite and anatase with low levels of chemical impurities. We also include new experimental results for natural cassiterite with low levels of chemical impurities and two natural rutile samples containing up to about 1.7 wt% of chemical impurities (mainly transition metals), new electron energy loss spectroscopy (EELS) results, and some additional atomistic modelling data.

EXPERIMENTAL PROCEDURES AND RESULTS

Samples were irradiated in situ at the IVEM-Tandem Facility at Argonne National Laboratory. TEM specimens were irradiated with 1.0 MeV Kr^{2+} ions in a Hitachi 300 keV electron microscope. All experiments were conducted at cryogenic temperatures (50-225 K) using a liquid He cooled sample holder with the electron beam turned off during the irradiations. Specimens were irradiated incrementally and observed in bright field and selected area diffraction modes after each step. For those specimens that become amorphous, we determined the critical temperature for amorphization using non-linear least squares fitting procedures.

The results of the irradiation experiments conducted at 50 K on the samples with low levels of chemical impurities have already been described in reference [3]. In summary, we found that synthetic rutile clearly remained crystalline up to a fluence of 5×10^{15} ions cm^{-2}, indicating that there is no significant accumulation of amorphous domains. In addition to rutile *sensu stricto*, we have recently irradiated a high-purity natural sample of cassiterite, SnO$_2$, which also has the rutile structure. Our results reveal that cassiterite remains crystalline up to the same fluence as synthetic rutile at a temperature of 50 K. Our previous investigation showed that brookite and anatase are both susceptible to amorphization under the conditions of the experiments and returned F$_c$ values of $8.1 \pm 1.8 \times 10^{14}$ ions cm^{-2} and $2.3 \pm 0.2 \times 10^{14}$ ions cm^{-2}, respectively.

In this work we conducted additional experiments on brookite and anatase in order to determine the fluence-temperature response. Irradiation at temperatures up to 225 K revealed T$_c$ = 170 K for brookite and 242 K for anatase. We have also examined two natural rutile samples from Graves Mountain, Georgia (sample GM), and from Alexander County, North Carolina (sample NC). Analytical data determined by SEM-EDX indicate that the GM rutile has about 1.7 wt% impurities (mainly Fe > V > Cr) whilst the NC rutile has about 1.2 wt% impurities (mainly Cr > V > Nb). Results of the irradiation experiments indicate that the GM and NC rutile samples become amorphous at critical fluences of $9.2 \pm 0.7 \times 10^{14}$ ions cm^{-2} and $8.6 \pm 0.5 \times 10^{14}$ ions cm^{-2}, respectively, at 50 K. The temperature response curve was determined for the GM rutile and gives a T$_c$ = 207 K. The experimental results acquired to date are summarized in Figure 1.

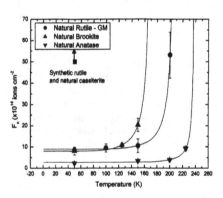

Figure 1. Ion irradiation data (1.0 MeV Kr) for the TiO$_2$ polymorphs and cassiterite (SnO$_2$) plotted as a function of temperature. Synthetic rutile and natural cassiterite with low levels of chemical impurities are radiation resistant at 50 K. The data at 50 K for synthetic rutile and natural brookite and anatase with low levels of impurities correlate with distortion and volume parameters. Natural rutile GM has approximately 1.7 wt% chemical impurities, mainly Fe, V, and Cr oxides, and is susceptible to amorphization at cryogenic temperatures. The critical dose (fluence) at 50 K is similar to that of brookite, but T$_c$ falls between brookite and anatase.

MODEL FOR AMORPHOUS TiO$_2$

Based on previous MD simulations and experimental results on glassy and amorphous TiO$_2$ discussed in reference [3], we proposed a model for the amorphous phase produced by ion irradiation. This model consists of a framework of distorted TiO$_{6-x}$ polyhedra, with x = 0.2-0.6, a slightly reduced mean Ti-O distance, and both edge and corner sharing between polyhedra with local configurations similar to those found in anatase and brookite. Our SAED data suggest that the amorphous states of all three polymorphs, produced by ion irradiation, have similar characteristics. To test this idea further, we collected EELS spectra for the Ti $L_{2,3}$ and O K edges for amorphous rutile and anatase. Preliminary results of this work are shown in Figure 2 and appear to support the proposed similarity of the amorphous structures of these polymorphs.

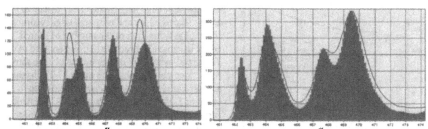

Figure 2. Experimental Ti $L_{2,3}$ EELS edges of crystalline (left) of amorphous (right) rutile and anatase. The rutile spectra are filled in and the anatase spectra are overlain as a smooth solid line. Differences in the crystalline states of rutile and anatase can be seen by their distinctive fine structures shown on the left. After irradiation to the amorphous state, the fine structures of both phases are similar, as shown on the right, and quite different from their respective crystalline states. Note that the raw data have been processed using the RL deconvolution procedure described by Gloter et al [5]. This may lead to some differences in the resolution of the corrected spectra.

CORRELATIONS WITH DISTORTION AND VOLUME PROPERTIES

For samples with low levels of chemical substitution, we have shown that the radiation response data of rutile, brookite, and anatase correlate inversely with the number of shared edges per TiO_6 octahedron in each structure: two for rutile (tetragonal, site symmetry mmm), three for brookite (orthorhombic, site symmetry 1), and four for anatase (tetragonal, site symmetry $\sum 4m2$). A characteristic feature of edge sharing in octahedral framework structures is an increase in the distortion of the coordination polyhedra as the shared edges become shorter and the unshared edges become longer. As the number of shared edges increases, the structures become more distorted, evidenced by increasing octahedral edge length distortion, bond angle variance, and quadratic elongation from rutile to brookite to anatase. Similar distortions are characteristic of amorphous oxide materials. Note that mean bond lengths and bond length distortion do not necessarily correlate with other distortion parameters.

Changes in the metal-metal distances of the TiO_2 polymorphs are shown in Figure 3. The average shared O-O distance decreases and the average unshared O-O distances increase with increasing number of shared edges. In contrast, the average Ti-Ti distances across shared and unshared octahedral edges both increase from rutile to brookite to anatase. Thus, the increase in Ti-Ti distances in the octahedral frameworks of the TiO_2 polymorphs lead to an increase in the molar volume from rutile to brookite to anatase [3]. This effect is accommodated by increased Ti-O-Ti angles across the shared corners, which change from 130.5° in rutile to an average of 139.0° in brookite to 155.4° in anatase. Using Hoang's [6] density value of 3.80 g cm^{-3} we obtain a molar volume of 21.03 cm^3 mol^{-1} for amorphous TiO_2 glass (from MD simulation). These data suggest that the difference in volume due to the crystalline-amorphous transformation (ΔV_{ca}) is approximately 11.8 % for rutile, 8.7 % for brookite, and 2.5 % for anatase.

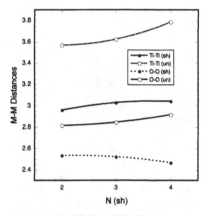

Figure 3. Plot of mean shared and unshared O-O octahedral edge lengths and Ti-Ti distances across shared edges and unshared edges (corner sharing) as a function of the number of shared edges in rutile (2), brookite (3), and anatase (4). The grand mean O-O distance is nearly the same in all three polymorphs. The mean Ti-Ti distance across shared edges approaches a maximum value of 3.04 Å in anatase, but the Ti-Ti distance across unshared edges (corners) increases more rapidly due to opening of the mean Ti-O-Ti angle from brookite to anatase. As a result, the mean Ti-Ti distance increases, accounting for the increased molar volume from left to right across the figure.

DISPLACEMENT ENERGIES

Atomistic simulations reported by Thomas et al. [7] for rutile show that the E_d for O displacements ranged from 50 to 85 eV depending upon direction, with an average value of 65 eV for a defect formation probability of 50%. For Ti displacements, E_d ranged from 80 to 160 eV depending upon direction, with an average value of 130 eV. In contrast to O, the E_d for Ti is highest along $\langle 111 \rangle$ and lowest along $\langle 001 \rangle$ directions. The higher average value for Ti is related to a) a higher lattice binding energy, b) greater electrostatic attraction between the vacancy and interstitial, and c) PKA energy transfer to O atoms. We have conducted additional calculations for anatase, finding that the E_d for O ranges from 20 to 100 eV with an average of about 55 eV. Due to difficulties in the statistics and determination of directional behaviour, only a minimum value of $E_d > 90$ eV has been determined for Ti in anatase. Similar difficulties were encountered for brookite, where we currently find E_d values to be > 60 eV for O and > 80 eV for Ti.

DEFECT FORMATION AND MIGRATION ENERGIES

Thomas et al. [7] also calculated formation energies of 9-11 eV for O Frenkel defects in rutile. These are of the split, channel, and channel wall type, with the split defects being the most common type. The O migration energy is < 0.1 eV as estimated from static and dynamic calculations and the interstitials migrate from split interstitial to split interstitial via channel sites, enabling movement in three dimensions, but are particularly active in the (001) plane. The Ti Frenkel defect energy was found to be approximately 18 eV and the Ti interstitials prefer the channel sites. The energy barrier for Ti migration was estimated to be 0.3 eV. Migration occurs from channel site to channel site along $\langle 001 \rangle$ and also by split interstitials and coordinated movements of many Ti atoms in the (001) plane, e.g., along $\langle 100 \rangle$ directions.

Additional static defect calculations were reported in reference [3], where we observed that the O Frenkel defect formation energies are very similar for all three polymorphs, with values of

9.0, 9.6, and 8.0 eV for rutile, brookite, and anatase, respectively. With regard to defect configurations, a common feature of all three polymorphs is the occurrence of the split oxygen interstitial. For Ti Frenkel defects, results of the calculations give formation energies of 18.1, 21.4, and 7.3 eV for rutile, brookite, and anatase, respectively. In both cases, the defect formation energies scale in the order anatase < rutile < brookite. We have also undertaken some preliminary modelling of O migration and our results suggest that the barriers scale in the order rutile < anatase < brookite.

EVOLUTION OF THERMAL SPIKES

In previous work [3,4], we performed MD simulations of the evolution of local regions of damage in rutile, brookite, and anatase using a thermal spike approach. Results of the simulations are summarized in Figure 4 wherein the evolution of the number of defects versus time is shown. This figure illustrates that the time required for peak development of the thermal spike and the maximum number of defects in the spike both follow the order rutile < brookite < anatase. Annealing of thermal spike damage occurs beyond a time frame of 0.1-0.2 ps. Rutile has completely annealed within 3 ps, brookite retains a few defects beyond 3 ps, and anatase retains many more defects out to 5 ps. In fact, the MD simulations reveal that a "stable" amorphous region is created in anatase that is slightly smaller that the intial spike dimension. Data analysis reported in reference [4] shows that, up to the peak level of spike damage, the numbers of O and Ti defects are close to the stoichiometric ratio of 2:1. Beyond the peak of damage, however, this is not the case and we see important differences in the dynamic defect behaviour as O defects migrate back to lattice sites more quickly than Ti defects.

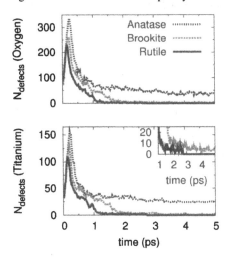

Figure 4. Plots showing the evolution of the number of O defects (top) and Ti defects (bottom) as a function of time during thermal spike formation and annealing in rutile, brookite, and anatase. In each diagram, rutile is represented by the lower solid line, brookite by the middle heavy dotted line, and anatase by the upper dotted line.

DISCUSSION AND CONCLUSIONS

Irradiation experiments conducted at 50 K with 1.0 MeV Kr ions reveal a correlation between structure and the critical fluence for amorphization in rutile, brookite, and anatase. We show that polyhedral distortion in these polymorphs leads to increased molar volume through electrostatic Ti-Ti repulsion in the octahedral framework. Hence the volume of open space available increases with distortion and provides a potential mechanism for lower defect formation energies at the preferred interstitial sites. Static defect calculations for anatase are consistent with this hypothesis; however, the calculated defect energies for brookite do not fit the observed trend, so other factors (e.g., defect migration pathways and energy barriers) must be invoked in order to explain the intermediate behaviour of this polymorph.

The MD simulations of small thermal spikes show good qualitative agreement with the experiments, suggesting damage recovery on picosecond time scales is related to mass transport within the spike. Undoubtedly, this involves the ability of the crystal to accommodate complex, coordinated multi-atomic motions. With the information at hand, we conclude that the radiation tolerance of rutile is due in part to the presence of low energy migration pathways for both O and Ti. At the other extreme, anatase is easily amorphized at 50 K and we speculate that this behavior may be influenced by fewer low energy migration pathways and significant energy barriers to defect migration and recombination.

ACKNOWLEDGEMENTS

We are grateful to Ed Ryan and Pete Baldo of Argonne National Laboratory for assistance with the IVEM Tandem experiments.

REFERENCES

1. M. Langlet, M. Burgos, C. Coutier, C. Jimenez, C. Morant, and M. Manso, *J. Sol-Gel Sci. Tech.* **22**, 139 (2001).
2. S. V. Stefanovsky, S. V. Yudintsev, R. Gieré, and G. R. Lumpkin, "Nuclear Waste Forms" *Energy, Waste, and the Environment: a Geochemical Perspective*, edited by R. Gieré and P. Stille, Geological Society, London, Special Publications, Vol. **236**, p36 (2004).
3. G. R. Lumpkin, K. L. Smith, M. G. Blackford, B. S. Thomas, K. R. Whittle, N. A. Marks, and N.J. Zaluzec, *Phys. Rev. B* **77**, 214201 (2008).
4. N. A. Marks, B. S. Thomas, K. L. Smith, and G. R. Lumpkin, *Nucl. Instr. Meth. Phys. Res. B* **266**, 2665 (2008).
5. A. Gloter, A. Douiri, M. Tencé and C. Colliex, *Ultramicroscopy* **96** 385 (2003)
6. V. V. Hoang, *Phys. Stat. Sol.* **244**, 1280 (2007).
7. B. S. Thomas, N. A. Marks, L. R. Corrales, and R. Devanathan, *Nucl. Instr. Meth. Phys. Res. B* **239**, 191 (2005).

Mater. Res. Soc. Symp. Proc. Vol. 1124 © 2009 Materials Research Society   1124-Q04-10

Porous Alumosilicates as a Filter for High-Temperature Trapping of Cesium-137 Vapour

A.S. Aloy and A.V. Strelnikov

RPA «V.G. Khlopin Radium Institute», 2-nd Murinsky pr. 28, St. Petersburg, Russia

ABSTRACT

High-temperature synthesis of glasses and ceramics containing ^{137}Cs is associated with volatilization of ^{137}Cs vapors.

Porous inorganic materials (PIMs) located in the synthesis area are proposed to effectively trap ^{137}Cs in a small volume of the filtering material.

Two types of porous inorganic materials were evaluated as filters for high-temperature irreversible ^{137}Cs chemosorption: «Gubka» based on alumosilicate hollow microspheres and commercially available porous Chamotte generated by using a foaming process.

The experiments show that, at above 700°C, the amorphous phase of these PIMs interacts with ^{137}Cs vapors, generating alumosilicate crystalline phases $CsAlSiO_4$ and $CsAlSi_2O_6$ (pollucite).

INTRODUCTION

Some of fission products have a significant vapour pressure at elevated temperatures. One of them is ^{137}Cs, which volatilization takes place during sintering of glass and ceramics.

The following two methods that differ in their mechanisms and location in the flow sheet are proposed for trapping ^{137}Cs vapors:

1. «Wet» trapping [1] is a "low-temperature" ^{137}Cs vapor condensation in the off-gas system (condensers, scrubbers, HEPA-filters, etc.) that contaminates gas lines and generates large volumes of secondary liquid radioactive waste.

2. «Dry» trapping [2] is a "high-temperature" chemosorption of ^{137}Cs vapors using filters that immobilize ^{137}Cs in stable crystalline phases.

It appears to be appropriate to use alumosilicate PIMs («Gubka», porous Chamotte, kaolin mineral wool, etc.) as filters for chemosorption of ^{137}Cs vapors. At high temperatures, these PIMs react with ^{137}Cs, generating $CsAlSiO_4$ and $CsAlSi_2O_6$ (pollucite). $CsAlSiO_4$ [3] and, especially, $CsAlSi_2O_6$ [4] are known to have a high chemical and thermal durability.

The «Gubka» material is defined as blocks of hollow (perforated and/or non-perforated) microspheres with the diameter of $0.1 - 0.4$ mm and the wall thickness of $5 - 10$ μm. These microspheres are part of fly ashes, resulted from incineration of mineral coal at fossil power plants. The «Gubka» material was developed at the Institute of Chemistry and Chemical Technology of the Siberian Branch of the Russian Academy of Science (Krasnoyarsk, Russia) and Mining and Chemical Combine (Zheleznogorsk, Russia) [5]. Porous Chamotte filters are the product of a mechanical treatment of the commercially available light-duty refractory brick ShL-0.4 (GOST RF 5040-96).

EXPERIMENT

Table I shows major characteristics of the «Gubka» and porous Chamotte materials.
Table I. Major characteristics of the «Gubka» and porous Chamotte

Characteristics	«Gubka»	Chamotte
Apparent density, g/cm^3	0.40 ± 0.02	0.40 ± 0.01
Open porosity, vol. %	50 ± 4	84 ± 4
Specific surface area, m^2/g	0.66 ± 0.04	19,84 ± 0,24
Chemical composition, wt.%:		
SiO_2	64.9-66.3	54.8-56.9
Al_2O_3	20.1-21.1	39.2-40.0
Fe_2O_3	3.1-4.6	0.8-0.9
MgO	1.9-2.2	0.1-0.2
CaO	1.8-2.7	0.7-0.8
Na_2O	0.3-0.6	0.1-0.2
K_2O	1.9-2.9	0.2-0.5
TiO_2	0.2-0.5	1.7-2.3
Phase composition, wt %	amorphous phase ~ 95 α-quartz ~ 5	amorphous phase ~90 α-quartz ~ 2.5 α-cristobalite ~ 5 mullite ~ 2.5

Figure 1 shows SEM photomicrographs of the «Gubka» and Chamotte materials illustrating a different morphology of their structures. «Gubka» has a honeycomb-organized structure, while Chamotte has a chaotic and unorganized structure.

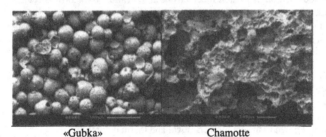

      «Gubka»   Chamotte
Figure 1. SEM microphotographs of the «Gubka» and Chamotte Materials

The ^{137}Cs vapor trapping using «Gubka» or Chamotte filters was conducted in a static mode (without a carrier gas) in a lab-scale system shown in Figure 2.

The lab-scale system contains the stainless steel can (5) with Al_2O_3 filling (6) where the crucible (4) with 2 g of $CsNO_3$ is installed. Two PIM disks (1), ~ 10 mm high and 35 mm in diameter each, are inserted into the upper part of the can. The lower disk (# 1) traps ^{137}Cs vapors and the upper disk (#2) collects the ^{137}Cs that escapes the lower disk (#1). The can (5) with Al$_2$O$_3$ filling (6) – crucible with CsNO$_3$ (4) – two disks (1) assembly is placed into the resistance furnace (7) that is heated to 1000°C. The disk (1) temperature ranges from 600 to 1000°C by varying the depth of the assembly immersion into the furnace (7). The thermocouple (2) monitors the temperature in the furnace (7), i.e., CsNO$_3$ calcination temperature and disk (1) temperature, and the other thermocouple (3) monitors temperature of the filters.

The assembly was held in the furnace for 50 hours, with the temperature of the crucible with CsNO$_3$ being 1000°C and the temperature of the «Gubka» disks varying from 600 to 1000° C by 100°C steps.

In each run, every five hours after cooling, the crucible (4) was replaced with a new one with a new portion of CsNO$_3$, and the disks (1) were weighed to measure both Cs saturation (disk #1) and escaped Cs (disk #2).

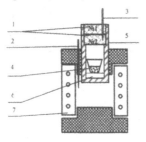

Figure 2. Schematic of Lab-Scale System for ^{137}Cs Vapor Trapping in a Static Mode

1 –	«Gubka» or Chamotte disks (#1 and #2)	4 –	crucible with CsNO$_3$
		5 –	stainless steel can
2 –	thermocouple	6 –	Al$_2$O$_3$ filling
3 –	thermocouple	7 –	resistance furnace

RESULTS AND DISCUSSION

Table II shows mass ratios of Cs-containing crystalline products, resulting from ^{137}Cs vapor interaction with «Gubka» or Chamotte as a function of temperature.

Table II. Phase Compositions Resulting from Interaction of ^{137}Cs Vapors with «Gubka» and Chamotte at Various Temperatures

t, °C	Ratios of Cs-containing crystalline phases, wt %	
	«Gubka»	Chamotte
600	CsNO$_3$=100	CsNO$_3$=100
700	CsAlSiO$_4$: CsAlSi$_2$O$_6$ =60:40	CsAlSiO$_4$: CsAlSi$_2$O$_6$ =20:80
800	CsAlSiO$_4$: CsAlSi$_2$O$_6$ =50:50	CsAlSi$_2$O$_6$ =100
900	CsAlSiO$_4$: CsAlSi$_2$O$_6$ =30:70	CsAlSi$_2$O$_6$ =100
1000	CsAlSi$_2$O$_6$ =100	CsAlSi$_2$O$_6$ =100

Table II indicates that, at 600°C, only $CsNO_3$ can be found on the filters, i.e., ^{137}Cs vapors do not interact with «Gubka» or Chamotte. Consequently, the mechanism of absorbing ^{137}Cs vapors by the «Gubka» or Chamotte filters at 600°C is a straightforward Cs_2O precipitation on the relatively cold surface of the filters followed by a reaction with NO_2.

At 700–1000°C, $CsAlSiO_4$ and $CsAlSi_2O_6$ (pollucite) crystallize in «Gubka», with the mass fraction of pollucite increasing with the temperature increase, and, at 1000°C, only pollucite crystallizes. Both phases crystallize in Chamotte at 700°C, and at 800 – 1000°C only pollucite crystallizes. «Gubka» and Chamotte absorb ^{137}Cs vapors at 700 – 1000°C due to a non-reversible high-temperature chemisorption, a chemical interaction of ^{137}Cs vapors with the «Gubka» or Chamotte amorphous alumosilicate phase.

Distribution of Cs_2O between crystalline and amorphous phases into PIM at different temperatures acoording to XRD data is shown in Table III.

Table III. Distribution of Cs_2O between crystalline and amorphous phases into PIM at different temperatures

T °C	Cs_2O in crystalline phases : Cs_2O in amorphous phase, %	
	«Gubka»	Chamotte
700	60 : 40	50 : 50
800	70 : 30	60 : 40
900	75 : 25	80 : 20
1000	85 : 15	80 : 20

It can be seen from Table III that amount of Cs_2O incorporated with crystalline phases bcome higher when temperature increase.

Figure 3 shows photomicrographs of the «Gubka» and Chamotte surface after absorption of ~0.11 g of Cs_2O/g of «Gubka»/Chamotte at 1000°C.

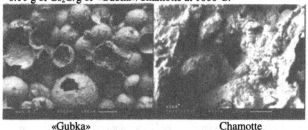

«Gubka»   Chamotte

Figure 3. SEM Photomicrographs of the «Gubka» and Chamotte Surface after Absorption of ~0.11 g of Cs_2O/g of «Gubka» or Chamotte at 1000°C

Table IV shows chemical compositions of the «Gubka» and Chamotte surface layers, depending on their saturation with cesium oxide (EPMA data).

Table IV. Chemical Compositions of the «Gubka» and Chamotte Surface layers

g of Cs₂O /g of «Gubka» or Chamotte	Oxides in the surface layers. wt %									
	Cs_2O	SiO_2	Al_2O_3	Fe_2O_3	MgO	CaO	Na_2O	K_2O	TiO_2	Total
«Gubka»										
0.05	35.20	36.43	19.05	2.48	0.00	1.10	2.34	3.40	0.00	100.00
0.11	44.16	32.85	15.29	1.58	0.00	0.98	1.96	3.18	0.00	100.00
0.18	45.82	31.11	15.10	2.64	0.00	0.77	1.36	3.20	0.00	100.00
Chamotte										
0.06	34.14	41.84	22.17	0.69	0.00	0.19	0.00	0.98	0.00	100.00
0.11	44.20	34.36	19.05	0.76	0.00	0.00	0.00	1.65	0.00	100.00
0.19	43.31	35.89	19.95	0.00	0.00	0.00	0.00	0.86	0.00	100.00

Table IV shows that the chemical compositions of the «Gubka» and Chamotte surface layers are similar to pollucite (Cs_2O – 45.2 wt. %; Al_2O – 16.3 wt. %; and SiO_2 – 45,2 wt. %).

Figure 4 shows photomicrographs of the «Gubka» and Chamotte vertical cross sections after absorption of ~0.11 g of Cs_2O/g of «Gubka» or Chamotte at 1000°C.

      «Gubka»   Chamotte

Figure 4. SEM Photomicrographs of the «Gubka» and Chamotte Cross Sections

Table V and VI provide data on ^{137}Cs penetration into «Gubka» and Chamotte.

Table V. «Gubka» Vertical Cross Section after Absorption of ~0.11 g of Cs_2O/g of «Gubka» at 1000°C. Analysis area – 500 x 500 µm

Measurement point	wt.%									
	Cs_2O	SiO_2	Al_2O_3	Fe_2O_3	MgO	CaO	Na_2O	K_2O	TiO_2	Total
A	46.94	34.23	14.41	1.91	0.00	1.03	0.00	1.48	0.00	100.00
B	43.11	37.22	14.08	3.59	0.00	0.87	0.00	1.12	0.00	100.00
C	37.04	41.61	15.58	2.78	0.00	1.10	0.00	1.89	0.00	100.00
D	11.13	52.29	19.85	5.97	4.05	1.71	0.00	5.00	0.00	100.00
E	2.06	55.24	23.71	6.14	3.52	1.24	3.81	3.41	0.88	100.00
F	0.88	61.23	22.33	6.93	2.41	1.45	0.00	3.73	1.04	100.00

Table VI. Chamotte Vertical Cross Section after Absorption of ~0.11 g if Cs_2O/g of Chamotte at 1000°C. Analysis Area – 500 x 500 µm

Measurement point	wt.%									
	Cs_2O	SiO_2	Al_2O_3	Fe_2O_3	MgO	CaO	Na_2O	K_2O	TiO_2	Total
A	46.67	31.78	17.12	1.17	0.00	0.80	0.00	1.37	1.09	100.00
B	40.37	36.27	19.76	1.55	0.00	0.57	0.00	1.48	0.00	100.00
C	10.08	56.22	29.11	1.53	0.90	0.38	0.00	1.78	0.00	100.00
D	0.88	58.77	34.33	2.66	1.04	0.58	0.00	1.74	0.00	100.00

Tables V and VI shown that [137]C penetrate «Gubka» deeper than Chamotte, i.e., Chamotte has a higher ^{137}Cs vapor capacity than «Gubka» because Chamotte has a higher surface area.

The high-temperature (700–1000°C) chemosorption of ^{137}Cs vapors using the PIMs had a linear character and generated a visible blurry separation line between saturated and non-saturated layers of the lower disk (#1), deformation of the lower disk and cracking of the «Gubka» disks. The Chamotte lower disks did not crack.

The «Gubka» and Chamotte lower disks cracked when their saturation of ^{137}Cs vapors exceeded 0.18 g Cs_2O /g of the PIM. The mass of the upper disks did not change, i.e., no ^{137}Cs vapors escaped through the lower disks (even if they cracked).

CONCLUSIONS

- Alumosilicate PIMs, such as «Gubka» and porous Chamotte, were found to be highly effective filters for a high-temperature irreversible chemosorption of ^{137}Cs vapors.
- The optimal temperature range for the ^{137}Cs irreversible chemosorption using «Gubka» or Chamotte was found to be 700 – 1,000°C.
- The use of Chamotte as a filter was found to be advantageous than use of «Gubka» because Chamotte has a higher specific surface area for ^{137}Cs vapors and a lower cost than «Gubka».

REFERENCES

1. Technical Reports Series 291, IAEA, Vienna, (1988).
2. J.M. Shin, J.J. Park, Korean J. Chem. Eng. 18 (6), 1010-1014 (2001)
3. S.A. Gallagher, G.J. McCarty and D.K. Smith, J. Mater. Res. Bull., 12(12), 1183-1190 (1977).
4. I. MacLaren, J. Cirre, and C B Ponton, J. Amer. Ceram. Soc., 82, 3242-3244, (1999).
5. N.N. Anshits, A.N. Salanov, Vereshchagina T.A., et. al. J. Nuclear Energy Science and Technology, 2 (1/2), 8-24 (2006).

Mater. Res. Soc. Symp. Proc. Vol. 1124 © 2009 Materials Research Society   1124-Q06-03-O07-03

Oxygen Lattice Distortions and U Oxidation States in UO_{2+x} Fluorite Structures

Lionel Desgranges[1] and Gianguido Baldinozzi[2]

[1]CEA, DEN, DEC, Centre de Cadarache, 13108 Saint-Paul-lez-Durance, France

[2] Matériaux Fonctionnels pour l'Energie, SPMS CNRS-Ecole Centrale Paris, 92295 Châtenay-Malabry, France & CEA/DEN/DANS/DMN/SRMA/LA2M, 91191 Gif-sur-Yvette, France

ABSTRACT

The structural changes induced by the changes of the uranium oxidation state in the ideal fluorite lattice of pure UO_2 are discussed. Experimental results evidence strong distortions of the oxygen sub-lattice due to dynamic (at high temperature) or static (at low temperature) fluctuations of the local charges in the cationic sublattice. These changes in the oxidation state are often described using the Vegard's law because a linear dependence of lattice parameter is observed over a wide range of compositions. Nevertheless, an ideal solid solution model cannot explain this behavior where the elastic effects are directly related to the ionic radii of the cations. Strong evidence is provided that enthalpy effects are relevant in these systems and that they are directly associated with the local structural changes observed during neutron scattering experiments.

INTRODUCTION

During its irradiation in power reactors, uranium dioxide, the most used nuclear fuel, undergoes not only displacive irradiation damage but also chemical changes because of the large range of oxidation states of U atoms interacting with the elements generated by the nuclear fissions. Amongst these fission products, lanthanides (Ln) are reported in literature [1-3] to form $(U,Ln)O_{2+/-x}$ solid solutions with x near to 0. It is generally admitted that the evolution of the lattice parameter of U based fluorite type structures follows a linear behaviour over a large range of dopant concentrations. For instance, the unit cell parameter of $(U,Ln)O_2$ binary systems, for Ln= Gd [1],La [2], Nd usually verifies the Vegard's law up to 50% [3]. From this experimental result, it was generally assumed that the incorporation of aliovalent cations (U with a different oxidation state or a Ln element) proceeds by a one to one substitution on the uranium site. Thus, the ionic radius of the aliovalent cation would be the most significant parameter to predict its actual substitution for a U atom in the cation sublattice of UO_2 [4,5]. However, from a chemical point of view, the assumption of uncorrelated vacancies in the oxygen sub-lattice over such a large range of compositions is questionable and the model of ideal solid solution is certainly not fulfilled for these binary systems. The coordination polyhedron of the lanthanide and the one of uranium, which may change its oxidation state to compensate lanthanide incorporation, are very likely to be distorted.

In UO_2, increasing temperature or changing the oxygen concentration modifies the oxidation state of uranium. In this paper, we demonstrate first that in the very simple case of stoichiometric UO_2 having uranium atoms with different oxidation states, ionic radii are not sufficient to describe the unit cell parameter. Distortions of the UO_2 oxygen sub-lattice associated with the incorporation of aliovalent cations have to be taken into account. Secondly, an illustration of such distortions is presented in the case of oxidised UO_2. And, at last, a possible distortion associated to lanthanide incorporation is discussed.

MODIFICATION OF URANIUM OXIDATION STATE IN UO_2

In UO_2, increasing temperature or changing the oxygen concentration modifies the oxidation state of uranium. The modification of the average UO_2 unit cell parameter is discussed in both cases from a structural point of view.

Defects and intrinsic electronic carriers at high temperature in stoichiometric UO_2

The electric conductivity of a UO_2 single crystal (with impurity content given in [6]) was measured as a function of temperature [6], and evidences a transition around 1300K. At low temperature (T<1300K) the electric conductivity was attributed to an extrinsic regime, the charge carriers being produced by impurity ionization. At high temperature (T>1300K), the electric conductivity was attributed to an intrinsic regime, the charge carriers being produced by the dismutation reaction:

$$2\ U^{4+} \Leftrightarrow U^{5+} + U^{3+} \tag{1}$$

Using the mass-action law the concentration of intrinsic carriers can be written as

$$n = \exp[-(\Delta H_p/2kT) + (\Delta S_p/2k)] \tag{2}$$

Where ΔH_p and ΔS_p are the enthalpy and the entropy of the reaction (1). ΔHp was evaluated to 2eV, ΔSp to $4.2k$. Therefore, the intrinsic carrier concentration is thermally activated and the activation energy is 1eV. UO_2 unit cell parameter, a, was measured by Ruello et al. as a function of temperature [7] (Figure 1).

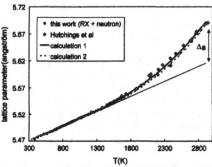

Figure 1. UO_2 unit cell parameter as a function of temperature from [5].

The thermal evolution of the lattice parameter can be decomposed as the sum of a regular linear term and an additional term. The additional term cannot be modeled by a 3rd order polynomial, as predicted by an anharmonic thermal expansion, nor by a power law that can be expected close to a phase transition. A thermally activated law best describes the evolution of the lattice parameter:

$$a = 5.4698 \left[1 + 1.015 \cdot 10^{-5} (T\text{-}300)\right] + 4.95 \exp(-1.06 \text{ eV}/kT) \qquad (3)$$

Where T is the absolute temperature of the sample. Within the experimental accuracy, this thermally activated term is proportional to the intrinsic carrier concentration, because they have the same activation energy 1eV. The evolution of the unit cell parameter can be used to evaluate the distortion imposed by intrinsic carriers on the crystalline lattice. This evaluation must be performed at temperature higher than 1300 K, where the extrinsic defect concentrations are smaller than the intrinsic ones as discussed in [6]; this is also consistent with the behavior of the additional thermally activated term, that is essentially negligible at temperatures lower than 1300K where the extrinsic defect concentrations dominate. Then, the measured unit cell parameter can be considered as a weighted average of a regular lattice and an ideal lattice whose unit cells contain an intrinsic carrier.

$$a_{measured} = (1\text{-}n)a_{regular} + n\ a_{intrinsic\ carrier} \qquad (4)$$

Making this assumption, the parameter of a unit cell containing an intrinsic carrier can be estimated as 6.07Å, compared to 5.47Å for the regular lattice. This corresponds to a strain of about 10%, which must be associated with atomic displacements. Whether the form of this average is close to the Vegard's law, the distortion associated with a $(U^{+3}:U^{+5})$ complex defect is much larger than what would be directly estimated by their ionic radii. In the literature [8], the ionic radii of U^{4+}, U^{3+} and U^{5+} are 1.0, 1.025 and 0.84 Å respectively. This distortion then involves a strong modification of the oxygen sub-lattice as it is also evidenced by in situ neutron diffraction experiments [9]: at high temperature oxygen atoms are displaced from their regular positions. The crystalline sites that were identified at high temperature are consistent with the ones already identified during UO_2 oxidation [10-13].

Defects and order at low temperature in undoped uranium oxide

At low temperature, the incorporation of oxygen in UO_2 results in the formation of new phases: U_4O_9 and U_3O_8. The analysis of their crystalline structures also evidences major changes in the oxygen sub-lattice, suggesting that the increase of the oxidation state of uranium atoms is always associated with a significant enthalpy effect. It should be pointed out that the formation of regular environments (like the oxygen cuboctahedra in U_4O_9) seems to be an effective way to accommodate a large defect concentration without inducing large strains. Obviously, it is not possible to apply Vegard's law in this framework since a phase change is observed. Nevertheless, understanding the local U environments may provide helpful hints for describing the effect of aliovalent cations, moreover the discussion of U_4O_9 and U_3O_8 crystalline structures is a mean not to consider impurities, that could have some influence at low oxygen concentration, but that can be neglected in the discussion of uranium average environment. In spite of the different crystal symmetries, the U sublattice in U_4O_9 and in U_3O_8 is still very similar to the one in UO_2 fluorite structure [13]. It is convenient to analyse both structures as a stacking of atomic layers. The fluorite structures can be seen as an alternate stacking along [111] of layers containing only U or O atoms. Each U layer appears as a hexagonal network where only one of the three sites is occupied. The fluorite structure consists of the regular repetition of three different layers according to the well-known ABC pattern of a face centred cubic structure. This hexagonal network of U atoms also exists in the U_4O_9 and U_3O_8 structure. Nevertheless, the layer atomic compositions and their staking sequences are modified.

For instance, the U layers in U_3O_8 are stacked according to an AA sequence but their atomic composition is also modified. While in UO_2 all the U and O atoms lie in separate layers, in U_3O_8 most of the O atoms now sit in the U layer forming a pentagonal pattern around each U atom. A small amount of O atoms still sits in the interlayer on top of each U atom creating pillars that connect the layers. These U-O-U bonds are shorter than the U-O bonds in UO_2 (2Å instead of 2.4Å).

The structure of U_4O_9 consists of a UO_2 lattice in which 12 clusters of 13 oxygen interstitials centred at an empty cube, the oxygen cuboctahedra, replace the original oxygen cubes in the fluorite structure. The unit cell parameter decreases from 5.47Å in UO_2 to 5.44Å in U_4O_9. Thus the significant changes of the structure of the oxygen sub-lattice associated with the clustering of the oxygen interstitials have only a limited impact on the contraction of the unit cell parameter.

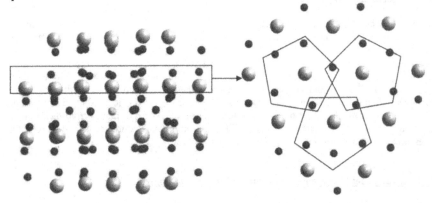

Figure 2. (left panel) Projection in the (**a+b+c** ; **a+b-c**) plane of U_4O_9 crystalline structure around a cuboctahedron (the oxygen at the centre of the cuboctahedron is displayed in yellow); (right panel) projection of the section taken from left image (represented by a black rectangle) on the (**a+b-c**; **a-b**) plane. The pentagonal arrangement of the oxygen atom is similar to the one observed in U_3O_8.

Each cuboctahedron consists of 12 O atoms (plus an additional O atom at the cuboctahedron centre) bonded to 6 U atoms that share the normal cation sublattice in the fluorite structure. These U atoms have a square antiprism coordination instead of the normal cubic one. The interstitial O atoms forming the cuboctahedron deplete the normal O sites in the fluorite lattice. As a result, the O layers are depleted and the interstitial O atoms lie on another layer closer to the U one. Moreover, the atom at the centre of the cuboctahedron lies midway between two U layers, in a position that can be identified with the O atoms forming the pillars connecting the layers in U_3O_8 structure. When a composite layer formed by the U layer and these O interstitial layers (figure 2) is projected along [111], the projected pattern is already very similar to the atomic configuration of the basal layer in U_3O_8 structure (a pentagonal environment for the U atoms). Therefore, the formation of cuboctahedrons can be seen as an intermediate step in the progressive transformation of the fluorite-type structure into the U_3O_8 one.

UO_2 oxidation evidences that defects may cluster and organize themselves in particular layers ; these defect organizations may accommodate in an efficient way the distortions produced by the changes in the uranium oxidation state.

DISCUSSION AND CONCLUSIONS

The experimental facts reviewed in this paper clearly suggest that the changes of the uranium oxidation state are always associated with a significant modification of the oxygen sub-lattice. These effects may also affect the behavior of oxide solid solutions between U and Ln atoms.

To this purpose, it is important to explore also alternative structural modification that may accommodate large defect concentrations without the onset of important long-range strain fields. The planar clusters of defects evidenced in oxidized UO_2 can give such an alternative; they are equivalent to extended defects like partial dislocations. According to Hyde [14], several oxide structures like UO_3, CaF_2, La_2O_3 and NaCl can be obtained maintaining the same basic cation sublattice consisting of a stacking along [111] of hexagonal layers but changing the characteristics of the oxygen layers stacking. This description points out that it is possible to move from one structure to the other one by the creation of lamellae made of oxygen vacancies parallel to the (111) plane. For instance, figure 3 displays the local transformation of a fluorite structure into a La_2O_3 one.

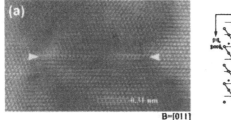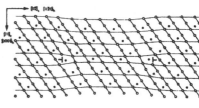

B=[011]

Figure 3. (left panel) High resolution TEM image [9] of a dislocation loop in irradiated CeO_2. (right panel) Coherent lamella of A type structure (La_2O_3) in fluorite type matrix from [8].

This structural modification can effectively accommodate trivalent cations in UO_2 when local compositional fluctuations or segregation occur. This assumption is also supported by recent high-resolution transmission electron microscopy results obtained in CeO_2 (Ce may also exhibit several oxidation states, depending on the oxygen partial pressure) in which edge dislocations where observed after electron irradiation [15].

Though the use of Vegard's law can be a valuable empirical tool for engineering nuclear materials, the structural features observed in uranium oxides where a complex reorganization of the oxygen sub-lattice and significant enthalpy effects are observed do not support the hypotheses of this description. These structural reorganizations are not always well documented because accurate information on the structural parameters describing the oxygen sub-lattice requires neutron diffraction experiments. EXAFS experiments can also provide valuable information on the local environments but only few experiments were performed on $(U,Ln)O_2$ compounds.

Therefore, new experiments should be planned to achieve a better understanding of $(U,Ln)O_2$ compounds, more specifically to understand whether a significant defect clustering also occurs in these systems. Such a behavior is expected to have significant impact for the design of radiation resistant fuels consisting of mixed oxides.

REFERENCES

1. H.G. Riella, M. Durazzo, M. Hirata, R.A. Nogueira, J. Nucl. Mater. **178**, 204 (1991).
2. M. P. Herrero, R. M. Rojas J. Solid State Chem., **73**, 536-543 (1988).
3. H. Weitzel and C. Keller, J. Solid State Chem., **13**, 136-141 (1975).
4. K. Kapoor, S.V. Ramana Rao, Sheela, T. Sanyal, A. Singh, J. Nucl. Mat. **321**, 331–334 (2003).
5. M. Kato and K. Konashi, J Nucl. Mat. (2009) doi:10.1016/j.jnucmat.2008.09.037.
6. P. Ruello & al., J. Am. Ceram. Soc. **88**, 604 (2005).
7. P. Ruello, L. Desgranges, G. Baldinozzi, G. Calvarin, T. Hansen, G. Petot-Ervas, C. Petot,. Journal of Phys. Chem. Solids **66**, 823–831 (2005).
8. R. D. Shannon Acta Cryst. A**32**, 751 (1976).
9. M.T. Hutchings J. Chem. Soc., Faraday Trans. 2, **83**, 1083-1103 (1987).
10. D. J. M Bevan, I. E. Grey, B. T. M. Willis, J. Solid State Chem. **61**, 1-7 (1986).
11. G. Baldinozzi, G. Rousseau, L. Desgranges, J.C. Nièpce, J.F. Bérar, Mat. Res. Soc. Symp. Proceedings **802** 3–8 (2003).
12. R.I. Cooper and B.T.M. Willis, Acta Cryst. A **60**, 322-325 (2004).
13. G.C. Allen and N.R. Holmes, J. Nucl. Mater. **223**, 231 (1995).
14. B.G. Hyde, Acta Cryst A **27**, 617-621 (1971).
15. K. Yasunaga et al., Nucl. Instr. and Meth. in Phys. Res. B **266**, 2877–2881 (2008).

Mater. Res. Soc. Symp. Proc. Vol. 1124 © 2009 Materials Research Society   1124-Q06-07-O07-07

Cerium Dioxide Surface Characterization and Determination of Surface Site Density by Potentiometric Fast Titration

N. Rodriguez Villagra [1], J.C. Marugan [1], E. Iglesias [1], J. Nieto [1], T. Missana[1], N. Albarran[1], J. Cobos[2], J. Quiñones[1]
[1] Ciemat. Avda. Complutense 22, 28040 – Madrid. SPAIN
[2] ITU-JRC European Commission Joint Research Centre Institute for TransU Elements. European Commission, Karlsruhe. Germany

ABSTRACT

Cerium dioxide has been used as a Pu analogous to study Pu-Th mixed oxide fuels behavior, (Th, Pu)O_2, known as "Th-MOX". They are considered as possible advanced nuclear fuels for Generation-IV Fast Reactors aiming the burning and reprocessing of Pu stocks generated on U cycle. The use of Pu in Mixed-Oxide Fuel (MOX) will generate less fission products than U dioxide fuels. Before performing the study of Th sorption behavior, it is necessary to identify essential parameters as surface coordination sites density, surface charge and the equilibrium constants of the functional surface groups of hydrous oxide. Ce oxide has been characterized by the BET (Brunauer–Emmett–Teller) method to measure the specific surface area using $N_2(g)$, by XRD (X-Ray Diffraction) to verify the crystalline structure, by SEM (Scanning Electron Microscopy) to analyze how well the pellets have been sintered, by AFM (Atomic Force Microscopy) to study the roughness, and by potentiometric titration to determine the surface sites density.

Potentiometric titrations were carried out both on a suspension of ceria colloids and on pellets. After the measurements, both proton intrinsic affinity constants and OH surface groups were compared. The experiments were developed in a wide range of pH and at different ionic strengths. The sites density values measured for ceria in 0.1M $NaClO_4$ media were 52.0 ± 5.5 and 2.3 ± 0.1 sites·nm^{-2}, for colloids and pellet, respectively. The hysteresis cycle obtained with pellets between acid/alkaline directions is rather important because indicates the tendency to retain other elements in his matrix.

INTRODUCTION

Nuclear fuels based on ThO_2 are reaching great importance in the Pu inventories reduction [1] as well as in minor actinides and long life fission products diminution [2] too. The development of this new generation of nuclear fuels requires their characterization before and after irradiation. ThO_2 has been chosen like an Inert Matrix Fuel (IMF), where the carrier of the plutonium is not uranium and it should be more or less transparent to neutrons [3], due to different reasons like: the Th abundance, around three times higher than U, its melt temperature (3300 °C) and its thermal conductivity higher for ThO_2 than UO_2, its high chemical stability (ThO_2 solubility is a thousand times less than UO_2 [4]), its better stability in the presence of radiation damage, and its higher actinide burnup. One option for this IMF after nuclear fuel cycle could be the Deep Geological Disposal and, for that purpose, it is essential to assess some criteria as its compatibility with the repository environment, chemical stability with time in the presence of known radiation fields [5].

The evaluation of Pu chemical behavior should be done hereby models with speciation data. Choppin [6] proposed Th(IV) as a Pu(IV) surrogate although Th(IV) has a weaker complexation and hydrolysis than Pu(IV). Th(IV) complexation constants can be adjusted multiplying $\log \beta(\beta^{Th})$ by Th(IV) and Pu(IV) ionic radium ratio. But this approach should be considered as a rude approximation only valid to 1:1 complexation ($log\ K°S(Th(OH)_4(cr)) = -9.4$ [7]). As happened with nuclear fuels based on U, it is assumed that radionuclide release from the matrix during the deep geological repository in Th-MOX will be controlled by the solubility of the mayor species (U and Th respectively). In ITU-JRC different experiments related with ThO_2, PuO_2 and $(Pu-Th)O_2$ fabrication sol-gel, from a liquid "sol" (colloidal) into a solid "gel" phase and characterization (lixiviation, chemical stability,...) are being performed. Thanks to the ITU-CIEMAT framework collaboration project, characterization and Th sorption experiments on CeO_2 have been performed to be compared with Pu results. The CIEMAT mayor goal was to get confidence on sorption processes (including sites density determination), and finally to be able to reproduce the Th sorption mechanism in natural waters. Pu is simulated with Ce and experiments were carried out using CeO_2 pellets.

The knowledge of radionuclide sorption mechanism in underground water conditions and behavior prediction model of retention are key issues in assessing the performance of a given nuclear waste disposal concept. An essential parameter for adsorption is the surface sites density that can be measured by acid-alkaline potentiometric surface titrations. Proton exchange sites density determination consists of two steps: first, the proton adsorption in acid media, and next, the proton desorption in alkaline media. Only one type of $\equiv$SOH sites has been considered because there is not knowledge about CeO_2 surface behavior [8]. In the present work, we attempt to quantify the sorption behavior of Th using CeO_2 pellets together with ITU focusing on the assessment of the long term safety of Th-based fuels. It is known that Th has a great tendency to adsorb colloidal particles and oxides on its surface, this affecting the migration of Th in natural groundwater [4, 9]. In this case, Th represents the main element of spent nuclear fuel by mass. Therefore, a continuous sorption test was carried out involving Th(IV) over Ce(IV) dioxide (as non irradiated chemical surrogate for Pu(IV)). The main objectives of the present work are (1) to compare the surface properties of colloidal and compacted CeO_2, and (2) to study Th adsorption on compacted CeO_2.

EXPERIMENTAL DETAILS

CeO_2 pellets were prepared by the sol-gel process in ITU. The specific surface area (BET method using both N_2 and Kr adsorbate) was measured with the ASAP 2020 equipment from Micromeritics. The values obtained were 84.74 ± 1.82 m^2/g and 0.22 ± 0.01 m^2/g, respectively [10]. AFM showed the roughness depth (Figure 1).

The crystalline structure was checked by XRD using an X Pert-MPD Philips diffractometer. The XRD pattern (Figure 2) is perfectly fit by the theoretical one, showing a cubic system with fluorite type structure, and the experimental value of the unit cell parameter(5.411 ± 0.012 Å) is almost the same as the theoretical value (5.412 Å)

SEM revealed surface morphology by using a FEG-SEM (Field-Emission gun Scanning Electron Microscope) at 10 kV in a JEOL 6335 equipment (Figure 3, Figure 4 and Figure 5). At low magnification Figure 3, first image, it is possible to appreciate a rough surface. At higher magnification, it is interesting to observe a possible grain boundary and marks with a width less than 0.1 μm

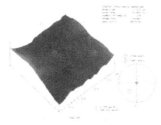

Figure 1 CeO₂ AFM surface image (scan size 2.175 µm).

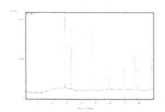

Figure 2 Characterization of CeO₂ pellet by X-ray diffraction pattern with a good agreement between experimental and theoretical results.

Figure 3 SEM X500 image pellet showing CeO₂ morphology.

Figure 4 SEM X2000 image CeO₂ pellet.

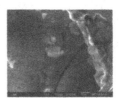

Figure 5 SEM X10000 image CeO₂ pellet

CeO₂ colloidal solution was employed to compare the sites density value to that of pellets. In this case, zeta potential ζ (colloids stability and mobility parameter) was measured with a MALVERN ZETAMASTER equipment through Laser Doppler Electrophoresis. This value will be dependent on the pH, from the point where protonation / deprotonation reactions take place with the correspondent intrinsic equilibrium acidic constant. The obtained result was 46.6 ± 3.5 mV. This potential corresponds to pH 6.4 in the titration curves. Colloidal concentration and particle size were measured in a Malvern 4700C equipment by Photon Correlation Spectrometry, obtaining an average particle size of 47.5 ± 4.9 nm. Knowing this value together with the concentration one, the colloid surface in suspension can be calculated.

Methods

The continuous fast titrations were performed under anoxic conditions inside glove box under argon atmosphere (<1ppm O₂) in order to avoid CO₂ presence; the background electrolyte was 0.1 M NaClO₄ at 298 K, and the pH has been adjusted by adding precise aliquots of HCl (acidic direction pH ≈ 1.5) or NaOH (alkaline direction until pH ≈ 12.5), both 0.1 M. This was done under stirring for reference (electrolytic media 0.1 NaClO₄), ceria colloids (1g/l) in 0.1M NaClO₄ (40 ml) and CeO₂ pellet in 0.1 M NaClO₄ (40 ml). Equilibration time for each solution before titration was 24 hours. The titration was carried out with a 808 Titrando equipment (Metrohm). The working conditions for each measurement were saved in a method ΔE (mV) <0.5 in 5 min as maximum time. Between measurements, a delay of 60 seconds for manual adding of the aliquot was set.

The selected pH region for dynamic sequential sorption experiments was 1.5-4.5 to avoid Th hydroxides formation, because this detail complicates the Th behavior modelling. Östhols [7] predicts those alkaline or neutral pH are enough for inducing Th desorption by hydrolysis on silica surface. As the added acid or base volume and taken samples was small opposite to the initial volume (330 ml), changes in the radionuclide concentration and solid area to liquid volume ratios (S:L) could be ignored. The S_{geom}/V_s initial value was 0.25 m^{-1}.

Firstly, a pellet was washed with MilliQ water until getting neutral pH. Th^{4+} solution $6.01 \cdot 10^{-7}M$ was prepared starting with 4.4 mg of ThO_2 in 49.7 ml suprapure HNO_3, 0.3 ml $HClO_4$ (70% , $\rho= 1.67 \cdot 10^3$ kg/m^{-3} or 11.643M) and 50 ml of 0.1M $HClO_4$ to get a final volume of 100 ml, electrolytic media 0.1 M $NaClO_4$ was added and adjusted around pH 1.5 with 0.1 M HCl under stirring. After equilibrium was reached, a CeO_2 pellet was added and one sample (1 ml filtrated with 0.22 μm filters and 1ml ultrafiltrated under ultracentrifugation) was taken to analyze by ICP-MS. This process was followed after one day of stabilization, after modify pH with 0.1M HCl or 0.1M NaOH. A new method was created, in this case ΔE (mV) < 0.1 between 60 and 150 s

RESULTS AND DISCUSSION

SEM of the CeO_2 pellet performed before the experiments (Figures 3- 5) showed no surface cracks and a non homogeneous porosity distribution. Potentiometric titration curves of colloids and pellet are shown in Figure 6 and Figure 7. Regarding to the acid/alkaline reversibility, a large hysteresis between the two directions is not observed. It was noted that the equilibrium in alkaline direction was reached more quickly than in the acid direction within the systematic equilibrium time used (approximately 2 minutes).

No consumption or release of protons takes place at lower pH, but above pH=4, the titration shows proton release. Additional with different ionic strengths are needed to adequately model the acid-base behavior and to calculate the active sites density.

The parameters calculated from CeO_2 are presented in Table I. For the pellet, we have obtained $(31.90 \pm 0.93) \cdot 10^{-2}$ $mol \cdot kg^{-1}$. Ceria pellets carry pH-dependent negative charge as it is shown in Figure 6 and Figure 7 as a consequence of the surface charge dependency on the pH, although it will be necessary to make other tests with other ionic strengths. In Figure 8 a swift decrease in Th concentration when pH increases is shown. There are two possible reasons: Th hydroxide formation could be take place (Figure 9) or, more reasonable, Th sorption on ceria coordination sites dominates through an exchange by protons in the equilibrium time (1-3 days).

Otherwise, the difference in Th concentration between filtered and ultrafiltered samples decreases as the pH increases, which indicates that at the beginning of the experiment there was Th in colloidal form. In future works, it would be necessary to develop experiments on Th sorption for constant pH, thus it would be possible to establish K_d on the basis of those conditions.

As expected, the Ce concentration levels in solution remain relatively constant since the beginning; it means that in the test conditions there have been no dissolution of Ce

Table I. Summary of the values calculated on Ce dioxide colloids and pellet.

	Colloids	Pellet
a / kg·l^{-1}	$(1.00 \pm 0.02)\cdot10^{-3}$	$(55.9 \pm 1.4)\cdot10^{-4}$
C_A / M	$(205.72 \pm 0.14)\cdot10^{-4}$	$(153.26 \pm 0.15)\cdot10^{-4}$
$[H^+]_{ref}$ / M	$(175.43 \pm 0.14)\cdot10^{-4}$	$(158.07 \pm 0.15)\cdot10^{-4}$
C_B / M	$(41.82 \pm 0.11)\cdot10^{-4}$	$(39.66 \pm 0.11)\cdot10^{-4}$
$[OH^-]_{ref}$ / M	$(26.63 \pm 0.12)\cdot10^{-4}$	$(26.63 \pm 0.12)\cdot10^{-4}$
Q / mol·kg^{-1}	1.51 ± 0.04	$(31.90 \pm 0.93)\cdot10^{-2}$
S_e / m^2·g^{-1}	17.5 ± 1.8	84.74 ± 1.82 (N$_2$)
		$(21.96 \pm 0.68)\cdot10^{-2}$ (Kr)
Total number of coordination sites	$(36.4 \pm 5.4)\cdot10^{18}$	$(43.0 \pm 1.8)\cdot10^{18}$ (N$_2$)
		$(43.0 \pm 2.3)\cdot10^{18}$ (Kr)
Coordination sites density / nm^{-2}	52.0 ± 5.5	2.3 ± 0.1 (N$_2$)
		875 ± 37 (Kr)

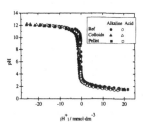

Figure 6 Experimental potentiometric titration curves (reference, suspension of dispersed CeO$_2$ colloidal and pellet) in 0.1M NaClO$_4$ M.

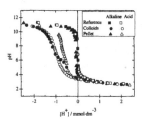

Figure 7 Zoom of the Figure 8 close to the zero charge point region.

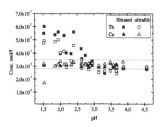

Figure 8 [Th] and [Ce] (filtered and ultrafiltered samples) in Th sorption experiments as a function of pH.

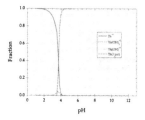

Figure 9 Aqueous Th(IV) speciation curves as a function of pH for ionic strength 0.1M and [Th] =6·10^{-7}M.

CONCLUSIONS

On one hand, the knowledge of surface sites density is essential because it is a controlling parameter in dissolution and precipitation processes and, on the other hand, surface complexation models require input this parameter, but no works related with CeO_2 pellet was found to compare our results. From these results is possible to conclude that ceria surface has certain tendency to retain protons on its surface. However, to refine the results, potentiometric titration at different ionic strengths should be made and use other techniques as well.

Because of Pu solubility is higher than Ce one in aqueous solutions, experiments with Pu should also be done to observe the effects on Th adsorption, as well as Th sorption essays at constant pH to get the distribution coefficient.

REFERENCES

1.   K. D. Weaver and J. S. Herring, *Nuclear Technology* 143, 22 (2003).
2.   L. C. Walters, D. L. Porter and D.C. Crawford, *Progress in Nuclear Energy* **40**, 513 (2002)
3.   R. P. C. Schram and F. C. Klaassen, *Progress in Nuclear Energy* **49**, 617 (2007).
4.   C. Kuetahyali, *Report No. JRC-ITU-TN-2007/79*, 2007.
5.   R. C. Ewing, *Progress in Nuclear Energy* **49**, 635 (2007).
6.   G .R. Choppin, *Marine Chemistry* **99**, 83 (2006)
7.   E. Östhols, PhD. Thesis, Royal Institute of Technology. Stockholm, Sweden, 1994
8.   M. H. Bradbury and B. Baeyens, *Geochim. Cosmochim. Acta* **69**, 875-892 (2005).
9.   E. Östhols, *Geochim. Cosmochim. Acta* **59**, 1235 (1995).
10.   E. Iglesias and J. Quiñones, *App. Surf. Science* **254,** 6890 (2008)

Mater. Res. Soc. Symp. Proc. Vol. 1124 © 2009 Materials Research Society　　　　　1124-Q10-08

In Situ Radiation Damage Studies of $Ca_3Zr_2FeAlSiO_{12}$ and $Ca_3Hf_2FeAlSiO_{12}$

Karl R. Whittle[1], Mark G. Blackford[1], Gregory R. Lumpkin[1], Katherine L. Smith[1], and Nestor J. Zaluzec[2]

[1] Institute of Materials Engineering, Australian Nuclear Science and Technology Organisation, PMB1, Menai, NSW 2234, Australia

[2] Argonne National Laboratory, 9700 South Cass Avenue, Argonne, IL 60439, USA

ABSTRACT

Garnets, $A_3B_2C_3O_{12}$, are considered to be potential host phases for the immobilization of high-level nuclear waste as they can accommodate a number of elements of interest, including Zr, Ti and Fe. The naturally occurring garnet, kimzeyite, $Ca_3(Zr,Ti)_2(Si,Al,Fe)_3O_{12}$, can contain ~30wt% Zr. An understanding of the radiation tolerance of these materials is crucial to their potential use in nuclear waste immobilization. In this study two synthetic analogues of kimzeyite of composition $Ca_3Zr_2FeAlSiO_{12}$ and $Ca_3Hf_2FeAlSiO_{12}$ were monitored in situ during irradiation with 1.0 MeV Kr ions using the intermediate voltage electron microscope-Tandem User Facility (IVEM) at Argonne National Laboratory. The structure of these materials was previously determined by neutron diffraction and ^{57}Fe Mössbauer spectroscopy. $Ca_3Zr_2FeAlSiO_{12}$ and $Ca_3Hf_2FeAlSiO_{12}$ have very similar structural properties with cubic Ia3d symmetry, the only significant difference being the presence of Zr and Hf, respectively, on the 6 coordinated B sites.

INTRODUCTION

Naturally occurring garnets are found with a large range of elemental compositions with the cations located across more than one site in the lattice, e.g., schorlomite. Garnets can be prepared containing U, Th and lanthanide elements; as such they make ideal candidates for the long-term storage of actinide waste. Current research in new materials for actinide waste tend to contain titanium and zirconium, this was originally due to the existence of compatible Ti phases for high-level waste, e.g., hollandite, pyrochlore and zirconolite. Zirconium has been added in recent research to enhance the stability to radiation damage. In nature there are two garnets that make ideal candidates for waste form research that have high levels of Ti^{4+} (schorlomite) or Zr^{4+} (kimzeyite), with kimzeyite and simplified analogues being the focus of this work. A schematic of the garnet structure is shown in Figure 1.

Kimzeyite $Ca_3(Zr,Ti)_2(Si,Al,Fe)_3O_{12}$ [1-5] was originally reported in 1961 for a sample from Magnet Cove in Arkansas, it is a naturally occurring garnet that contains high levels of Zr (~30wt% ZrO_2). There have been two subsequent analyses published based on samples from the Aeolian Islands, and the Sabatini Volcanic district, both in Italy. All three analyses give similar values for elemental composition and crystal parameters. As these garnets are naturally occurring with good long-term stability, they are ideal to study as potential storage media for radioactive nuclear waste.

In this work we have irradiated $Ca_3Zr_2FeAlSiO_{12}$ and $Ca_3Hf_2FeAlSiO_{12}$, which have previously been studied to determine the cation location within the lattice using combined Mossbauer and neutron diffraction techniques. This work found that the Zr/Hf located on the B-site (in octahedral coordination with oxygen) with Fe/Al/Si locating on the C-site (in tetrahedral coordination with oxygen). The results from the synthetic samples were then compared with electron channeling analysis of naturally occurring kimzeyite.

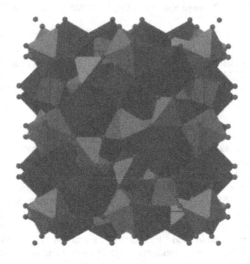

Figure 1. Schematic of the unit cell of garnet, the A-site cations are the dark blue polyhedra, the B-site cations are grey, and the C-site cations are green. The image is projected down the [100] direction.

EXPERIMENTAL

Stoichiometric mixtures of CaCO$_3$ (Alfa-Aesar, 99.9%), SiO$_2$ (Aldrich, 99.9%), Al$_2$O$_3$ (Aldrich, 99.9%), Fe$_2$O$_3$ (Aldrich, 99.9%), ZrO$_2$ (Alfa-Aesar, 99.5%) and HfO$_2$ (Alfa-Aesar, 99.5%), were ground together in an agate ballmill for 60min. The samples were dried and heated as pellets for 24 h at 850 °C to decarbonate the CaCO$_3$ in situ. After heating the samples were ground and calcined as pellets at 1400 °C for 48 h, placed on α-Al$_2$O$_3$ plates. After cooling, the samples were ground to a fine powder sufficient to pass through a 38 μm sieve. Once ground the phase purity was checked using laboratory X-ray powder diffraction. The chemical and structural analysis are reported in Whittle et al [1].

TEM samples were prepared by crushing small ceramic fragments in methanol and collecting the suspension on holey carbon coated copper grids. Samples were characterized prior to the irradiation experiments using a JEOL 2000FXII TEM operated at 200 kV and calibrated for selected area diffraction over a range of objective lens currents using a gold film standard. The compositions of the grains were checked by thin film EDX analysis procedures using a Link ISIS energy dispersive spectrometer attached to the TEM. Ion irradiation experiments with 1.0 MeV Kr ions were carried out *in situ* at the IVEM-Tandem User Facility at Argonne National Laboratory using a Hitachi TEM interfaced to a NEC ion accelerator [6]. All TEM observations were carried out using an accelerating potential of 300 kV. Ion irradiations were performed at variable temperature, both cryogenic and raised, with the electron beam of the TEM turned off, using a flux of 3.125×10^{14} ions cm^{-2} s^{-1} per count within a beam of ~ 2 mm diameter. Each sample was irradiated using incremental irradiation steps and selected grains were observed using bright field imaging and selected area electron diffraction after each irradiation step. The critical amorphization fluence (F_c) was determined from the last fluence increment in which weak Bragg diffraction spots were observed and the next increment for which only diffuse rings occur in the diffraction pattern. For each sample, we determined F_c from the average of several grains. For a more detailed description of the irradiation procedures, readers should consult recently published work. [7]

RESULTS

The numerical results for each sample are shown in Table 1, and Figure 2.

Table I - Results obtained from the irradiation of $Ca_3Zr_2FeAlSiO_{12}$ and $Ca_3Hf_2FeAlSiO_{12}$ at different temperatures.

$Ca_3Zr_2FeAlSiO_{12}$			$Ca_3Hf_2FeAlSiO_{12}$		
Temp. (K)	Fluence (x10^{14} ions cm^{-2})	Error (x10^{14} ions cm^{-2})	Temp. (K)	Fluence (x10^{14} ions cm^{-2})	Error (x10^{14} ions cm^{-2})
302	2.6	0.4	302	3.0	0.5
500	3.3	0.2	500	3.1	0.6
750	3.4	0.3	850	4.1	0.3
850	3.9	0.3	1050	7.7	2.0
1000	7.1	1.5			

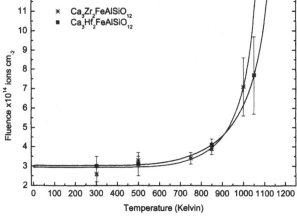

Figure 2. Critical fluence against temperature for $Ca_3Zr_2FeAlSiO_{12}$ and $Ca_3Hf_2FeAlSiO_{12}$, the curves are fitted using Equation 1.

239

The critical temperature (T_c) for each sample was determined using Equation (1),

$$Fc = \frac{Fc_0}{1 - \exp\left[\left(\frac{Ea}{k_b}\right)\left(\frac{1}{Tc} - \frac{1}{T}\right)\right]} \qquad (1)$$

where F_c is the critical amorphization fluence, F_{c0} is the critical fluence at 0 K, E_a is the activation energy for recovery of radiation damage, k_b is Boltzmann's constant, T_c is the critical temperature above which the specimen remains crystalline, and T is the temperature of the experiment. The critical fluence extrapolated to 0 K, critical temperature, and activation energy for recovery of damage were determined by non-linear least squares refinement of the data, weighted according to $1/\sigma^2$, where σ is the reported error on the critical fluence. Allowing all three parameters to vary, the numerical results of these refinements are reported in Table 1. This equation is known to underestimate the expected energy of activation for the recovery process [8], as a result a second equation is used to determine Ea, shown in Equation 2.

$$Tc = \frac{Ea}{k_b \ln\left[\frac{Fc_0 \nu}{\phi}\right]} \qquad (2)$$

where ν is the effective attempt frequency, F_{c0} is the reciprocal of the cross section for amorphization ($\sigma_a = 1/F_{c0}$), and ϕ is the ion flux. This equation can be rearranged to calculate E_a using the data listed in Table 1. Thus we have calculated a range of E_a values for the samples by assuming a range of ν values from 10^9 to 10^{12} s^{-1}. The results obtained using this procedure suggest that the activation energies range from about 0.4 to 1.6 eV.

The analyses of data collected from the irradiations are shown in Table 2 for both $Ca_3Zr_2FeAlSiO_{12}$ and $Ca_3Hf_2FeAlSiO_{12}$.

Table II - Results obtained from analysis of the numerical data in Table 1, using equations 1 and 2. The values in brackets are the errors from analysis using equation 1. The range of values for Ea using Eqn 2, derives from frequencies of 10^9-10^{12}.

System	T_c (K)	F_{c0} (ions cm^{-2})	E_a (eV) – Eqn 1	E_a (eV) – Eqn 2
$Ca_3Zr_2FeAlSiO_{12}$	1129(49)	2.9 (0.25)	0.40 (0.15)	3.2-3.9
$Ca_3Hf_2FeAlSiO_{12}$	1221(9)	3.0 (0.34)	0.32 (0.02)	3.5-4.2

The values obtained from the analysis of the recorded data are in broad agreement with those previously published by Utsunomiya et al [9-11], who found in their studies of natural garnets and synthetic Fe-containing garnets that the critical temperatures for amorphization range from 1030 to 1130 K. There are some differences, but they are in broad agreement.

As can be seen in both Table 2 and Figure 1 there is a difference between both systems, the Tc's differ by ~100 K, and the energy of activation is also higher in the $Ca_3Hf_2FeAlSiO_{12}$. This implies that both systems are similar in tolerance but that the Hf-based sample is slightly less resistant to damage. As both structures are essentially identical [1], there is likely to be no structural implications on stability. Therefore the small changes in radiation resistance are likely linked to the change in composition from Zr-based to Hf-based.

If the literature is examined for radiation resistance studies of system where Zr has been exchanged for Hf, a similar set of results can be seen. For example in the systems $La_2Zr_2O_7$ and $La_2Hf_2O_7$, exchange of Zr by Hf increased [12] the Tc from 339(49) K to 563(10) K, an increase of 224 K. Since again in this system the structural changes were minimal, and as such could be considered identical, the change from Zr to Hf has induced less resistance to damage.

An explanation for this change can be attributed to changes within the orbital overlaps between the Zr/Hf and O octahedra. In the $Ln_2Ti_2O_7$ systems, it can be seen that the degree of orbital overlap between the Ti-3d/Ln-5d and O-2p changes across the period[13]. When the systems have a high T_c, e.g. $Gd_2Ti_2O_7$, there is no driving force for the two networks to be separate, i.e. the Ti-3d/O-2p and Ln-5d/O-2p orbitals are similar in energy, where as in systems where there is a driving force, e.g. $Lu_2Ti_2O_7$ where there is an energy gap the result is a low Tc.

If this approach is applied to the garnet structure, a similar explanation can be postulated. When Hf replaces Zr, the orbital overlaps will change as the number of orbitals is increased. Since such overlaps not only dictate structural stability but also drive the structure that is formed, where there is an energy gain from being ordered the structure is likely to be ordered, where there is not, there is unlikely to be order. Since the effects observed here match those observed when $La_2Zr_2O_7$ and $La_2Hf_2O_7$ are irradiated, this is a possible explanation. The degree of difference in tolerance is likely to be smaller between $Ca_3Zr_2FeAlSiO_{12}$ and $Ca_3Hf_2FeAlSiO_{12}$, than between $La_2Zr_2O_7$ and $La_2Hf_2O_7$ as the composition change is smaller in the garnets than in the pyrochlores. In the garnets, there are also the additional effects of Ca on the dodecahedral site and Fe, Al, Si on the tetrahedral site. The small change in T_c between the two garnet compounds is potentially within the experimental error of measurements, and is the subject of further examinations.

CONCLUSIONS

The critical temperatures for amorphization of the synthetic garnets $Ca_3Zr_2FeAlSiO_{12}$ and $Ca_3Zr_2FeAlSiO_{12}$ were found to be in general agreement with previous studies of natural garnets and polyphase, garnet bearing nuclear waste form materials. The effect of replacing Zr in $Ca_3Zr_2FeAlSiO_{12}$ by Hf has increased the critical temperature for amorphisation by ~ 100 K. This small change can be explained by the change from Zr to Hf and the inherent change in the internal chemical bonding. This change agrees with previously published work on $La_2Zr_2O_7$ and $La_2Hf_2O_7$.

ACKNOWLEDGMENTS

The authors thank the IVEM-Tandem Facility staff at Argonne National Laboratory for assistance during the ion irradiation work. The IVEM-Tandem Facility is supported as a User Facility by the U.S. DOE, Basic Energy Sciences, under contract W-31-10-ENG-38. We acknowledge financial support from the Access to Major Research Facilities Programme (a component of the International Science Linkages Programme established under the Australian Government's innovation statement, Backing Australia's Ability). Part of this work was supported by the Cambridge-MIT Institute (CMI), British Nuclear Fuels Limited (BNFL), and EPSRC Grant (EP/C510259/1) to G.R. Lumpkin.

REFERENCES

1. K. R. Whittle; G. R. Lumpkin; F. J. Berry; G. Oates; K. L. Smith; S. Yudintsev; N. J. Zaluzec, *Journal Of Solid State Chemistry* **2007**, 180, (2), 797-803.
2. E. Schingaro; F. Scordari; F. Capitanio; G. Parodi; D. C. Smith; A. Mottana, *European Journal of Mineralogy* **2001**, 13, (4), 749-759.
3. L. Lupini; C. T. Williams; A. R. Woolley, *Mineralogical Magazine* **1992**, 56, (385), 581-586.
4. R. Munno; G. Rossi; C. Tadini, *American Mineralogist* **1980**, 65, 188-191.
5. C. Milton; B. L. Ingram; L. V. Blade, *American Mineralogist* **1961**, 46, 533-548.
6. C. W. Allen; L. L. Funk; E. A. Ryan in: *New Instrumentation in Argonne's HVEM-Tandem, facility: expanded capability for in situ beam studies*, Ion-Solid Interactions for Materials Modification and Processing, 1996; D. B. Poker; D. Ila; Y. T. Cheng; L. R. Harriott; T. W. Sigmon, (Eds.) Materials Research Society: 1996; pp 641-646.
7. G. R. Lumpkin; K. L. Smith; M. G. Blackford; B. S. Thomas; K. R. Whittle; N. A. Marks; N. J. Zaluzec, *Physical Review B* **2008**, 77, (21), 212401.
8. W. J. Weber, *Nuclear Instruments & Methods in Physics Research Section B- Beam Interactions with Materials and Atoms* **2000**, 166-167, 98-106.
9. S. Utsunomiya; S. Yudintsev; R. C. Ewing, *Journal of Nuclear Materials* **2005**, 336, (2-3), 251.
10. S. Utsunomiya; L. M. Wang; R. C. Ewing, *Nuclear Instruments & Methods in Physics Research Section B-Beam Interactions with Materials and Atoms* **2002**, 191, 600-605.
11. S. Utsunomiya; L. M. Wang; S. Yudintsev; R. C. Ewing, *Journal of Nuclear Materials* **2002**, 303, (2-3), 177-187.
12. G. R. Lumpkin; K. R. Whittle; S. Rios; K. L. Smith; N. J. Zaluzec, *Journal Of Physics-Condensed Matter* **2004**, 16, (47), 8557-8570.
13. K. R. Whittle; G. R. Lumpkin; K. L. Smith; M. G. Blackford; N. J. Zaluzec; E. J. Harvey in: *Radiation Tolerance of A₂Ti₂O₇ Materials - A Question of Bonding?*, Scientific Basis For Nuclear Waste Management XXX, Boston, US, 2007, 2006; D. S. Dunn; C. Poinssot; B. Begg, (Eds.) Materials Research Society: Boston, US, 2006; pp 0985-NN09-02.

Mater. Res. Soc. Symp. Proc. Vol. 1124 © 2009 Materials Research Society   1124-Q10-09

Characterisation of Ion Beam Irradiated Zirconolite for Pu Disposition

Martin C. Stennett[1], Neil C. Hyatt[1], Daniel P. Reid[1], Ewan R. Maddrell[2], Nianhua Peng[3], Chris Jeynes[3], Karen J. Kirkby[3], and Joseph C. Woicik[4]
[1]Immobilization Science Laboratory, Department of Engineering Materials, University of Sheffield, Sheffield, South Yorkshire, UK
[2]National Nuclear Laboratory, Sellafield, Cumbria, UK
[3]Ion Beam Centre, University of Surrey, Guildford, Surrey, UK
[4]Ceramics Division, National Institute of Standards and Technology, Gaithersburg, Maryland, USA

ABSTRACT

A number of mineral systems retain significant actinide fractions over geological timescales, rendering them metamict as a result of α-decay events. These systems can be viewed as natural actinide wasteforms and are a useful guide when designing synthetic materials for the disposition of actinide species arising from civilian and military nuclear activities. The effect of radiation damage on synthetic materials can be simulated by a number of different techniques including incorporation of short lived actinide species into the structures and irradiating samples with heavy charged particles. In this study a polycrystalline zirconolite sample was irradiated with 2 MeV Kr^+ ions at a fluence of 5 x 10^{15} ions cm^{-2}. The irradiated surface area of the sample was analysed using grazing incidence X-ray diffraction (GIXRD) and grazing incidence X-ray absorption spectroscopy (GIXAS). Pristine samples were also analysed for comparative purposes. GIXRD gave information on the amorphous character of the irradiated surface and GIXAS was used to investigate the effect of irradiation on valence and coordination of key elements in the zirconolite structure.

INTRODUCTION

Research programmes have been underway for several decades, in both the USA [1] and Russia [2], to develop immobilisation matrices for separated plutonium. This work has resulting in number of crystalline ceramic phases being proposed as promising candidates. Civilian nuclear programmes in the UK over the last 50 years have generated over 100 tHm of separated plutonium some of which has been deemed surplus to UK nuclear fuel requirements and therefore requires immobilisation. A Nuclear Decommissioning Authority (NDA) funded programme has been ongoing for several years to underpin scientifically the case for immobilisation of UK separated plutonium [3-6]. This required the development of a suitable immobilisation matrix (or wasteform), a suitable process for wasteform manufacture and demonstration of the long term integrity of the wasteform product.

During the lifetime of the wasteform in the disposal environment, a substantial amount of plutonium will undergo α-decay, resulting in several thousand atomic displacements per decay event and ultimately leading to amorphisation of the wasteforms crystalline matrix. It is therefore important to assess the effect amorphisation has on key wasteform properties and product integrity. One method of simulating radiation damage over disposal timescales is to accelerate the rate of damage, by the incorporation of short half life actinide species such as ^{238}Pu or ^{244}Cm [7, 8]. Another method is *in-situ* ion-beam irradiation on thin specimens coupled with transmission electron microscopy. The critical ion fluence (ions cm^{-2}) required to cause amorphisation of a TEM specimen under a heavy ion beam flux, at a given temperature, may be determined by inspection of selected area diffraction patterns [9-11]. Zirconolite has been shown to become amorphous at room temperature, when irradiated with 1.5 MeV Kr$^+$ ions at a fluence of approximately 4 x 10^{14} ions cm^{-2} [10]. A third alternative is to irradiate the surface of bulk samples with heavy ions to a known fluence and then analyse the specimens *ex-situ*. Interpretation of the results can however be problematic due to the technique only generating thin ($\sim$ 100 to 1000 nm) amorphous surface layers.

The effects generated in these accelerated tests can be compared with the effects of radiation damage in naturally occurring uranium and thorium bearing mineral specimens which has occurred over geological timescales. Structural studies on natural zirconolites have shown that specimens having experienced α-decay doses in the range of 10^{18}-10^{20} α cm^{-3} to be X-ray amorphous and exhibit diffuse rings rather than discrete spots when examined by selected area electron diffraction [12, 13]. These doses are consistent with studies of synthetic zirconolites doped with ^{238}Pu or ^{244}Cm [7, 8].

EXPERIMENTAL

The zirconolite (CaZrTi$_2$O$_7$) powder sample was synthesised by conventional solid state sintering using oxide (TiO$_2$, ZrO$_2$) and carbonate (CaCO$_3$) precursors. A 50 g powder batch was planetary milled (Retsch PM100) at 250 rpm for 2 hours in a 125 ml stainless steel pot using hardened Cr-steel balls and iso-propanol as the carrier fluid. The powder slurry was then dried for 16 hours at $\sim$ 100 °C. The resulting dry powder cake was sieved (250 μm mesh) and calcined in an alumina crucible, in air, for 10 hours at 1300 °C. The calcine was then sieved again and colliodally milled (Retsch PM100) at 500 rpm in a 125 ml stabilised zirconia pot with 3mm stabilised zirconia media and isopropanol as the carrier fluid. The slurry was dried as above and 20mm diameter pellets approximately 30 mm high were uniaxially pressed at 60 MPa in a hardened stainless steel die. The green pellets were then vacuum sealed in latex gloves and cold isostatically pressed at 200 MPa. The pellets were transferred onto stabilised zirconia setter plates and sintered at 1450 °C for 4 hours. The sintered pellets were sectioned using a low speed diamond saw to afford 1 mm thick cylindrical sections. These sections were then polished to an optical finish (1 μm) and thermally etched at 1300 °C for one hour, to reveal the grain boundaries, and relax out any surface stresses associated with the cutting proceedure.

Several of the samples were then ion beam irradiated with 2 MeV Kr$^+$ ions at a fluence of 5 x 10^{15} ions cm^{-2}. The ion beam irradiation experiments were performed in the Surrey Ion Beam Centre at the University of Surrey, Guildford. Ceramic sections were mounted onto a clean 4 inch Si wafer with conducting carbon tape. This Si wafer was loaded into the implantation chamber of 2 MeV van der Graaf implanter with a standard 4 inch carousel sample holder and a 4 inch aperture. The beam current of 2 MeV Kr$^+$ ions was maintained at $\sim$ 0.25 μA during the ion irradiation process.

GIXRD was performed on a Siemens D5000 diffractometer using Cu K$_\alpha$ (λ = 1.540 Å) radiation. Patterns were collected at fixed incidence angles between 0.5 and 5 ° θ over an angular range of 10 – 70 ° 2θ in steps of 0.02 ° with a count time of 0.5 ° min^{-1}.

GIXAS was performed on beamline X23A2 of the National Synchrotron Light Source (NSLS), Brookhaven National Laboratory (BNL), USA. Beamline X32A2 has a bending magnet source with an upwards reflecting, fixed exit Golovchenko-Cowan design; piezo-feedback stabilized single bounce harmonic rejection mirror monochromator. The experimental apparatus consisted of a radiation hutch, ion chambers for transmission and fluorescence measurements and adjustable horizontal and vertical exit slits. The beamline can operate at 4.9 – 30 keV, using a Si (311) monochromator crystal, with a resolution of 2×10^{-4} and a flux of 1×10^{10} photons per second (monochromator bandpass at 10 keV, 100mA, 2.5 GeV), a spotsize of 25 mm (horizontal) x 1.0 mm (vertical) and a total angular acceptance of 4 mrad. The beamline was tuned to the Ti K edge energy of 4.9664 keV for analysis of the zirconolite samples. In order to facilitate measurement at grazing incidence a specially designed sample stage was used which in addition to being able to translate in all three directions could also tilt the sample in relation to the plane of the incoming beam to within an accuracy of 0.1 ° (denoted by θ in the schematic diagram on the far right of Figure 1).

Figure 1. GIXAS setup on NSLS beamline X23A2: A) reference detector; B) transmitted beam detector; C) sample stage; D) fluorescence detector; sample indicated by arrow in centre image. Schematic diagram of experimental setup shown on far right hand side.

RESULTS AND DISCUSSION

The surfaces of the pristine and irradiated zirconolite samples were examined using an optical microscope. No obvious differences were observed in the microstructure; both consisted of equiaxed grains between 5 and 50 μm in diameter, Figure 2.

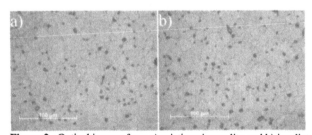

Figure 2. Optical images from a) pristine zirconolite and b) irradiated zirconolite samples.

The nature of the pristine and irradiated surface layers were probed using GIXRD. Calculation performed using SUSPRE [14] at the University of Surrey suggested the depth of the damaged surface layer from irradiation with 2 MeV Kr$^+$ ions to be 0.7 μm. Figure 3 shows diffraction patterns obtained from the pristine and irradiated samples as a function of angle of incidence. The presence of reflections in the patterns acquired at low incidence angle from the pristine zirconolite sample, although rather weak and diffuse in character, indicate the surface layer to be crystalline in nature. The absence of any reflections in the pattern obtained from the irradiated zirconolite sample indicate that the surface layer is amorphous in nature and lacks any long range crystalline order.

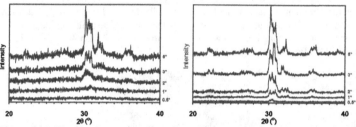

Figure 3. Diffraction patterns, collected at various grazing incidence, from irradiated (left) and pristine (right) zirconolite samples.

The X-ray absorption data in Figure 4 shows clear differences in the spectra of the pristine and irradiated samples. The two edge bands in the irradiated sample are broadened compared to the bands in the spectrum of the pristine sample. This indicates a reduction in long and medium range order in the irradiated sample [15]. This is due to the surface of the sample becoming amorphous as a consequence of the Kr$^+$ ion bombardment. The other obvious difference is an increase in the normalized height of the pre-edge feature in the irradiated sample, compared to the pre-edge feature in the spectra of the pristine sample. There is also a shift in energy of the pre-edge feature from 4971.0 eV for the pristine sample, to 4970.5 eV for the irradiated sample. Previous XAS studies of titanium oxides [16, 17] and titanium containing minerals [18] have shown that such changes in the pre-edge feature can be accounted for by changes in the co-ordination environment of the Ti atoms.

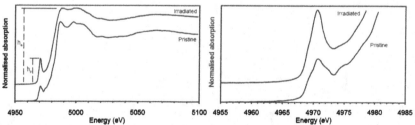

Figure 4. Ti K-edge X-ray absorption near edge spectra (XANES) of irradiated and pristine zirconolite samples. h_e = height of edge, h_p = height of pre-edge peak. Close up of pre-edge shown on left hand side.

A graph relating the normalized height of the pre-edge feature *versus* the shift in energy, to the co-ordination environment of the Ti atoms, is shown in Figure 5. This graph shows that in the pristine sample (filled circle no. 1) the Ti atoms are mainly in a 6 co-ordination environment. However, in the irradiated sample (filled circle no. 2), the datum is more consistent with Ti in a co-ordination environment of 5. Comparison of the Ti-edges obtained from metamict natural zirconolite sample show that the local-structure of the metamict state produced by fast ion irradiation is essentially identical to that derived from accumulated damage in natural U and Th bearing mineral analogues [19-21].

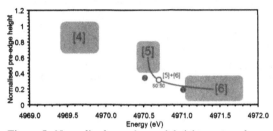

Figure 5. Normalised pre-edge peak height vs. pre-edge energy. Grey regions indicate bounding regions for different Ti^{4+} environments; the number in the square brackets indicates co-ordination number.

CONCLUSIONS

The XANES spectra obtained for the pristine and irradiated zirconolite samples show a reduction in long and medium range order of the irradiated sample, suggesting a more amorphous structure. This can be accounted for by the bombardment of the sample with Kr^+ ions. There is also the change in pre-edge feature which points towards a dominant 5-fold co-ordination environment in the irradiated sample compared to 6-fold co-ordination in the pristine sample. Work is currently underway to further investigate these issues.

ACKNOWLEDGMENTS

This work was carried out as part of the TSEC programme KNOO and as such we are grateful to the EPSRC for funding under grant EP/C549465/1.

REFERENCES
1. R. C. Ewing, W. J. Weber and W. Lutze, in *Disposal of Weapons Plutonium*, edited by E. R. Merz and C. E. Walter (Kluwer Academic Publishing, The Netherlands, 1996), p 65.
2. E. B. Anderson and B. E. Burakov in *Scientific Basis for Nuclear Waste Management XXVII*, edited by V. M. Oversby and L. O. Werme (Mater. Res. Soc. Symp. Proc. 807, Warrendale, PA, 2004), p. 207.
3. M. C. Stennett, N. C. Hyatt, W. E. Lee, and E. R. Maddrell in *Environmental Issues and Waste Management Technologies in the Ceramic and Nuclear Industries XI*, edited by C. C.

Herman, S. Marra, D. R. Spearing, L. Vance, and J. D. Vienna (Ceram. Trans. 176, Westerville, 2006), p. 81.

4. M. C. Stennett, N. C. Hyatt, E. R. Maddrell, F. G. F. Gibb, G. Moebus and W. E. Lee in *Scientific Basis for Nuclear Waste Management XXIX*, edited by P. Van Iseghem (Mater. Res. Soc. Symp. Proc. 932, Warrendale, PA, 2006), p 623.

5. M. C. Stennett, E. R. Maddrell, C. R. Scales, F. R. Livens, M. Gilbert and N. C. Hyatt in *Scientific Basis for Nuclear Waste Management XXX*, edited by D. S. Dunn, C. Poinssot, B. Begg (Mater. Res. Soc. Symp. Proc. 985, Warrendale, PA, 2007), p 145.

6. M. C. Stennett, N. C. Hyatt, M. Gilbert, F. R. Livens and E. R. Maddrell in *Scientific Basis for Nuclear Waste Management XXXI*, edited by W. E. Lee, J. W. Roberts, N. C. Hyatt and R. W. Grimes (Mater. Res. Soc. Symp. Proc. 1107, Warrendale, PA, 2008), p 413.

7. F. W. Clinard Jr., D. L. Rohr and R. B. Roof. J Nucl. Mater. 126, 245 (1984).

8. W. J. Weber, L. W. Wald and Hj. Matzke. J. Nucl. Mater. 138, 196 (1986).

9. R. C. Ewing and L. M. Wang. Nucl. Inst. Meth. Phys. Res. B65, 319 (1992).

10. K. L. Smith, N. J. Zaluzec and G. R. Lumpkin. J. Nucl. Mater. 250, 36 (1997).

11. S. X. Wang, G. R. Lumpkin, L. M. Wang and R. C. Ewing. Nucl. Inst. Meth. Phys. Res. B166-167, 293 (2000).

12. G. R. Lumpkin, K. L. Smith and R. Giere. Micron. 28, 57 (1997).

13. G. R. Lumpkin, R. C. Ewing, B. C. Chakoumakos, R. B. Greegor, F. W. Lytle, E. M. Foltyn, F. W. Clinard Jr, L. A. Boatner and M. M. Abraham. J. Mater. Res. 1, 564 (1986).

14. R. P. Webb in *Practical Surface Analysis*, edited by D. Briggs and M. P. Seah (John Wiley & Sons, Inc., New York, 1983), p. 604.

15. N. J. Hess, B. D. Begg, S. D. Conradson, D. E. McCready, P. L. Gassman and W. J. Weber. J. Phys. Chem. B. 106, 4663 (2002).

16. K. E. R. England, J. M. Charnock, R. A. D. Patrick and D. J. Vaughan. Mineral. Mag. 63, 559 (1999).

17. G. N. Greaves, N. T. Barrett, G. M. Antonini, F. R. Thornley, B. T. M. Willis and A. Steel. J. Am. Chem. Soc. 111, 4313 (1989).

18. F. Farges. Am. Mineral. 82, 44 (1997).

19. F. Farges. Geo. Cosmochim. Acta. 60, 3023 (1996).

20. F. Farges, G. E. Brown and J. J. Rehr. Phys. Rev. 56, 1809 (1997).

21. G. A. Waychunas. Am. Mineral. 72, 89 (1987).

Engineered Barrier Systems
and the Near Field

Mater. Res. Soc. Symp. Proc. Vol. 1124 © 2009 Materials Research Society   1124-Q05-02

Identification of Oxygen-Depleting Components in MX-80 Bentonite

Torbjörn Carlsson and Arto Muurinen
VTT Technical Research Centre of Finland, Otakaari 3J, Espoo, FI-02044 VTT, Finland

ABSTRACT

After closure, the near-field of a nuclear waste repository contains large amounts of oxygen in tunnels and deposition holes. The bentonite buffer/backfill will contain oxygen as a gas phase in unsaturated pores as well as dissolved gas in porewater. The redox conditions in the bentonite filling after post-closure will change towards reducing conditions. In the initial stage, the development of the redox state is mainly governed by the depletion of oxygen. The main mechanisms of oxygen depletion in the bentonite are: 1) diffusion into the surrounding rock and 2) reactions with accessory minerals and by microbial aerobic consumption of organic matter [1,2]. The reactions leading to oxygen depletion are not, however, well understood. The objective of this work was to gather new information concerning oxygen depletion in MX-80. This was done by measuring oxygen depletion and changes in the redox state in suspensions of 1) MX-80, 2) a heavy fraction of MX-80, or 3) a light fraction of MX-80.

INTRODUCTION

The KBS-3 concept for the long-term disposal of spent nuclear fuel aims to keep the waste isolated in deep crystalline bedrock. In the Finnish case, the depth of the planned repository is between 400 and 700 meters [3]. Briefly, the spent fuel is placed in copper-coated iron canisters, which are placed in bentonite-filled holes in the bedrock. Details concerning the KBS-3 concept are found in, e.g. [3]. The final choice has not yet been made concerning the bentonite to be used, but MX-80 bentonite has long been used as a reference material in scientific research and it has also been used in the present study.

One topic of concern in the safety assessments of nuclear waste repositories is the redox condition, which may have an impact on, e.g., waste canister corrosion and radionuclide sorption and migration [4]. The conditions in the repository will be oxidizing directly after closure, but the redox conditions will thereafter slowly change towards being less oxidizing and, finally, reducing.

Lazo et al. [5] mention that the redox changes are closely related to questions like: What happens to O_2 initially present in the backfill/buffer at repository closure? How long does it take before oxygen is consumed in the buffer? Oxygen depletion in a repository had been modelled previously, with the results indicating almost complete oxygen depletion after 7-290 years [6]. However, since experimental confirmation was lacking, Lazo et al. started a series of experiments where oxygen depletion was studied in MX-80 and Montigel suspensions. One of the findings was that dissolved oxygen (DO) disappears within a few days.

The objective of the present study was to take a first step towards the identification of those components that are involved in oxygen consumption in MX-80 suspensions. This was done by dividing MX-80 into a heavy fraction and a light fraction by using centrifugation and sedimentation procedures. Known amounts of these fractions were subsequently placed in aerated water. The DO concentration and the redox potential were thereafter measured as functions of time.

EXPERIMENT

This section discusses two types of experiments carried out in various solutions or suspensions: oxygen measurements and redox measurements. The former experiments focussed on studying the possible depletion of oxygen, while the latter experiments focussed on possible redox changes. In both cases, the suspensions were made using Milli-Q water containing dissolved oxygen and either MX-80 bentonite or a sub-fraction of this bentonite.

Bentonite materials

The bentonite materials used were 1) untreated, commercially available MX-80 Wyoming bentonite, 2) a light MX-80 fraction (LF), and 3) a heavy MX-80 fraction (HF). The LF and the HF materials were obtained from anaerobic MX-80 by repeated centrifugation and sedimentation under N_2 atmosphere The anaerobic MX-80 used was obtained by storage under N_2 atmosphere for several months prior to use and was for the present purposes considered as free from gaseous oxygen. The composition of MX-80 was recently thoroughly determined by analyzing XRD diffractograms with Siroquant quantitative software [7]. Seventeen minerals were identified in the MX-80, with montmorillonite being the dominant mineral. In this context it is, however, sufficient to give the shortened version of Karnland's et al. data as found in [8]: montmorillonite 87%, quartz 5%, feldspar + mica 7 %, pyrite 0.07%, and gypsum 0.7%.

The composition of the HF material was analyzed in the same way as described above, but with another program (Panalytical X'Pert software). A semi-quantitative analysis of the identified phases was made by means of experimentally measured absorption coefficients (accuracy about ±5%). The composition of the HF material was smectite 10%, quartz 30%, K-feldspar 10%, plagioclase 40%, pyrite 5%, and calcite 5%. Finally, the mineral in the LF material was almost completely montmorillonite.

Solutions and suspensions

All solutions were made using Milli-Q water. Solutions devoid of dissolved oxygen were prepared by N_2 flushing and used, e.g., during centrifugation and sedimentation of MX-80. Air-saturated water was put into the vessels containing MX-80, HF, or LF by means of a syringe. The bentonite and water were mixed by magnetic stirring, but 15-20 minutes were sometimes needed before a more or less homogeneous suspension could be observed. However, the period during which the sample was clearly inhomogeneous was short in comparison to the duration of the whole experiment and is therefore of no importance here.

Electrodes

The pH was measured with IrO_x electrodes of the type described by Yao et al. [9]. The redox potentials were measured with Au and Pt wires. The reference electrodes used were commercially available Ag/AgCl electrodes (LF-2) from Innovative Instruments, Inc. Further details can be found in [10].

Experimental setup

The oxygen measurements took place in a closed polypropylene vessel containing initially a stirring magnet, solid material and an Orbisphere 3600 oxygen analyzer (see Figure 1). The experiments were started by quickly inserting aerated Milli-Q water with a syringe through a small hole in the vessel. The syringe was then removed and the hole was carefully closed.

The liquid phase occupied in most cases less than half of the vessel, while the rest of the vessel volume was occupied by N_2, water vapour and the oxygen meter. The latter was entirely kept in the gas phase, with its oxygen-sensitive membrane placed about 1 cm above the surface of the liquid. It should be explicitly pointed out that the meter was placed above the liquid, but measured the concentration of dissolved oxygen in the liquid. This was possible since the gas in the closed system was fully saturated with water vapour and the oxygen activity, therefore, was the same in both gas and liquid. A minor disadvantage with the experimental setup was that most of the oxygen was present in the gaseous phase, in accordance with Henry's Law, and only a fraction in the liquid. However, this was not a problem, since the Orbisphere analyzer could measure the DO concentrations ranging from 0.1 ppb to 8 ppm, which was sufficient for the present purposes. The advantage of the design was that it offered a way to avoid any direct contact between the bentonite suspension and the sensitive membrane in the sensor. Such contact was judged to have possible negative effects on the results.

The redox measurements were performed in a straightforward manner in vessels containing four electrodes attached to a data logger (see Figure 1). Details concerning the electrodes are found elsewhere [10].

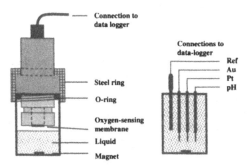

Figure 1. Schematic drawings of the experimental setup during oxygen measurements (left) and redox measurements (right). Vessel dimensions: height 63 mm, diameter 36 mm.

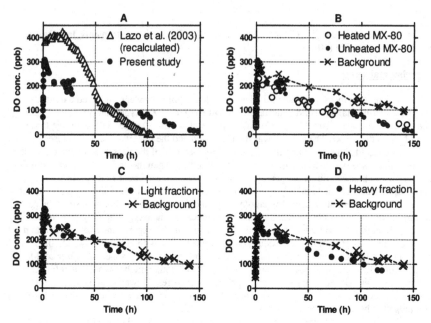

Figure 2. Depletion of dissolved oxygen vs. time in four aqueous suspensions: A) MX-80 as delivered compared with re-calculated data from Lazo et al. (2003), B) MX-80 heated to 120 °C, C) Light fraction of MX-80, D) Heavy fraction of MX-80. Background data refer to aerated water without solids.

RESULTS AND DISCUSSION

DO measurements

The experimental setup was initially tested against a study described in [5]. The chosen test case concerned the depletion of dissolved oxygen (DO) in an MX-80 suspension containing 18.3 g/L MX-80. However, the published data referred to measurements performed in a vessel completely filled with bentonite suspension, while in our case the vessel contained both liquid and gas. In order to enable comparison, the data in [5] were straightforwardly converted to what they should have been, had the experiment been performed with our setup. The test in this study was run in a glove-box under N_2 under conditions similar to those reported in [5]. The two studies gave roughly the same results (Figure 2A). It was found that DO was consumed after 100 h, while the corresponding figure achieved in this study was 150 h. The agreement was considered satisfactory for the present purposes and indicated that the experimental arrangement functioned properly.

The effect of heating with regard to the bentonite's DO-depleting capability was subsequently studied in an experiment that copied the one described above. The material used this time was MX-80 that had been heated under N_2 at 120 °C for 65 h. It was found that the DO depletion was the same, irrespective of whether the MX-80 had been heated (Figure 2B). This indicates that the heating did not have any significant effect on the MX-80. The observation is supported by XRD-diffractograms of the heated and unheated MX-80 which were almost identical, although the former exhibited a slightly higher crystallinity. The bacterial activity in the two samples may or may not have been the same. Experiments by Masurat et al. [11] indicate that the heat tolerance limit for sulphate reducing bacteria is somewhere between 100 °C and 120 °C.

The slope of the background curve (Figure 2) deserves some comments. The curve shows the DO concentration of water containing DO but no solid. If no oxygen consumption took place and if the vessel (polypropylene) was absolutely tight, there should be no change in measured DO. However, the material of the vessel is known to be slightly permeable. This contributes to the appearance of the background curve. However, as long as all experiments are performed in the same manner, and the results are designed for comparative studies as in the present case, the background behaviour is not a serious problem.

Finally, the DO depletion was also studied in light fractions (LF) and heavy fractions (HF) of MX-80. The results indicate that no significant DO depletion takes place in the LF material (Figure 2C), while on the other hand there is a clear indication of DO depletion in the HF material (Figure 2D).

Redox measurements

The redox conditions were studied in three suspensions containing DO, Milli-Q water and MX-80 (1.1 g/60 mL), HF (0.4 g/60 mL) or LF (0.3 g/L), and compared with a background sample consisting of DO and water (60 mL). Figure 3 shows the redox potential, Eh, in the four samples at start and after six days. The background sample exhibited a constant Eh during this time, while the samples containing solids all showed a marked decrease. The reason for this is not known to us. A comparison between Figures 2C and 3 shows that the Eh decreased also when no oxygen was consumed, which indicates that the decrease in Eh might be coupled to the presence of more than one redox-pair in MX-80.

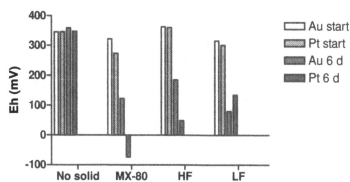

Figure 3. Eh values at the start and after six days in the following DO containing systems: 1) water, 2) MX-80 suspension, 3) HF suspension, and 4) LF suspension. "Au" and "Pt" in the legend refers to results achieved with Au- and Pt-electrodes, respectively.

CONCLUSIONS

This study shows data on dissolved oxygen depletion and on redox potential development in suspensions containing 1) MX-80, 2) a heavy fraction of MX-80, or 3) a light fraction of MX-80. The solids in the heavy fraction were smectite, quartz, K-feldspar, plagioclase, pyrite, and calcite, while the light fraction mainly consisted of montmorillonite. Oxygen depletion was observable in the suspensions with MX-80 and the heavy fraction, but not in those containing the light fraction. The Eh values decreased in the three suspensions, but not in the background sample containing only DO and water. The reason for the decrease in Eh is presently not fully known.

ACKNOWLEDGMENTS

The work was financed by Posiva Oy and by the Swedish Nuclear Fuel and Waste Management Co. (SKB). Petri Korkeakoski (Posiva Oy) is thanked for useful comments. The encouragement by Lars Werme (SKB) and Marjut Vähänen (Posiva Oy) is appreciated. Kristian Lindqvist (GTK) is thanked for XRD analyses and Nigel Kimberley (IN-Kontext) for correcting the language.

REFERENCES

1. F. Grandia, C. Domènech, D. Arcos, and L. Duro, SKB Report R-06-106 (2006).
2. B. Pastina and P. Hellä, Posiva Report 2006-5 (2006).
3. Posiva Oy, Safety Case Plan, Posiva Report 2008-05 (2008).
4. M.J. Danielson, O.H. Koski, and J. Myers in *Scientific Basis for Nuclear Waste Management VII*, edited by G.L. McVay, (Mater. Res. Soc. Proc. **26**, North-Holland, New York, 1984) pp.153-160.
5. C. Lazo, O. Karnland, E.-L. Tullborg, and I. Puigdomenech in *Scientific Basis for Nuclear Waste Management XXVI*, edited by R.J. Finch and D.B. Bullen, (Mater. Res. Soc. Proc. 757, Warrendale, PA, 2003) pp. 643-648.
6. P. Wersin, K. Spahiu, and J. Bruno, SKB Report TR-94-02 (1994).
7. O. Karnland, S. Olsson, and U. Nilsson, SKB Report TR-06-30 (2006).
8. D. Arcos, F. Grandia, C. Domènech, A.M. Fernández, M.V. Villar, A. Muurinen, T. Carlsson, P. Sellin, and P. Hernán, *J. Contam. Hydrol.* **102**, 196-209 (2008).
9. S. Yao, M. Wang, and M. Madou, *J. Electrochem. Soc.* **148(4)**, H29-H36 (2001).
10. A. Muurinen and T. Carlsson, *Phys. Chem. Earth* **32**, 241-246 (2007).
11. P. Masurat, S. Eriksson, and K. Pedersen, *Appl. Clay Sci.* (2008) doi:10.1016/j.clay.2008.07.002

Mater. Res. Soc. Symp. Proc. Vol. 1124 © 2009 Materials Research Society   1124-Q05-03

Soft X-ray Spectroscopic Characterization of Montmorillonite

J-E Rubensson[1], F Hennies[2], L O Werme[1,3], O Karnland[4]
[1] Department of Physics and Materials Science, Uppsala University, Box 530, SE-75121 Uppsala, Sweden
[2] MAX-lab, Lund University, Box 118, SE-22100 Lund, Sweden
[3] Swedish Nuclear Fuel and Waste Management Co, Box 250, SE-10124 Stockholm, Sweden
[4] Clay Technology AB, IDEON Research Center, SE-22370 Lund, Sweden

ABSTRACT

Soft X-ray spectroscopy was applied to study a calcium bentonite from the Kutch area in India. We recorded the X-ray absorption spectra from the L-edge of calcium, silicon, and aluminum, and from K-edge of oxygen. The Ca absorption spectrum shows a quasi-atomic behavior, while the Si spectrum closely simulates the absorption spectrum of a pure silicon oxide. The O K spectrum shows a pre-peak, which is absent in the spectra of both the pure, bulk aluminum and silicon oxides. The Al L spectrum is complex and shows almost no resemblance to the absorption spectrum of aluminum oxides. The chemical state of the Al atoms (in octahedral coordination) must, thus, be quite different from what is common in the oxides. The obtained data show that soft X-ray spectroscopy is a promising technique for studying clay minerals. It is capable of supplying unique information that is complementary to information accessible using other techniques; especially, it can be used to determine the local electronic structure at various atomic sites in the complex samples.

INTRODUCTION

Bentonite is proposed in many countries, including Sweden, to be used as a buffer material in nuclear waste repositories. Bentonite is a geological term for soil materials with a high content of a swelling mineral, which usually is montmorillonite. The montmorillonite belongs to the smectite group, in which all members have an articulated layer structure. The desirable physical bentonite properties of the buffer, mainly swelling pressure and low hydraulic conductivity, are due to interaction between water and the montmorillonite in the bentonite. This interaction is affected by changes in the ion concentration in the groundwater and by changes in the mineral structure of the montmorillonite. The mineralogical stability of the montmorillonite is, therefore, of crucial importance for the performance of the buffer. Montmorillonite can be stable for hundreds of millions of years in its formation environment, but changes in the geochemical environment can lead to a relatively rapid change of the mineral structure.

Montmorillonite consists of nanostructures of silicon and aluminum oxide sheets, where typically one monolayer of octahedrally coordinated aluminum atoms with oxygen as ligands, is sandwiched between two monolayers of silicon oxide. The aluminum in the octahedral sheet is partly substituted, principally by Mg. The silicon in the tetrahedral sheet may partly be substituted, principally by aluminum. The substitutions result in a net negative charge of the montmorillonite layer in the range of 0.4 to 1.2 elementary charges per unit cell ($O_{20}(OH)_4$), and the octahedral charge is larger than the tetrahedral charge. (If the tetrahedral charge is larger than the octahedral charge, the mineral is referred to as beidellite.) The induced negative layer charge is balanced by cations (M), e.g., Ca and Na with the valence v, located in the interlayer space. The ideal montmorillonite formula may consequently be written:

$Si_{8-x} Al_x$ $Al_{4-y} Mg_y (Fe) O_{20} (OH)_4$ $M_{(x+y)/v} n(H_2O)$ (x<y) and 0.4 <x+y<1.2
tetrahedral   octahedral   anions   interlayer cations

and the structure may schematically be illustrated as in Figure 1 [1].

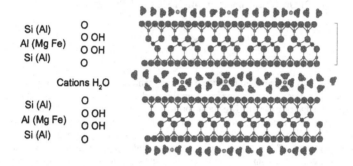

Si (Al)   O
Al (Mg Fe)   O OH
   O OH
Si (Al)   O

Cations H_2O

Si (Al)   O
Al (Mg Fe)   O OH
   O OH
Si (Al)   O

Figure 1. Schematic structure of montmorillonite [1]. A central layer of aluminum oxide is surrounded by a silicon oxide layer. Substitutions are frequent. The negatively charged layers are electrostatically bound by cations. The layers separate as water molecules enter between them, and the material swells.

The bentonite water interaction has been studied extensively, both experimentally and through modelling. Among the experimental techniques used are NMR [see, e.g., 2], IR [see, e.g., 3, 4]. The theoretical studies have involved Monte Carlo studies [5, 6, 7] as well as first-principles calculations [8]. A limited number of X-ray absorption studies [7, 9] have been performed, and the need for further efforts in that direction is identified [9].

Soft X-ray emission spectroscopy with monochromatized synchrotron radiation has unique features, which make it ideal for studies of samples with complex structures. First, it focuses on transitions between the outermost levels and the quasi-atomic core levels. Spectra resulting from such transitions carry information about local electronic structure at specific atomic sites. Alternative methods can often only provide average electronic structure information, which is ungainly for the understanding complex structures. Second, the penetration length of soft X-ray photons is typically several thousand Ångström, sufficient for measuring at depth in a complex sample. Most important in this context is that they enable the use of ultra-thin windows, separating the sample and the ultra-high vacuum which is a prerequisite for generating and analyzing the radiation. Hence, it is straightforward to study gases, liquids, and samples under ambient pressure.

EXPERIMENTAL

This paper presents an account of a preliminary study of feasibility of applying soft X-ray spectroscopy to the study of a complex natural material such as bentonite. In this investigation, a purified calcium conditioned bentonite sample from the Kutch area in India was used. The purification method is described in ref. 1 and the montmorillonite content was > 95 %. The sample was studied dry in vacuum. The spectrometer can, however, be modified for the study of wet samples. This has successfully been applied to studies of uranium [10]. The present cell for wet samples has, however, a silicon nitride window that makes it unsuitable for the study of silicates. A new window of diamond-like carbon has been developed and has been tested successfully, but has not yet been applied to bentonite studies.

The soft X-ray absorption spectra from oxygen, calcium, and silicon were recorded at the I511 undulator beam-line at MAX-lab synchrotron light source in Lund, Sweden. By changing the undulator gap, the usable energy range can be varied from about 90 eV to about 1200 eV. The beamline has a plane-grating monochromator with Kirkpartick-Baez mounted elliptical mirrors for refocusing. The the standard grating is gold coated with 1221 lines/mm [11]. At the beamline, a Gammadata Scienta XES-300 grazing incidence grating spectrometer is also available for X-ray emission measurements [12].

RESULTS AND DISCUSSION

We recorded the X-ray absorption spectra from the L-edge of calcium, silicon, and aluminum, and from K-edge of oxygen. The spectra are shown in Figure 2. The calcium, silicon and oxygen absorption spectra were collected in the total fluorescence yield (TFY) mode. The aluminum spectrum was recorded in the partial fluorescence yield (PFY) mode using the XES-300 grating spectrometer with the detector gated to the energy interval of the Al L emission. The total count rate of the detector signal was recorded as a function of excitation energy.

Calcium is the main inter-layer counter ion in the Kutch bentonite. The absorption spectrum shows a typical quasi-atomic behavior. Small deviations from the quasi-atomic behavior can be analyzed in terms of coordination and ligand interactions. These ligand interactions lead to ligand field multiplet splittings and manifest themselves as additional peaks in the X-ray absorption spectrum. At the present level of accuracy, however, we do not see any effects of the ligand field, which are prominent in all ionic Ca compounds [see e.g. 13]. The absence of ligand field multiplets could indicate that the calcium cations are loosely bound to the negatively charged mineral sheets surfaces and can move relatively freely along the sheet surface.

The O K spectrum shows a pre-peak at around 532 eV, which is absent in the spectra of both the pure, bulk aluminum [see e.g. ref. 14] and silicon [see e.g. ref. 15] oxides. Such a pre-peak has previously normally been associated with impurities and/or lattice distortions. Here, the reduced dimensionality in the layers suggests that the pre-peak is associated with the 'dangling oxygens' in the outermost atomic layer, since the level of impurities in the sample is low. Other interpretations cannot, however, be excluded at this stage.

Surprisingly, the Si L absorption closely simulates the absorption spectrum of a pure silicon oxide [16]. This indicates that the local electronic structure at the Si sites is affected very little, and that, therefore, the SiO_4 tetrahedrons in the clay must be essentially undisturbed as compared to the pure silica.

The Al L spectrum, on the other hand, is complex and shows almost no resemblance to the absorption spectrum of aluminum oxides. The spectrum has, however, many features in common with the spectra of tetrahedrally coordinated aluminum silicates [17]. Although the physical significance of these results is not fully understood and must be further analyzed, this qualitative discussion illustrates the large potential of the methods and indicates some specific research directions.

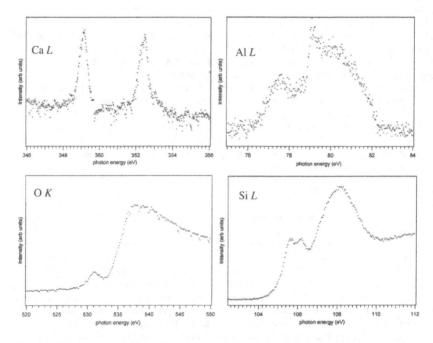

Figure 2. Fluorescence yield soft X-ray absorption spectra of montmorillonite from the Kutch area in India, with counter ions exchanged to Ca $^{2+}$.

CONCLUSIONS

The obtained data show that soft X-ray spectroscopy is a promising technique for studying clay minerals. It is capable of supplying unique information that is complementary to information accessible using other techniques; especially, it can be used to determine the local electronic structure at various atomic sites in the complex samples. We aim at future systematic studies where X-ray absorption and emission spectroscopy is applied to various montmorillonite samples, of different composition as a function of humidity. The goal is to increase the understanding of the microscopic origin of the fascinating properties of montmomorillonite.

REFERENCES

1. O. Karnland, S. Olsson, U. Nilsson, Mineralogy and sealing properties of various bentonites and smectite-rich clay materials, Svensk kärnbränslehantering AB, TR-06-30 (2006).

2. Y. Nakashima, J. Nucl. Sci. Technol. **41**, 981-992 (2004).

3. N. I. E. Shewring, T. G. J. Jones, G. Maitland, and J. Yarwood, J. Colloid Interface Sci. **176**, 308-317 (1995).

4. J. Madejová, M. Janek, P. Komadel, H.-J. Herbert, H. C. Moog, Appl. Clay Sci. **20**, 255-271 (2002).

5. A. Meleshyn and C. Bunnenberg, J. Chem. Phys. **122**, 034705 (2005).

6. A. Meleshyn and C. Bunnenberg, J. Chem. Phys. **123**, 074706 (2005).

7. F-R. Chang, N. T. Skipper, G. Sposito, Langmuir **11**, 2734-2741 (1995).

8. L. Stixrude and D. R. Peacor, Nature **420**, 165-167 (2002).

9. G. E. Brown, Jr., V. E. Henrich, W. H. Casey, D. L. Clark, C. Eggleston, A. Felmy, D. W. Goodman, M. Grätzel, G. Maciel, M. I. McCarthy, K. H. Nealson, D. A. Sverjensky, M. F. Toney, and J. M. Zachara, Chem. Rev. **99**, 77-174 (1999).

10. S. M. Butorin, D. K. Shuh, K. Kvashnina, I. L. Soroka, K. Ollila, K. E. Roberts, J.-H. Guo, L. Werme and J. Nordgren, Resonant inelastic soft X-ray scattering studies of U(VI) reduction on iron surfaces, Mat. Res. Soc. Symp. Proc. Vol. 807, 113-118 (2004).

11. R. Denecke, P. Väterlein, M. Bässler, N. Wassdahl, S. Butorin, A. Nilsson, J-E. Rubensson, J. Nordgren, N. Mårtensson, R. Nyholm, J. Electron Spectrosc. **101-103**, 971-977 (1999).

12. J. Nordgren, G. Bray, S. Cramm, R. Nyholm, J-E. Rubensson, N. Wassdahl, Rev. Sci. Instrum. **60**, 1690-1696 (1989).

13. F. M. F. de Groot, J. Electron Spectrosc. **67**, 529-622 (1994).

14. W. L. O'Brien, J. Jia, Q-Y, Dong, T. A. Callcott, D. L. Mueller, D. L. Ederer, N. D. Shinn, S. C. Woronick, Phys. Rev. 44, 13 277-13 282 (1991).

15. P. Lagarde, A. M. Flank, G. Tourillon, R. C. Liebermann and J. P. Itie, J. Phys. I France **2**, 1043-1050 (1992).

16. W. L. O'Brien, J. Jia, Q-Y, Dong, T. A. Callcott, J-E. Rubensson, D. L. Mueller, D. L. Ederer, Phys. Rev. **44**, 1013-1017 (1991).

17. C. Weigel, G. Calas, L. Cormier, L. Galoisy, G. S. Henderson. J. Phys.: Condens. Matter **20**, 135219 (8pp) (2008).

Mater. Res. Soc. Symp. Proc. Vol. 1124 © 2009 Materials Research Society 1124-Q05-04

Diffusion Behavior of Humic Acid in Compacted Bentonite: Effect of Ionic Strength, Dry Density and Molecular Weight of Humic Acid

Kazuki Iijima[1], Seiichi Kurosawa[2], Minoru Tobita[2], Satoshi Kibe[2] and Yuji Ouchi[2]
[1]Geological Isolation Research and Development Directorate, Japan Atomic Energy Agency, 4-33, Muramatsu, Tokai, Ibaraki 319-1194, Japan
[2]Inspection Development Co. Ltd., 4-33, Muramatsu, Tokai, Ibaraki 319-1112, Japan

ABSTRACT

Diffusion behavior of humic acid (HA) in the compacted bentonite was investigated by through diffusion method under various ionic strength and dry density conditions for 1 - 2 years. The breakthrough of HA is observed in case of 1.2-1.6 Mg/m^3 dry density in 1M NaCl, 1.2-1.5 Mg/m^3 in 0.1M NaCl, however, no breakthrough is observed even with 1.2 Mg/m^3 in lower ionic strength than 0.01M within this experimental duration. Molecular weight of HA was evaluated by size exclusion chromatography (SEC). The HA diffusing through bentonite is of lower molecular weight than 5kD in 1M NaCl or 3kD in 0.1M NaCl, those are independent of dry density of bentonite. It indicates that the structure of diffusing path for HA in compacted bentonite is dominated by ionic strength rather than dry density. Diffusion behavior of Nd(III) is also investigated in the presence and absence of HA, and concluded that it is facilitated by HA. These diffusion behaviors are modeled by one dimensional diffusion equation taking the equilibrium Nd + HA↔Nd-HA simultaneously into account. This model can explain the behavior of Nd in the presence of HA. The effective diffusion coefficient and distribution coefficient for Nd-HA are considered to be similar to those of HA.

INTRODUCTION

Humic acid (HA) is one of the natural organic matters. Since HA forms strong complexes with multivalent radionuclides (RNs), it is expected to affect migration behavior of RNs. Diffusion behavior of colloidal particles through compacted bentonite was studied and it was shown that gold colloids with mean size 15 nm were filtered by compacted bentonite with dry density 1.8 Mg/m^3 [1]. Therefore, it is essential to evaluate the size of colloids in the performance assessment [2].

Recently, some studies show that macromolecule organics can migrate through compacted bentonite. Wold and Eriksen [3] studied diffusion behavior of lignosulfonate colloids with molecular weight 30kD through compacted bentonite and obtained apparent diffusivities in the order of 10^{-12} m^2/s, which was independent to ionic strength (0.1-0.01 M) and dry density (0.6-1.8 Mg/m^3). Through diffusion experiments of Eu(III) and Co(II) in the compacted bentonite were also carried out in the presence of HA under same ionic strength and dry density conditions [4]. It was shown that HA diffused through compacted bentonite regardless of the conditions and apparent diffusivities significantly increased for both Co(II) and Eu(III) when HA were present. On the other hand, Wang et al. [5] reported that HA formed precipitation or complexation with Eu at the surface of compacted bentonite and reduced the diffusion of Eu in the compacted bentonite. Difference of the behavior of HA and RN-HA complex between these studies might be caused by the difference of molecular weight of RN-HA complex. Therefore, the influence of molecular weight of HA on the diffusion behavior of HA should be evaluated.

In this study, diffusion behavior of HA was investigated under various ionic strength and dry density conditions. The influence of molecular weight of HA was also evaluated by using size exclusion chromatography (SEC). The diffusion behavior of Nd(III), which was used as an chemical analogue of tetravalent actinides, was studied in the presence and absence of HA and the influence of HA was evaluated.

EXPERIMENTAL

Material

The bentonite used in this study was Kunigel V1®, supplied by Kunimine Industry Co. Ltd. This bentonite contains about 50-60wt% of smectite, and accessory minerals such as quartz, chalcedony, plagioclase, calcite, dolomite, analcime, and pyrite. Its detailed properties were given elsewhere [2]. The commercial HA purchased by Aldlich Co. Ltd was purified based on the International Humic Substance Society Method [6]. All other chemicals are reagent grade and used without purification. Doubly distilled water supplied by ADVANTEC CPW-200 was used to prepare all the solutions.

Diffusion and sorption experiments

Experiments are carried out in the through-diffusion method under aerobic condition. The schematic drawing of the diffusion cell is shown in figure 1. Experimental conditions are summarized in table I. In this paper, experimental conditions for each data point are indicated by corresponding sample number shown in italic in this table. Dried bentonite was compacted into the acrylic cell. After saturation with NaCl solution for one month, the tracer cell solution was replaced to the experimental solution, in which 500 mg/L HA and / or 1×10^{-6} M Nd was added, according to the experimental conditions. The pH of the experimental solution was adjusted at 8, which has been kept within 7.7-8.2 during whole experiments. The 5 mL aliquot of sample cell solution was taken at time intervals and same volume of NaCl solution was added. Absorbance, molecular weight distribution and Nd concentration of the sample were measured by UV-VIS spectrophotometer (Hitachi, U-3300), size exclusion chromatography (TOSOH, HPLC-8020) and ICP-MS (GV instruments, Plasma Trace 2), respectively.

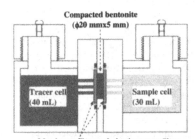

Membrane filter and plastic porous filter

Figure 1. Schematic drawing of diffusion cell.

Table I. Experimental conditions for through-diffusion.

Dry density [Mg/m³], content of HA and Nd	NaCl concentration [M]			
	1	0.1	0.01	10⁻³
1.2, HA, Nd	2056	2156		
1.2, HA	2050	2150	2250	2350
1.2, Nd	2006	2106		
1.2, (without HA, Nd)	2000	2100	2200	
1.3, HA		3150		
1.4, HA	4050	4150	4250	
1.5, HA		5150		
1.6, HA	6050	6150		

The conditions for SEC are as follows: column; TSKgel G3000, SuperSW2000, solution; 10^{-3}M KH_2PO_4 – 2×10^{-3}M Na_2HPO_4 – 1M NaCl, elution rate; 1 mL/min (G3000), 0.35 mL/min (SW2000), detector; spectrophotometer (270 nm). The sample for total organic carbon concentration (TOC) was diluted 11 times with 0.1M KH_2PO_4, bubbled with N_2 gas for 15 minutes to remove carbonate and measured by TOC analyzer (SHIMADZU TOC-5000A). After 1-2 years, post-diffusion extraction was carried out. Bentonite sample was sliced into thin pieces, then each slice was immersed into 5 mL of 1M NaCl solution and shaken for two weeks to extract HA and Nd (if added) from bentonite. The extraction solution was centrifuged at 6,000 rpm for 100 min and supernatant was taken for measurements.

Batch type sorption experiments were also carried out to evaluate distribution behavior of HA and Nd in the post-diffusion extraction. Experimental conditions are as follows: Solid liquid ratio (S/L); 2, 20 g/L, HA concentration; 0, 34 mg/L, Nd concentration; 0, 7.7×10^{-7} M, solution; 1M NaCl, pH;8, replicate; 3. The 40 mL of batch sorption sample was prepared in the PTFE tube and shaken end-over-end for one month. The aliquots of the sample were taken at time intervals for the measurements and treated same as the samples of post-diffusion extraction. To confirm sorption reversibility of HA, desorption experiments were also carried out for some sorption samples. The residual sorption samples were centrifuged at 6,000 rpm for 60 min, followed by taking 4.5 mL of supernatant and adding same volume of 1M NaCl solution. Sampling was carried out same as sorption experiments.

<u>Data evaluation</u>

One-dimensional diffusion of diffusant in the compacted bentonite can be described by the Fick's second law:

$$\frac{\partial C}{\partial t} = \frac{De}{\alpha} \frac{\partial^2 C}{\partial x^2} \qquad (1)$$

where De is the effective diffusivity (m^2/s), α is the rock capacity factor, C is the concentration of diffusant in solution (mol/m³), t is the diffusion time (s) and x is the distance (m). In this study, three species, HA, Nd-HA complex, and free Nd, are considered as dominant species. One of the models to describe complexation reaction of a metal ion with humic acid is the charge neutralization model [7]. In this model, the stability constant for Nd + HA↔Nd-HA is described as follows:

$$\beta = \frac{C_{Nd-HA}}{C_{Nd-free}C_{HA}} \approx \frac{C_{Nd-HA}}{C_{Nd-free}} \frac{3}{[HA]_0 \cdot PEC \cdot LC} \tag{2}$$

where β is the stability constant of Nd-HA complex, C_{Nd-HA}, $C_{Nd-free}$ and C_{HA} are the concentration of Nd-HA complex, free Nd and free HA, respectively, $[HA]_0$ is the initial concentration of HA (g/L), PEC is the proton exchange capacity (eq/g) and LC is the loading capacity. It is noticed that $C_{Nd-free}$ is the concentration of Nd^{3+} without complexation in this model. However, it is difficult to evaluate the concentration of Nd^{3+} in the compacted bentonite. Therefore, the total concentration of soluble Nd without complexation with HA was used for $C_{Nd-free}$ in this study, since it has a linear relationship to C_{Nd3+} if the solution chemistry is defined. Taking account for the equilibrium among three species described by equation (2), equation (1) is modified for each species as follows:

$$\frac{\partial C_i}{\partial t} = \frac{De_i}{\alpha_i}\frac{\partial^2 C_i}{\partial x^2} + \Delta(Nd-HA) \quad (i = Nd-free, HA)$$

$$\frac{\partial C_j}{\partial t} = \frac{De_j}{\alpha_j}\frac{\partial^2 C_j}{\partial x^2} - \Delta(Nd-HA) \quad (j = Nd-HA) \tag{3}$$

where $\Delta(Nd-HA)$ is the concentration of Nd-HA dissociated to free Nd and HA to keep equilibrium. The C_{HA} is usually higher than total concentration of Nd so that $\Delta(Nd-HA)$ is negligible for HA. Then the equations for $C_{Nd-free}$ and C_{HA} can be obtained as follows:

$$\frac{\partial}{\partial t}\left(\frac{C_{Nd-HA}}{\beta C_{HA}}\right) + \frac{\partial C_{Nd-HA}}{\partial t} = \frac{De_{Nd-free}}{\alpha_{Nd-free}}\frac{\partial^2}{\partial x^2}\left(\frac{C_{Nd-HA}}{\beta C_{HA}}\right) + \frac{De_{Nd-HA}}{\alpha_{Nd-HA}}\frac{\partial^2 C_{Nd-HA}}{\partial x^2}$$

$$\frac{\partial C_{HA}}{\partial t} = \frac{De_{HA}}{\alpha_{HA}}\frac{\partial^2 C_{HA}}{\partial x^2} \tag{4}$$

The difference equations were obtained from equation (4). In this study, the C_{Nd-HA} and C_{HA} were calculated numerically using equation (4), then $C_{Nd-free}$ using equation (2).

RESULTS and DISCUSSION

Effect of ionic strength, dry density and molecular weight of HA

Figure 2 shows the examples of UV-VIS absorption spectra and SEC chromatograms of the sample and the tracer cell solution. The tracer cell solution shows typical absorption spectrum of HA, that is, continuous absorption increasing with decreasing wavelength. The sample cell solution in the presence of HA (#2050) also shows similar spectrum, however, slightly steep slope in the region 250-400 nm. It means that the composition of functional groups of HA diffusing through the compacted bentonite is different from that of initial HA. It is also pointed out that absorption is observed even in the absence of HA (#2000) in the region 250-300 nm, indicating that the bentonite contains some species which show absorption in this region.

266

Therefore, HA should be distinguished from the species originally contained in the bentonite. In SEC chromatograms in figure 2, the sample cell solution in the presence of HA shows clear peak compared with that of in the presence of HA. Thus, the breakthrough of HA should be clarified by comparison of SEC.

Figure 3 a)-c) shows the SEC chromatograms of the sample cell solutions under various experimental conditions. All samples in 1M NaCl show breakthrough of HA, while no significant breakthrough can be observed in 10^{-2}M NaCl within this experimental duration. In 0.1M NaCl, no significant breakthrough can be observed in case of 1.6 Mg/m^3 dry density (#6150) but in lower dry density than 1.5 Mg/m^3 (#2150 and #5150). Thus the diffusion of HA through the compacted bentonite is dependent on both ionic strength and dry density.

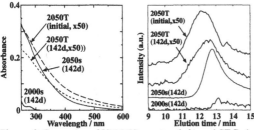

Figure 2. Examples of UV-VIS spectra (left) and SEC chromatograms (right). Sample numbers are corresponding to those of table I. Following "s" indicates the sample cell solution, while "T" indicates the tracer cell solution diluted 50 times by 1M NaCl. SEC column: SuperSW2000.

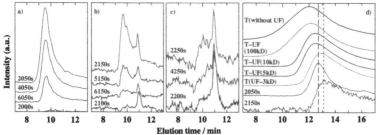

Figure 3. SEC chromatograms of the sample cell solutions in a)1M, b)0.1M and c)10^{-2}M NaCl. Chromatograms of #2050 and #2150 are compared with those of the filtered tracer solution by various ultrafilters (d). Peak height is normalized by maximum in d). Diffusing durations: 2050;299d, 4050;406d, 6050;422d, 2000;413d, 2150;383d, 5150;478d, 6150;755d, 2100;483d, 2250;405d, 4250;421d, 2200;450d. SEC column: G3000(a-c), SuperSW2000(d).

Since the peak arrival time in SEC is later, as shown in figure 2, the molecular weight of HA in the sample cell solution is lower than that of in the tracer cell. It indicates that only lower molecular weight fraction can diffuse through the compacted bentonite. In order to evaluate the molecular weight of diffusing HA, the chromatograms of the sample cell solution are compared with those of filtrates of the tracer cell solution by ultrafilters from 3kD to 100kD MWCO, as

267

shown in figure 3 d). The peak arrival time for the sample in 1M NaCl (dashed line) is intermediate between those of the filtrates by 3kD and 5kD, while that of 0.1M NaCl (short dashed line) is later than that of 3kD. Therefore, it can be considered that the molecular weight of diffusing HA is lower than 5kD in 1M NaCl and 3kD in 0.1M NaCl, respectively.

Wold and Eriksen [4] reported that diffusion of HA took place in the major part of the pore space of compacted bentonite and ion exclusion was not significant. It was also pointed out that higher ionic strength in the pore water probably contributed to size reduction in humic colloids, while the molecular weight of diffusing HA slightly increases with increasing ionic strength in this study. On the other hand, the molecular weight of diffusing HA is independent of dry density. It indicates that the pore structure dominating molecular size of diffusing HA in compacted bentonite is highly affected by ionic strength rather than dry density in this experimental condition.

Diffusion behavior of HA and Nd

Figure 4 shows breakthrough curves of HA and Nd for #2056 as an example. The concentration of HA was evaluated by TOC. Breakthrough of Nd in the presence of HA is similar to that of HA as shown in this figure, while no breakthrough is observed in the absence of HA by 300 d. It indicates that diffusion of Nd is facilitated by HA under this condition.

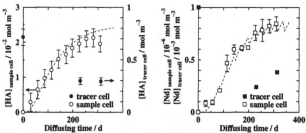

Figure 4. Breakthrough curves of HA (left) and Nd (right) in the sample (open marks) and the tracer (closed marks) cell solution for #2056 (1M NaCl, 1.2 Mg/m^3 dry density, 500 mg/L HA, 1×10^{-6}M Nd). Fitting curves for the sample cell are also shown by dashed lines.

Table II shows results of batch sorption and desorption experiments in 1M NaCl solution. The distribution coefficient, Kd (m^3/kg), is calculated as follows:

$$Kd = \frac{C_{init} - C_{eq}}{C_{eq}} \frac{1}{S/L} \qquad (5)$$

where C_{init} and C_{eq} is the initial and the final concentration, S/L is the solid liquid ratio (kg/m^3). The Kd value for HA is not affected by the presence of Nd. It is considered that sorption behavior of Nd-HA complex is similar to that of HA itself, or, the concentration of Nd-HA is too low to affect on the distribution coefficient. The Kd in sorption is also similar to that of desorption, indicating that sorption of HA onto the bentonite is reversible. On the other hand, the Kd for Nd is highly dependent on the presence of HA due to high stability of Nd-HA complex.

Table II. Results of batch sorption and desorption experiments.

Experimental conditions			Results			
S/L (g/L)	$[HA]_{init}$ (mg/L)	$[Nd]_{init}$ (M)	$Kd_{HA,\ sorption}$ (m^3/kg)	$Kd_{HA,\ desorption}$ (m^3/kg)	$Kd_{Nd,\ sorption}$ (m^3/kg)	β (L/mol)
2	34.2±1.5	$(7.7\pm0.5)\times10^{-7}$	$(1.9\pm0.6)\times10^{-1}$	-	$(7.3\pm0.9)\times10^{-1}$	10.4
2	0	$(7.7\pm0.5)\times10^{-7}$	-	-	$\mathbf{(2.0\pm0.2)\times10^{2}}$	-
20	34.2±1.5	0	$(7.7\pm0.9)\times10^{-2}$	$(3.9\pm1.2)\times10^{-2}$	-	-
20	34.2±1.5	$(7.7\pm0.5)\times10^{-7}$	$\mathbf{(7.7\pm1.0)\times10^{-2}}$	$(4.0\pm1.2)\times10^{-2}$	$(4.7\pm1.1)\times10^{0}$	**2.3**

The stability constant of HA defined by equation (2) is calculated from the results of sorption experiments. It is assumed that the distribution coefficient of Nd-HA complex is similar to that of HA, as mentioned above. The *PEC* is determined by pH-titrarion as 6.0×10^{-3} eq/g at pH 8, and *LC* is assumed as 0.72 taking only deprotonation rate at pH 8 into account [8]. Using experimentally obtained C_{Nd-HA}, $C_{Nd-free}$, β values are calculated as listed in table II. The values emphasized in bold are used to evaluate the distribution of HA and Nd in the extraction from slice sample.

Figure 5 shows the profiles of HA and Nd in the bentonite for the same sample as shown in figure 4. In both profiles, two regions can be cleary observed. The first region is from tracer source to 1.3 mm, showing drastic decrease of the concentration. The second region is from 1.3 mm to the sample cell solution, with gradual decrease. The SEC chromatograms of extracted solution in the first region (0-1.3 mm) show relatively earlier arising, indicating that it contains higher molecular weight fraction. On the other hand, those of in the second region (1.3-5 mm) are almost similar to that of the sample cell solution. Therefore, higher molecular weight fraction than 5kD diffuses up to 1.3 mm, while lower fraction can diffuse through compacted bentonite.

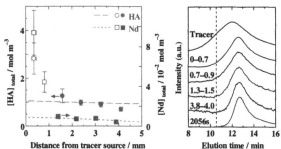

Figure 5. Concentration profiles (left) of HA and Nd in the bentonite for #2056 (1M NaCl, 1.2 Mg/m^3 dry density, 500 mg/L HA, 1×10^{-6}M Nd) calculated from the concentration in the extraction solution from the bentonite slice. Fitting curves to closed marks are shown by dashed lines. SEC chromatograms (right) of corresponding extraction solution are also shown, compared with those of the tracer and the sample cell solution. SEC column: SuperSW2000.

Numerical calculation based on equation (4) in the implicit scheme are carried out to fit to breakthrough curves and profiles in the bentonite from 1.3 to 5 mm (shown as closed marks).

In the calculation, it is assumed that Nd-HA complex has similar De and α to HA and β obtained in the sorption experiments can be applied. Fitting parameters are the initial concentration of lower molecular weight fraction of HA and Nd, $[HA]_0$, $[Nd]_0$, respectively, De_{HA}, Kd_{HA} and Kd_{Nd}. The De_{Nd} was not sensitive in this calculation so that it is fixed at 10^{-13} m^3/kg. The effect of the filters in which the compacted bentonite is sandwiched is also considered.

Obtained curves by the least square fitting are also shown in the figures 4 and 5 as dashed lines. Parameters are evaluated as follows: $[HA]_0$=45 mg/L, $[Nd]_0$=5.5x10^{-9} mol/L,, De_{HA}=2.5x10^{-11} m^2/s,, Kd_{HA}=3.3x10^{-2} m^3/kg, and Kd_{Nd}=2.3x10^{2} m^3/kg. The De value is comparable with the values in ref.[4], which are in the order of 10^{-11} m^2/s. As shown in figure 5, the calculated profile is slightly different from experimental data. Part of the larger molecular weight fraction might also contribute to the profile in the region 1.3-5.0 mm. The diffusion behavior of HA and Nd are likely to be explained by this model and it seems reasonable to apply diffusion parameters of HA to Nd-HA complex.

CONCLUSIONS

Diffusion behavior of HA in the compacted bentonite was investigated under various ionic strength and dry density conditions. The breakthrough of HA is observed in case of 1.2-1.6 Mg/m^3 dry density in 1M NaCl, 1.2-1.5 Mg/m^3 in 0.1M NaCl. Lower fraction than 5kD in 1M NaCl or 3kD in 0.1M NaCl can diffuse and the structure of diffusing path for HA in compacted bentonite is dominated by ionic strength rather than dry density. Diffusion of Nd(III) is facilitated by HA. These diffusion behaviors are modeled by the diffusion equation taking equilibrium Nd + HA$\leftrightarrow$Nd-HA into account. This model can explain the behavior of Nd in the presence of HA and De and α for Nd-HA are considered to be similar to those of HA.

ACKNOWLEDGEMENT

The Authors thank Yoshimi Seida and Masahiro Shibata for their valuable comments and suggestion on this work.

REFERENCES

1. S. Kurosawa, M. Yui, and H. Yoshikawa in *Scientific Basis for Nuclear Waste Management XX,* edited by W.J. Gray, and I.R. Triay, (Mater. Res. Soc. Symp. Proc. **465**, Pittsburgh, PA, 1997) pp.963-970.
2. Japan Nuclear Cycle Development Institute, JNC Technical Report, JNC TN1410 2000-004, 2000.
3. S. Wold, and T.E. Eriksen, *Appl. Clay Sci.* **23**, 43 (2003).
4. S. Wold, and T. Eriksen, *Phys. Chem. Earth* **32**, 477 (2007).
5. X. Wang, Y. Chen, and Y. Wu, *Appl. Radiat. Isot.* **60**, 963 (2004).
6. International Humic Substance Society, *http://www.ihss.gatech.edu/soilhafa.html.*
7. J.I. Kim, and K.R. Czerwinski, *Radiochim. Acta* **73**, 5 (1996).
8. K. Iijima, S. Tobitsuka, and Y. Kohara, JNC Technical Report, JNC TN8400 2004-005, 2004 (in Japanese).

Mater. Res. Soc. Symp. Proc. Vol. 1124 © 2009 Materials Research Society   1124-Q05-05

Experimental Work Conducted on MgO Characterization and Hydration

Haoran Deng[1], Yongliang Xiong[1], Martin Nemer[2] and Shelly Johnsen[1]
[1]Repository Performance Dept. 6712
[2]Performance Assessment and Decision Analysis Dept. 6711
Sandia National Laboratories Carlsbad Programs Group, Carlsbad, NM 88220

ABSTRACT

Magnesium oxide (MgO) is the only engineered barrier certified by the EPA for emplacement in the Waste Isolation Pilot Plant (WIPP), a U.S. Department of Energy repository for transuranic waste. MgO will reduce actinide solubilities by sequestering CO_2 generated by the biodegradation of cellulosic, plastic, and rubber materials. Demonstration of the effectiveness of MgO is essential to meet the U.S Environmental Protection Agency's requirement for multiple natural and engineered barriers. In the past, a series of experiments was conducted at Sandia National Laboratories to verify the efficacy of Premier Chemicals LLC (Premier) MgO as a chemical-control agent in the WIPP. Since December 2004, Premier MgO is no longer available for emplacement in the WIPP. Martin Marietta Magnesia Specialties LLC is the new MgO supplier. MgO characterization, including chemical, mineralogic, and reactivity analysis, has been performed to address uncertainties concerning the amount of reactive constituents in Martin Marietta MgO. Characterization results of Premier MgO will be reported for comparison.

Particle size, solid-to-liquid ratio, and stir speed could affect the rate of carbonation of MgO slurries. Thus, it's reasonable to hypothesize that these factors will also affect the rate of hydration. Accelerated MgO hydration experiments were carried out at two or three levels for each of the above factors in deionized water at 70 °C. The Minitab statistical software package was used to design a fractional-factorial experimental matrix and analyze the test results. We also fitted the accelerated inundated hydration data to four different kinetic models and calculated the hydration rates. As a result of this study we have determined that different mechanisms may be important for different particle sizes, surface control for large particles and diffusion for small particles.

INTRODUCTION

The Waste Isolation Pilot Plant (WIPP) is a U.S. Department of Energy repository for defense-related transuranic waste. It is located in southeast New Mexico at a depth of 655 m in the Salado Formation, a Permian bedded-salt formation. Cellulosic, plastic, and rubber materials are included in the waste containers as actinide-contaminated waste. This includes lab coats, plastic bottles, rubber gloves, as well as waste-emplacement materials such as wood palettes, plastic wrap, and slip sheets. Possible microbial degradation of cellulosic, plastic, and rubber materials could generate carbon dioxide (CO_2). Carbon dioxide dissolution would acidify the brine, which would directly affect actinide speciation and solubilities. Therefore, MgO is emplaced in the WIPP repository to reduce actinide solubilities by sequestering CO_2. In addition, after the WIPP is filled and sealed, the periclase in the MgO will react with water in the gas phase or in brine to form brucite ($Mg(OH)_2$). The brucite dissolution reaction will increase the pH to about 9 [1]. A series of experiments was conducted at Sandia National Laboratories to quantify the efficacy of MgO from Martin Marietta Magnesia Specialties LLC. Martin Marietta (MM) MgO WTS-60 is the MgO currently being emplaced in the WIPP. In this paper we will present the experimental results from MgO characterization and accelerated inundated hydration experiments.

The objectives of MgO characterization were to determine the mineralogic composition and particle-size distribution, and to address uncertainties concerning the amount of reactive constituents, periclase (MgO) plus lime (CaO).

Particle size, solid-to-liquid ratio, and stir speed all affect the rate of carbonation of MgO slurries [2]. It is reasonable to hypothesize that these factors will also affect the rate of hydration. The objectives of the accelerated inundated hydration experiments were to gauge the effect of each of the factors above. We then fitted the accelerated inundated hydration data to various reaction models assuming surface or diffusive control of the hydration process.

EXPERIMENTAL

All experiments discussed herein were conducted using MM MgO WTS-60. The MgO particle-size distribution was determined as a function of mass fraction by passing MM MgO through Fisher Scientific sieves ranging from 75 μm (200 mesh) to 2.0 mm (10 mesh). The fractions between each pair of sieves (or above and below the sieve for the largest and smallest) were then collected and weighed.

Lower bounds on the amounts of Mg, Al, Si, Ca, and Fe were determined by dissolving samples of MM MgO in trace-metal-grade nitric acid followed by analysis using a Perkin Elmer Optima 3300 DV inductively coupled plasma atomic emission spectrometer (ICP-AES). Scanning-electron-microscope (SEM) images and energy dispersive spectra (EDS) were taken of as-received MM MgO, and of the insoluble portion after dissolution in nitric acid.

To determine the reactive content of MM MgO, the material was first hydrated in deionized (DI) water at 90 °C for at least 3 days, the time required for most of the periclase and lime in MgO to convert to brucite and portlandite ($Ca(OH)_2$). Next, loss-on-ignition (LOI) tests and thermogravimetric analysis (TGA) were performed to quantify the amount of brucite and portlandite in the hydrated MgO, based on the mass of water lost at the decomposition temperatures for $Mg(OH)_2$ and $Ca(OH)_2$. By assuming that all of the brucite and portlandite will eventually carbonate, the hydration followed by LOI testing allowed us to estimate the amount of MM MgO that is capable of reacting with CO_2.

A fractional-factorial experimental matrix (Table I) was designed for the accelerated inundated hydration experiment to determine the factors that are important to MgO hydration. Wheaton serum bottles (125 mL) or Nalgene high density polyethylene centrifuge tubes (30 mL) containing DI water and MM MgO were placed either in a VWR oven, where they were not agitated, or in a New Brunswick Scientific water-bath shaker set to a shaking speed of 150 rpm. Both the oven and water-bath shaker were set to 70 °C. The two MgO particle sizes with the highest particle-size fraction were used in the accelerated inundated hydration experiments. The larger particle-size range had particles with diameters between 1.0 and 2.0 mm, which accounted for 32 wt % of the as-received MM MgO. The smaller range had particles with diameters <75 μm, 18 wt % of the MM MgO. Three WIPP-relevant MgO-to-liquid ratios, 1 g/mL, 0.4 g/mL, 0.05 g/mL were used in the accelerated inundated hydration experiments. Nemer [3] showed that a range of 10^{-3} to 10^1 g/mL brackets the expected range of MgO-to-liquid ratios in the WIPP. Experiments were performed on MM MgO in DI water at 70 °C for 43 days. Duplicate samples were prepared for each experiment. The solid portion of each hydrated sample was filtered using Whatman #40 filter paper and then dried in lab air. A LOI test was then performed on the dried sample, from which the brucite concentration was calculated.

Table I. Accelerated Inundated Hydration Experimental Matrix.

	ID	Motion[1]	Particle Size[2]	MgO/water Ratio[3]
1	Ahy-m-sml-tu	1	1	1
2	Ahy-m(8gmgo)-sml-tu	1	1	2
3	Ahy-m-sml-bo	1	1	3
4	Ahy-m-big-tu	1	2	1
5	Ahy-m(8gmgo)-big-tu	1	2	2
6	Ahy-m-big-bo	1	2	3
7	Ahy-s-sml-tu	2	1	1
8	Ahy-s(8gmgo)-sml-tu	2	1	2
9	Ahy-s-sml-bo	2	1	3
10	Ahy-s-big-tu	2	2	1
11	Ahy-s(8gmgo)-big-tu	2	2	2
12	Ahy-s-big-bo	2	2	3

1. Samples with motion level 1 were placed inside an oven, where they were not agitated. Samples with motion level 2 were shaken at 150 rpm in a water-bath shaker.
2. Particle-size level 1 represents MgO with diameters from 1.0-2.0 mm. Particle-size level 2 has diameters <75 μm.
3. Samples with MgO/water ratio level 1 contained 10 g of MM MgO and 10 mL of DI water. Samples with MgO/water ratio level 2 contained 8 g of MM MgO and 20 mL of DI water. Samples with MgO/water ratio level 3 contained 5 g of MM MgO and 100 mL of DI water.

DISCUSSION

MgO Characterization

The particle-size distribution of one lot of Martin Marietta (MM) MgO WTS-60 was determined by the method described in the experimental section. On average, 7 wt % of the MgO particles have diameters bigger than 2.0 mm, and 18 wt % of the MgO particles have diameters smaller than 75 μm. Of the remaining 75 wt %, 32, 20,13, 5, and 3 wt % of MgO particles have diameters between 1.0-2.0 mm, 600 μm-1.0 mm, 300-600 μm,150-300 μm, and 75-150 μm respectively. The resulting particle-size distribution is similar to the manufacturer's analysis sheet that accompanied the tested lot of MM MgO. Lower bounds on the amounts of Mg, Al, Si, Ca and Fe were determined by ICP-AES as discussed in the experimental section. The results are listed in Table II. Table II represents lower bounds because a small fraction of nitric-acid insolubles remained after dissolution and were not quantitatively analyzed. The insolubles remaining after dissolution amounted to about 0.7 wt % of the MM MgO. SEM and EDS spectra of the insolubles indicated the solids were (1) spinel ($MgAl_2O_4$) or a solid solution of several spinels, such as ($FeCr_2O_4$), hercynite ($FeAl_2O_4$), magnesiochromite ($MgCr_2O_4$); (2) Fe_2O_3; and (3) SiO_2 (polymorph yet to be determined).

273

Table II. Weight percent of Mg, Al, Si, Ca and Fe (reported as oxides) that dissolved in nitric acid. MgO, Al_2O_3, SiO_2, CaO and Fe_2O_3 are reported here as oxides, which aren't necessarily representative of the actual phases in the MM MgO.

	MgO (wt %)	CaO (wt %)	Al_2O_3 (wt %)	Fe_2O_3 (wt %)	SiO_2 (wt %)
Average	98.46	0.87	0.13	0.12	0.31
Standard Deviation (1σ)	2.54	0.03	0.018	0.01	0.01

The concentration of reactive constituents (periclase and time) in MM MgO was determined as described in the experimental section. The results are shown in Table III. Detailed calculations can be found in reference [4]. The concentration of periclase and lime in MM MgO, 96 ± 2 (1σ) wt %, is higher than that of Premier MgO, 92 wt % [5]. In addition, we have tested 21 lots of MM MgO supplied to WIPP. In these 21 lots, the average concentration of periclase plus lime is 98 ± 0.5 (1σ) wt %.

Table III. Weight fraction of periclase and lime in (unhydrated) MM MgO.

	Average (wt %)	Standard Deviation (wt %)	Reported Value mean $\pm \sigma$
periclase	94.8	1.72	95 ± 2
lime	0.874	0.0256	0.87 ± 0.03
periclase + lime	95.7	1.74	96 ± 2

Accelerated Inundated Hydration Experiments

Inundated hydration experiments were performed on MM MgO in DI water at 70 °C for 43 days. The hydration products were examined using X-ray diffraction (XRD) analysis. The XRD patterns showed that brucite was the only hydration product. Portlandite may have formed but the amount was below the XRD detection limit. After 43 days, the large MgO particles were completely converted to brucite, while a small amount of periclase remained in the small MgO particles. Given more time, we expect that the small MgO particles would completely hydrate.

We used the fraction of periclase converted to brucite, W, as the measure of reaction progress.

$$W = \frac{X_{brucite}(t)}{X_{periclase}}, \tag{1}$$

Here $X_{brucite}(t)$ is the mole fraction of brucite in the hydrated sample at time t, and $X_{periclase}$ is the initial mole fraction of periclase. MM MgO contains minerals other than periclase and lime. These include spinels and iron oxides as discussed in the MgO characterization section. During this experiment, those unreactive components may not have hydrated. Therefore, the average hydration of the large MgO particles at 43 days was used as a measure of complete hydration

$X_{brucite}(\infty) = X_{periclase}$. At the end of 43 days, the average W of the small MgO particles was W = 0.9735 and by our above definition the large particles reached W=1.

Figure 1 shows the reaction progress versus time for the different factors and levels listed in Table I. Figure 1 shows that the small MgO particles hydrated faster than the large MgO particle during the first few days, which is probably due to the larger specific surface area (m²/g) of the small particles. However for the remainder of the experiment, the large-MgO particles hydrated faster than the small particles. There are no obvious differences between the results of the stirred and unstirred experiments. The MgO/water ratio did not significantly influence the hydration rate until later stages of hydration. A Minitab analysis of variance (ANOVA) was performed on the reaction progress (W) as a function of the effect variables (motion, particle size, and MgO/water ratio) at each point in time by multiple linear regression. The Minitab linear regression model confirmed our visual conclusions from Figure 1.

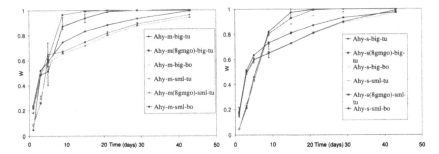

Figure 1.   MM-MgO hydration progress (W) versus time (days) for experiments carried out in DI water at 70 °C. The results in the left graph are from the unstirred experiments; those on the right were shaken at 150 rpm. The conditions for each sample identified in the legend can be found in Table I.

In dilute solutions, researchers have found that MgO hydration proceeds by the dissolution of MgO followed by precipitation of brucite crystals, usually attached to the MgO surface [6]. Rocha [7] concluded that at high temperature (≥70 °C), the hydration of MgO is at first governed by MgO dissolution (surface-area control), then as the reaction progresses both the surface and pores of MgO are covered by Mg(OH)₂, changing the porosity of the solid. As a result, transport becomes hindered inside particles (diffusive control). Rocha observed that the kinetics of MgO hydration change from surface-area control to diffusive control at W >0.6. We fitted the W versus time data to both surface area and diffusion-controlled kinetic models to determine if a particular model or class of models fit the data better than another. The surface-area controlled models assume that the hydration kinetics are controlled by the number of active sites on the reacting surface of a sphere that is shrinking as the reaction proceeds [8]. The diffusion-controlled models assume the kinetics are controlled by diffusion through the brucite layer to the periclase reaction interface [9-10]. Our study shows that different mechanisms may be important for different particle sizes, surface-control for larger particles and diffusion for small particles.

CONCLUSIONS

MgO characterization and accelerated inundated MgO hydration have been reported in this study. Characterization has shown that the concentration of reactive components (periclase plus lime) in MM MgO is 96 ± 2 wt %. The calculation included conservative assumptions, which decreased the result.

The accelerated inundated hydration experiments were performed to gauge the effects of various factors that could affect the hydration rate, including particle size, solid-to-liquid ratio, and stir speed. As a result of this study we have determined that particle size is the most important factor. In addition, we found that different mechanisms may be important for different particle sizes, surface control for larger particles and diffusion for small particles.

ACKNOWLEDGMENTS

Sandia is a multi-program laboratory operated by Sandia Corporation, a Lockheed Martin Company, for the United States Department of Energy's National Nuclear Security Administration under Contract DE-AC04-94AL85000. This research is funded by WIPP programs administered by the Office of Environmental Management of the U.S Department of Energy.

REFERENCES

1. U.S. DOE, *Title 40 CFR Part 191 Compliance Recertification Application for the Waste Isolation Pilot Plant, Appendix BARRIERS* (DOE/WIPP 2004-3231, U.S. Department of Energy Carlsbad Area Office, Carlsbad, NM, (2004).
2. A.I. Fernandez, J.M. Chimenos, M. Segarra, M.A. Fernandez and F. Espiell, *Hydrometallurgy.* **53**, 155-167 (1999).
3. M.B. Nemer, Memo to the records center, ERMS 542612, Sandia National Laboratories, Carlsbad, NM, 2006.
4. H. Deng, Y. Xiong, and M.B. Nemer, Report, ERMS 546570, Sandia National Laboratories, Carlsbad, NM, 2007.
5. A.C. Snider, and Y. Xiong, Analysis report, ERMS 537188, Sandia National Laboratories, Carlsbad, NM, 2004.
6. O. Fruhwirth, G.W. Herzog, I. Hollerer and A. Rachetti, *Surf. Technol.* **24**, 293-300 (1985).
7. S.D. Rocha, M.B. Mansur and V.S. Ciminelli, *J. Chem. Technol. Biotechnol.* **79**, 816-821 (2004).
8. G.K. Layden, and G. W. Brindley, *J. Am. Ceram. Soc.* **46**, 518-522 (1963).
9. R.J. Raymond and G.W. Brindley, *Trans. Faraday Soc.* **61**, 1017-1025 (1964).
10. R.E. Carter, *J. Chem. Phys.* **34**, 2010-2015 (1961).

Mater. Res. Soc. Symp. Proc. Vol. 1124 © 2009 Materials Research Society   1124-Q07-06

Diffusion of ^{60}Co, ^{137}Cs and ^{152}Eu in Opalinus Clay

Miguel García-Gutiérrez[1], José L. Cormenzana[2], Tiziana Missana[1], Manuel Mingarro[1], Ursula Alonso[1]
[1]CIEMAT, Departamento de Medioambiente, Av. Complutense 22, 28040 Madrid, SPAIN
[2]Empresarios Agrupados, Magallanes 3, 28015 Madrid, SPAIN

ABSTRACT

This study addresses the diffusion of representative sorbing elements, cobalt, cesium and europium in the Opalinus Clay (OPA). The methodology used here to determine diffusion coefficients is the "instantaneous planar source" method. In this setup, a paper filter impregnated with tracer is introduced between two clay samples, avoiding contact between the tracer and the experimental vessels.

The apparent diffusion coefficients (D_a) perpendicular to the bedding plane, obtained with this experimental method and fitting the experimental results with an analytical solution, were $D_a(Co) = (2.4-3.5) \cdot 10^{-14}$ m^2/s, $D_a(Cs) = (5.9-8.0) \cdot 10^{-14}$ m^2/s, and $D_a(Eu) = (1.0-2.1) \cdot 10^{-15}$ m^2/s. With cobalt and cesium, classical in-diffusion experiments were also performed for comparison, and similar D_a values were obtained but with a large dispersion.

To analyze the possible effects of the paper filter impregnated with the tracer on the determinations of D_a with the analytical solution, one experiment was also analyzed using a detailed stochastic model of the setup. The good agreement between the two modeling approaches confirms the validity of this experimental setup and the analytical model fitting procedure.

INTRODUCTION

Clay formations are being considered as potential host rocks for radioactive waste disposal in many countries. In these materials, which have low hydraulic conductivity, diffusion is the main transport mechanism for radionuclides accidentally released from the canisters.

The Mont Terri Underground Research Laboratory (URL) is located in a service gallery of the Mont Terri motorway tunnel, near St. Ursanne in north-western Switzerland, and is excavated in the Opalinus Clay (OPA). The OPA is currently under investigation to demonstrate the basic feasibility of disposing of spent fuel, vitrified high-level waste, and long-lived intermediate-level waste in Switzerland [1].

Diffusion is a process by which mass is transported from one part of a system to another by random molecular motion. The mathematical description of diffusion is based on Fick's laws, which have to be modified for porous media such as clays, because several interactions between solid phases and solutes exist. The nature of the ions and the porous medium, as well as the specific geochemical conditions, are all elements influencing the diffusion process [2].

Several studies of diffusion behavior of neutral (HTO), anionic (Br⁻, I⁻, Cl⁻), and weakly sorbing cationic elements (Na⁺, Sr²⁺) on clay formations exist [3-5], but only a few studies are available for medium sorbing elements such as Cs [6], and no studies for Eu, a highly sorbing element.

Diffusion studies with sorbing elements are not straightforward to carry out, because it is necessary to take into account their sorption on the diffusion cells and/or stainless-steel filters used in the experimental set-ups. Strongly sorbing elements such as europium cannot be easily studied using classical through-diffusion or in-diffusion methods because of their strong adsorption onto laboratory materials (e.g., vessels, cells, and tubing), such that their contact with these materials has to be minimized.

To avoid contact of the tracer with the cell materials, the "instantaneous planar source" or "thin source" method was used in this study, in which a paper filter impregnated with the tracer is placed between two samples of clay. About 200 days were necessary to obtain a good concentration profile for diffusion of cobalt and cesium, and more than 400 days were needed for europium diffusion.

EXPERIMENTAL

The OPA samples were collected in the Mont Terri URL, at a depth between 200 and 300 m below the surface from the overdrilling of the DI-A2 in situ diffusion experiment [7], in which several non-reactive and reactive tracers were injected in a packed-off borehole. The OPA was deposited about 180 million years ago as marine sediment consisting of fine mud particles and is ~100 m thick. The consolidated material has a density of $2.3 - 2.4$ g/cm^3. The OPA contains between 40 and 80% clay minerals (illite, illite/smectite mixed-layer, chlorite, kaolinite), 10-40% quartz, 6-45% carbonates, 0-8% feldspars, 0-1.7% pyrite and 0.1-0.5% organic carbon [8].

When a filter paper tagged with a tracer is located between two samples of clay sandwiched in a cell, the tracer can diffuse into each sample. The apparent diffusion coefficient (D_a) can be obtained by fitting the tracer concentration profile in the samples at the end of the experiment.

Cylindrical samples of OPA clay approximately 2 cm in diameter are drilled perpendicular to the bedding plane and introduced in stainless-steel rings. Whatman n° 54 filter paper saturated with 0.025 mL of tracer solution is placed between the two clay samples. The ring is closed with 2 end-pieces and two sintered filters and the diffusion cell is placed in a closed vessel to avoid humidity loss. At the end of the experiment, the cell is disassembled, the clay is sliced and the tracer activity is measured in each slice to obtain the concentration profile within the consolidated clay.

An instantaneous planar source or constant model can be applied to model experimental data, depending on the solubility limiting phase [9]. The experiments were performed using ^{60}Co ($3.5 \cdot 10^{-8}$ M), ^{137}Cs ($1.1 \cdot 10^{-7}$ M) and ^{152}Eu ($4.1 \cdot 10^{-7}$ M) as tracers. The elements come from CoCl$_2$, CsCl and EuCl$_3$ acidified solutions. For Co and Cs the concentration is well under the solubility limits in the OPA pore water; for Eu the concentration is slightly lower. The instantaneous planar source method has been selected in this work. In all cases, radiotracer activity was measured directly in the solid samples by gamma counting (in the ^{137}Cs case by the gamma emitter ^{137m}Ba in equilibrium) using a Packard Cobra II auto-gamma counter.

For Cs and Co, in-diffusion (ID) experiments by immersion in the tracer solution were also performed. In these experiments the cell with the sample is immersed into a large volume of synthetic OPA pore water [10]. After water saturation of the sample, the tracer is added to the reservoir and it can enter through one side of the clay sample. After a given time, the diffusion cell is disassembled, the sample cut into slices, and the activity in each slice measured to obtain a concentration profile in the clay. More details can be found in [11].

MODELING

The results of the experiments are modeled using the analytical solution for the instantaneous injection of a solute in a one-dimensional (1D) semi-infinite medium. If the mass of tracer M is injected uniformly across the cross-section of area A, and the initial width of the tracer source is infinitesimally small, the apparent diffusion coefficient is evaluated from the following analytical solution [12]:

$$C(x,t) = \frac{M}{2 \cdot A \sqrt{\pi \cdot D_a t}} \exp\left(-\frac{x^2}{4 D_a t}\right) \tag{1}$$

where C is the tracer concentration in the clay, and x and t the diffusion length and time, respectively.

The analytical solution (Equation 1) does not consider any potential effects related to the finite thickness and sorption properties of the paper filter. To analyze the effect of the filter, a 1D model of the experimental setup and detailed calculations were performed for a cesium experiment confirming the validity of using the analytical solution.

RESULTS AND DISCUSSION

Four experiments with each tracer were performed. A first group of 6 experiments was performed with diffusion times of approximately 120, 180 and 200 days, for Cs, Co and Eu respectively. The diffusion time was enough for obtaining an evolved profile for Cs and Co, but not for Eu. The second group of 6 experiments lasted 200 days for Cs and Co and 433 days for Eu. Good concentration profiles were obtained, even for Eu. All the experiments were fitted and D_a was determined in all cases using Equation 1.

Examples of the concentration profiles are shown in Figure 1 [Co after 200 days (left) and Eu after 433 days (right)].

The apparent diffusion coefficients (D_a) obtained fitting the experimental results with the Equation 1 are: $D_a(Cs) = (5.9 - 8.0) \cdot 10^{-14}$ m^2/s, $D_a(Co) = (2.4 - 3.5) \cdot 10^{-14}$ m^2/s, and $D_a(Eu) = (1.0 - 2.1) \cdot 10^{-15}$ m^2/s. The variation of the calculated D_a is very small in all the cases.

ID experiments with Cs and Co provided similar values but within a wider range: for cesium, $D_a(Cs) = (2.9 - 12) \cdot 10^{-14}$ m^2/s and for cobalt, $D_a(Co) = (1.2 - 4.6) \cdot 10^{-14}$ m^2/s.

Van Loon and Eikenberg [6], using a high-resolution abrasive method for determining a Cs diffusion profile, obtained a $D_a(Cs)$ value of $5.6 \cdot 10^{-14}$ m^2/s (calculated from D_e and K_d values available in their work), which is in good agreement with the ranges of $D_a(Cs)$ presented here.

The "instantaneous planar source" method, followed by a fit with the analytical solution, is therefore a simple and precise method to determine diffusion coefficients for sorbing elements.

279

In order to analyze completely the experiment accounting for the possible role of the paper filter in which the tracer is spiked, a detailed, stochastic 1D model of one experiment with Cs was done using the GoldSim 9.60 computer code [13].

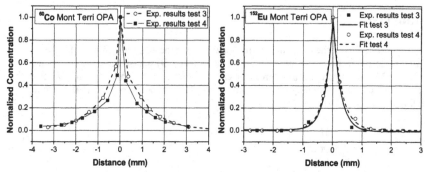

Figure 1. Experimental results of the instantaneous planar source method. Left: Co concentration profiles after 200 days. Right: Eu concentration profiles and fits after 433 days of diffusion.

In the stochastic 1D model wide probability distributions were assigned to the transport parameters of the clay and the paper filter: porosity (θ) and dry density (ρ_d) of the medium, tracer pore water diffusion coefficient (D_p), and tracer distribution coefficient water/solid (K_d). The apparent diffusion coefficient (D_a) is a function of the previous parameters:

$$D_a = \frac{D_p}{\theta + \rho_d(1-\theta)K_d} \qquad (2)$$

The probability distributions for the 8 stochastic parameters were sampled 2000 times, and for each set of values a complete diffusion calculation was done and the activity in each slice was calculated. Of the 2000 calculations performed, only those in which the final concentration in the filter was similar to the measured value were retained. Around 500 individual calculations that fulfilled this condition in the Cs experiment are presented here.

Ten XY graphics were created representing, for a given slice, the calculated activity (Y axis) vs. one of the 8 stochastic parameters, D_a(clay) or D_a(filter) from the approximately 500 calculations selected. Only for D_a(clay) was a clear correlation observed, as shown in Figure 2 (left) for a clay slice near the filter. For the other 9 parameters, correlation with the calculated activity is weak, as can be seen in Figure 2 (right) for K_d(clay) in the same slice. As can be seen in Figure 2 (left) the activity in the slice depends nearly exclusively on D_a(clay). In the other slices the same behavior was observed. As a consequence, comparison of calculated and measured activities in a slice allows us to make a quite precise estimation of D_a(clay).

The clear correlation between D_a(clay) and activity in a given slice has been used to make estimations of D_a(clay). For each slice, a graphic of the calculated activity vs. D_a(clay) was done. From these graphs, and considering an uncertainty of ±5% in the activity measured, a range of D_a(clay) consistent with the activity measured in each slice was obtained. Figure 3 shows the application for a slice, the uncertainty band around the measured activity and the corresponding range of D_a(clay) values. Combining the ranges of D_a(clay) values obtained for all the slices of the Cs experiment, a range of D_a(Cs) = (6.0 - 9.1)·10^{-14} m^2/s was calculated. The value obtained using the simple analytical solution in the same experiment was D_a(Cs) = 8·10^{-14} m^2/s, well centered in the range obtained with the detailed stochastic 1D model.

The results obtained with the detailed 1D model confirm that it is appropriate to use the simple analytical solution to model the experiments and that the effect of the paper filter is of small relevance.

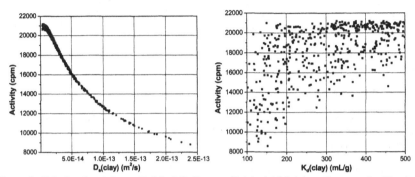

Figure 2. Calculated activity *vs.* D_a(clay) (left) or *vs.* K_d(clay) (right) in a slice near the filter in a Cs experiments using the detailed 1D model of the experiment.

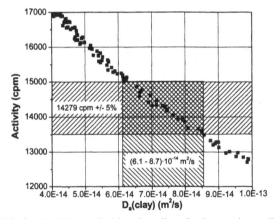

Figure 3. Calculated activity vs. D_a(clay) in a slice of a Cs experiment. The ranges of experimental values (measured value ±5%) and the consistent D_a(clay) value are represented by bands.

CONCLUSIONS

The "instantaneous planar source" was shown to be a simple and useful method to obtain apparent diffusion coefficients for sorbing species (such as Cs and Co) and even for strongly sorbing elements such as Eu in the Opalinus Clay (OPA).

The experiments performed with the instantaneous source model can be modeled using the analytical solution for the instantaneous injection of solute in a 1D semi-infinite medium.

The range of values for the apparent diffusion coefficients perpendicular to the bedding in OPA are: $D_a(Cs) = (5.9 - 8.0) \cdot 10^{-14} \text{ m}^2/\text{s}$, $D_a(Co) = (2.4 - 3.5) \cdot 10^{-14} \text{ m}^2/\text{s}$, and $D_a(Eu) = (1.0 - 2.1) \cdot 10^{-15} \text{ m}^2/\text{s}$. The method seems to be more precise than "conventional" in-diffusion tests where, for Cs and Co, D_a in the following ranges were obtained: $D_a(Cs) = (2.9 - 12) \cdot 10^{-14} \text{ m}^2/\text{s}$ and $D_a(Co) = (1.2 - 4.6) \cdot 10^{-14} \text{ m}^2/\text{s}$.

$D_a(Cs)$ values obtained in this work are in good agreement with the value of $5.6 \cdot 10^{-14}$ m^2/s from Van Loon and Eikenberg [6].

ACKNOWLEDGMENTS

This work has been carried out in the frame of the ENRESA-CIEMAT association and partially funded by the EU within the FUNMIG (Fundamental Processes of radionuclide Migration) Project (Ref. FP6-516514).

REFERENCES

1. NAGRA, *Project Opalinus Clay: Safety report. Demonstration of disposal feasibility for spent fuel, vitrified high-level waste and long-lived intermediate-level waste (Entsorgungsnachweis)*. Technical Report NTB 02-05 (NAGRA, Wettingen, Switzerland, 2002).
2. P. Grathwohl, *Diffusion in Natural Porous Media: Contaminant Transport, Sorption/Desorption and Disolution Kinetics* (Kluver Academic Publishers, 1998).
3. M. Descostes, V. Blin, F. Bazer-Bachi, P. Meier, B. Grenut, J. Radwan, M.L. Schlegel, S. Buschaert, D. Coelho and E. Tevissen, *Appl. Geochem.* **23**, 655 (2008).
4. C. A. J. Appelo and P. Wersin, *Environ. Sci. Technol.* **41**, 5002 (2007).
5. L.R. Van Loon, J.M Soler, A. Jakob and M.H. Bradbury, *App. Geochem.* **18**, 1653 (2003).
6. L.R. Van Loon and J. Eikenberg, *App. Radiation and Isotopes* **63**, 11 (2005).
7. P. Wersin, J.M. Soler, L.R. Van Loon, J. Eikenberg, B. Baeyens, D. Grolimund, T. Gimmi and S. Dewonck, *App. Geochem.* **23**, 678 (2008).
8. M. Mazurek *Mineralogical composition of Opalinus Clay at Mont Terri. A laboratory intercomparison*. Mont Terri Project Technical Note 98-41 (University of Bern, Switzerland, 1998).
9. B. Torstenfelt, B.Allard, H. Kipatsi, *Soil Sci.* **139(6)**, 512 (1985).
10. F.J. Pearson *Opalinus Clay experimental water: A1 type, version 980318*. Technical Report TM-44-98-07 (Paul Scherrer Institut, Villigen, Switzerland, 1998).
11. M. García-Gutiérrez, J.L. Cormenzana, T. Missana, M. Mingarro, J. Molinero, *J. of Iberian Geology* **32(1)** 37 (2006).
12. J. Crank, *The Mathematics of Diffusion*, 2nd edition (Clarendon Press, Oxford, 1975).
13. GoldSim Technology Group. GoldSim User's Guide (2007). http://www.goldsim.com/.

Mater. Res. Soc. Symp. Proc. Vol. 1124 © 2009 Materials Research Society　　　　1124-Q07-07

Migration Behavior of Plutonium in Compacted Bentonite Under Reducing Conditions Controlled With Potentiostat

Kazuya Idemitsu[1], Hirotomo Ikeuchi[1], Syeda Afsarun Nessa[1], Yaohiro Inagaki[1], Tatsumi Arima[1], Shigeru Yamazaki[2], Toshiaki Mitsugashira[3], Mitsuo Hara[3], Yoshimitsu Suzuki[3]
[1]Dept. of Applied Quantum Physics and Nuclear Engineering, Kyushu Univ., Fukuoka, JAPAN
[2]Kobelco Research Institute, Kobe, JAPAN
[3]Tohoku Univ., Oarai, JAPAN

ABSTRACT

The electro-migration method was applied to study the migration behavior of plutonium in compacted bentonite under reducing condition. The Reducing environment was controlled with iron ions supplied by anode corrosion of iron coupon using potentiostat. Ten micro liter of tracer solution containing 1 kBq of ^{238}Pu was spiked on the interface between an iron coupon and bentonite before assembling. The iron coupon was connected as the working electrode to the potentiostat and was held at a constant supplied potential between - 500 to 0 mV vs. Ag/AgCl reference electrode for up to 7 days. Plutonium migrated from anode toward cathode as far as 1mm from the interface, and the penetration length of plutonium grew deeper as the higher potential supplied. This result indicated plutonium migrated as a chemical form with positive charge through bentonite. Concentration profiles were fitted by convection-dispersion equation to obtain two migration parameters, D_a, apparent dispersion coefficient, and V_a, apparent migration velocity. Apparent migration velocity of plutonium was one-tenth of that of ferrous ion at the potential condition of -500 to +0mV vs. Ag/AgCl. Apparent diffusion coefficient of plutonium was estimated as $\sim 10^{-13}$ m^2/s by comparing the values of migration velocity of Pu and Fe. These results indicate that the reducing condition changes the chemical form of plutonium, and accelerate the migration of plutonium in compacted bentonite.

INTRODUCTION

Carbon steel is one of the candidate overpack materials for high-level waste disposal and is expected to assure complete containment of vitrified waste glass during an initial period of 1000 years in Japan [1]. Carbon steel overpack is corroded after closure of the repository. Corrosion products diffuse into buffer materials and then maintain the reducing environment in the vicinity of the repository [2]. The reducing condition is expected to retard the migration of redox-sensitive radionuclides. For example, a rare study on plutonium diffusion in compacted bentonite is available. It has been reported that no movement could be measured for plutonium diffusion in a concrete-bentonite system under an oxidizing condition in an experimental period as long as 5 years [3]. Authors have developed and carried out electro-migration experiments with source of iron ions supplied by anode corrosion of iron coupon in compacted bentonite [4]. Our research group succeeded in migrating plutonium about 1mm in bentonite within a week under a reducing condition by using the electro-migration technique [5]. The result shows that plutonium has a higher diffusion coefficient under the reducing condition in bentonite with iron corrosion products than under oxidizing conditions.

In this study, the electro-migration experiment was carried out varying an electrical potential supplied at iron coupon. It is reported in the previous study that the larger amount of ferrous ion was introduced into bentonite as the higher potential is supplied to iron coupon [6]. The main objective of this study is to accumulate the knowledge about migration behavior of plutonium in compacted bentonite under the reducing conditions at different potentials with this electro-migration technique.

EXPERIMENT

A typical Japanese sodium bentonite, Kunipia-F, was used in this experiment. Bentonite powder was compacted into cylinders with a diameter of 10 mm and a height of 10 mm with a dry density of around 1.4 Mg/m^3 (porosity 0.46). Each compacted bentonite was inserted in an acrylic resin column and saturated with water including 0.01 M of NaCl for one month.

A carbon steel coupon was assembled with bentonite saturated by contacting with water including 0.01M of NaCl into an apparatus for electromigration as shown in Figure 1. Ten micro liter of tracer solution containing 1kBq ($7x10^{-12}$mol) of ^{238}Pu was spiked on the interface between carbon steel coupon and bentonite before assembling. There was a reference electrode of Ag/AgCl and a counter electrode of platinum foil in the upper part of the apparatus with 0.01 M of NaCl solution. The carbon steel coupon was connected to a potentiostat as a work electrode and was supplied electrical potential at -500, -400, -300 or 0 mV vs. Ag/AgCl electrode at 25°C for up to 168 hours. After supplying electrical potential, the bentonite specimen was pushed out from the column and was sliced in steps of 0.3 to 2mm. Each slice was submerged in 1N HCl solution to extract plutonium and iron, and the liquid phase was separated by the centrifugal method. Then the supernatant was taken to be measured with alpha liquid scintillation counter for plutonium and with Atomic Absorption Spectrometry for iron.

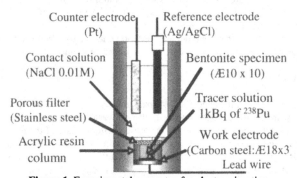

Counter electrode (Pt)
Reference electrode (Ag/AgCl)
Contact solution (NaCl 0.01M)
Bentonite specimen (Æ10 x 10)
Porous filter (Stainless steel)
Tracer solution 1kBq of ^{238}Pu
Acrylic resin column
Work electrode (Carbon steel:Æ18x3)
Lead wire

Figure 1. Experimental apparatus for electromigration.

RESULTS AND DISCUSSION

Profiles of plutonium and iron in bentonite specimens

The concentration profiles of plutonium infiltrated in bentonite specimens at different potentials are shown in Figure 2. The result from previous study [5] with the supplied potential of +300mV is also shown. Plutonium migrated as far as 1 mm from the interface between carbon steel and bentonite, and penetrated into deeper part of bentonite as the higher potential was supplied. This indicated that plutonium migrated as chemical forms of cations.

The concentration profiles of iron in bentonite specimens at different potentials are shown in Figure 3. Iron migrated as far as 8 mm from the interface between carbon steel and bentonite, and penetrated into deeper part of bentonite as the higher potential was supplied.

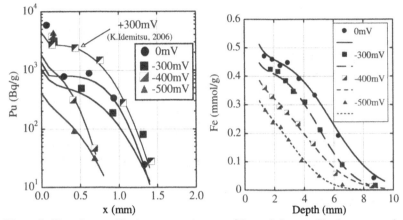

Figure 2. Plutonium profiles in compacted bentonites by electromigration up to 168 h.

Figure 3. Iron profiles in compacted bentonites by electromigration up to 168 h.

Amount of plutonium and iron in bentonite specimen as a function of quantity of electricity

Both penetration depths of plutonium and iron and corrosion current depended on the potential supplied to the work electrode. There should be correlation between the penetration depth and corrosion current as shown in Figure 4. As a matter of fact amount of iron infiltrated into bentonite was a function of quantity of electricity calculated by integrating the corrosion current with respect to time [6]. Therefore iron corroded to ferrous ions at the interface between carbon steel and the bentonite specimen as the following reaction.

$$Fe \rightarrow Fe^{2+} + 2e^- \tag{1}$$

The amounts of iron and plutonium calculated from profiles were plotted as a function of quantity of electricity as shown in Figure 5. Amount of iron migrated into bentonite was a linear function. On the other hand amount of plutonium migrated into bentonite was not a linear function. Beside that, amounts of plutonium migrated into bentonite were less than the amount of plutonium spiked, 1kBq, in all cases. This shows that not all plutonium could migrate into bentonite specimen, however, the total amount of plutonium migrated was increased depending on the amount of ferrous ions introduced into bentonite. Thus we assumed the formation of diffusive trivalent plutonium converted from non-diffusive tetravalent by ferrous ions introduced into bentonite at the interface between carbon steel and bentonite.

$$Pu^{4+} + Fe^{2+} \leftrightarrow Pu^{3+} + Fe^{3+} \tag{2}$$

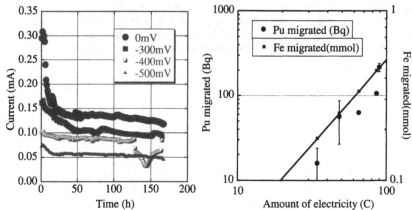

Figure 4. Corrosion currents as a function of time up to 168 h.

Figure 5. Correlation between the amounts of plutonium and iron as a function of quantity of electricity.

Model of electromigration

The movement of ions under the influence of the electric potential gradient, hence electromigration, can be described with the dispersion-convection equation,

$$\frac{\partial C}{\partial t} = D_a \frac{\partial^2 C}{\partial x^2} - V_a \frac{\partial C}{\partial x} \tag{3}$$

where D_a is an apparent dispersion coefficient, V_a is an apparent migration velocity including mostly electromigration of iron and negligible electro-osmotic flow of water. To solve Eq.(3) with the boundary conditions in this experimental configuration, we considered successive spikes of ion in each time step, $M(t)$, which is a function of the corrosion current. Then we can obtain the solution as follows (see ref.[6]in detail) .

$$C(x,t) = \int_0^t \frac{M(t-\tau)}{2\sqrt{\pi D_a \tau}} \left[\exp\left\{ -\frac{(x-V_a\tau)^2}{4D_a\tau} \right\} + \exp\left\{ -\frac{(x+V_a\tau)^2}{4D_a\tau} \right\} \right] d\tau \tag{4}$$

where τ is the migration period of ions introduced into the specimen at '$t - \tau$'. We emphasize that this solution would be correct when the dispersion coefficient is independent of concentration. The amount of ions introduced into bentonite, $M(t - \tau)$, in each time step can be calculated from corrosion current at the time step with assumption of Eq.(1) for iron migration. Though the amount of diffusive plutonium was not linear function of quantity of electricity or the amount of iron introduced, we assume that the amount of plutonium introduced into bentonite in each time step is proportional to the corrosion current to obtain the migration parameters for plutonium.

Apparent advection velocity and dispersion coefficient of plutonium

Fitted curves are also plotted in Figure 1 for plutonium and Figure 2 for iron. The obtained values of V_a and D_a were almost constant in each supplied potential as ca. 10^{-9} m/s and 10^{-13} to 10^{-14} m^2/s for plutonium and ca. 10^{-9} m/s and 10^{-12} m^2/s for iron, respectively. Those parameters are plotted as a function of applied potential in Figure 6.

If a value of apparent mobility of plutonium in bentonite is known, apparent diffusion coefficient of plutonium in bentonite can be calculated by the Einstein's relation.

$$D = \frac{uRT}{ZF} \tag{5}$$

where T is the temperature, Z is the valence of diffusing cation, and F is the Faraday constant, C. The mobility, u, could be obtained from following relation,

$$V_a = uE, \tag{6}$$

where E is an electrical gradient in bentonite specimen. However, the electrical gradient in bentonite cannot be measured directly. So we tried to estimate the apparent diffusion coefficient of plutonium by comparing the migration velocity with iron ions. Applying Eq.(5) and Eq.(6) for both ions, following relation is obtained.

$$\frac{D_{Pu}}{D_{Fe}} = \frac{V_{a,Pu}}{V_{a,Fe}} \cdot \frac{Z_{Fe}}{Z_{Pu}} \tag{7}$$

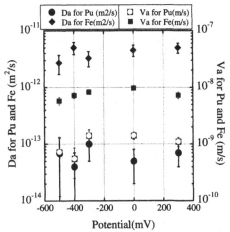

Figure 6. Comparison of apparent migration velocity and dispersion coefficients of plutonium with those of iron.

Here we assumed the ratio of valence described as second term in Eq.(7) is approximately unity. Then the ratio of diffusion coefficients is simply described as the ratio of migration velocities. Figure 6 shows the comparison of migration velocity of both ions. It is found that the migration velocity of plutonium is almost one-tenth of the velocity of iron in each condition. On the other hand it is shown that the dispersion coefficient of plutonium is several-tenth of that of iron in each condition. The dispersion coefficient of iron could include large apparent diffusion coefficient as large as mechanical dispersion coefficient [6]. The dispersion coefficient of plutonium could include much larger apparent diffusion coefficient than mechanical dispersion coefficient but not larger than 10^{-13} m^2/s.

CONCLUSIONS

(1) The total amount of diffusive plutonium in bentonite was increased with the supplied potential. The formation of diffusive plutonium is expected to have been depending on the amount of dissolved ferrous ions introduced into bentonite. It is possible that non-diffusive tetravalent plutonium could be reduced to diffusive trivalent plutonium at the interface between carbon steel and bentonite.

(2) Based on the above assumption, the migration model was constructed, and the two migration parameters, dispersion coefficient and migration velocity, gave best fitting curves for plutonium profiles. The apparent diffusion coefficient of plutonium was estimated as not larger than 10^{-13} m^2/s by the comparison of migration velocity of plutonium and iron.

ACKNOWLEDGMENTS

This research is partly financed by a Grant-in-Aid for scientific research of Japan (contract B19360430) and by the Japan Atomic Energy Agency.

REFERENCES

1. JNC, H12: *Project of Establish the Scientific and Technical Basis for HLW Disposal in JAPAN* (JNC, Tokai Japan, 2000).
2. K. Idemitsu, S. Yano, X. Xia, Y. Inagaki, T. Arima, T. Mitsugashira, M. Hara, Y. Suzuki in *Scientific Basis for Nuclear Waste Management XXV*, edited by B.P. McGrail and G. A. Cragnolono (Mater. Res. Soc. Proc. **713**, Pittsburgh, PA, 2001) pp.113-120.
3. Albinsson, Y., K. Andersson, S. Börjesson, B. Allard, J. Contaminant Hydrology 12, 189 (1996).
4. K. Idemitsu, X. Xia, Y. Kikuchi, Y. Inagaki, T. Arima in *Scientific Basis for Nuclear Waste Management XXVIII*, edited by John M. Hanchar, Simcha Stroes-Gascoyne, Lauren Browning (Mater. Res. Soc. Proc. **824**, Pittsburgh, PA, 2004) pp.491-496.
5. K. Idemitsu, Y. Yamasaki, S. A. Nessa, Y. Inagaki, T. Arima, T. Mitsugashira, M. Hara, Y. Suzuki in *Scientific Basis for Nuclear Waste Management XXX*, edited by D.S. Dunn, C. Poinssot, B. Begg (Mater. Res. Soc. Proc. **985**, Pittsburgh, PA, 2007), NN11-7, pp.443-448.
6. K. Idemitsu, S. A. Nessa, S. Yamazaki, H. Ikeuchi, Y. Inagaki and T. Arima in *Scientific Basis for Nuclear Waste Management XXXI*, edited by W. E. Lee, J. W. Roberts, N. C. Hyatt and R. W. Grimes (Mater. Res. Soc. Proc. **1107**, 2008), pp.501-508.

Mater. Res. Soc. Symp. Proc. Vol. 1124 © 2009 Materials Research Society   1124-Q07-08

State of Compacted MX-80 Bentonite After Simulation of Thermo-Hydraulic Conditions in Deep Geological Storage

Roberto Gómez-Espina and María Victoria Villar
Centro de Investigaciones Energéticas Medioambientales y Tecnológicas (CIEMAT), Avd. Complutense 22. Madrid, 28040, Spain.

ABSTRACT

These studies were developed in the framework of high-level nuclear waste (HLW) disposal in deep geological formations, in which the waste is protected by a group of geological and engineering barriers. In many disposal concepts, one of those engineering barriers is formed by compacted bentonite blocks. Bentonite was chosen as sealing material because of its expansive capacity, low permeability and high plasticity.

The conditions of the bentonite in an engineered barrier for HLW disposal have been simulated in a laboratory test, using a 20-cm long cylindrical block of MX-80 clay that was heated and hydrated for 496 days. This bentonite is from Wyoming (USA), and has been selected in many disposal concepts as backfilling and sealing material (Sweden, Finland, Germany, France).

This paper summarises some results obtained after the test was dismantled. The dry density and water content were measured in sections along the column. Also, soluble salts content, exchangeable cations and the cation exchange complex of the smectite were determined in the same sections.

INTRODUCTION

This study was developed in the framework of the Temperature Buffer Test (TBT project), which is a full-scale test for HLW disposal that aims at improving the current understanding of the thermo-hydro-mechanical (THM) behaviour of buffers with a temperature around and above 100°C during the water saturation transient. The principle of the TBT test is to observe, understand and model the behaviour of the components in the deposition hole, starting from an initial unsaturated state under thermal transient and ending with a final saturated state with a stable heat gradient. The French organisation ANDRA is running this test in cooperation with SKB (Svensk Kärnbränslehantering AB 2005) at the Äspö Underground Research Laboratory (Sweden).

The participation of CIEMAT has consisted of the design, manufacturing and launching of laboratory infiltration tests, and the study of the bentonite after completion of the experiments, in order to support the modelling work.

EXPERIMENT

The experiment consists of a cylindrical cell with two 10-cm long blocks of MX-80 bentonite compacted at a nominal dry density of 1.70 Mg/m^3 with a water content of 16.9 %. The MX-80 bentonite is composed mainly of montmorillonite (65-82%) and it also contains quartz (4-12%), feldspars (5-8%), and smaller quantities of cristobalite, calcite and pyrite [1].

Deionised water was injected at a pressure of 0.01 MPa through the upper lid of the cell. The bottom part of the cell was a flat stainless steel heater, set during the test at a temperature of 140°C, which is the temperature on the surface of the heater of the large-scale TBT test. Over the upper lid of the cell, water was circulated at 30°C. The thermal gradient along the bentonite is not linear, but sharpest towards the heater, the steady temperature at 6 and 16 cm from it being about 77°C and 41°C, respectively. A schematic diagram of the setup and the final appearance of the mounted cell are shown in Figure 1.

The cell was instrumented with capacitive-type sensors (VAISALA HMP233), placed inside the clay at three different levels, at 4, 9 and 14 cm from the hydration surface. The cell was placed over a balance to measure the water intake and the test run for 496 days before dismantling.

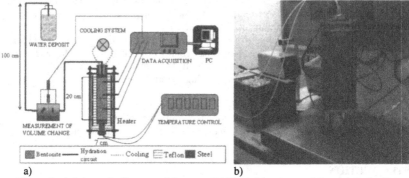

a) b)

Figure 1. a) Schematic diagram of the setup. b) Final appearance of the mounted cell.

The gravimetric water content (w) was determined by oven drying at 110°C for 24 hours, and is defined as the ratio between the weight of water and the weight of dry solid expressed as a percentage. Dry density (ρ_d) is defined as the ratio between the weight of the dry sample and the volume occupied by it prior to drying. The volume of the specimens was determined by immersing them in a receptacle containing mercury and by weighing the mercury displaced.

The soluble salts and pH were analysed by means of aqueous extracts, with a solid/liquid ratio of 1:8. Immediately, the pH was measured with a digital pH-meter. To determine the carbonate and bicarbonate content an automatic titration instrument (Metrohm 702 SM Titrino) was used. The rest of the ions were analyzed by means of a chromatograph (Metrohm 861 Advanced Compact IC).

The exchangeable cations were analyzed by means of aqueous extracts, obtained by adding to the bentonite a 0.5 N solution of $CsNO_3$ with a solid/liquid ratio of 1:8. The Cs displaced the cations in the interlayer and these could be measured in the aqueous extract. The Na^+ and K^+ were measured with a Perkin Elmer 2280 spectrophotometer and the Ca^{2+} and Mg^{2+} with an ICP mass spectrometer (Jobin Yvon JY48-JY38). The value obtained for the soluble cations was subtracted from each exchangeable cation to obtain a reliable value of these.

RESULTS AND DISCUSSION

Final physical state

A steady state was not reached before dismantling. The final water content and dry density of the bentonite are shown in Figure 2. The average value of final water content is 18.3 %, gradually decreasing from the hydration zone (31.1%) towards the heater (0.4%). The final average is not much higher than the initial value, which was 16.9 %. There is also a gradient of dry density caused by the swelling of the bentonite in the hydrated zone (1.47 Mg/m³), and there is an increase near the heater, where it reaches values as high as 1.88 Mg/m³. There has been an overall decrease of the dry density of the bentonite (from 1.70 to 1.66 Mg/m³) caused by the swelling of the clay, and its expansion on dismantling and trimming of the subsamples. The final overall degree of saturation was 67 %.

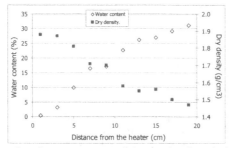

Figure 2. Final gravimetric water content and dry density along the bentonite column.

Geochemistry

The cation exchange capacity has been determined in every section. There are not important changes in the number of exchangeable positions due to the treatment, except in the 3 cm closest to the heater, where this capacity is reduced by more than 10 meq/100g. This phenomenon could be enhanced by the reduction in the number of stacked TOT layers in smectite particles in this zone. On the other hand, there are important changes in the distribution of cations, mainly between Na^+ and Ca^{2+} (Figure 3a).

The cation exchange capacity measured in the initial bentonite is 77 meq/100g, and the average along the column after the treatment is 74 meq/100g. The initial value is similar to that obtained by other authors [1, 2].

The concentration of soluble ions has been measured in aqueous extracts. The chemical composition of the aqueous extracts reflects the bentonite-pore water interactions and the resulting chemical reactions, which affect the distribution profile of each ion. Mass balance calculations, made by adding the average results obtained for the soluble cations and the value of each exchangeable cation indicate a very good fit for all the cations, especially Na^+ (Figure 3b).

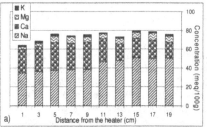 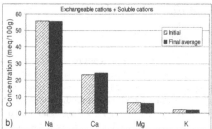

Figure 3. a) Changes in the cation exchange capacity along the column and distribution of exchangeable cations. b) Mass balance calculations for the main exchangeable cations.

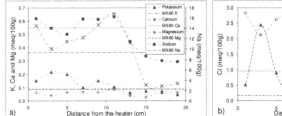

Figure 4. a) Concentration of principal soluble cations along the bentonite column. b) Concentration of soluble sulphate and chloride along the column. The horizontal lines indicate the values for the original material.

Overall, we can appreciate that the concentration of soluble potassium, sodium and calcium increases towards the heater. Meanwhile, the magnesium concentration remains constant along the column (Figure 4a). The main anions are sulphate, chloride, carbonate and bicarbonate (Figure 4b and Figure 5a).

The mass balance indicates that the final average content in sulphate and chloride are greater than the initial values. These increases can be attributed to mineral dissolution processes. In the case of sulphate the presence of pyrite could explain its increase. Sulphate and chloride move towards the heater along the column, the latter being much more mobile and generating a front at 3 cm from the heater. Villar *et al.* [3] observed in a similar series of tests performed with a different bentonite, that the chloride front moved quickly and increased its concentration over time.

The contents of carbonate and bicarbonate are highly dependent on pH. The pH decreases from the hydration zone to the heater, possibly due to the increase in the CO_2 partial pressure toward the hottest zones. Likewise, the pH in the aqueous extract could change during the analysis with respect to the pH in the bentonite during the TH treatment due to changes in the CO_2 partial pressure, changing the carbonate and bicarbonate relations. For this reason, it seems better to evaluate alkalinity changes along the column (Figure 5b). There is a clear decrease in alkalinity with respect to the initial value, possibly due to precipitation of calcite, dolomite or magnesite.

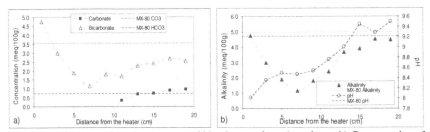

Figure 5. a) Concentration of carbonate and bicarbonate along the column. b) Concentration of alkalinity and pH along the column.

Figure 6 shows the value measured in each section for several soluble cations and also the initial value of the soluble cation modified by taking into account the changes in the cationic exchange complex measured in each section (calculated). The latter would reflect the changes expected in the soluble cations values if only exchange reactions would have taken place. There is an increase of soluble Na^+ with respect to the initial value towards the heater, although it is lower than expected due to the exchange reactions, what could point to a process of backward diffusion.

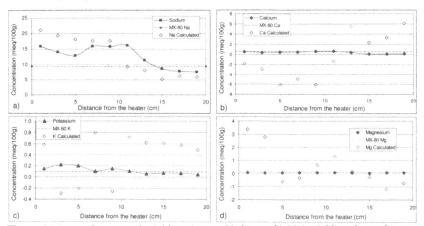

Figure 6. Measured content of soluble cations and balance of initial soluble cation and measured exchangeable cation for (a) Na^+, (b) Ca^{2+}, (c) K^+ and (d) Mg^{2+}.

Calcium is washed from the hydrated zone towards the heater. Since the final average content of Ca2+ in the system is higher than the initial one, some dissolution of calcite must have taken place. However, the main processes controlling the Ca2+ concentration are the interlayer cationic exchange and the movement of the cation towards the heater.

Potassium does not move along the column, but there must be precipitation of some mineral species containing it, since its content shows an overall reduction after the experiment. Precipitation of magnesium minerals must have also taken place near the heater.

293

CONCLUSIONS

The thermo-hydraulic treatment of the compacted bentonite generates water content and dry density gradients along the column, as well as changes in the geochemistry of the material.

The cation exchange capacity does not suffer changes with the treatment except in the hottest zone of the column, where there is a decrease in its value. The main change in the cation exchange complex is the increase in Ca^{2+} with respect to Na^+, especially in the hottest half of the column.

The content of soluble ions in the bentonite is conditioned by different processes, such as advection, cation exchange and dissolution/precipitation.

The overall salinity of the pore water increases towards the heater. The contents of soluble potassium, sodium and calcium, as well as those of chloride and sulphate, are below the original value near the hydration surface and increase towards the heater. In the case of sodium, this increase is due to exchange reactions by which sodium in the interlayer of the smectite is replaced by calcium, sodium being released into the pore water. There seems to be a movement of sodium from the heater towards the hydration zone by diffusion. Chloride and sulphate probably move by advection from the hydrated zones towards the heater, as well as calcium.

Dissolution and precipitation processes have been also detected after the TH treatment. It is possible that the major precipitation processes (K- and Mg- rich minerals, carbonates) occur in the heater zone due to movement of the water from this zone to cooler zones. On the other hand, dissolution of chlorides, sulphates and carbonates could occur in the wet zones due to hydration.

ACKNOWLEDGMENTS

The research agreement CIEMAT/CIMNE 04/113 has financed this work. The collaboration of A. Ledesma (Technical University of Catalonia, Barcelona) and Manuel Velasco (DM-Iberia, Madrid) in the design of the infiltration tests is gratefully acknowledged. The laboratory work has been performed at the Soil Mechanics Laboratory of CIEMAT (Madrid, Spain) by R. Campos and J. Aroz. The software for the data acquisition of the tests has been developed by J.M. Barcala from CIEMAT.

REFERENCES

1. M.V. Villar, R. Gómez, and P.L. Martín, *Behaviour of Mx-80 Bentonite at Unsaturated Conditions and Under Thermo-Hydraulic Gradient –Work Performed by Ciemat in the Context of the TBT Project*, Technical Report CIEMAT/DIAE/M2146/1/06v1 (CIEMAT, Madrid, 2006).

2. O. Karnland, S.Olsson, U. Nilsson and P. Sellin, *Mineralogy and sealing properties of various bentonites and smectite-rich clay materials*, SKBTR-06-30 (Swedish Nuclear Fuel and Waste Management Co, Stockholm, 2006).

3. M.V. Villar, A.M. Fernández and R.Gómez, *Effect of Heating/Hydration on Bentonite: Test in a 60-cm Long Cell*. NF-PRO Project, Deliverables 2.2.7 and 3.2.10. CIEMAT Technical Report (CIEMAT, Madrid, 2006).

Mater. Res. Soc. Symp. Proc. Vol. 1124 © 2009 Materials Research Society   1124-Q07-09

Temporal Evolution of the Concrete-Bentonite System Under Repository Conditions

Elena Torres, Alicia Escribano, María J. Turrero, Pedro L. Martin, Javier Peña, María V. Villar
Division of Engineered and Geological Barriers, CIEMAT, Avda. Complutense 22, 28040,
Madrid, Spain

ABSTRACT

Concrete could be part of the barrier system and/or be needed during repository construction for the management of the High Level Radioactive Waste (HLW) in a Deep Geological Repository (DGR). Depending upon the design, concrete could be physically in contact with the bentonite, or be sufficiently close that alkaline pore fluid from the cement may interact with the bentonite, affecting its properties. A comprehensive study based on a series of short term experiments is being performed to provide experimental evidences on the physical, chemical and mineralogical changes during the concrete-compacted bentonite interaction. Concrete and bentonite samples were analyzed by means of XRD, FTIR spectroscopy and SEM-EDS. In addition, in bentonite sections, swelling capacity measurements, specific surface area (BET) and chemical analysis for cation exchange capacity and soluble salts were performed.

INTRODUCTION

A Deep Geological Repository (DGR), based on the concept of multibarriers, is the solution proposed by most countries for the management of High Level Radioactive Waste (HLW). In order to isolate the wastes, a series of barriers are inserted between the wastes and the biosphere.

The Spanish repository design contains concrete to support the access galleries and in the final sealing of access routes. Concrete will be in contact with the geological clay formation and the bentonite barriers. Cement in contact with the bentonite barrier will be a source of alkaline fluids in wet conditions. Concrete pore fluids are in quasi-equilibrium with the solid hydrated phases of the matrix [1]. Ca solubility is principally controlled by $Ca(OH)_2$. Sulphate solubility is limited by ettringite. The Mg contents are controlled by $Mg(OH)_2$ and Al solubility by C_3AH_6. Si concentrations are determined by the solubility of alkali-containing C-S-H gels. Alkaline cement pore water enhances the solubility of many radioactive elements.

Furthermore, concrete degradation generates a diffusive alkaline plume which can affect the swelling and transport properties of the bentonite, as well as the properties of the adjacent host rock and groundwater. Many studies have been performed in order to study the influence of alkaline media on bentonite. Eberl et al. [2] reported that the formation of illite or illite/smectite depends on the solution concentration rather than on temperature and time of reaction.

Cuevas et al. [3] have carried out a set of batch (alkaline solution and bentonite, 25-200°C) and column experiments (granitic and alkaline water-mortar-bentonite, 25-120°C) experiments in order to study the influence of alkaline media on FEBEX bentonite. The main phases identified in the alkaline reaction of FEBEX bentonite are phillipsite, Mg-clays, analcime, tobermorite and C-S-H gels. The formation of a new smectitic trioctahedral phase was confirmed, as well.

The scope of the present study is to provide experimental data about alteration products at the concrete/bentonite interface and the related changes of the bentonite and concrete properties under these conditions.

EXPERIMENTAL

Materials

The tests were performed with the Spanish reference bentonite (FEBEX bentonite) from the Cortijo de Archidona deposit (Almería, Spain). FEBEX bentonite[4] has a content of dioctahedral smectite of the montmorillonite type of 92±3%. It contains variable quantities of quartz (2±1%), plagioclase, cristobalite (2±1%), potassium feldspar (traces), calcite (traces) and trydimite (traces). The cation exchange capacity is 102 meq/100 g, and the exchangeable cations are Ca (35±3 meq/100g), Mg (31±3 meq/100g), Na (27.1±0.2 meq/100g) and K (2.6±0.4 meq/100g). The cement used in the tests is a Sulfate-Resistant Portland Cement CEM I 42.5 R/SR [5]: its chemical composition is shown in Table I.

Table I. Chemical composition of the CEM I 42.5 R/SR cement.

CHEMICAL COMPOSITION (WT. %)	SIO_2	AL_2O_3	FE_2O_3	CAO(TOT AL)	MGO	SO_3	NA_2O	K_2O	CAO(FR EE)
	19.6	4.43	4.27	64.5	0.95	3.29	0.11	0.28	1.92

Experimental Set up

The tests were carried out in hermetic cylindrical cells in which a concrete slab was placed on the top of the compacted bentonite block. For hydration, a saline solution, whose composition is described in Table II, was used. The hydration solution is injected under a pressure of 12 bars. A plane heater constitutes the bottom of the cell and on the top of the cell a chamber allows the circulation of water at a controlled temperature (around 22°C), lower than that of the heater (100°C), so a gradient of temperature is established. A hydration channel crosses the upper chamber and allows the hydration of the sample through a stainless steel sinter. The body of the cell is made out of Teflon, although an external steel cylinder prevents its deformation swelling. The clay, with its water content at equilibrium with the laboratory conditions, is uniaxially compacted outside the cell to a dry density of 1.65 g/cm^3. The cells are inside a methacrylate chamber, under vacuum, to prevent oxidation. Six medium cells were assembled. In the context of NF-PRO project, three cells were dismantled after 6, 12 and 18 months. The remaining cells are planned to operate for at least two years more. Two sensors were installed in each cell to monitor the evolution of water content and temperature with time. Sensors were located 50 and 95 mm from the top of the cell, at the hydration zone and the interface, respectively.

Table II. Chemical composition of the saline solution.

Na(M)	Ca(M)	Mg (M)	SO_4^{2-} (M)	Cl⁻(M)	HCO_3^-(M)	pH = 7.54 pE = -3.16 log P CO_2 = -2.65
1.3 E-01	1.1 E-02	8.2 E-02	7.0 E-02	2.3 E-02	1.8 E-03	

Methods

Three types of samples were analyzed: (1) the precipitates found at the concrete/bentonite interface and some millimetres far away from interface; (2) bentonite samples at different distances from the interface, named sections 1, 2 and 3; and (3) concrete sections representative of the whole slab. The samples collected were analyzed by using different techniques: X-ray diffraction (XRD) [Philips Xpert-MPD diffractometer], Fourier transform-infrared spectroscopy (FT-IR) [Nicolet 4700 in the range of 4000 to 400 cm^{-1}], Scanning Electron Microscopy (SEM) with microanalysis (XEDS) [JEOL JSM 6400], and N_2-BET specific surface area methods. Cation exchange capacity (Cu-triethylenetetramine method) [6], and exchangeable cations (CsNO$_3$ method) [7] analysis were also performed.

RESULTS

Concrete degradation

Carbonation represents the main degradation process observed in the concrete slabs from the dismantled cells. Calcite precipitates in the hydration zone to a greater extend than in the rest of the slab and increases with time (Figure 1). Leaching of the most soluble elements, such as alkali hydroxides, leads to the reprecipitation of portlandite (Figure 2) and brucite on the bottom part of the concrete blocks. The effects of temperature, humidity and hydration with a solution rich in sulphates favour the formation of ettringite [$Ca_6Al_2(SO_4)_3(OH)_{12} 26(H_2O)$] and gypsum ($CaSO_4 \cdot H_2O$). Ettringite was found in the 12-month test (Figure 3). In concrete tested for 18 months, together with ettingite, gypsum was also formed (Figure 4), possibly resulting from the substitution reaction between portlandite and sulfate, either from the saline solution or from the dissolution of initial gypsum. Although gypsum is a common component of Portland cement, the amount measured, by means of XRD and TGA analysis, in the 18-month reacted concrete was much greater than the average content of gypsum found in the unaltered sample.

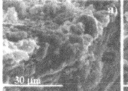

Figure 1. Calcite: (a) in the hydration zone of the cell after 6 months and (b) inside of a pore after 12 months.

Concrete/bentonite interface

A continuous layer of portlandite was found in the three cells (Figure 5a). According to the EDS line profile, in the 18 months test, the layer is 2 μm thick (Figure 5b and 5c). Portlandite formation is a consequence of the previous leaching of the matrix concrete by diluted groundwaters providing pore water oversaturated with respect to portlandite.

297

Figure 2. Portlandite: (a) inside of a pore after 6 months, (b) covering the sample surface after 12 months, and (c) detail of the crystals.

Figure 3. Ettringite: in a cell dismantled after 12 months.

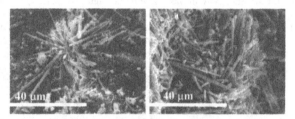

Figure 4. Gypsum crystals after 18 months.

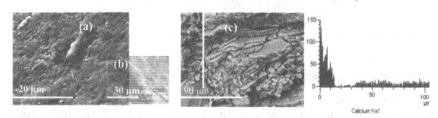

Figure 5. (a) Portlandite precipitate layer on the bentonite surface from the 18 month test. (b) Thickness of the precipitate layer. (c) Bentonite interface profile and variation of the Ca concentration (cps) from interface.

Below the portlandite layer, CSH gels were formed. Tobermorite-type phases (Ca/Si molar ratio of 0.6), were identified together with brucite and ettringite (Figure 6). Also, in the 18-month test, marks of dissolution of quartz were detected. This process enables the precipitation of CSH minerals such as tobermorite.

X-ray diffraction confirmed the formation of a smectitic trioctahedral phase. The peak found at 1.53 Å is typical of saponite (Figure 7a). In the FTIR spectrum of the sample collected at the interface, a small band was observed at 669 cm⁻¹. This band could correspond to the Mg_3OH bending vibration of saponite. XRD analysis of the oriented aggregates of the fraction below 2 μm shows an increase of the d-spacing. The increase of the interlayer space in the smectite could result from the formation of a brucite-saponite-smectite mixed phase. This process does not depend on time, as it can be observed in figure 7b.

Figure 6. New phases formed at the concrete-bentonite interface.

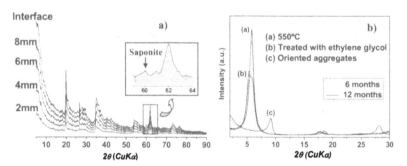

Figure 7. (a) XRD patterns of the bentonite at different depths in the 18 month test, (b) XRD diffraction profile of the oriented aggregates.

Bentonite

Regarding exchangeable cations (Figure 8), Ca^{2+} and K^+ do not manifest consistent variations along of the compacted bentonite block. Nevertheless, an increase of Na^+ and a drop in Mg^{2+} is observed at the interface. At the interface, Mg^{2+} is released from exchangeable positions and is introduced into octahedral sites. Because of this Mg^{2+} mobility, saponite can be formed.

Cation Exchange Capacity (CEC) in bentonite tends to rise as we approach the interface. As a result of CSH precipitation, there is a marked increase in the CEC values.

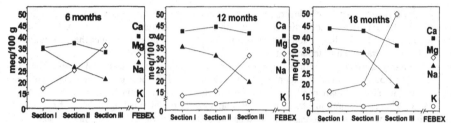

Figure 8. Distribution of the exchangeable cations in bentonite in the different experiments (from the interface, section 1: 3 to 25 mm; section 2: 25 to 50 mm; section 3: 50 to 78 mm).

The temperature gradient induces the movement of soluble salts towards the heater. Cl^-, SO_4^{2-} and Na^+ increase in the bentonite section closer to the heater.

CONCLUSIONS

Due to hydration in the cells, the leaching of salts contained in the porous and soluble concrete compounds induces the precipitation of new mineral phases at the interface of both materials. The secondary mineral phases formed are: CSH-gel tobermorite-type (Ca/Si = 0.6), portlandite, brucite and ettringite. The interlayer space in the smectite at the interface corresponds to a brucite-saponite-smectite mixture phase. Chloride, sulfate and sodium contents increase near the heater, resulting in the precipitation of NaCl near the heater.

Mineralogical and chemical changes were observed in the bentonite in contact with concrete. CEC values increase because of the formation of calcium silicate hydrate gel. A significant variation in the magnesium content and sodium at the exchange positions was measured.

ACKNOWLEDGEMENTS

This work has been financially supported by ENRESA and the EU NF-PRO project under contract FIGW-CT-2003-02389.

REFERENCES

1 M. García-Gutiérrez, U. Alonso, T. Missana, M. Mingarro, N. Granizo, M. Grivé, X. Gaona, E. Colás, L. Duro and J. Bruno, *Estudio bibliográfico sobre sorción y difusión de radionucleidos en cementos, hormigones y productos de corrosión en presencia de cementos*, Report PT-02/2007, (ENRESA, Madrid, 2007).
2 D.D. Eberl, B. Velde and T. McCormick, *Clay Miner.* **28**, 49 (1993).
3 J. Cuevas, R. Vigil de la Villa, S. Ramírez, L. Sánchez, R. Fernández, and S. Leguey, *J. of Iberian Geology* **32**, 151 (2006).
4 A.M. Fernández, B. Baeyens, M. Bradbury and P. Rivas, *Phys. Chem. Earth, Parts A/B/C*, **29(1)**, 105 (2004).
5 A. Hidalgo, S. Petit, C. Domingo, C. Alonso, C. Andrade. *Cem. Concr. Res.* 37, 67 (2007).
6 L.P. Meier and G. Kahr, *Clays and Clay Miner.* **47(3)**, 386-388 (1999).
7 B.L. Shawhney. *Clays and Clay Miner.* **18**, 47 (1970).

Mater. Res. Soc. Symp. Proc. Vol. 1124 © 2009 Materials Research Society   1124-Q07-10

Evolution of the Geochemical Conditions in the Bentonite Barrier and Its Influence on the Corrosion of the Carbon Steel Canister

Elena Torres [1], Alicia Escribano [1], Juan L. Baldonedo [2], María J. Turrero [1], Pedro L. Martín[1], Javier Peña [1], María V. Villar [1]

[1] Division of Engineered and Geological Barriers, CIEMAT, Avda. Complutense 22, 28040, Madrid, Spain
[2] Centro de Microscopía Electrónica, Universidad Complutense de Madrid, Avda. Complutense s/n, 28040, Madrid, Spain

ABSTRACT

This work has been focused on characterization of geochemical processes occurring in the bentonite barrier and their influence on corrosion of the carbon steel container. In the cells used for these tests, a block of compacted bentonite was subjected simultaneously to constant hydration and heating, in opposite directions, in order to simulate the conditions of the clay barrier in the repository and better understand the coupled THMC (Thermo-Hydro-Mechanical-Chemical) processes that can affect the performance of the bentonite barrier or the metallic container. In these tests, carbon steel was substituted by Fe powder in order to enhance corrosion phenomena. At present, only two out of the six cells assembled have been dismantled after 6 and 15 months. In both cases, an advective movement of salts towards the heater has been observed. CEC values increased in saturated areas. In the exchange complex, Mg was replaced by Na in saturated areas, whereas in unsaturated zones close to the heater, Mg is the prevalent exchangeable cation. Goethite and hematite were the main corrosion products found in the 6-month and 15-month tests, respectively.

INTRODUCTION

Carbon steel is one of the candidate materials for the waste package in the Spanish High-Level Waste (HLW) Disposal Project. These waste packages will be emplaced in a granitic host formation. As buffer material, FEBEX bentonite (Na-Ca-Mg bentonite) has been chosen by the Spanish radioactive waste management agency as the reference bentonite.

The behavior of a HLW repository is determined by the characteristics of design of the engineered barrier system. The combined effect of the heat generated by radioactive decay and the ingress of groundwater can provoke changes in the mechanical, hydraulic and geochemical properties of the bentonite barrier. Geochemical conditions in the bentonite barrier surrounding the carbon steel container can compromise canister performance.

Chloride and sulfate move by advective transport towards the metallic container due to the constant hydration and heating of the clay barrier. Both chlorides and sulfates are hygroscopic salts. They are known to enhance corrosion of mild steel at relative humidity (RH) below 100% [1]. The aqueous salt films formed on the steel surface allow electrochemical corrosion to occur. Bulk aqueous solutions will form at values equal or greater than the equilibrium RH of the saturated salt solution. However, at RH values below the equilibrium value there is an increase in the weight gain when compared to the corrosion in the absence of the deposited salt [2].

The aim of this work is, on one hand, to identify the different geochemical processes taking place in compacted bentonite as a function of the variations of the Thermo-Hydro-Mechanical (THM) parameters and, on the other hand, to characterize the corrosion products formed at an Fe/bentonite interface during the unsaturated state of bentonite.

EXPERIMENT

Materials

The experimental set-up consists on hermetic cells (figure 1a) where a compacted block of FEBEX bentonite (1.65 g/cm^3) in contact with Fe powder at its bottom is hydrated with reduced granitic water (Grimsel, Switzerland) (see table I) on the top while a thermal load is applied from the bottom. The body of the cell is made out of Teflon, although an external steel cylinder prevents its deformation by swelling. A plane heater (100°C) constitutes the bottom of the cell while, on the top of the cell, a chamber allows the circulation of water at a controlled temperature (around 22°C), so a thermal gradient is established. Two sensors, placed at 25 and 80 mm from the top of the bentonite block, record the evolution of RH and temperature as the hydration front advances.

Table I. Chemical composition of the granitic water used in the tests (Grimsel Test Site, Switzerland).

Chemical species	Na	Ca	Mg	K	SO_4^{2-}	Cl^-
Concentration (ppm)	15.0	5.9	<0.4	---	6.3	5.7
pH 9.72		E_h -200 mV		Alk (HCO_3^-)		23 ppm

FEBEX bentonite is composed of 92±3% dioctahedral smectite of the montmorillonite type. It contains variable quantities of quartz (2±1%), plagioclase (2±1%), cristobalite (2±1%), potassium feldspar (traces), calcite (traces) and trydimite (traces). The cation exchange capacity is 102 meq/100g, and the exchangeable cations are Ca (35±3 meq/100g), Mg (31±3 meq/100g), Na (27.1±0.2 meq/100g) and K (2.6±0.4 meq/100g). The water content of the clay at laboratory conditions is about 13.7 ± 1.3 %.

Methods

To determine the exchangeable cations of the samples, a $CsNO_3$ solution was used to displace the exchangeable cations [3]. Soluble elements were analysed in aqueous extract solutions at a solid to liquid ratio of 1:8 (5 g of clay in 40 ml of deionized water reacted for 24 h.). Additional measures of soluble salts were realized at the interface. The sampling of the bentonite block is shown in figure 1b.

Iron powder was analysed by means of Transmission Mössbauer Spectroscopy (TMS), Scanning Electron Microscopy (SEM) coupled to Energy Dispersive X-ray Spectroscopy (EDS), Transmission Electron Microscopy (TEM) and Scattered Area Electron Diffraction (SAED).

RESULTS AND DISCUSSION

THM behavior

The distribution of water content along the bentonite columns in the tests occurs soon after the beginning of the experiment. As observed in figure 2, RH increases in the coldest end of the block, where hydration is applied. However, close to the Fe/bentonite interface, where temperature is about 100°C, bentonite loses its absorbed water. Desiccation seems to take place rather quickly and affects especially the section closest to the heater, where RH values are lower than the initial values.

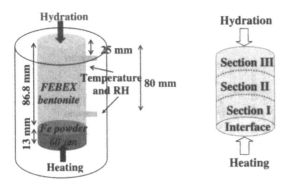

Figure 1. *a) Scheme of the cylindrical cells used; b) Sampling of the bentonite block.*

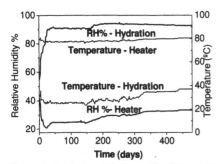

Figure 2. Evolution of Relative Humidity and Temperature recorded by the sensors installed at 0.7 (heater) and 6.2 (hydration) cm from the Fe/bentonite interface.

Geochemical characterization

The hydration of bentonite leads to the dissolution of the soluble accessory minerals in the bentonite (sulfates, carbonates and chlorides). The anions can have a significant influence on corrosion processes. Carbonates can precipitate at the interface as siderite ($FeCO_3$), once the barrier is fully saturated. Chlorides and sulfates are hygroscopic salts and their precipitation can favour initiation or enhancement of corrosion processes, even at low RH. As expected, soluble carbonates were only detected in saturated areas (934 and 980 mg of CO_3^{2-} per kilogram of dry bentonite, for the cell dismantled after 6 and 15 months, respectively). Figure 3 shows the advance of the chloride and sulfate fronts towards the heater. In the cell dismantled after 15 months, it is observed that most of soluble chloride is concentrated at the interface. So, it seems that after the initial precipitation of chlorides when heating started, there was a later precipitation that resulted from the advective transport of chloride towards the heater. Sulfate concentration is controlled by gypsum solubility. So, in both tests, the sulfate concentration at the interface is much smaller than that of chloride.

Soluble cations, measured in the aqueous extracts, increased as well with time in all the sections, especially in section I and at the Fe/bentonite interface. Cations move much more slowly than anions. Sodium is the main counterion (figure 4). Ca increases with longer times due to dissolution of calcite in the hydrated zones.

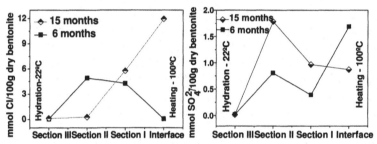

Figure 3. Movement of chlorides (left) and sulfates (right) along the bentonite block (aqueous extract solid:liquid 1:8).

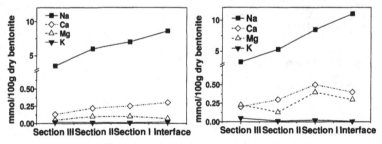

Figure 4. Na, Mg, Ca and K measured in the aqueous extracts: (left) 6-month, (right) 15-month test.

The distribution of the exchangeable cations is modified by the thermal gradient (figure 5). Na seems to substitute Mg in saturated areas (sections III and II), whereas near the Fe/bentonite interface, Na is replaced by Mg in the exchange complex (section I). Magnesium complexes are thought to be more stable at high temperature than Na, Ca or K-complexes [4]. This fact would be in good agreement with the experimental data obtained in both experiments.

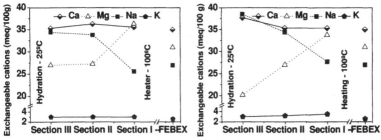

Figure 5. Distribution of exchangeable cations along the bentonite block: (left) 6 months (right) 15 months.

Corrosion products

Different corrosion products were found in the Fe powder depending on the distance from the interface. Oxides were found in both experiments at the interface (figure 6). In the case of the cell dismantled after 6 months, room temperature Mössbauer spectra of the bentonite collected at the interface shows a sextet (quadrupolar splitting: -0.27 mm/s; isomeric shift: 0.25 mm/s; hyperfine field: 100-350 KOe) typical of goethite. In the FTIR spectra recorded for the Fe oxide found at the interface after 15 months, there is a doublet placed at 570 and 478 cm^{-1}. These bands are characteristic of hematite. EDS analysis detected high contents of chlorine (up to 5% at. in some cases), and traces of common elements present in bentonite either in goethite or in hematite.

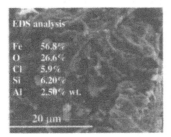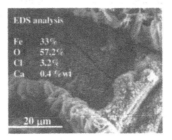

Figure 6. SEM images of Fe oxides found at the Fe/bentonite interface: (Left) goethite in the six-month test; (Right) hematite formed after 15 months of experiment. (EDS analysis was performed over the whole image).

Iron powder that was not in direct contact with bentonite seemed to undergo slight corrosion. In the cell dismantled after 6 months, iron powder away from the bentonite interface kept its metallic luster and no corrosion products could be identified in it. In the cell dismantled after 15 months, corroded and non-corroded areas were observed throughout the iron powder. EDS analysis of corroded Fe powder detected, in most cases, traces of chlorine (0.6% at.) and Ca

(0.3% at.). Precipitation of chloride was not homogeneous and where it was found, a thin layer of hematite (α-Fe$_2$O$_3$) (figure 7 left) and maghemite (γ-Fe$_2$O$_3$), was grown (below 1 μm in all cases). At this stage, corrosion may continue, however to a lesser extent, as RH at the interface is below 50%. Although the chloride content measured at the interface in the 15-month test is very high, close to 7% wt. (figure 7 right), no Cl-containing phases have been identified.

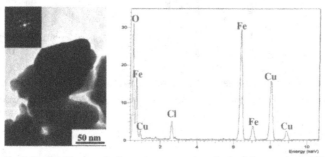

Figure 7. (Left) TEM image of a hematite suspension obtained from corroded Fe powder and SAED pattern of one of the crystals; (right) EDS analysis of the diffracted crystal (Cu lines corresponds to the copper grid in which the hematite suspension was deposited for TEM observation).

CONCLUSIONS

Initial precipitation of chloride plays a relevant role in the first stages of the corrosion process, as it helps to initiate the nucleation of goethite. When bentonite at the interface gets desiccated, goethite can transform into hematite. Once the chloride front reaches the interface and precipitates onto the Fe powder, it seems that corrosion is enhanced again. Results obtained in these tests are preliminary and will be completed after the dismantling of the four experiments on-going at the moment.

ACKNOWLEDGEMENTS

This is a contribution to the NF-Pro project IP number FIGW- CT-2003-02389 financed by the EU and the CIEMAT/ENRESA association.

REFERENCES

1.   V. Kucera and E. Mattson, Atmospheric Corrosion in *Corrosion Mechanisms*, edited by F. Mansfeld (Marcel Dekker, Inc., New York, 1987), p.211-252.
2.   J. C. Estill and G.E. Gdowski, in *Proceedings of the Seventh International Conference on High Level Radioactive Waste Management*, edited by Holly A. Dockery, (ASCE, New York, 1996), pp. 457-458.
3.   B.L. Shawney, Clays and Clay Miner, 18, 47 (1970).
4.   J. Cuevas, M.V. Villar, A.M. Fernández, P. Gómez and P.L. Martín, Appl. Geochem, 12, 473 (1997).

Mater. Res. Soc. Symp. Proc. Vol. 1124 © 2009 Materials Research Society   1124-Q07-11

A Thermodynamic Approach on the Effect of Salt Concentration on Swelling Pressure of Water-Saturated Bentonite

Haruo Sato[1]
[1]Japan Atomic Energy Agency, 432-2 Hokushin, Horonobe-cho, Hokkaido 098-3224, Japan

ABSTRACT

The effect of salt concentration on swelling pressure (dP_{ext}) of water-saturated bentonite was calculated based on the thermodynamic data of water at montmorillonite surface and of water in solutions of various salinities coming in contact with the montmorillonite. Activities (a_w) of water at montmorillonite surface were obtained as a function of water content and temperature by a vapor pressure method, and the relative partial molar Gibbs free energies (dG_w) of the water were determined.

Water affected from the montmorillonite surface was estimated to be almost all interlayer water. dP_{ext} versus montmorillonite partial density was estimated for solutions of various salinities and compared to data measured for various montmorillonite contents and silica sand contents. The calculated dP_{ext} decreased with increasing salinity. The effect of salinity on dP_{ext} was not clear in the measured data due to the scattering in the measured data. This cause is presumed to be due to the increase of the ionic strength of porewater by dissolution of soluble minerals.

INTRODUCTION

In the safety assessment of the geological disposal for high-level radioactive waste in Japan, bentonite is used for its physical and chemical functions as backfilling and buffer material of the engineered barrier system. Since the major clay mineral constituent of bentonite is montmorillonite which has the nature of swelling, bentonite swells by contacting with groundwater and restricts the groundwater flow. And diffusion field is finally formed. Since swelling pressure (dP_{ext}) develops by hydration of the interlayer of montmorillonite, understanding the thermodynamic properties of water on the montmorillonite surface, especially of interlayer water, is important.

It has been reported that activity (a_w) of the interlayer water of montmorillonite and the chemical potential are smaller than those of free water [1-3]. It also has been reported that the moisture potential and suction increased with decreasing water content (W_c) of the bentonite [4, 5], being indicated that water at montmorillonite surface has a high-moisture potential and a high suction. Thus, in view of thermodynamics, water at montmorillonite surface and water existing in ultra-narrow spaces such as interlayer water should not be treated similarly to free water.

The moisture potential of montmorillonite is closely related to dP_{ext}. Kahr et al. [6] have predicted dP_{ext} values for 2 kinds of bentonites (Montigel, MX-80) versus respective dry densities based on the water vapor adsorption-desorption isotherms, and shown that predictions are consistent with experimental results. Kanno and Wakamatsu [5] also have predicted dP_{ext} values for the Kunigel-V1 bentonite based on the moisture potential and arrived to similar conclusions. In a recent study, Tanaka et al. [7] have calculated dP_{ext} values for several kinds of bentonites (Kunigel-V1, Volclay, Kunibond, Neokunibond) versus concentration of artificial saline water and respective effective bentonite densities, based on the repulsion force by osmotic pressure derived from electric potential distribution between montmorillonite stacks, the gravitation in the van der Waals force and the repulsion force that originates in peculiar rigidity to montmorillonite crystal, and roughly explained the measured data excepting Ca-bentonite (Kunibond). Many dP_{ext} data have been reported for various bentonites with different montmorillonite contents (Kunigel-V1, Kunipia-F, MX-80, Kunigel OT-9607, Volclay, Kunibond, Neokunibond) so far [4, 6-10]. It is known from past studies that volumetric swelling ratio and dP_{ext} of bentonite decrease with increasing salinity and that they are affected by montmorillonite content, interlayer cation, dry density of bentonite, silica sand content, saturation degree and temperature [7-10].

Summarizing earlier studies, dP_{ext} values against dry density and salinity have been reported for various bentonites, but few data on the effect of temperature have been reported. The model concerning dP_{ext} is divided into what based on electric potential distribution from the surface of clay mineral and the one based on water adsorption-desorption characteristics at the surface of the clay mineral, and previous measured dP_{ext} values can be explained approx. by both models. However, such conventional models depend on the origin of bentonite, montmorillonite content, etc., and the generality to other kinds of bentonites and different conditions is scarce. Thereon, the author has developed a general model in a recent study which can calculate dP_{ext} for various montmorillonite contents and dry densities and different silica sand contents based on the thermodynamic data of water at montmorillonite surface [3].

In this study, the effect of salinity on dP_{ext} of bentonite was further calculated based on the thermodynamic data of water at montmorillonite surface and of water in solutions of various salinities. This paper presents a general model which can quantitatively calculate dP_{ext} for various montmorillonite contents of bentonites saturated with solutions of various salinities, in addition to measurements of a_w values of water at the montmorillonite surface.

EXPERIMENTAL

Preparation of purified Na-montmorillonite and experimental conditions

Purified Na-montmorillonite, with interlayer cation completely exchanged with Na^+ ions, was prepared in this study. The reason that bentonite was purified is because the ionic strength of porewater changes by dissolution of soluble impurities such as calcite contained in the bentonite and because the nature of exchangeable cation in the interlayer of montmorillonite affects swelling properties. It is generally known that dP_{ext} and swelling volume of bentonite decrease with an increase of the ionic strength of solution [7-9]. Therefore, controlling the ionic strength of porewater and interlayer cation is important for interpreting the mechanisms. The starting material, the Kunipia-P bentonite (Kunimine Industries Co. Ltd.), contains approx. 100wt.% montmorillonite, was used. Details of the purification process of the Na-montmorillonite are described in [3, 11]. The cation exchange capacity (CEC) of the Na-montmorillonite was 109.6±16.2meq/100g [12]. This CEC is at similar level as that (116.5meq/100g) of the Kunipia-F bentonite, for which montmorillonite content is approx. 100wt.% [13].

The measurements were performed in triplicate. Montmorillonite samples with a size of ϕ10x5mm were used.

Vapor pressure measurements

Figure 1 illustrates an experimental apparatus for measuring vapor pressure. The apparatus consists of a digital manometer (DM2, Shibata Scientific Technol. Ltd.) and a cylindrical stainless steel container, into which a thermocouple is inserted. Purified montmorillonite powder dried at 110°C was filled into an acrylic sample holder in a sample cell to obtain dry densities of 0.6 and 0.9Mg/m³. Totally 3 samples (1 for 0.6 and 2 for 0.9Mg/m³) were prepared. The montmorillonite in the sample holder was initially moisturized with distilled water and the sample

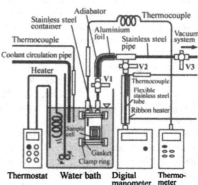

Fig. 1. Experimental apparatus for the measurement of water vapor pressure.

cell was disassembled. The surface of the sample holder was wiped off and again assembled by using dried parts. This sample cell was put in a stainless steel container and then placed in a constant-temperature water bath. The inner diameter of the stainless steel pipe is about 4mm, which is substantially longer than the mean free path of water vapor ~5x10^{-5}m at the low-temperature region of the vapor pressure measurements (ca. 0.1kPa) [2]. All valves (1, 2 and 3) were initially opened. The container was evacuated to reach vacuum for about 10min by a rotary pump, and then valve 3 was closed. Vapor pressure and temperature in the container were periodically measured with accuracies of 0.1kPa and 0.1°C, respectively. These measurements were repeated until the vapor pressure reached a constant. Thus, water vapor pressure in equilibrium with porewater was measured at a temperature interval of 3 to 5°C as a function of W_c (0-83%) and temperature (15-40°C).

At the end of a series of vapor pressure measurements, the sample cell was removed from the container and weighed. And the sample cell was placed again in the container and evacuated to lower the W_c for a while (ca. 10min). The aforementioned experimental procedure was repeated. Vapor pressure during cooling was also measured to check for the reversibility.

After all vapor pressure measurements, the montmorillonite samples were dried at 110°C for obtaining the dry density and for determining W_c. No vapor pressure was detected for this dried sample over the range of temperature.

Determinations of thermodynamic parameters

The activity of porewater is determined by the following relation;

$$a_w = P_w / P^\circ_w, \tag{1}$$

where a_w is the activity of the porewater, P_w the measured vapor pressure of the porewater at 25°C, and P°_w the vapor pressure of pure water at 25°C (3.168kPa) [14].

The relative partial molar Gibbs free energy of porewater is determined by the following relation;

$$dG_w = RT \ln(P_w / P^\circ_w), \tag{2}$$

where dG_w is the relative partial molar Gibbs free energy, R the gas constant (8.314J/mol/K), and T the absolute temperature.

Since the enthalpy of vaporization of porewater is obtained from the temperature dependence of vapor pressure by the following Clausius-Clapeyron equation;

$$d\ln P_w / d(1/T) = dH_V(s)/RT, \tag{3}$$

the relative partial molar enthalpy of the porewater is determined by the following relation;

$$dH_w = dH^\circ_V(H_2O) - dH_V(s), \tag{4}$$

where $dH_V(s)$ is the enthalpy of vaporization, dH_w the relative partial molar enthalpy, and $dH^\circ_V(H_2O)$ the enthalpy of vaporization of pure water (44.0kJ/mol) [14].

Moreover, the relative partial molar entropy is determined from the following relation;

$$dG_w = dH_w - TdS_w, \tag{5}$$

where dS_w is the relative partial molar entropy.

RESULTS AND DISCUSSION
dG_w of porewater

Figure 2 shows the dG_w values of porewater at montmorillonite surface against W_c obtained in this study together with those of water on the Kunipia-F bentonite [2] at 25°C. The dG_w showed a tendency to decrease with a lowering of W_c in the region where W_c is lower than about

40%. Similar results have been obtained for the Kunipia-F bentonite reported by Torikai *et al.* [2]. This trend indicates that the porewater changed into water of low-chemical potential, which is more stable in the region where W_c is lower than about 40%. Representative data of the measured results of the water vapor pressure, the calculated results of a_w and dG_w of the porewater, and a_w and dG_w values for some representative electrolytes at 25°C are described in [3, 11]. The a_w and dG_w values for those electrolytes were calculated based on respective osmotic coefficients [15]. Moreover, from the correlation between a_w versus W_c and the specific surface area of montmorillonite (800m²/g), water affected from montmorillonite surface was estimated to be almost all interlayer

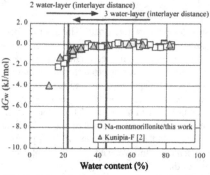

Fig. 2. dG_w values of porewater at the surface of montmorillonite vs. W_c obtained in this study and those of water on the Kunipia-F bentonite [2] at 25°C.

water. The W_c values (22.4 and 44.8%) corresponding to the left side and right side of bold lines in Fig. 2 are respectively equivalent to when montmorillonite surface was hydrated by 1-layer (2 water-layer in interlayer) and 2-layer (4 water-layer in interlayer) of water molecules on the average.

Modelling of swelling pressure and its validation

Figure 3 illustrates a concept of the chemical potential balance of water in equilibrium state between a solution of an arbitrary salinity and bentonite saturated with the solution through filter. Where, the α phase is the solution of an arbitrary salinity in equilibrium with its vapor, the β phase is the solution-saturated bentonite in equilibrium with its vapor, and the salinities in both phases are assumed to be a constant. If the α and the β phases reached equilibrium by penetration of the solution to the bentonite, the chemical potentials in both phases are equivalent.

In this system, the free energy change (dG) between both phases in equilibrium state acts as swelling energy to take the energy balance in both phases. The dG can be calculated as below assuming that the relative partial molar Gibbs free energy of water in the α phase is dG_s and that that in the β phase is dG_w.

$$dG = dG_s - dG_w \qquad (6)$$

Where, dG_s and dG_w are the relative partial molar Gibbs free energy in the α and the β phases, respectively.

The chemical potentials of waters in respective phases are expressed as follows;

$$\mu°(W, \alpha) = RT \ln(P^\alpha_w / P°_w) \qquad (7)$$

$$\mu°(W, \beta) = RT \ln(P^\beta_w / P°_w) + \int_{P°_{ext}}^{P_{ext}} V_w dP, \qquad (8)$$

where, $\mu°(W, \alpha)$ and $\mu°(W, \beta)$ are the chemical potentials of waters in the α and the β phases, respectively, P^α_w and P^β_w the vapor pressures of the waters in the α and the β phases, respectively, P_{ext} and $P°_{ext}$ the external pressures at saturated and

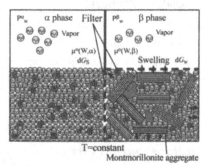

Fig. 3. Concept of the chemical potential balance of water in equilibrium state between a solution of an arbitrary salinity and bentonite saturated with the solution through filter at a constant temperature.

310

dried conditions, respectively, and V_w the specific volume of water (18.0686cm^3/mol at 25°C).

Since the chemical potentials of waters in both phases are equivalent and $P°_{ext}$=0 at dried condition, dP_{ext} is finally calculated from the following relations assuming that V_w can be regarded as a constant between P_{ext} and $P°_{ext}$;

$$dP_{ext}=P_{ext}-P°_{ext}= -RT/V_w \ln(P^\alpha_w /P°_w)-RT/V_w \ln(P^\beta_w /P°_w)=(dG_s-dG_w)/V_w. \qquad (9)$$

Swelling pressure against montmorillonite partial density under saturated condition was calculated for solutions of various salinities ([NaCl]=0.5-3.4M) based on Eq. (9), and compared to data measured for various montmorillonite contents and silica sand contents of bentonites (Kunigel-V1, Kunipia-F and MX-80, Kunigel-V1 and Kunipia-F with silica sand) [4, 8-10]. In the model calculation, the dG_w values, calculated based on the best fit curve derived as the correlation between dG_w and W_c, were used. The best fit curve, derived by the least squares method, is as follows;

$$dG_w(W_c)=m_1[0.5erf\{(W_c+m_2)/m_2\}+0.5erf\{(W_c-m_2)/m_4\}-1], \qquad (10)$$

where d$G_w(W_c)$ is the dG_w at W_c (%) derived from the least fit curve, m_1=2.3786, m_2=21.081, m_3=52.857, and m_4=8.1004. The correlation coefficient in this case is 0.98007.

Montmorillonite partial density was calculated assuming montmorillonite contents of 48wt.% for Kunigel-V1, 75wt.% for MX-80, and 99wt.% for Kunipia-F. Water-saturated montmorillonite partial density was converted by the following relation [3, 11];

$$\rho_m=100\rho_{th}\rho_w/(W_c\rho_{th}+100\rho_w)|_{fm=1}, \qquad (11)$$

where ρ_m is the montmorillonite partial density, ρ_{th} the solid density of montmorillonite (2.7Mg/m^3), ρ_w the density of water (0.997044Mg/m^3 at 25°C), and f_m the montmorillonite content in the bentonite ($0 \leq f \leq 1$ and montmorillonite content is 100wt.% at f_m=1).

In addition, the montmorillonite partial density for various montmorillonite contents and different silica sand contents was calculated based on relation given in [16]. Figure 4 shows the comparison of the calculated and measured dP_{ext} values [2, 4, 8-10] for solutions of various salinities versus montmorillonite partial density. The montmorillonite partial density of 0.9Mg/m^3 is just equivalent to the dry density of 1.6Mg/m^3 with a silica sand content of 30wt.% for Kunigel-V1, which is the reference condition in the 2nd progress report [17]. The calculated dP_{ext} clearly decreased with increasing salinity, while little difference

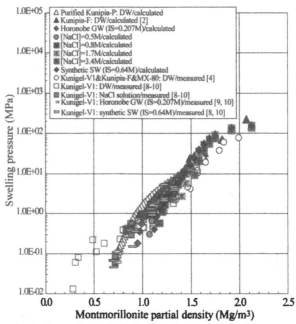

Fig. 4. Comparison of calculated and measured swelling pressures [2, 4, 8-10] for solutions of various salinities vs. montmorillonite partial density (IS: ionic strength, DW: distilled water, GW: groundwater, SW: saline water).

was found in a range of high montmorillonite partial density where is higher than $1.6Mg/m^3$. This cause is considered to be because the dG_w of interlayer water which is predominant in the region of high montmorillonite partial density is quite low. The effect of salinity on dP_{ext} of bentonite was not clear in the measured data due to the scattering in the measured data. This cause is presumed to be due to the increase of the ionic strength (salinity) of porewater by dissolution of soluble minerals such as calcite contained in the bentonite. The effect of salinity on dP_{ext} of bentonite can be regarded to be at most within the range of the scattering in the measured data.

CONCLUSIONS

The a_w values of water at montmorillonite surface were obtained as a function of W_c and temperature, and dG_w values of the water were determined. Water affected from the montmorillonite surface was estimated to be almost all interlayer water. In addition, an empirical correlation between dG_w and W_c was derived by the least squares fitting to plots of dG_w versus W_c.

Based on the thermodynamic data (dG_w and dG_s) of interlayer water and of water in solutions of various salinities and the empirical correlation between dG_w and W_c, a general model, which can quantitatively calculate swelling pressure for different montmorillonite contents of bentonites saturated with solutions of various salinities, a given dry density of bentonite and different silica sand contents of bentonites, was developed.

REFERENCES

1. Y. Torikai, S. Sato, and H. Ohashi, in Scientific Basis for Nucl. Waste Manag. XVIII, edited by T. Murakami and R. C. Ewing, (Mater. Res. Soc. Symp. Proc., **353**, Pittsburgh, PA, 1995), pp.321-328.
2. Y. Torikai, S. Sato, and H. Ohashi, *Nucl. Technol.*, **115**, 73 (1996).
3. H. Sato, in Proc. 15th Int'l Conf. on Nucl. Eng., April 22-26, 2007, Nagoya, Japan, Paper No.: ICONE15-10207, 7 pages (in CD-ROM format) (2007).
4. H. Suzuki, T. Fujita, and T. Kanno, PNC Tech. Rep., PNC TN8410 92-057 (1992).
5. T. Kanno and H. Wakamatsu, in Scientific Basis for Nucl. Waste Manag. XVI, edited by C. G. Interrante and R. T. Pabalan, (Mater. Res. Soc. Symp. Proc., **294**, Pittsburgh, PA, 1993), pp.425-430.
6. G. Kahr, F. Bucher, and P. A. Mayor, in Scientific Basis for Nucl. Waste Manag. XII, edited by W. Lutze and R. C. Ewing, (Mater. Res. Soc. Symp. Proc., **127**, Pittsburgh, PA, 1989), pp.683-689.
7. Y. Tanaka, T. Hasegawa, and K. Nakamura, Civil Eng. Res. Lab. Rep., Central Res. Inst. of Electric Power Industry, No. N07008 (2007).
8. H. Suzuki and T. Fujita, JNC Tech. Rep., JNC TN8410 99-038 (1999).
9. H. Kikuchi and K. Tanai, JNC Tech. Rep., JNC TN8430 2004-005 (2005).
10. Japan Atom. Ener. Agency, Buffer Material Database, http://bufferdb.jaea.go.jp/bmdb/ (2006).
11. H. Sato, *Physics and Chemistry of the Earth*, **33**, S538 (2008).
12. H. Sato, in Proc. 2007 Fall Meeting of the Atom. Ener. Soc. of Japan, Sep. 27-29, 2007, Kitakyushu, I34 (2007).
13. M. Ito, M. Okamoto, M. Shibata, Y. Sasaki, T. Danbara, K. Suzuki, and T. Watanabe, PNC Tech. Rep., PNC TN8430 93-003 (1993).
14. R. A. Robinson and R. H. Stokes, *Electrolyte Solutions*, 2nd ed. (Butterworths, London, 1959).
15. Nihon-Kagakukai (Chemical Society of Japan) ed., *Kagaku-Binran, Kisohen II (Handbook of Chemistry, Basic version II)*, 2nd ed. (Maruzen, Tokyo, 1975).
16. H. Sato and S. Miyamoto, *Appl. Clay Sci.*, **26**, 47 (2004).
17. Japan Nucl. Cycle Develop. Inst., JNC Tech. Rep., JNC TN1410 2000-004 (2000).

Mater. Res. Soc. Symp. Proc. Vol. 1124 © 2009 Materials Research Society 1124-Q08-07

Limitations on Radionuclide Release From Partially Failed Containers

Lubna K. Hamdan and John C. Walton
Environmental Science and Engineering, University of Texas at El Paso, 500 W. University Ave., El Paso, TX, 79968-0626, U.S.A.

ABSTRACT

Over time, due to physical and chemical disturbances, nuclear waste packages at the Yucca Mountain repository are anticipated to fail gradually and evolve into a mixture of corroded locations and discrete objects of intact Alloy-22 and other waste package materials. Radionuclides leak from waste packages by dissolution and transport in water.

In this paper, we address the condition of a partially failed waste container in which multiple penetrations potentially allow water to flow through it. In this situation, residual heat release in the waste is expected to create a capillary pressure gradient and set up flow systems in the relict protected areas. In these systems water flows into the protected area toward the warmest region and evaporates, effectively sequestering radionuclides in the relict sheltered areas. We explore analytically the impacts of these processes on radionuclide release from the waste packages.

INTRODUCTION

To ensure the future protection of the public health, it is essential to understand the processes actually controlling radionuclide release at Yucca Mountain (YM). Once radionuclides are released from the Engineered Barrier System (EBS) at YM into the host rock, over time they will make it out into the accessible environment. Since source term – the transport mechanisms determining the amounts and types of radionuclides released – is an area with high conceptual uncertainty, the models used to estimate radionuclide release from YM may be largely based upon processes that do not in general occur.

Here we develop a simple conceptual model of radionuclide release from a partially failed waste container and explore the implications of this model on the performance of the proposed repository at YM. This model addresses a serious failure of a waste package, where multiple penetrations potentially allow water to flow through the waste package.

THEORY

One of the biggest challenges facing radionuclide release rate analysis is the fact that the geometry and properties of the inside of a failed waste container are unknown, and basically unknowable. For example, the corrosion rate of Alloy-22, stainless steel, and Zircaloy, may or may not proceed in the estimated order, since the corrosion rate predictions are based on short term data in perhaps non representative environments. In general, small cells of evaporation and condensation, inside a failed waste container, are likely to leave some portions of the waste package metals in more benign and others in more harsh corrosion environments. As a result, it will be difficult to predict which parts will endure the longest. However, in the long run, and under such conditions, the waste package is likely to evolve into a mixture of corroded locations (i.e., porous media) and relict discrete objects. Inside such a heterogeneous medium, capillary breaks (i.e., discontinuous liquid transport pathways) may or may not exist, further complicating the radionuclide transport issue.

Given all these uncertainties, it will be possible to further understand the processes controlling radionuclide release rate only if these processes are robust and not tightly related to many of the evolving material properties. One potential controlling process is heat release from radioactive waste over time. Figure 1 shows the average rate of thermal energy release per volume over time from radioactive waste packages (pressurized water reactor [21PWR], and defense glass logs [DHLW-L]) [1]. Figure 1 indicates that heat release continues, although at a very low rate, beyond one million years.

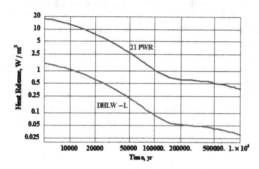

Figure 1. The rate of heat release per volume from radioactive waste packages over time, plotted on a logarithmic scale. Small, but significant, heat release continues up to one million years.

Stagnant zone conceptual model

Based on the above information about the geometry and environment inside a waste package, we develop a conceptual model to evaluate the behavior of radionuclides in the stagnant portions of a partially failed waste container. Residual heat release in the waste package is anticipated to set up flow systems in the relict protected areas (areas with no seepage). A simple schematic for this model is shown in Figure 2. Evaporation at hotter areas, typically where the greatest concentration of heavy metal is present, creates a capillary pressure gradient that drives water (in the absence of direct seepage or condensate), with its dissolved and suspended constituents, to flow or wick toward the warmest region; that is, advective transport for radionuclides occurs toward heat source. As the water evaporation continues, the concentration of radionuclides in these regions increases and radionuclides start to precipitate. Meanwhile, the resultant gradient in radionuclide concentration causes diffusion transport of radionuclides in the opposite direction (away from hotter regions; see Figure 2).

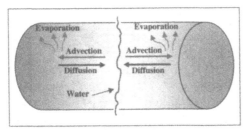

Figure 2. A schematic drawing illustrates the physical processes in a partially failed waste package. Evaporation in the hotter regions causes water to flow toward these regions (advective transport), while the increase in radionuclide concentration (due to evaporation) causes a diffusion transport in the opposite direction.

Since the liquid flow is toward the heat source, radionuclide release from these areas not only must happen by diffusion transport (i.e., swimming upstream), but also the rate of diffusion transport should be greater than the rate of advective transport. The ratio of advection to diffusion is given by the dimensionless Peclet number:

$$Pe = \frac{VL}{D} \qquad (1)$$

Where V is inward advective velocity (m/s), L is length of non drip region (m), and D is the effective diffusion coefficient (m²/s).

In the case of direct seepage or condensate, diffusion and advective transport are in the same direction; therefore $Pe > 0$ and there will be a rapid release for radionuclides. On the other hand, in the relict protected areas, if $0 > Pe > -1$, diffusion transport is greater than advective transport and a gradual release will occur, while if $Pe < -1$, the advective transport is greater and sequestration process for radionuclides will take place instead of release.

NATURAL ANALOG

Nature provides us with a straightforward demonstration of the flow system and sequestration process mentioned in the flow-through conceptual model. Figure 3 shows a picture taken at the 1880's era silver mine on North Percha Creek, NM. To slow the release of contaminated water from the mine tunnels, the mouth of the mine was blocked with two berms of mine tailings. Although there is a significant hydraulic head difference (~20 cm/100 cm) that drives water to flow through the berms, efflorescent crusts form on the top of the wet tailings (see Figure 3). When water evaporates at the upper surfaces of the wet tailings, it creates a gradient in the capillary pressure that causes water to rise upward in the berms. As the water continues to evaporate, the concentration of dissolved species increases near the tailings/air interface and minerals begin to precipitate, forming the efflorescent crusts.

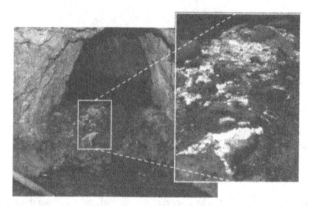

Figure 3. North Percha Creek mine. Efflorescent crusts form on the top of the wet tailings, when water evaporates at the upper surfaces. Water evaporation leads to a gradient in capillary pressure that causes water to rise upward in the berms.

In this system, the advective transport for the minerals from the tailings forming the berms is greater than the diffusion transport (i.e. Pe < -1). As a result, a sequestration process occurs instead of release and the minerals accumulate at the top. This situation is similar to that in a failed waste package, where liquid water flows toward the heat source, except that water evaporation in the mine is driven by the low humidity air near the drift entrance, rather than residual heat release (as in the waste packages).

DISCUSSION

Steady state, one-dimensional calculations were made for the concentration profile of an infinitely soluble salt over distance to give a quantitative perspective for radionuclide transport in stagnant portions of the inside of a partially failed waste package (see Figure 4). These calculations are based on equating inward transport by advection with outward transport by diffusion [2].

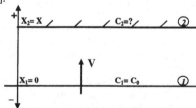

Figure 4. Schematic draw shows the boundary conditions used for the one dimensional calculation of salt concentration over distance.

Advective velocity is estimated from the heat generation of the waste assuming half of this heat transfer leads to water evaporation. In these calculations, the first order continuity equation for advective and dispersive transport in groundwater (soil is the closest representation for the rubble inside a waste package) is used, considering the steady state [3]. The effective diffusion is estimated using Archie's Law, which assumes that the porosity and tortuosity corrections are equal to volumetric water content squared [4].

Figure 5 shows the concentration profile of the salt over distance, in terms of the ratio of salt concentration at any point (x; downstream) to the initial point (x = 0; see Figure 4). These ratios are calculated at different times. Figure 5 indicates that the concentration of salt increases

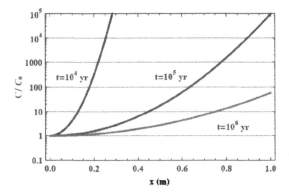

Figure 5. The ratio of salt concentration (C(x) / C(x=0)) vs. distance, calculated at three different times.

as we move inside, toward the heat source (downstream). As expected, the concentration ratios are greater in earlier times, since the heat release is greater. The distance (x) can be viewed as the size of the area sheltered from seepage. Figure 5 indicates that during early time periods, even very small sheltered areas prevent leaching of radionuclides. The concentration ratio will not reach these large values, shown in Figure 5, since salts and compounds have limited solubility and start to precipitate after critical concentration. Nevertheless, these calculations demonstrate the transfer process of salts toward the heat source, under the conditions illustrated in the flow-through conceptual model.

Figure 6 shows the concentration ratio of the salt (at x = 1 m) over time, calculated for several volumetric water contents. Figure 6 indicates that the sequestration process is more effective in lower moisture environments (lower volumetric water contents). These results are reasonable, since radionuclides are released by water transport.

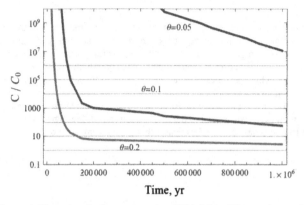

Figure 6. The ratio of salt concentration (C(x) / C(x=0)) over time, calculated for different values of volumetric water content.

CONCLUSIONS

Conceptual model calculations showed that heat-generating waste stored in a partially saturated environment is inherently safe, in the sense that, in the absence of strong advection, radionuclides tend to concentrate rather than disperse. Regions of partially failed waste packages which have been historically assumed to lead to diffusional release [5], instead mostly result in sequestration. The sequestration process is stronger at higher rates of heat generation and lower moisture conditions.

ACKNOWLEDGMENTS

The Nye County Nuclear Waste Repository Project Office funded this work through funding provided by the U.S. Department of Energy Office of Civilian Radioactive Waste Management. We also thank the Center for Environmental Resource Management of The University of Texas at El Paso for their support.

REFERENCES

1. Bechtel SAIC Company, 2005. Multiscale Thermohydrologic Model. ANL-EBS-MD-000049, Rev. 03, Bechtel SAIC Company.
2. J. C. Walton, "Fate and Transport of Contaminants in the Environment", First edition, College Publishing, Glen Allen, USA (2008).
3. J. C. Walton, "Fate and Transport of Contaminants in the Environment", First edition, College Publishing, Glen Allen, USA (2008).
4. S. Roy and S. Tarafdar, The American Physical Society 55(13), 8038 (1997).
5. CRWMS M&O 2008. Total System Performance Assessment - 2008: Model/Analysis for the License Application. MDL-WIS-PA-000005 REV00. Las Vegas, Nevada: CRWMS M&O.

Mater. Res. Soc. Symp. Proc. Vol. 1124 © 2009 Materials Research Society 1124-Q10-07

Dependence of the Dynamic Behavior of Supersaturated Silicic Acid on the Surface Area of the Solid Phase

Yuichi Niibori[1], Yasunori Kasuga[1], Hiroshi Kokubun[1], Kazuki Iijima[2] and Hitoshi Mimura[1]

[1]Dept of Quantum Science and Energy Engineering, Tohoku University, Aoba-ku, Sendai, 980-8589 Japan.

[2]Geological Isolation Research and Development Directorate, Japan Atomic Energy Agency, Tokai-mura, Naka-gun, Japan.

ABSTRACT

Cement is an essential material to construct a geological disposal system. Such a material (which is stable in high pH) may alter the groundwater up to 13 in pH. In Japan, the water table is predicted to be shallow compared to the depth level of the repository system (deeper than 300 m). The backfill will condition the groundwater entering the repository to high pH. Since the solubility of silicic acid is very large in pH>9, its mixing with natural groundwater would cause silicic acid supersaturation. So far, the authors have reported that the deposition layer (amorphous), resulting from the supersaturated silicic acid, strongly affect the sorption of RNs. The current study examined the precipitation rates of the silicic acid, in order to evaluate the area of altered surface surrounding the repository.

In the results, when the initial supersaturated concentration was set to 2.5 mM or 5.6 mM, the initial precipitation rates of the silicic acid depended on the surface area of the solid phase. However, the kinds of the solid phase did not affect the precipitation rates. With the increment of temperature, the precipitation rates increased. Moreover, the precipitation rate constant, k, was evaluated in the initial precipitation rates. The value of k was around 2.0×10^{-10} (m/s) in this experiment condition. This value was suggested to be sufficiently large to limit the altered surface area, even if the advection effect of groundwater occurs.

INTRODUCTION

The use of cement to construct the repository system (e.g., for so-called TRU waste) would bring high pH plume of groundwater [1-4]. Since the solubility of silicic acid is very large in pH>9 [5], the silicic acid becomes supersaturated in the mixing zone of high pH plume with the surrounding groundwater (pH=8). Then, in the downstream zone, the supersaturated silicic acid alters the surface of flow-path around the repository. However, if the altered area is limited, its effect on the migration of RNs (radionuclides) is negligibly small. Such altered area depends on flow rate of groundwater and the precipitation rate of supersaturated silicic acid.

This study examined the precipitation rate of the silicic acid onto the solid surface with relatively large surface area. Iler [6] deduced the precipitation behaviors of supersaturated silicic acid in the presence of solid phase such as silica or silicate minerals. When the silicic acid concentration is supersaturated and above 3.3 mM to 5.0 mM (amorphous silica solubility at pH 8 is around 1.9 mM), the polymerization occur in the addition to the precipitation of the monomeric silicic acid onto the surface. Such dynamic behavior of silicic acid may depend also on the specific surface area of solid phase. Recently, Conrad et al. [7] summarized the polymerization of silicic acid and examined the kinetics of silica nanocolloid formation and precipitation (with no addition of solid phase). However, such processes seem different from that including the solid phase [2-3]. Moreover, the underground environment (natural barrier) has large surface area in the flow paths of underground water such as fractures in rock matrix [8]. In this study the precipitation rates of supersaturated silicic acid onto the solid surface were evaluated, in order to understand the behavior of silicic acid in the mixing zone of high pH plume with the surrounding groundwater.

EXPERIMENTAL

Mallinckrodt silica powder (amorphous silica), opal-CT and quartz were prepared as the solid sample. Minerals were ground and classified into a size fraction of 74–149 μm particle diameters by sieving. Table I shows the Brunauer-Emmett-Teller (BET) specific surface areas (use of N_2 gas) of the solid samples. Na_2SiO_3 solution (250 ml, pH>10, 298 K) was poured into a polyethylene vessel containing the solid sample, HNO_3 and a buffer solution (a mixture of MES (2-morpholinoethanesulfonic acid, monohydrate) and THAM (tris(hydroxymethyl) aminomethane)). The pH of the solution was set to 8. (At the time, the silicic acid initially in a soluble form at pH >10 becomes supersaturated.) Both concentrations of the soluble silicic acid and the polymeric silicic acid were respectively monitored by using silicomolybdenum-yellow method and ICP at each predetermined time. That is, soluble silicic acid was defined as silicic acid reacting with molybdate reagent and coloring yellow, and polymeric silicic acid was defined as silicic acid in liquid phase except for soluble silicic acid. Its total concentration of silicic acid was measured by ICP-AES. It was confirmed in pre-experiments that the polymeric silicic acid can pass through the filters of not only 0.45 μm but also 0.21 μm [4].

The initial supersaturated concentration of the silicic acid was initially set to a given value of 2.5 mM, 4.8 mM or 5.6 mM, by considering the dependence of the solubility (pH=8) on the temperature. The temperature was kept constant within ±0.5 K in the range of from 288 K to 323 K. The solid sample was prepared with a given weight within ±0.1 mg in the range of from 0.0 g (no addition of solid phase) to 1.0 g. In these experiments, nitrogen gas continuously passed through the polyethylene vessel to avoid contact with air. The silicic acid solution including the solid sample was mechanically stirred with a polypropylene stirrer.

Table I. BET (N_2 gas) specific surface area of the solid samples.

Solid phase (SiO_2)	Specific surface area /$10^3 m^2 kg$
Mallinckrodt silica powder	350
Quartz	1.00
Opal-CT	16.9

RESULTS AND DISCUSSION

As the results under the given experimental conditions, the concentration of the polymeric silicic acid was negligibly small compared to that of the soluble silicic acid in the presence of the solid phase except for quartz. Figure 1 shows the behaviors of the soluble silicic acid in the conditions of 0.0 g (no addition of solid sample) and 0.5 g in weight of solid sample (Mallinckrodt silica powder). The initial supersaturated concentration, C_{ini}-C_e, was set to 4.8 mM, where C_{ini} is the initial concentration of soluble silicic acid (pH>10) and C_e is the solubility of soluble silicic acid (pH=8) in the presence of amorphous silica. (Since the solubility depends on temperature, the value of C_{ini} prepared initially was different.) As shown in figure 1(a) and figure 1(b), the dependence of the kinetics on temperature became quite opposite. Generally, it is well-known that the polymerization needs nucleus formation at the initial process [e.g., 6,9]. However, the presence of the solid phase does not always need such nucleus for the supersaturated silicic acid. In fact, the polymeric silicic acid was not detected in the condition of figure 1(b). In other words, the supersaturated silicic acid gradually deposited onto the solid surface.

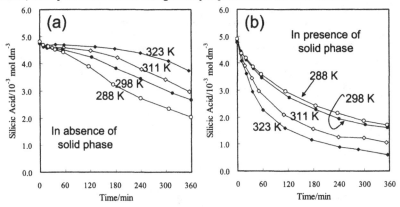

Figure 1. The dynamic behaviors of soluble silicic acid in the cases without solid phase and with solid phase. (Initial supersaturated concentration: 4.8 mM, Solid sample: Mallinckrodt silica powder, 0.5 g.)

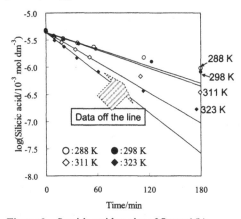

Figure 2. Semi-logarithm plot of figure 1(b).

Figure 2 is the data of figure 1(b) in semi-logarithmic form. The experimental data thus represent a slowing of the initial-rate, as a result of the decrease in surface area. This would suggest the change of specific surface area caused by the precipitation of supersaturated silicic acid. After the experiment, the BET (N_2 gas) specific surface area of the solid sample decreased down to one-third or two-third. If the surface of the solid particles were covered with the precipitation layer, the BET specific surface area cannot always reflect the net surface area mainly contacting the solution. However, at least the change of specific surface area is not small. Therefore, this study focused on the initial precipitation rate. Here, the initial precipitation rates were evaluated by using second-order curve with three points of data in each experimental condition. These apparent rates include the contribution of polymerization as shown in the cases of 0.0 g (no solid phase) of figure 2.

Figure 3 shows the effects of the amount of solid phase (Mallinckrodt silica powder) on the precipitation rates at two temperatures, where (a) is at 298 K and (b) is at 323 K. The supersaturated silicic concentration was set to 4.8 mM in both cases. As shown in figure 3, the surface area of the solid phase significantly affects the precipitation rates. Moreover, the effect is more evident in the case of 323 K is large compared to that in the case at the lower temperature of 298 K.

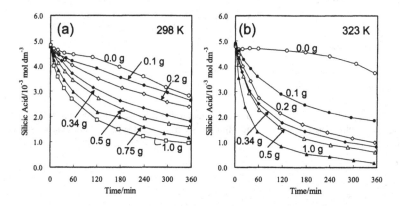

Figure 3. Dependence of the precipitation rates on the weight amount of Mallinckrodt silica. ((a) is the case of 298 K and (b) is the case of 323 K in temperature of thermostat. The initial supersaturated concentration of silicic acid was 4.8 mM.)

Figure 4 compares the precipitation rates evaluated from the data of figure 3 and the other cases of 288 K and 311 K. The precipitation rates are linearly proportional to the specific surface area of the solid phase with the correlation coefficient > 0.93. The specific surface area (the surface area in figure 4), a, was defined by $a = w a_B / V$ (1/m), where w is the weight amount of the solid phase (kg), a_B is the BET (N_2 gas) specific surface area (m^2/kg) of each solid phase as shown in table I, and V is the volume of the solution. If the aperture of parallel flat boards is 0.1 mm, the specific surface area is 2.0×10^4 (1/m). Moreover, Soler & Mäder [8] reported that the specific surface area of the flow-path (fracture) in the host rock of natural barrier was estimated to be around 10^6 (1/m), where the surface area of secondary minerals packed in fracture was considered in the addition of the fracture surface area. These both values are within the range of a (up to 1.4×10^6 (1/m)) given in this study.

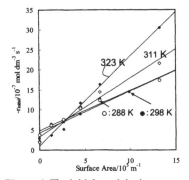
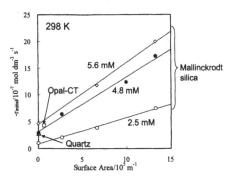

Figure 4. The initial precipitation rate $-r_{\text{ini}}$ vs. surface area of the solid phase, a (Solid phase: Mallinckrodt silica, Initial supersaturated concentration of silicic acid: 4.8 mM.)

Figure 5. Effects of the initial supersaturated concentration on the relation of $-r_{\text{ini}}$ to a (Solid phase: Mallinckrodt silica, opal-CT or quartz, Thermostat temperature: 298 K.)

The current study assumed a first order equation to describe the initial process. To confirm its validity, the different initial concentrations (i.e., 2.5 mM and 5.6 mM) of the supersaturated silicic acid were given. (So far, this paper displayed only data of 4.8 mM as shown in figures 1-4.) Figure 5 shows the results. Any case takes the linearity against the specific surface area, a. Moreover, the other kinds of solid samples such as opal-CT and quartz indicate the same tendency as Mallinckrodt silica (amorphous silica) in the relation of the initial precipitation rate, $-r_{\text{ini}}$, to the specific surface area, a. This suggests that the surfaces of opal-CT and quartz were also covered with thin amorphous layer caused by the precipitation [4]. (Note that the value of 2.5 mM, 4.8 mM or 5.6 mM in figure 5 was evaluated by $C_{\text{ini}}-C_{\text{e}}$, where C_{ini} is the initial condition of soluble silicic acid, and C_{e} is assumed to be the solubility of soluble silicic acid controlled by the surface of amorphous silica, even if the solid sample is opal-CT or quartz.)

The rate equation that summarizes the initial precipitation rates of this study was defined as:

$$-r_{ini} = ka(C - C_e) + r_{ini,p}, \tag{1}$$

where k is rate constant, a is the specific surface area mentioned above, C_e is the solubility of the soluble silicic acid (at pH 8) depending on temperature, and $r_{\text{ini,p}}$ is the initial polymerization rate (i.e., the intercept on $(-r_{\text{ini}})$-axis in figure 5). Table II is the k-value of each initial supersaturated concentration of silicic acid evaluated by using equation (1) and the data of figure 5. While it is hard to explain the dependence of r_p on the saturation index, i.e., $(C/C_e)-1$ [7, 9], the dependence of initial precipitation rate (except for $r_{\text{ini,p}}$) on specific surface area can be approximately described by equation (1).

Table II. Rate-constants, k, evaluated from the results of figure 5.

Initial supersaturated concentration / mM	2.5	4.8	5.6
Rate constant, k / m s^{-1}	2.0×10^{-10}	2.1×10^{-10}	2.1×10^{-10}

CONCLUSIONS

This study examined the initial precipitation rate of soluble silicic acid, considering the relatively large surface area of the natural barrier (up to 1.4×10^6 (1/m)). The results showed the strong dependence of the precipitation rates on specific surface area, suggesting that the solid surface is covered with a thin amorphous layer. In this case the rate constant was estimated to be 2.0×10^{-10} (m/s). This value may be applicable to estimate the altered surface-area by the supersaturated silicic acid around the repository. For example, consider the one-dimensional, advection-dispersion model including the precipitation rate of soluble silicic acid. The model has a key non-dimensional parameter, i.e., Damköhler number, $D_a = x_1 ak/u_0$, where x_1 is the characteristic length (length of flow-path), and u_0 is the fluid flow velocity of underground water. Assuming $u_0 = 1$ (m/year) $= 3.2 \times 10^{-8}$ (m/s), $x_1 = 10$ m, $a = 10^6$ (1/m) [8] and $k = 2.0 \times 10^{-10}$, D_a is estimated to be larger than 10^4. Such a large value of D_a suggests that the altered range is sufficiently limited surrounding the repository.

However, when the concentration of supersaturated silicic acid is large compared to its value given by this study and the specific surface area is limited (e.g., like open-fracture), we need more reliable model including the precipitation of polymeric silicic acid in order to estimate the altered surface area.

REFERENCES

1. Y. Niibori, Y. Kasuga, H. Yoshikawa, K. Tanaka, O. Tochiyama and H. Mimura in *Scientific Basis for Nuclear Waste Management XXIX*, edited by P. Van Isheghem, (Mater. Res. Soc. Symp. Proc. **932**, Warrendale, PA, 2006) pp. 951-958.
2. T. Chida, Y. Niibori, O. Toshiyama, H. Mimura, K. Tanaka, Applied Geochemistry **22**, 2810 (2007).
3. T. Chida, Y. Niibori, O. Tochiyama, and K. Tanaka in *Scientific Basis for Nuclear Waste Management XXVI*, edited by J. R. Fintch and B. D. Bullen, (Mater. Res. Soc. Symp. Proc. **757**, Pittsburgh, PA, 2003) pp. 497-502.
4. H. Kokubun, Y. Niibori, K. Iijima, and H. Mimura, Proc. of 16th Pacific Basin Nuclear Conference (CD-ROM), Paper ID P16P1154 (2008).
5. W. Stumm, and J. J. Morgan, *Aquatic Chemistry*, 3rd ed. (John Wiley & Sons, New York, 1996) p. 782.
6. R. K. Iler, *The Chemistry of Silica, Solubility, Polymerization, Colloid and Surface Properties, and Biochemistry* (John Wiley & Sons, New York, 1979) pp. 83-84.
7. C. F. Conrad, G. A. Icopini, H. Yasuhara, J. Z. Bandstra, S. L. Brantley, P. J. Heaney, Geochimica et Cosmochimica Acta **71**, 531 (2007).
8. J. M. Soler and U. K. Mäder, Applied Geochemistry **22**, 17 (2007).
9. B. A. Fleming, Journal of Colloid and Interface Science **110**, 40 (1986).

Cement-Based Systems

Mater. Res. Soc. Symp. Proc. Vol. 1124 © 2009 Materials Research Society　　　1124-Q05-06

Modelling the spatial and temporal evolution of pH in the cementitious backfill of a geological disposal facility

J.S. Small and O.R. Thompson
National Nuclear Laboratory, Risley, Warrington WA3 6AS UK

ABSTRACT

The UK disposal concept for intermediate-level and low-level radioactive waste (ILW/LLW) utilizes a cementitious material for backfilling vaults containing cementitious waste packages. The backfill is porous and is designed to promote an alkaline environment in which the mobility of radionuclides is reduced. This paper presents an initial investigation into the use of multi-dimensional reactive-transport geochemical models to examine the spatial and temporal effects that may occur as cementitious backfill and waste packages interact with groundwater. Models have been developed, verified against experimental data, and used to examine the mineralogical reactions in the cement materials and their effect on pH buffering and porosity.

INTRODUCTION

The geological disposal facility (GDF) concept for the long-term management of radioactive wastes in the UK utilizes a cementitious backfill (Nirex Reference Vault Backfill, NRVB) for filling void space around packages of conditioned ILW and LLW. The NRVB is composed of ordinary Portland cement (OPC), limestone flour, hydrated lime and water [1]. This mixture provides a high porosity material that promotes chemical homogeneity, permits gas migration and provides a substrate for the sorption of radioelements. The NRVB formulation produces strongly alkaline conditions in which the solubility of many radioelements is low and their sorption is enhanced [1]. ILW is typically encapsulated in grout composed of OPC blended with pulverized fuel ash or blast furnace slag. The evolution of the pH buffering behaviour of the NRVB and encapsulation grouts will be determined by leaching of calcium from $Ca(OH)_2$ and calcium silicate hydrate (CSH) gel by groundwater and also by reaction with groundwater solutes. Chemical interactions may also occur with the wastes and their encapsulation grouts.

Previous calculations of the volume of NRVB required to maintain suitably alkaline conditions (pH >10) have considered the vault contents to be mixed uniformly with all chemical reactions assumed to be complete prior to leaching by homogeneous groundwater flow [1]. Such simple calculations using robust assumptions for a homogeneous model are useful in determining the required amount of NRVB for a given volume of conditioned waste. However, it is necessary to develop more detailed models in order to understand the potential spatial and temporal variations in pH and the formation of reaction products. Leaching and secondary mineral precipitation will also influence the physical properties of the NRVB, including mass transport properties such as porosity and hydraulic conductivity.

This study presents results from initial geochemical reactive-transport modelling to examine the key pH buffering and mineral reactions occurring in a cementitious ILW/LLW disposal vault. The modelling considers, in one or two dimensions, (1) the overall vault scale evolution of pH in the presence of representative hard rock and clay rock groundwaters and (2) the pH buffering and mineral volume changes occurring through interactions between groundwater, NRVB and waste package grout that occur at a spatial scale of 1 to 10 m.

DESCRIPTION AND VERIFICATION OF THE pH EVOLUTION MODEL

A modelling approach to examine pH evolution of NRVB has been developed using the PHREEQC geochemical speciation and reaction path code [2]. PHREEQC is able to represent a wide variety of chemical processes of interest, including equilibrium pure phase and solid solution dissolution and precipitation, together with the capability to include kinetic processes that can represent waste degradation processes. PHREEQC enables 1-dimensional reactive-transport (advection and diffusion) effects to be considered. The PHAST code [3] incorporates PHREEQC in a 3 dimensional groundwater flow and transport model to examine a wider range of coupling of groundwater flow and transport effects with reactive chemistry.

The principal pH buffering processes occurring in NRVB and cement-containing waste packages are considered to be the dissolution of free $Ca(OH)_2$ and CSH gel materials, which comprise the main volume fraction of NRVB. $Ca(OH)_2$ is considered as a pure phase and effectively buffers at pH ~12.5. In cement compositions of lower Ca/Si ratio (~0.8 to 1.8), CSH phases dissolve incongruently and buffer pH in the range ~ 9 to 12.5 [4]. The pH buffering effects of CSH undergoing incongruent dissolution have been represented in PHREEQC as an ideal solid solution model using the approach and thermodynamic data of Kulik and Kersten [5]. This represents the CSH as two solid solution series: CSH-(I) considers a pure silica phase (SiO_2) and a tobermorite-like phase (Tob-I, $5Ca(OH)_2:6SiO_2:5H_2O$); CSH-(II) comprises solid solution between an identical tobermorite-like phase and a jennite phase ($10Ca(OH)_2:6SiO_2:6H_2O$).

To test the implementation of the $Ca(OH)_2$ and CSH dissolution model in PHREEQC the model was used to reproduce the results of an experimental study of $Ca(OH)_2$ and CSH leaching in carbonate free deionized water undertaken by Greenberg and Chang [4]. Figure 1 compares the experimentally measured pH and the solubility of Ca- and Si-bearing phases as a function of varying Ca/Si ratio of the solid with the modelled pH and Ca and Si concentrations. The PHREEQC model provides a good representation of the experimental data for this important pH buffer and this indicates that the CSH solid solution model [5] is correctly implemented.

Additional pure mineral phases; calcite, $CaCO_3$; brucite, $Mg(OH)_2$; and ettringite, $Ca_6Al_2(SO_4)_3(OH)_{12}:26H_2O$, were included in the PHREEQC model. These are important primary and secondary cementitious phases and their reaction may affect pH and the porosity of the system. The HATCHES thermodynamic database [6] was used in all calculations and this includes solubility products for these phases and also defines the aqueous species considered.

To represent the initial compositions of NRVB and waste encapsulant grout, mineral concentrations were calculated to represent the bulk composition of the two materials (Table I). This method represents a simplified approach to represent the cement mineral assemblage to examine the effects of the main chemical components. The enhanced Ca content of the NRVB is apparent by the presence of $Ca(OH)_2$ in the NRVB assemblage. The waste grout, which has a lower Ca/Si ratio, has no free $Ca(OH)_2$ according to this calculation.

Table I. Composition of NRVB and waste encapsulation grouts considered in the models.

	Bulk concentration, mol m^{-3}			Initial mineral concentration, mol kgw^{-1}			
	Ca	*Si*	*Al*	*Ca(OH)₂*	*CSH: Tob-I*	*CSH: SiO₂*	*ettringite*
NRVB	7398	1516	622	2.59	0.38	absent	0.19
Waste grout	5500	5300	1465	absent	1.53	13.25	2.44

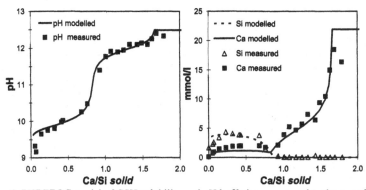

Figure 1. PHREEQC model of CSH solubility and pH buffering, compared to the experimental data of Greenberg and Chang [4].

Application to examine pH evolution in a geological disposal facility

To examine the effect of groundwater interaction and transport effects with a cementitious disposal vault PHREEQC and PHAST models have been constructed that examine processes at the scale of a disposal vault (length 300 m) and also to examine interactions at the smaller waste package scale (6 m). Two reference groundwater compositions have been examined (Table II); a hard rock groundwater composition (DET 5), considered in previous studies of geological disposal in the UK, and a clay rock groundwater (COx), representing the composition of a potential host rock formation in France [7].

Table II. Composition of reference hard rock (DET 5) and clay rock (COx) groundwaters.

	Master species concentration, mol kgw^{-1}								
	Ca	Mg	Na	K	Si	Cl	CO_3^{2-}	SO_4^{2-}	pH
DET 5	2.9×10^{-2}	5.7×10^{-3}	3.7×10^{-1}	4.4×10^{-3}	2.5×10^{-4}	4.2×10^{-1}	1.0×10^{-3}	1.2×10^{-2}	7.22
COx	1.5×10^{-2}	1.4×10^{-2}	3.2×10^{-2}	7.1×10^{-3}	9.4×10^{-5}	3.0×10^{-2}	2.5×10^{-3}	3.4×10^{-2}	7.00

The vault scale model of groundwater interaction was comprised of 30 transport cells each of 10 m length. For the purposes of this comparison, the vault was subject to leaching by 40 column pore volumes of each groundwater over a nominal period of 0.5 Ma. A lower volumetric flow rate might be expected in a clay rock than in a hard rock environment and a model variant has examined the case where the same volumetric flow took place over 60 Ma. A dispersivity of 30 m was considered in the large scale model.

PHREEQC and PHAST models have examined the chemical interactions of the NRVB and an 'averaged' waste package grout (Table I) and their effect on transport properties (e.g. porosity). These models considered the interaction of a 1m region of waste grout (porosity 0.25) surrounded by NRVB (porosity 0.55) 2 m upstream from the groundwater interface. For the PHAST model the grout was considered to have a hydraulic conductivity of 1e-7 ms^{-1}, one hundredth that of the NRVB to simulate flow around the waste package. The effects of variations in hydraulic conductivity in regions of the NRVB have also been represented. The presence of the waste container has been ignored at this stage of model development.

EXAMPLE RESULTS OF pH EVOLUTION AT THE VAULT SCALE

Figure 2 compares the modelled pH variation along a 300 m long cementitious filled disposal vault subject to leaching by the DET 5 and COx groundwaters. The clay rock groundwater (COx), results in a significant reduction in pH buffering, with the reaction front being advanced by around 50 m compared to leaching by the equivalent volume of DET 5 groundwater. This is due mainly to the higher magnesium concentration in the COx groundwater. Magnesium exchanges with $Ca(OH)_2$, resulting in $Mg(OH)_2$ formation which buffers between pH 9.5 and pH 10. Similarly, the higher carbonate concentration of the COx results in enhanced reaction with calcium. In the case of the lower flow rate model, where flow occurs over a period of 60 Ma, the effects of diffusive transport of species along chemical gradients result in a broader reaction front and higher pH at the groundwater interface.

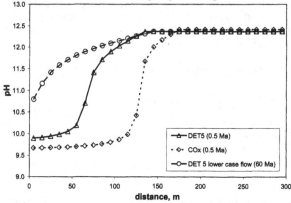

Figure 2. PHREEQC 1-dimensional transport model of pH evolution of a cementitious vault.

EXAMPLE RESULTS OF NRVB – WASTE PACKAGE INTERACTION

Figure 3 presents mineral volume and porosity changes calculated from a PHREEQC 1-dimensional transport model, where a waste package is situated 2 m from the interface of NRVB with groundwater. The pH of the model varies from pH 12.4, where $Ca(OH)_2$ is present, to pH 10.3 at the groundwater interface, where brucite and calcite precipitate. In the region of the waste package, pH was initially buffered at around pH 10 as a consequence of the absence of $Ca(OH)_2$. During the simulation, calcium leached from the NRVB reacts with the waste package grout and this increases the waste package pH. Ettringite forms at the margins of the waste package (Figure 3a). The leaching of Ca from $Ca(OH)_2$ and CSH results in an increase in porosity in most regions (Figure 3b). However, at the groundwater interface, calcite and brucite precipitation reduces porosity. In the case of COx groundwater, a 1% reduction in porosity occurs in the first cell, which represents the first 20 cm of NRVB. Similar effects have been noted in previous studies [8,9], which have modelled such systems at a finer level of discretisation, where significant porosity reduction may result in sealing by small scale (cm) precipitation of carbonate. More significant reduction in porosity occurs at the downstream NRVB-waste package interface as a result of ettringite precipitation for the DET 5 case.

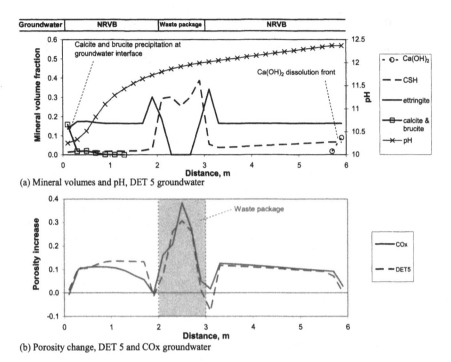

(a) Mineral volumes and pH, DET 5 groundwater

(b) Porosity change, DET 5 and COx groundwater

Figure 3. PHREEQC transport model of mineral evolution at the waste package scale, after reaction with 40 column pore volumes of groundwater.

Figure 4 presents results of a 2-dimensional PHAST model that examines groundwater-NRVB-waste package interactions, including the effects of water flow around the waste package. Dissolution of $Ca(OH)_2$ in the NRVB surrounding the waste package is enhanced by reaction with the lower Ca content waste package grout. At the groundwater interface leaching of $Ca(OH)_2$ is enhanced in the regions of increased flow, between regions of lower permeability.

CONCLUSIONS

A modelling methodology has been developed that is capable of examining the spatial and temporal evolution of pH buffering and mineral reactions in the near field of a cementitious geological disposal facility as it interacts with groundwater. A simplified representation of the mineral reactions in NRVB and waste package cement grouts has been developed at this stage including an ideal solid solution model to represent CSH dissolution, along with discrete phases $Ca(OH)_2$, calcite, brucite and ettringite.

A groundwater representing a clay host rock with higher Mg and carbonate concentration has been observed to result in an acceleration of the pH reaction front through the NRVB compared to the equivalent volume of a hard rock groundwater. Although the clay groundwater has a greater chemical effect, the flow rate would be expected to be lower for such a low permeability clay rock. In the case of lower flow rates, diffusive transport of chemical species results in a broadening of the pH front and higher a pH at the groundwater interface.

The modelling also provides information concerning the porosity evolution of the backfill. Results indicate an increase in porosity in the main body of the NRVB but decreases in porosity at the interfaces of the NRVB with groundwater and with the waste packages as a result of secondary precipitation.

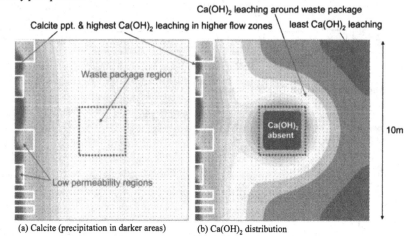

(a) Calcite (precipitation in darker areas) (b) Ca(OH)$_2$ distribution

Figure 4. PHAST model output (monochrome print of colored contour figure) of the mineral reactions between groundwater, NRVB and waste package (flow from left side) after 10,000 y.

ACKNOWLEDGMENT

This work has been undertaken within the Nuclear Decommissioning Authority (NDA) Radioactive Waste Management Directorate Research and Development Programme.

REFERENCES

1. I. Crossland and S.P. Vines, Nirex Report N/034 (2001).
2. D.L. Parkhurst and C.A.J. Appelo, U.S. Geological Survey, Water Resources Investigations Report 99-4259 (1999).
3. D.L. Parkhurst, K.L. Kipp, P. Engesgaard and S.R. Charlton, U.S. Geological Survey, Techniques and Methods 6-A8, 154p (2004).
4. S.A. Greenberg and T.N. Chang, *J. Phys. Chem.*, **69**, 182-188 (1965).
5. D.A. Kulik and M. Kersten, *Am. Ceram. Soc.*, **84**, 3017-3026 (2001).
6. K.A. Bond, T.G. Heath and C.J. Tweed, Nirex Report NSS/R379 (1997).
7. E.C. Gaucher, *et al.*, *C.R. Geoscience*, **338**, 917-930 (2006).
8. W. Pfingsten, PSI Report 01-09, Paul Scherrer Institut, Villigen, Switzerland (2001).
9. J.W. Carey and P.C. Lichtner, *Transport Properties and Concrete Quality: Materials Science of Concrete, Special Vol.* Am. Ceram. Soc.; (John Wiley & Sons, 2007) pp. 73-106.

Mater. Res. Soc. Symp. Proc. Vol. 1124 © 2009 Materials Research Society   1124-Q05-07

Properties and composition of cemented radioactive wastes extracted from the mound-type repository

Galina A. Varlakova, Zoya I. Golubeva, Alexander S. Barinov and Igor A. Sobolev
Moscow SIA "Radon", 2/14, 7-th Rostovsky per., Moscow, 119121, Russian Federation

Michael I. Ojovan
Immobilisation Science Laboratory, Department of Engineering Materials, University of Sheffield, Mappin Street, Sheffield, S1 3JD, United Kingdom

ABSTRACT

The results of several decade long-term field tests of cemented radioactive wastes are described with the focus on physical properties and composition. Cementitious wasteform in form of cubic blocks with specific radioactivity from 0.34 to 1.5·MBq/kg were manufactured and disposed of in 1965 in a mound-type repository covered by loam soil. Properties of aged cementitious wasteforms were assessed analyzing the density, the porosity, the compressive strength, the normalized radionuclide leaching rates and their structure using SEM/EDS/XRD techniques. Microbiological examination of cementitious wasteforms evidenced presence of viable silicate minerals destructing bacteria. The cementitious wasteform after 40 year is homogeneous and has a stone-like structure, retained a good physical state without visual mechanical damage, wall saltpetre, structural or colour changes caused by freeze-thaw cycles.

INTRODUCTION

Most of cemented low- and intermediate-level radioactive wastes are currently stored or disposed of in near-surface storage facilities and repositories. Performance of cementitious wasteforms in such conditions is of paramount importance. Modern Portland cements have been in use for about one and a half century and $Ca(OH)_2$-based compositions have been successfully used in civil engineering for more than two millennia [1]. Although the experience of cement utilisation demonstrates its prominent long-term durability, data obtained via extended field experiments with real cementitious wasteforms are most relevant to improve our confidence in radionuclide retention. This paper analyses data obtained during the long-term field tests of cemented radioactive waste in a mound type repository and describes properties, composition and structure of cemented radioactive wastes extracted from the repository.

Long-term field tests of cemented aqueous radioactive wastes in an experimental mound-type surface repository were carried out at Moscow Scientific and Industrial Association "Radon" since 1965 [2,3]. Aqueous radioactive waste of different compositions containing short-lived radionuclides including ^{90}Sr and ^{137}Cs at concentrations from 0.34 to 1.8 MBq/l were cemented using ordinary Portland cement (OPC) Russian trade mark M500, water solution to cement ratio 0.66, grout mixing time 10-15 minutes, and cement paste hardening time 7 days. 73 cementitious wasteform in form of cubic blocks of sizes 30*30*30 cm were disposed for long-term field tests in a mound-type surface repository in 1965. Cementitious wasteforms were placed on a concrete plate inside of steel trays. The blocks were covered by a 0.2 m thick layer of sand and gravel, 0.5 - 0.7 m thick layer of loam and then 0.1 - 0.3 m thick layer of sod. In August 2004 the experimental repository was opened, cemented blocks, underlying and covering materials were retrieved for analyses. It was found that cementitious wasteforms retained their integrity, have no visual mechanical damage, wall saltpetre, structural or colour changes caused by freeze-thaw cycles and that the cement paste has retained waste radionuclides [2,3]. This paper gives details on such properties of aged cementitious wasteforms as density, porosity, compressive strength, normalized radionuclide leaching rates and their structure using SEM/EDS/XRD techniques. Moreover we present data on microbiological examination of cementitious wasteforms which evidenced presence of viable silicate minerals destructing bacteria.

EXPERIMENTAL

Properties of aged cementitious wasteforms were assessed analysing the density, the porosity, the compression strength, the humidity and the normalized radionuclide leaching rates (NR). All cementitious samples were equivalent except the waste loadings [2,3]. The real (true) density of wasteform ρ_t (g/cm^3) was measured using the water-displacement method with a picnometer and the average density ρ_a (g/cm^3) using the hydrostatic weighing [4]. The porosity of wasteform P (%) was found using the expression

$$P = \left(1 - \frac{\rho_a}{\rho_t}\right) \cdot 100\%$$

The compression strength of wasteforms was determined using the tensile-testing machine IR-5047-50C. The normalized leaching rate NR (g/cm^2 day) was assessed using the IEAE testing protocol ISO 6961-1982.

SEM/EDX and XRD analyses were used to assess the structure of wasteforms. Scanning electron microscopes JSM-5300 (JEOL, Japan) with spectrometer Link-ISIS (Oxford, UK) and JSM-5610LV (JEOL, Japan) using the low vacuum regime and energy dispersive analyzer JED-2300 were used. Identification of crystalline phases was done using the X-ray diffractometer D/MAX-2200 (Rigaku).

In order to carry out necessary analyses samples of aged cementitious wasteform were required in form of powders, chips, slices. These were prepared using a specialized corer and polishing machine. A schematic of analytical samples preparatory works is shown in Fig. 1.

The microbiological stability of wasteforms was examined using standard microbiological standards. The bacteria count and selection of clean cultures have been done on liquid elective nutrient media using the method of ten-fold breeding in two retests with consequent statistical processing of results using McCready's tables [5]. Microscopic mushrooms were grown on solid Czapek nutrient medium. Clean cultures were obtained selecting from individual colony on classical meat-peptone broth and broth-wort-agar. Identification of selected strains was done based on morphological and physiology-biochemical signs [6,7].

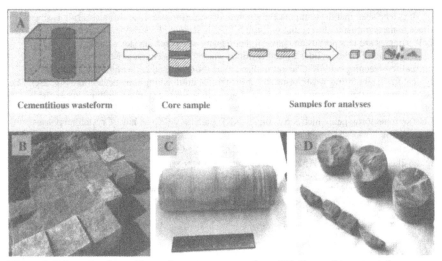

Figure 1. Schematic of sampling and preparatory procedures (A). Cementitious wasteforms 30·30·30 cm (B), a core sample (C) and samples for analytical procedures (D).

RESULTS AND DISCUSSION

Physico-chemical properties of aged cementitious wasteforms analysed are given in Table 1.

Table 1. Main physical and chemical properties of wasteforms.

Sample and sampling place		Compressive strength, MPa	Humidity, %	Density, g/cm³		Porosity, %
				Real	Average	
CC-1	Surface	5.1	32.5	2.50	1.79	27.5
	Bulk	20	21.1	2.51	1.98	24.1
CC-2	Surface	Not measured	25.5	2.54	1.80	28.9
	Bulk	17	22.1	2.56	1.84	28.0
CC-3	Surface	Not measured	18.7	2.52	1.78	29.0
	Bulk	18	15.9	2.43	2.00	17.3
CC-4	Surface	4.9	23.5	2.41	1.65	31.5
	Bulk	14	22.7	2.74	1.85	32.3

Measurements were carried out on samples taken from both the surface of wasteforms and core samples from the bulk of blocks. The identification marks in this table correspond to liquid radioactive waste (LRW) batches used to prepare the wasteforms in 1965 as follows: CC-1 – LRW from the Tank No. 5, CC-2 – LRW from the Tank No. 4, CC-3 – simulant LRW with ^{90}Sr, CC-4 – simulant LRW with [137]Cs [2,3].

It can be seen that the compressive strength of wasteforms 5.1 -20 MPa as a rule is several times higher compared the regulatory limit 5 MPa in Russian Federation. It can be also seen that bulk samples are less porous and have higher density compared surface samples.

Retention of radionuclides by cementitious wasteforms is characterised here by the normalised leaching rate [8]. The normalised rates (NR) of aged cementitious wasteforms are typical for cementitious wasteforms and show typical diminishing time behaviour (Fig. 2A). The XRD patterns of wasteforms demonstrate along with typical cement crystalline phases the presence of typical humps characteristic to the amorphous phase (Fig. 2B). This evidences on CSH or tobemorite gel, which forms on the XRD patterns a diffuse halo. Crystalline phases identified in the wasteform were Portlandite – $Ca(OH)_2$; Alite – $54Ca·16SiO_2·Al_2O_3·MgO$; Calcite – $CaCO_3$; Ettringite – $3CaO·Al_2O_3·3CaSO_4·32H_2O$; Two-water-three-calcium hydrosilicate $3CaO·SiO_2·2H_2O$. The calcite present in the wasteform is formed on $Ca(OH)_2$ carbonisation. The carbonisation process occurs first in the near surface layers of wasteform blocks whilst deeper volumes are carbonised slower and to a lower degree. Note that carbonisation lead to an increased mechanical and water durability [9-11].

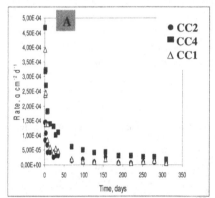 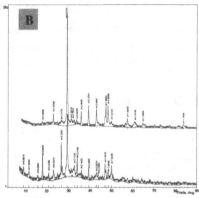

Figure 2. Normalised leaching rate ($\sum\beta$ on ^{137}Cs) time dependence of aged cementitious wasteform (A) and XRD pattern of CC-1 wasteform. The highest NR are characteristic for the CC-4 samples prepared with simulant LRW with ^{137}Cs.

The structure of aged cementitious wasteforms is illustrated by SEM images (Fig. 3A, B and C). These images evidence that the aged cementitious wasteform has a dense grainy structure with grain sizes about 10 – 100 μm. Some of pores are relative large up to 200 μm. Crystalline needle and rhombic formations can be seen on most of the internal pore surfaces (Fig. 3C, D). Most probable that needle-like crystals are ettringite whereas rhombic – calcite.

Needle shaped crystals are also present within intergrain spaces. It was noted that such structures result in an enhanced strength of cements [9]. Recent studies [12] have shown that pore filling with secondary minerals (i.e. calcite) in cementitious materials suppresses radionuclide migration.

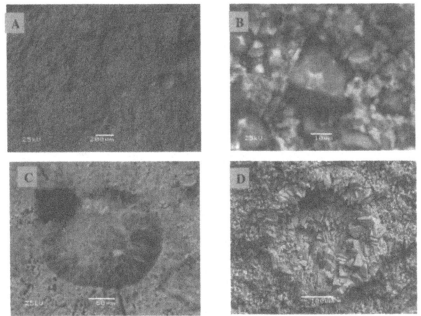

Figure 3. SEM image evidencing on dense grainy structure of cementitious wasteform (A, B) and crystalline needle (C) and rhombic (D) neoformations on the pore surface.

Biocorrosion of cementitious wasteforms is currently not well investigated although it can be important in the degradation of cements [13]. The destructive effects in the biocorrosion are caused by microorganism activities and the products of their metabolism such as acids, bases, ferments and other aggressive media. Microbiological examination of aged cementitious wasteforms from the mound type repository evidenced the presence of viable bacteria of different physiological groups: denitrification, fermenting, iron oxidizing and microscopic mushrooms. The micro-organism population was low. Cell count has shown that the denitrification group is present in the range from 5 to 9 cells/g, fermenting – from 25 to 256 cells/g. The total number of bacteria is 10 colony-formation units (CFU) per g, mushrooms 10 CFU/g. There were identified bacteria of physiological groups related to *Bacillus, Pseudomonas, Mycobacterium* species. These micro-organisms are strong destructors of silicate minerals including those present in the cement stone [14]. Although their population is low their activity can significantly grow [15].

CONCLUSIONS

The cementitious wasteform retained acceptable immobilisation properties after 40 years storage in the mound-type repository. The wasteform has a homogeneous stone-like structure and retained a good physical state without visual mechanical damage, wall saltpetre, structural and colour changes caused by freeze-thaw cycles.

ACKNOWLEDGEMENTS

Authors acknowledge the assistance of Prof. L.B. Prozorov, A.Yu. Yurchenko, Dr. A.V. Mokhov and Dr. O.A. Gorbunova.

REFERENCES

[1] F.P. Glasser. *Mat. Res. Soc. Symp. Proc.*, **713**, JJ9.1.1-12 (2002).
[2] I.A. Sobolev, S.A. Dmitriev, A.S. Barinov, G.A. Varlackova, Z.I. Golubeva, M.I. Ojovan. Reliability of cementation technology proved via long-term field tests. *Proc. ICEM'05*, September 4-8, Glasgow, Scotland, ICEM05-1216, ASME (2005).
[3] I.A. Sobolev, S.A. Dmitriev, A.S. Barinov, G.A. Varlakova, Z.I. Golubeva, I.V. Startceva, M.I. Ojovan. 39-years Performance of Cemented Radioactive Waste in a Mound Type Repository. *Mater. Res. Soc. Symp. Proc.* **932**, Warrendale, PA, 721-726 (2006).
[4] State Standard GOST 12730.0-78 – GOST 12730.4-78. Concrete. Methods to etermine the density, moisture, water absorbing, porosity and water permeability. Standards, Moscow (1985).
[5] N.S. Egorov. *Handbook on practical training on microbiology.* Moscow State University, Moscow, 224pp (1995).
[6] *Bergey's Manual of Determinative Bacteriology*, 9th ed. J.G. Holt, ed., Williams & Wilkins, Lippincott, 500 pp. (1993).
[7] T.G. Dobrovolskaya, I.N. Skvortcova, L.V. Lysak. *Method to determine and identify soil bacteria.* Moscow State University, Moscow, 72pp. (1990).
[8] M. I. Ojovan, W.E. Lee. *An Introduction to Nuclear Waste Immobilisation*, Elsevier Science Publishers B.V., Amsterdam, 315pp. (2005).
[9] *Formation and cement stone microstructure genesis (Electron stereomicroscopy of cement stone)*. Ed. L.G. Shpynova. Vischa Shcola, Lyvov, 158pp. (1975).
[10] G.A. Balayaev, V.M. Medvedev, N.A. Moshchansky. *Protection of building units from corrosion.* Stroyiztad, Moscow (1966).
[11] V.B. Volkonsky, S.D. Myakishev, N.P. Schteiert. *Technical, physical and mechanical, physical and chemical studies of cementitious materials.* Stroyizdat, Leningrad (1972).
[12] M. Toyohara, M. Kaneko, F. Matsumura, N. Kobayashi and M. Imamura. Study on Effects of Hydraulic Transport of Groundwater in Cement. *Mater. Res. Soc. Symp. Proc.* **608**, 337-342 (2000).
[13] J.M. West. A review of progress in the geomicrobiology of radioactive waste disposal. *Radioact. Waste Manag. and Environ. Restoration. Radioact. Waste Manag. and Nucl. fuel Cycle* **19** (4) 263–283(1995).
[14] G.I. Karavaiko. Microbial destruction of silicate materials. *Proceedings of S.N. Vinogradsky S.N. Institute of Microbiology*, **XII**, Collected articles on 70 jubilee of institute, Nauka, Moscow, 172-196 (2004).
[15] K. Pedersen. Investigation of subterranean in deep crystalline bedrock and their importance for the disposal of nuclear waste. *Can. J. Microbiol.* **42**, 382-391 (1996).

Mater. Res. Soc. Symp. Proc. Vol. 1124 © 2009 Materials Research Society   1124-Q05-08

RBS and μPIXE study of I and Cs heterogeneous retention on concrete

Ursula Alonso[1]*, Tiziana Missana[1], Miguel García-Gutiérrez[1], Alessandro Patelli[2], Daniele Ceccato[3], Valentino Rigato[3], Nairoby Albarran[1], Henar Rojo[1], Trinidad Lopez[1]

[1] CIEMAT, Avda. Complutense, 22 Madrid, 28040 Spain
[2] CIVEN, Via delle Industrie 5, Venezia-Marghera, 30175 Italy
[3] LNL-INFN, Viale dell' Universitá 2, Legnaro-Padova, 35020 Italy
*Corresponding author: ursula.alonso@ciemat.es

ABSTRACT

A combination of two nuclear ion beam techniques, Rutherford Backscattering Spectrometry (RBS) and μ-Particle Induced X-Ray Emission (μPIXE) was tested to evaluate both diffusion profiles and radionuclide spatial distribution of radionuclides (RN) onto cement-based materials. The methodology was tested on a Spanish reference backfill concrete, used as engineering barrier in low-level radioactive waste repositories, using two elements (Cs and I) with different sorption behaviour onto the cement. The applicability and limitations of the selected methodology is discussed for both elements.

INTRODUCTION

Cement-based materials, like concrete and mortar, are being used as major component in the barriers of low- level and high-level radioactive waste repositories, because they are well suited to retain radionuclides (RN) both by physical and chemical processes. In the Spanish low level repository, concrete is mainly used as engineered barrier in the vaults (main structures) and in the containers where the metal canisters with the waste are placed [1].

To assess the long-term safety of cement barriers, the transport or retention of critical RN have to be studied. Cements are composed by different phases and RN migration would be highly conditioned by their heterogeneity. A methodology based on combining the RBS and μPIXE techniques was tested here for the first time in cement-based materials, to evaluate diffusion coefficients, and to quantify tracer surface retention. Surface retention was experimentally measured, through the determination of surface distribution coefficients (Ka), at a mineral scale, on the heterogeneous concrete. A similar methodology has been previously applied in granite [2-3] with very good results. Due to the big difference within the two materials it was considered necessary this previous study to evaluate the applicability of the techniques for cements as well as for the optimization of the experimental methodology.

EXPERIMENTAL

The concrete selected for this study is the reference material used as engineered barrier at the Spanish low-level radioactive waste repository. The concrete composition is given in Table 1. It has a low C3A and alkali content to prevent sulphate or alkali-aggregate attack. The density of this material is 2.31 g/cm^3. The BET specific surface area is 2.60 ± 0.20 m^2/g.

Table 1. Main composition of the selected concrete in components (kg/m^3) and in oxides (%).

Concrete	(kg for 1 m^3)
Cement I/42,5/R/SR	400
Sand 0 – 2 mm	353
Sand 0 – 4 mm	484
Aggregate 4 – 16 mm	1023
Superplastifier Melcret 222	6.5
Water	175

Oxides	(%)
SiO$_2$	19.90
Al$_2$O$_3$	2.86
Fe$_2$O$_3$	4.40
CaO	65.04
MgO	2.13
SO$_3$	3.70
Na$_2$O	0.11
K$_2$O	0.45
Cl$^-$	0.02

For sorption and diffusion experiments, samples were cut with an area of around 1 cm^2 and thickness of few millimeters, since micro - analyses are performed on the near surface region. Prior to tracer addition, samples were immersed in deionised water during two days to saturate, allowing sufficient time to equilibrate the pH to alkaline conditions. Cs and I solutions were prepared in deionised water, from high purity salts of ClCs and IK to a final concentration of $5·10^{-3}$ M. The final pH was adjusted to pH 12.5.

After saturation, for RBS diffusion experiments, concrete samples were immersed in 3 ml of Cs or I solutions during 6 hours, for μPIXE analyses, during 8 days. After the immersion, the surface of the samples was cleaned with deionised water, trying to eliminate tracer deposited on the surface. Ion beam techniques measurements (RBS and uPIXE) were carried out at the nuclear microprobe facility at the Laboratori Nazionali di Legnaro (INFN-LNL, Italy).

Diffusion experiments: RBS measurements

RBS measurements were performed on different concrete phases (1 mm^2), in particular on a portlandite-type area (Ca(OH)$_2$) and on a phase containing Ca, Si, Al and Mg (alite-type area). Measurements were carried with ^{4}He$^+$ particles at 2 MeV and with a scattering angle of 160°. Within the experimental conditions, heavy tracers at depths of around 5 microns, with detection limits of 10^{12} atoms/cm^2 can be detected. The natural distance RBS units are given in atoms /cm^2, so that to convert the RBS units to distance scale (x, diffusion length in m) the density of the concrete must be considered (2.31 g/cm^3). RBS spectra were analyzed with the X-RUMP code [4]. Spectra simulation is performed iteratively, by defining the sample composition and introducing tracer concentration profiles by using a RUMP modelling subroutine. The equation used to simulate the concentration profiles within the solid was a solution of Fick's second law which satisfies the boundary conditions C=C$_B$, x=0, t>0 and the initial conditions C=0, x>0, t = 0 is [5]:

$$C (x, t) = C_B \mathrm{erfc}\, \frac{x}{2\sqrt{Dt}} \qquad (E.1)$$

μPIXE measurements

Several 2*2 mm^2 areas were analyzed on each sample by the nuclear ion beam technique micro-Particle Induced X- Ray Emission (μPIXE) [6]. Samples were irradiated with 1.9 MeV protons with a beam size of around 4 μm², at perpendicular incidence. The typical beam currents were 1.5 nA with a the spatial resolution to 2 μm. A Camberra Ge detector with a polymer window of 0.4 μm thickness and a maximum nominal resolution of 160 eV was used to enhance the sensibility to heavy elements, all measurements were carried out with an Al filter 80 μm thick with a hole in the centre that allows passing no more than 8% of the total signal. The μPIXE technique includes a scanning system that allows mapping a studied area (2*2 mm²), obtaining elemental distribution maps. Therefore, the main mineral phases present can be identified.

To perform an accurate individuation of trace elements in the studied area it is necessary to analyze the individual PIXE spectra. To do so, the μPIXE images were processed with the Mappix code (developed at the LNL, Italy). Individual PIXE spectra were obtained on selected areas, and quantitative analyses were performed with the windows version of the GUPIX code [7]. Therefore, both the area composition (in elemental atomic %) and the tracer concentration retained (Tracer $_{RETAINED}$ in ng/cm²) can be obtained.

The distribution coefficients (Ka, in m), that represents the degree of retention, can be calculated with the following equation:

$$K_a \, (m) = \frac{Tracer_{RETAINED}}{Tracer_{EQUILIBRIUM}} \qquad (E.2)$$

where Tracer$_{RETAINED}$ is the concentration measured on the selected area (in g/m²) and Tracer$_{IEQUILIBRIUM}$ is the tracer concentration in the liquid phase upon equilibrium (in g/m³).

RESULTS

Diffusion experiments: RBS measurements

Figure 1 (left) shows experimental RBS spectra of two different concrete phases (portlandite and alite-type) kept in contact for 6 hours with Cs. The spectra are shown in the region of interest for Cs (1.5-1.8 MeV). In both cases, peaks attributable to the presence of Cs are detected, and also pointed out in the Figure. The Cs signal was never detected either in the reference alite and portlandite RBS spectra (also included in Figure 1, for comparison). Simulations were performed considering, depending on the concrete phase, either the alite or the portlandite composition and fitting the tracer concentration profiles, considering the equation 1. The Cs concentration profiles are shown in Figure 1 (right) both for alite and portlandite. The Cs concentration is clearly higher in the alite area. The apparent diffusion coefficients measured for Cs were D_a = (2.5 ± 0.5) E-15 m²/s in alite and Da = (1.1 ± 0.5) E-15 m²/s in portlandite. The RBS methodology is able to detect differences between different areas in a heterogeneous material as concrete, allowing to experimentally evaluating possible variations.

RBS measurements were also performed on the samples kept in with I. In this case, no peaks attributable to I presence could be observed in any area. This indicated that I concentration inside the concrete is below the RBS detection limit (around 1 ppm), probably due to his very low retention at the surface and even smaller diffusion coefficient than the Cs one. Thus, in the case of I, larger experimental times are needed.

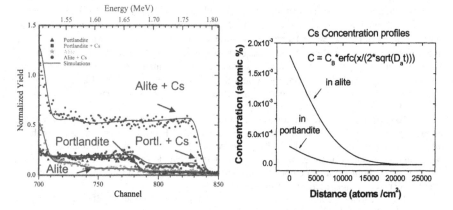

Figure 1. (Left) RBS spectra of obtained after immersion in Cs solution during 6 hours on (■) portlandite-type area and (★) alite-type area. The reference alite and portlandite RBS spectrum, where no Cs is detected, are also included for comparison. Simulations are plotted as continuous lines. **(Right)** Cs concentration profiles used to simulate the RBS spectra on each area.

μPIXE measurements

The μPIXE analyses were first measured directly on the samples after the immersion on the Cs or I solutions. In those preliminary measurements, it was observed that the elemental distribution maps were rather scattered, indicating the presence of particles spread all over the surface. During the measurements, this was attributed to Ca phases that were formed upon immersion on the solutions that were later precipitated over the concrete surface. This assumption hypothesis was later confirmed by SEM analyses. To avoid mistaken the tracer retained on the precipitated phases, with the tracer really retained over the concrete sample, the surface of the samples was slightly polished and measured again.

As an example, Figure 2 shows the elemental distribution maps of the Ca, Si, K and Fe content of two different $2*2$ mm^2 concrete areas after polishing the surface. It can be seen that the different areas, mainly related to Ca-phases or to quartz minerals can be perfectly appreciated. Only the Ca, Si, K and Fe maps are shown, but the distribution maps of elements with atomic number higher than Al can, in principle, be detected within our experimental conditions, provided their X-ray lines are not to close to that of the main elements composing the concrete, or their concentration too low to be detected.

The presence of Cs could be detected in most of the concrete areas, mostly equally distributed over the area. But actually, the distribution maps of elements at low concentration need to be taken with care, since the background noise of other elements exist. This problem was solved by selecting different zones inside the studied area (of $2*2$ mm^2) and obtaining the individual PIXE spectrum, for further quantitative analyses. Typical size of the quantitatively analyzed zones is around 500 μm^2.

From the quantitative analyses of the individual PIXE spectra it was shown that no Cs was generally retained (or it was below detection limit) on quartz areas and on Ca areas with no Al, or K content. Figure 3 shows the PIXE spectrum of the zone were higher Cs was detected, Apart

342

from the X-ray characteristics of the main elements (Al, Si, S, K, Ca, Fe and Cu) the Cs Lα, Lα and Lγ peaks were clearly detected. The simulation estimated a Cs concentration of 5000 ng/cm^2 in this area. The Cs surface distribution coefficients (Ka) were determined with E.2. In this concrete the range of values vary from $K_a = 0$ m in quartz areas to $K_a = 7.5 \cdot 10^{-4}$ m in the area where higher Cs was detected. Again, contrary to the Cs case, no peak attributable to I could be measured in any of the samples immersed into the I solution. We cannot ensure that I is not retained on the surface because, the Ca X-ray emission lines (very strong because of the high Ca concentration in the concrete samples) are very close to that of I, so that, as mentioned, the Ca signal hinders the detection of I as trace element.

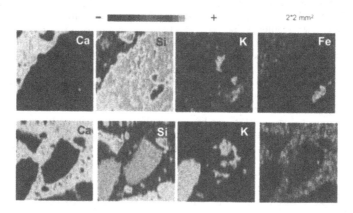

Figure 2. Elemental distribution maps (Ca, Si, K and Fe) obtained by µPIXE on two different concrete areas 2*2 mm^2.

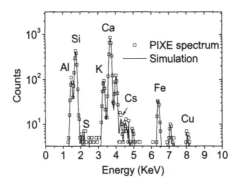

Figure 3. PIXE spectrum and simulation obtained on a concrete zone where higher Cs was detected.

CONCLUSIONS

A methodology based on combining the RBS and μPIXE techniques was tested here for the first time, to evaluate diffusion coefficients and to identify main retention areas on the heterogeneous concrete surface.

The experimental diffusion times are clearly shorter (days) than those required by conventional diffusion experiments (months/years). It was possible to determine diffusion coefficients for Cs, while no I access within the concrete was detected.

The RBS diffusion methodology was able to measure differences on the Cs diffusion on different phases of the heterogeneous concrete surface (allows independent measurements in areas of 1 mm^2). The alkaline chemical conditions of the cements restrict the RBS application to elements with low solubility, because high tracer concentration is required (around 10^{-3} M).

Moreover, the μPIXE nuclear ion beam technique, allowed both visualizing and quantifying the main phases composing the concrete. By PIXE quantitative analyses it was possible to quantify the Cs retention on the main identified areas. Surface distribution coefficients (Ka) could experimentally determined in a heterogeneous concrete surface. The methodology is sensitive enough to measure differences amongst the different phases. In the case of I, no retention was observed (bellow detection limit).

The proposed methodology can be in principle applied to other relevant radionuclide, if they are heavy elements and if the concrete geochemical conditions do not restrict their solubility, or their concentration being lower than the RBS and μPIXE detection limits.

ACKNOWLEDGMENTS

This work was partially supported by ENRESA (E) under the FISQUIA project (05/190/007/800070) and by the EU under the Research Infrastructure Action under FP6 Contract n. 506065 EURONS- EUROpean Nuclear Structure Research.

REFERENCES

1. C. Andrade, I. Martinez, M. Castellote, P. Zuloaga. J. of Nuclear Materials 358 82–95 (2006).
2. U. Alonso, T. Missana, A. Patelli, V. Rigato, J. Ravagnan. Nuclear Instruments and Methods in Physics Research B 207/2, 195-204 (2003).
3. U. Alonso, T. Missana, M. García-Gutiérrez, A. Patelli, J. Ravagnan, V.Rigato. in Scientific Basis For Nuclear Waste Management XXVII. MRS Symposium Proceedings 807, 621-626 (2004). Eds. V.M. Oversby, L.O. Werme. Materials Research Society, Warrandale, Pennsylvania, USA.
4. L.R. Doolittle. Nucl. Instrum. Methods in Physics Research B, v. 15, p. 227-231 (1986).
5. J. Crank, J., 1956, The Mathematics of Diffusion: Oxford, Claredon Press.
6. S.A.E. Johansson, J.L. Campbell, K.G.E. Malmqvist, Particle Induced X-Ray Emission Espectrometry (PIXE) Chemical Analysis, a series of monographs on analytical chemistry and its applications, John Wiley& Sons, Ltd, 1995.
7. J.A. Maxwell, W. Teesdale, J.L. Campbell. Nuclear Instruments and Methods in Physics Research B 95 (1995) 407.

Mater. Res. Soc. Symp. Proc. Vol. 1124 © 2009 Materials Research Society 1124-Q05-09

The Inhibition of Aluminium Corrosion in OPC Based Composite Cements

Nicholas C. Collier, Neil B. Milestone and Paul D. Swift
Immobilisation Science Laboratory, Department of Engineering Materials, The University of
Sheffield, Mappin Street, Sheffield, S1 3JD, UK.

ABSTRACT

The pore solution pH in composite cements based on ordinary Portland cement (OPC) is
lower than in pure OPC pastes but is still high enough to corrode encapsulated aluminium
generating hydrogen and forming expansive corrosion products, both of which may affect the
integrity of the hardened wasteform. The encapsulation of aluminium metal in blastfurnace slag
(BFS)/OPC and pulverised fuel ash (PFA)/OPC composite cements containing added anhydrite
and gypsum has been studied and results compared with the same systems without additions. It
was found that while aluminium corrodes in all systems yielding hydrogen, the amount was
lowest in the samples containing added anhydrite or gypsum. The corrosion layer in the samples
containing the additives consisted of a combination of aluminium hydroxides (gibbsite and
bayerite), C-S-H, and significant quantities of ettringite, whilst in the samples without additives
little ettringite was present and strätlingite was observed. It is suggested that this formation of
ettringite after the initial corrosion removes free water from the paste providing a protection to
the aluminium from significant further corrosion.

INTRODUCTION

The UK currently uses composite blends of OPC with BFS or PFA to encapsulate or
immobilise intermediate and low level radioactive wastes (ILW and LLW). OPC has a high
internal pH, typically above 13 [1], which is beneficial for immobilising heavy metals since the
solubility of many hydroxides is reduced in the highly alkaline pore solution. However, this high
pH can lead to reactions of the matrix with the waste, causing expansive corrosion of metals such
as aluminium and excess generation of hydrogen. Although current formulations replace OPC
with up to 90 wt% BFS, the pH still remains high at approximately 12.5 [2] and as a result, metal
corrosion and hydrogen generation is still a major concern [3].

In these composite systems, the underlying reaction is the activation of the BFS or PFA
by the OPC, utilising the $Ca(OH)_2$ and alkalis generated during hydration to produce a form of
C-S-H (an ill defined calcium silicate hydrate where C = CaO, S = SiO_2 and H = H_2O) similar to
that from OPC but with additional Al substitution and a lower Ca:Si ratio. However, these
systems with their low OPC contents usually contain large amounts of unreacted BFS or PFA
[4]. While other strongly alkaline activators such as sodium hydroxide and sodium silicate can
be used with both BFS or PFA, recent work has demonstrated that weakly alkaline salts such as
sodium carbonate or near neutral salts such as calcium and sodium sulfates will also function the
same way for slag activation [5]. This paper reports initial results from investigations into the
use of various calcium sulfates to provide further activation of BFS and PFA in composite
cement systems and their potential as immobilisation matrices for aluminium.

EXPERIMENTAL DETAILS

All materials were supplied by the National Nuclear Laboratory (NNL, formally Nexia Solutions) with the OPC, BFS and PFA supplied by Castle Cement Ltd, Civil and Marine and Ash Resources Ltd respectively. The BFS and PFA were both largely amorphous but the former contained a small amount of gehlenite (C_2AS, $2CaO.Al_2O_3.SiO_2$) and the latter small amounts of quartz (SiO_2), mullite ($Al_6Si_2O_{13}$), hematite (Fe_2O_3) and magnetite (Fe_3O_4). Anhydrite ($CaSO_4$) and gypsum ($CaSO_4.2H_2O$) were used as the additives and were supplied by Chance & Hunt Limited UK.

Samples were mixed at a water:cement solids ratio (w:s) of 0.4. Powders of 9:1 BFS/OPC and 3:1 PFA/OPC were pre-blended with the additives at 0, 10, and 20 wt% of the anhydrous composite cement powders before mixing with water. All formulations were mixed by adding the combined powders to a known mass of distilled water over a total powder addition plus mixing time of 20 minutes and then immersing a strip of grade 1050 aluminium ($\approx$ 99.0 wt% purity) into each polypot of cement paste. The pots were sealed with lids and hydrated at 20°C and 95% RH for 3, 7 and 28 days.

After hydration the physical condition of the sample pots was noted before the hardened cement paste was removed from the pots and split to remove the corroded aluminium which was stored in a vacuum desiccator (1×10^{-2} mBar).

The composition of the aluminium corrosion products was determined by analysing pieces of the corroded aluminium strip by XRD. The corrosion product was scraped from the aluminium strip, ground and analysed by TGA. A Siemens D500 diffractometer (2°/min, 0.02° step size, 5-65° 2θ) and a Perkin Elmer Pyris 1 Thermogravimetric Analyzer (30 - 1000°C, 10°C/minute, flowing nitrogen) were used. For examination by scanning electronic microscopy (SEM), a piece of the corroded aluminium was analysed in secondary electron imaging (SEI) mode using a Jeol JSM 6400 electron microscope fitted with a Link EDS (energy dispersive spectroscopy) analyser.

RESULTS AND DISCUSSION

The generation of hydrogen gas caused by aluminium corrosion was qualitatively estimated by its effect on the containers. With no additions to the BFS/OPC mixes, the production of gas caused the lids of the containers to burst after only a few hours hydration. With the addition of anhydrite or gypsum, while corrosion was still apparent (Figure 1), the amount of hydrogen generated was reduced, and the lids did not burst. though there was still considerable pot distortion. The PFA/OPC system showed similar

Figure. 1. Corrosion of aluminium in 9:1 BFS/OPC with no additions for 7 days (left) and with 20 wt% gypsum for 28 days (right)

trends with the lids of the pots containing mixes with no additions bursting within a few hours of mixing. Adding anhydrite or gypsum reduced the severity of lid bursting and pot distortion, but the effect of gas evolution was more apparent than with the BFS/OPC samples due to the higher OPC content.

There was a significant difference between the crystalline phases formed in the samples with and without additions (Figure 2). The main crystalline phase formed in both the BFS/OPC and PFA/OPC samples with no additions was aluminium hydroxide (both bayerite and gibbsite were detected) together with production of a large quantity of hydrogen gas. A small quantity of strätlingite, formed by $Al(OH)_4^-$ attack on the C-S-H [3], was also detected along with reflections for the aluminium metal in the XRD trace. Upon adding either anhydrite or gypsum to the cement powders, the quantity of aluminium hydroxide formed was reduced significantly, a large amount of ettringite ($C_3A.3CaSO_4.H_{32}$, $3CaO.Al_2O_3.3CaSO_4.32H_2O$) was present and no strätlingite was formed. Ettringite contains 32 molecules of water per molar formula and as such it binds large amounts of free water, reducing the quantity available for corrosion of the aluminium. The quantity of ettringite to aluminium hydroxide formed was less in the PFA/OPC samples than in the BFS/OPC samples but still considerably more than seen with additional calcium sulfate addition.

The XRD results were corroborated by those from TGA (Figure 3). The DTG curve for the BFS/OPC system with no additions was dominated by a peak at approximately 300°C due to the dehydroxylation of aluminium hydroxide. A small shoulder on the lower temperature side of this peak was due to the release of water from strätlingite while the broad peak between approximately 50 and 550°C showed the presence of C-S-H. The DTG traces for the samples with added anhydrite and gypsum were dominated by a peak at approximately 100°C due to the release of water from ettringite. The temperature of this peak was slightly lower than reported in the literature, but the presence of so many ettringite XRD reflections confirmed the presence of large amounts of ettringite. The small dehydroxylation peak for aluminium hydroxide (shifted to a slightly lower temperature than in the samples with no additions) indicated that, even though aluminium corrodes in the samples containing sulfate additions, the amount of corrosion is significantly reduced. The TGA/DTG traces for the PFA/OPC system were very similar to those for BFS/OPC and again showed that by adding anhydrite or gypsum, the corrosion of the aluminium was significantly reduced and a substantial quantity of ettringite was produced. For all samples, the weight losses at approximately 700°C and 950°C were due to phases in the BFS and PFA.

SEM of the aluminium surface exposed in both the BFS/OPC and PFA/OPC composites, (Figures 4 and 5) showed the formation of a number of areas of distinct corrosion product (A in Figure 4) which EDS confirmed as aluminium hydroxide. For all samples made with no additions, C-S-H was observed adhered to the surface of the aluminium/hydroxide (B in Figure 5), whereas upon adding anhydrite or gypsum, much fine needle-like ettringite was observed on the surface of the aluminium hydroxide (C in Figures 4 and 5) forming a layer which could contribute to the reduced aluminium corrosion.

CONCLUSIONS

• Aluminium corrodes when encapsulated in BFS/OPC and PFA/OPC cement matrices generating hydrogen and giving aluminium hydroxide as the main corrosion product (both bayerite and gibbsite were detected) and forming strätlingite due to attack on the C-S-H.

- When anhydrite and gypsum were added to the encapsulating matrices, aluminium corrosion was reduced and significant quantities of ettringite were detected in the corrosion product
- This suggests that the ettringite may form a protective layer around the aluminium limiting the amount of long term aluminium corrosion and reducing the likelihood of radionuclide discharge due to loss of wasteform integrity from expansion cracking.

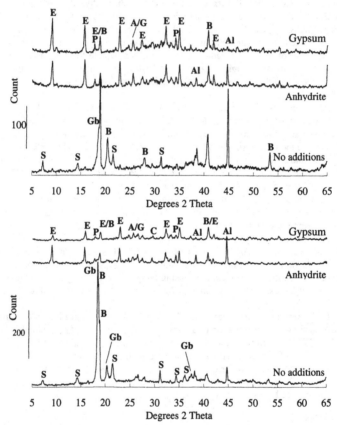

Figure. 2. XRD traces of the aluminium corrosion layer for 9:1 BFS/OPC samples (upper) and 3:1 PFA/OPC samples (lower) with no additions and additions of 20 wt% anhydrite and gypsum after hydration for 28 days.

Notes: E - ettringite ($C_3A.3CaSO_4.H_{32}$, $3CaO.Al_2O_3.3CaSO_4.32H_2O$), P - portlandite ($Ca(OH)_2$), B - bayerite ($Al(OH)_3$), A - anhydrite ($CaSO_4$), G - gypsum ($CaSO_4.2H_2O$), Al - aluminium, S - strätlingite (C_2ASH_8, $2CaO.Al_2O_3.SiO_2.8H_2O$), Gb - gibbsite ($Al(OH)_3$).

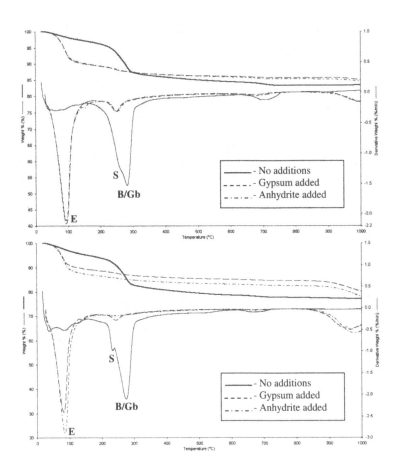

Figure 3. TGA/DTG traces of the aluminium corrosion layer for 9:1 BFS/OPC samples (upper) and 3:1 PFA/OPC samples (lower) with no additions and additions of 20 wt% anhydrite and gypsum after hydration for 7 days.

Notes: E - ettringite ($C_3A.3CaSO_4.H_{32}$, $3CaO.Al_2O_3.3CaSO_4.32H_2O$), S - strätlingite ($C_2ASH_8$, $2CaO.Al_2O_3.SiO_2.8H_2O$), B - bayerite ($Al(OH)_3$), Gb - gibbite ($Al(OH)_3$).

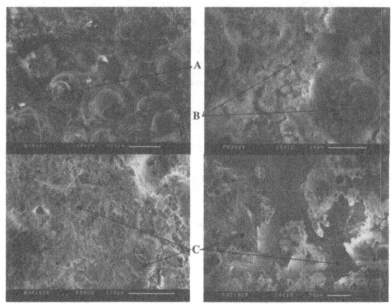

Figure 4. SEM micrographs of the surface of the corroded aluminium strip encapsulated in 9:1 BFS/OPC with no addition (upper) and 10 wt% gypsum (lower) hydrated for 28 days

Figure 5. SEM micrographs of the surface of the corroded aluminium strip encapsulated in 3:1 PFA/OPC with no addition (upper) and 10 wt% gypsum (lower) hydrated for 28 days

ACKNOWLEDGMENTS

The authors thank the National Nuclear Laboratory (NNL) for the supply of materials and for useful discussions.

REFERENCES

1. H. F. W. Taylor, *Cement Chemistry*, 1st ed (Academic Press Limited, London, 1997) p. 199.
2. M. Hayes (private communication).
3. A. Setiadi, N. B. Milestone, J. Hill and M. Hayes, Adv. Appl. Ceram. 105, 191 (2006).
4. C. A. Utton, PhD. Thesis, The University of Sheffield, 2006.
5. N. C. Collier, X. Li, Y. Bai and N. B. Milestone (submitted to Cem. Concr. Res., July 2008).

Mater. Res. Soc. Symp. Proc. Vol. 1124 © 2009 Materials Research Society   1124-Q07-19

Assessing Radionuclide Solubility Limits for Cement-Based, Near-Surface Disposal

David A. Pickett,[1] Karen E. Pinkston,[2] and James L. Myers[1]
[1]Center for Nuclear Waste Regulatory Analyses, Southwest Research Institute®, San Antonio, TX
[2]Division of Waste Management and Environmental Protection, U.S. Nuclear Regulatory Commission (NRC), Washington, DC

ABSTRACT

Recommended solubility limits for Se, Tc, U, Np, and Pu are developed for evaluating performance assessments for shallow, cement-based disposal systems. These recommendations consider existing literature models and experimental data and new equilibrium models.

INTRODUCTION

NRC has a consultative role in U.S. Department of Energy (DOE) plans to dispose of radioactive waste in two types of near-surface, cement-based facilities: (i) Saltstone grout monoliths in Savannah River Site (SRS) vaults [1] and (ii) grouted waste tanks at the Idaho National Laboratory (INL) and SRS [e.g., 2]. As part of this role, NRC staff reviews DOE 10,000-year performance assessments demonstrating whether these systems can meet the performance objectives in 10 CFR Part 61.

This study provides ranges of radionuclide solubility limits that NRC may consider when evaluating DOE release models, focusing on the risk-significant (from a performance perspective) radioelements selenium (Se), technetium (Tc), uranium (U), neptunium (Np), and plutonium (Pu). Key reactive components of the materials are ordinary Portland cement (OPC), blast furnace slag (BFS), and fly ash (FA). Our approach is to establish a range of expected chemical conditions during degradation stages, survey the existing literature for relevant solubility experiments and models, model solubilities using the anticipated range of water compositions, and establish recommended ranges, addressing uncertainty by conservatively favoring higher values. The conservative approach meant, for example, that low-solubility actinide oxides not observed in experiments were excluded because their long-term stability could not be confidently assumed. This paper is a condensed version of a portion of a report prepared for NRC [3], which contains the literature references used.

CHEMICAL CONDITIONS AND DEVELOPMENT OF SOLUBILITY MODELS

During the evolution of cement-based disposal systems, key aqueous chemical parameters affecting the five redox-sensitive elements include pH and Eh [3]. Fresh cement maintains pH > 12.5. Once alkalis are largely leached, portlandite [$Ca(OH)_2$] will buffer pH to 12.5; the presence of BFS and FA may shorten this stage. After portlandite exhaustion, pH will evolve from 12.5 to ~ 10.5, controlled by dissolution of calcium silicate hydrates (C-S-H). During degradation of C-S-H and carbonation, pH will continue to decrease toward 8, reflecting calcite buffering. In the long term, reactive materials are exhausted and pH reflects the infiltrating water. Eh tends to increase with degradation. Early on, BFS imposes reducing conditions, with Eh as low as −350 mV (SHE). If the BFS reducing capacity is exhausted during the portlandite stage, Eh may rise to +100 mV or higher. Upon further degradation, atmospheric oxygen will eventually control solution Eh to high values. Other solution characteristics include (i) a decrease in Ca/Si as degradation proceeds, (ii) relatively low carbonate species during C-S-H evolution, and (iii) increasing carbonate under the influence of atmospheric CO_2.

For the solubility models, 14 water chemistries relevant to the INL and SRS systems were identified based on literature models and data on pore waters in cement-based materials, ranging in pH from 13.4 to 8.3 and in Eh from −350 mV to +730 mV [3]. As potential end-members to chemical evolution when reactivity is exhausted, models were developed for groundwater chemistries from INL and SRS. Equilibrium solubility models for Se, Tc, U, Np, and Pu were conducted using Geochemist's Workbench® Professional 6.0 with the thermo.com.v8.r6+.dat thermodynamic database [4]. For each model water, Eh was varied for the given pH to cover the chemical range delineated in the previous paragraph. Results were compared to models using an alternative provisional database [5] based on the Nuclear Energy Agency (NEA) thermodynamic data [e.g., 6]. For data display, a concentration of 0.5 M was assigned for cases in which no solubility limit was indicated.

DISCUSSION AND SOLUBILITY LIMIT RECOMMENDATIONS

Selenium

Se solubility model results are shown in Figure 1. For the solubility-limited Eh–pH regions, the controlling solids were the tetravalent phase $CaSeO_3:2H_2O$ (Eh < 200 mV) and metallic Se (lowest Eh). Runs using the provisional NEA database differed mainly in yielding lower limits at Eh < 200 mV. Literature data and models are consistent with the overall picture of control to very low Se concentrations by metallic Se only at lowest Eh and pH > 11. Concentrations of as low as 10^{-2} M, controlled by $CaSeO_3:2H_2O$, are

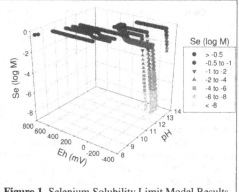

Figure 1. Selenium Solubility Limit Model Results

possible at Eh < 200 mV and pH 10–12.5; these concentrations, however, are strong functions of dissolved Ca. Table I shows the important aqueous chemical effects on Se solubility limits. Consideration of our models and literature information leads to these recommendations:

- Se is not solubility limited over much of the Eh–pH range (Figure 1).
- Very low Se solubility limits should be used only in cases in which very low Eh control can be confidently demonstrated.
- $CaSeO_3:2H_2O$ control to 10^{-2} M or higher may be used at Eh < 200 mV, but only if dissolved Ca can be confidently predicted.

Technetium

Tc model results are shown in Figure 2. Only Tc^{IV} and Tc^{VII} solids were included in the model; other oxidation state solids were suppressed because they are not typically observed in

Table I. Qualitative rating of effects of solution characteristics on modeled solubility limits in cement-based systems (L = low, M = moderate, H = high). Arrows indicate whether the correlation is positive or negative, or shows both positive and negative effects.

	pH	Eh	Carbonate	Other
Se	L	H ↑	–	Calcium – M (pH 10 to 12.5 only) ↓
Tc	L	H ↑	–	Potassium? (pH 13.4 only)
U	M ↕	H ↑	H (pH < 12 only) ↑	Sodium – H (pH ≥ 11 only) ↓ Silicon – H (pH < 11 only) ↓
Np	L	H ↑	H ↑	
Pu	L	M ↕	H (pH < 12 only) ↑	

relevant aqueous systems. $KTcO_4$ was suppressed because it has not been observed in cementitious systems; its inclusion would decrease predicted Tc to 0.05 M in only two high-K waters at pH 13.4. Tc is controlled by solubility limits only at lowest Eh values at pH > 11, where concentrations as low as 10^{-7} M were controlled by $TcO_2:2H_2O(am)$. Runs using the provisional NEA database differed in that TcO_2 controlled Tc to lower concentrations at lowest Eh.

Literature data and models, though revealing a great deal of uncertainty with respect to

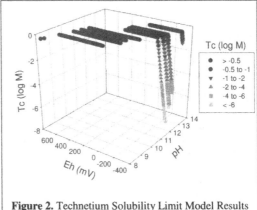

Figure 2. Technetium Solubility Limit Model Results

identification and quantification of precipitating Tc solids, are consistent with the overall picture in Figure 2. Table I shows the important aqueous chemical effects on Tc solubility limits. Consideration of our models and literature information leads to these recommendations:

- Tc is solubility limited only at the very margins of the Eh–pH region of interest, representing the most reducing conditions.
- Unless the reducing capacity of the grout can be confidently predicted to endure during the modeled period of radionuclide release, solubility limits should not be applied to Tc.

Uranium

U model results are shown in Figure 3. There are no overall systematic variations with pH. Because sodium uranate was an important model phase at high pH and uranyl silicates (particularly soddyite) governed solubility at lower pH, Figure 3 excludes waters that did not report concentrations of sodium or silicon. (In addition, the phases $CaUO_4$ and haiweeite were excluded because they imposed U concentrations—< 10^{-11} M—much lower than observed in experiments). The high U concentrations in two high-pH waters reflect their low Na content because solubility is limited by $Na_2U_2O_7$. The tendency to lower U concentrations as pH decreases reflects decreasing soddyite solubility as aqueous Si increases. The steep drops in U at lowest Eh result from precipitation of the U^{IV} phase uraninite (UO_2) and mixed-valent U^{IV}/U^{VI} oxides. Sensitivity analyses showed that equilibrium with atmospheric CO_2 could result in significantly higher U concentrations at high Eh. Models using the NEA-derived database predicted U control by schoepite at pH < 11, with U concentrations higher than those in Figure 3 by two to three orders of magnitude (up to 10^{-5} M). This result reflects the higher soddyite solubility constant Guillaumont, et al. [6] recommended compared to the value in thermo.com.v8.r6+.dat.

Experimental studies suggest that, for cementitious systems with pH 9 to 13, U is limited to maximum values of 10^{-8} to 10^{-6} M. Lower and higher limits are suggested under certain conditions, but are not consistently observed.

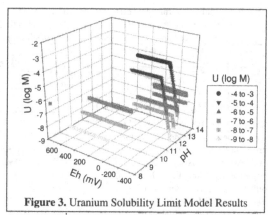

Figure 3. Uranium Solubility Limit Model Results

Modeled U solubilities have varied considerably over the years because of differences among and evolution of thermodynamic databases. The broad outlines of the new model results are consistent with the literature laboratory results. Table I shows the important aqueous chemical effects on U solubility limits. Consideration of our models and literature information leads to these recommendations:

- Over the conditions of interest, U solubility limits are between 10^{-8} and 10^{-5} M, with potential deviations to lower values under the most reducing conditions and to higher values if Na is especially low.
- U solubility limits must take into particular consideration aqueous Na, Si, and carbonate.
- Choices of U mineral thermodynamic data can strongly affect solubility predictions.

Neptunium

Np solubility model results are shown in Figure 4. $Np^{IV}O_2$ was suppressed in these models because it imposed markedly low Np concentrations compared to experiments and has not been identified in relevant solubility studies. In the models, solubility was controlled by $Np^{IV}(OH)_4$ at lower Eh (sloped portions of curves) and by $Np^{V}O_2OH(am)$ at high Eh (flat portions of the curves); the majority of model Np concentrations lie between 10^{-9} M and 10^{-4} M, with no

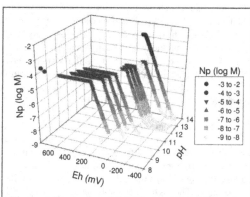

Figure 4. Neptunium Solubility Limit Model Results

systematic pH dependence. The one exception to the overall pattern, with notably higher Np at pH 13.4, has a possibly unrealistically high carbonate content of 0.04 M. Carbonate sensitivity was also seen in the groundwaters. Models using the NEA-derived database yielded Np concentrations up to four orders of magnitude lower than with thermo.com.v8.r6+.dat. This contrast is due to (i) much lower $Np^V_2O_5$ solubility and (ii) much lower stability of the $Np^VO_2OH(aq)$ species in the NEA database.

Relevant literature experimental data are rare and show Np^{IV} and Np^V solubility limits varying between 10^{-4} and 10^{-9} M over the pH range of interest, with contradictory data on pH dependence. Literature Np^V models, focused on $Np^VO_2OH(am)$ solubility, have given widely varying results due to thermodynamic data differences. Modeled Np^{IV} solubilities tend to lower values than Np^V. Table I shows the important aqueous chemical effects on Np solubility limits. Consideration of our models and literature information leads to these recommendations:

- The differences in thermodynamic data and modeled solubilities at higher Eh noted in our new model results are not easily reconciled with the existing literature.
- Taking a conservative approach that favors higher solubility limits in light of uncertainties suggests that in oxidizing systems, Np solubility varies from approximately 10^{-4} M at pH 8 to 10^{-5} M at pH 11–12. Concentrations may be higher at pH > 13, particularly in the unlikely event carbonate is abundant in solution.
- Under more reducing conditions, 10^{-8} M appears to represent a reasonable Np solubility limit at all pH, though values as low as 10^{-9} M are possible at Eh < –100 mV.

Plutonium

Pu model results are shown in Figure 5. $Pu^{IV}O_2$ was suppressed in these models because it imposed Pu concentrations (< 10^{-11} M at pH < 9 and < 10^{-14} M at pH > 9) much lower than observed in the laboratory and because it has not been identified in experiments. Dissolved Pu is constant at 2×10^{-9} M over most of the Eh– pH space, controlled by $Pu^{IV}(OH)_4$. Between pH 9 and 11, Pu solubility rises with increasing Eh as dissolved Pu^V species become more stable, but decreases

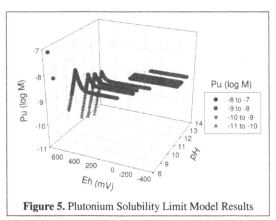

Figure 5. Plutonium Solubility Limit Model Results

markedly to 10^{-11} to 10^{-10} M at even higher Eh when hexavalent $PuO_2(OH)_2$ becomes the stable solid phase. The two lowest pH waters resulted in higher modeled Pu because they have high carbonate species contents. Models using the NEA-derived database gave lower Pu under most conditions due to different solubility data for the hydrous Pu^{IV} solid. At high Eh, however, Pu concentrations are higher in the NEA models as a result of differing solid and aqueous species stabilities for Pu^V and Pu^{VI}.

Literature experimental and model results for Pu in cementitious systems are schematically depicted in Figure 6 (which does not account for potentially high carbonate species at lower pH). This figure also depicts an apparently equilibrium 10^{-8} M concentration for colloidal Pu^{IV} polymer Neck, et al. [7] proposed. Table I shows the important aqueous chemical effects on Pu solubility limits. Consideration of our models and literature information leads to these recommendations:

- Under most Eh–pH conditions, 10^{-9} M is a reasonably conservative solubility limit for dissolved Pu.
- As pH decreases to ≤ 9, the limit on Pu concentrations is more appropriately defined as around 10^{-8} M.
- Elevation of dissolved carbonate due to atmospheric CO_2 could stabilize Pu concentrations greater than 10^{-6} M. Aqueous carbonate content needs to be predicted with confidence if the Pu solubility limit is to be better defined.
- A 10^{-8} M concentration for colloidal Pu^{IV} may more realistically reflect a maximum Pu concentration for most Eh values and should be considered unless the potential for colloid mobilization can be excluded.

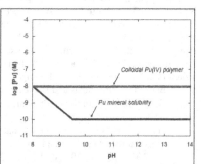

Figure 6. Literature Pu concentration limit trends for carbonate-poor, cement-based systems [3,7].

ACKNOWLEDGMENTS

We thank R. Pabalan and D. Turner for discussions and constructive reviews. This paper is an independent product of the Center for Nuclear Waste Regulatory Analyses and does not necessarily reflect the views or regulatory positions of the NRC. The NRC staff views expressed herein are preliminary and do not constitute a final judgment or determination.

REFERENCES

1. DOE, *Basis for Section 3116 Determination for Salt Waste Disposal at the Savannah River Site*, DOE–WD–2005–001 (DOE, Washington, DC, 2006).
2. DOE–ID, *Basis for Section 3116 Determination for the Idaho Nuclear Technology and Engineering Center Tank Farm Facility*, DOE/NE–ID–11226 (DOE, Idaho Falls, ID, 2006).
3. R.T. Pabalan, F.P. Glasser, D.A. Pickett, G.R. Walter, S. Biswas, M.R. Juckett, L.M. Sabido, and J.L. Myers, *Review of Literature and Assessment of Factors Relevant to Performance of Grouted Systems for Radioactive Waste Disposal* CNWRA 2009-001 (Center for Nuclear Waste Regulatory Analyses, Southwest Research Institute®, San Antonio, TX, 2009).
4. C.M. Bethke, *Geochemical Reaction Modeling, Concepts, and Applications* (Oxford University Press, New York, 1996).
5. Japan Atomic Energy Agency, "Thermodynamic Database," File 050000g0.tdb <http://migrationdb.jaea.go.jp/english.html> (28 March 2008).
6. R. Guillaumont, T. Fanghänel, J. Fuger, I. Grenthe, V. Neck, D.A. Palmer, and M.H. Rand, *Chemical Thermodynamics, Volume 5: Update on the Chemical Thermodynamics of Uranium, Neptunium, Plutonium, Americium and Technetium* (Elsevier, New York, 2003).
7. V. Neck, M. Altmaier, A. Seibert, J.I. Yun, C.M. Marquardt, and T. Fanghänel, *Radiochim. Acta* **95**, 193 (2007).

Mater. Res. Soc. Symp. Proc. Vol. 1124 © 2009 Materials Research Society 1124-Q07-21

Acoustic emission characterisation of cementitious wasteforms under three-point bending and compression

Lyubka M. Spasova and Michael I. Ojovan
Immobilisation Science Laboratory, Department of Engineering Materials,
University of Sheffield, Mappin Street, Sheffield S1 3JD, UK

ABSTRACT

Correlation was observed between load of failure, curing age and energy of recorded acoustic emission (AE) signals from OPC and BFS/OPC (at mass ratio 7:3) samples in three-point bending tests. The released stress energy was concentrated in single distinctive, high amplitude signals (up to 10 V at the piezoelectric sensor output) with primary frequencies below 50 kHz and an amplitude spectrum spread from 20 to 1000 kHz. The fracture surfaces showed brittle-like failure with a significant role of pre-existing microcracks.

AE responses of OPC and 7:3 BFS/OPC samples showed Kaiser's effect during three consecutive loading-unloading cycles under compression before the ultimate strength of the structures to be reached. The strains measured during the loading-unloading cycles were used to determine the elastic modulus and Poisson ratio of the monitored wasteforms enabling calculation of the velocities of the longitudinal and transverse ultrasonic waves.

INTRODUCTION

One of the general requirements to the safe storage and disposal of cementitious structures encapsulating radioactive wastes is to preserve their mechanical integrity over long periods (tens and hundreds of years) [1]. As a result of electrochemical interactions between the waste and the encapsulating cementitious matrix corrosion of metals such as Al and Magnox occurs [2]. Previous studies [2-3] report that the accumulation of corrosion products and hydrogen gas release cause nucleation and propagation of cracks (visible to the naked eye) on the surface of cementitious wasteforms up to 2 years of curing. The mechanisms of formation of those cracks involve accumulation of diagonal tension stresses along the encapsulated metal due to the expanding corrosion products and the low adhesion provided by the porous zone formed between the encapsulated metal and the hardened cement. Real issues are to detect nucleation and propagation of these cracks which is difficult to be predicted.

The AE technique, based on detection and processing of ultrasonic waves generated and released within materials, has been shown as a suitable tool to detect and characterise in terms of signal rate and characteristics in time and frequency domains the damage development within OPC and 7:3 BFS/OPC structures encapsulating Al [3-5]. To extend and support these results from passive AE monitoring we carried out three-point bending and compression tests on OPC and BFS/OPC samples.

EXPERIMENTAL DETAILS

In this paper the results from three-point bending tests on two OPC and two 7:3 BFS/OPC specimens are reported. The samples were prepared following the mixing procedure in [3-5] and casted as prisms 180 mm long, 50 mm wide and 20 mm high. Two uniaxial strain gauges (Kyowa KFG-10-120-C1-11) were attached symmetrically on the top and bottom of the samples as shown in Figure 1a (arrowed). Two of those samples were cured for 14 days and two for 28 days at 20°C and 95% relative humidity (RH). A broadband piezoelectric sensor type WD, calibrated and supplied by Physical Acoustics Corporation, was attached to each of the samples during the tests via a thin layer of grease as shown in Figure 1b. The load was applied manually and recorded by a computer-based acquisition system. At the same time the AE was processed and recorded by a PCI-2 based system with input settings for data collection as reported in [3,5]. The AE parameters determined were signal durations, amplitudes, absolute (ABS) energies, amplitude spectra by fast Fourier transformation (FFT) and primary frequencies also defined in [3,5].

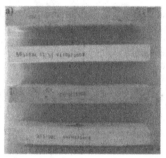

Figure 1. (a) Strain gauges attached to the cementitious samples and (b) experimental setup for a three-point bending test.

For compression tests OPC and 7:3 BFS/OPC cylindrical samples (prepared as in [3,4]) cured for 90 days at 20°C and 95% RH with a diameter and a height of 60 mm were used. Since the region of the stress-strain curve corresponding to elastic deformation was needed to determine the elastic constants (elastic modulus and Poisson ratio) each sample was loaded with a rate of 10 kN/min to ~100 kN. This was chosen after measurement of the load under compression when failure occurred of another pair of OPC and BFS/OPC samples with the same geometry and curing age. The loads of failure were 155 kN for the OPC and 144 kN for the BFS/OPC sample. The longitudinal strains were measured by three symmetrically attached strain gauges on the outer surfaces of the samples. The transverse strain was measured by one strain gauge with a length of 30 mm (Figure 2a). All strain gauges were the same type as those for three-point bending tests. The BFS/OPC sample before to be loaded is shown in Figure 2b. Despite the top and bottom surfaces of the samples having been polished to be as flat as possible, the loading equipment included a specially designed "free" moving component (see Figure 2b) to minimize the effect of an irregular surface during loading.

Figure 2. (a) Strain gauges attached to the cementitious samples and (b) experimental setup for a compression test.

RESULTS AND DISCUSSION

Three-point bending tests

AE hit rate in relation to the applied load for the OPC and BFS/OPC samples cured for 28 days are presented in Figure 3. For the OPC sample significant AE (with a total ABS energy of 8.57×10^4 aJ) was detected during the first 100 s of the loading up to 60 N (Figure 3a). The failure of the sample under loading was associated with a significant increase in the number of hits and their total ABS energy of 1.87×10^9 aJ. The load of failure of the OPC sample was 510 N whereas this was 275 N for the BFS/OPC specimen. AE was also recorded from the composite cement sample from the beginning of the loading up to a load of ~200 N (Figure 3b). The total ABS energy for those signals was 4.19×10^3 aJ. The sudden failure of the sample corresponded to an increase in the number of hits and their ABS energy being 6.08×10^8 aJ. This energy was concentrated at single distinctive signal (Figure 4). Very similar signal with an amplitude spectrum as that shown in Figure 4b was recorded during the failure of the OPC sample.

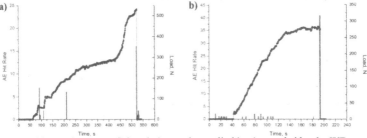

Figure 3. AE hit rate (per second) in relation to the applied load recorded by the WD sensor for (a) the OPC and (b) the BFS/OPC samples cured or 28 days.

As it can be seen in Figure 4b the amplitude spectrum of the signal was characterised by low frequency peak (below 50 kHz) and high frequency components spread between 200 and 1000 kHz.

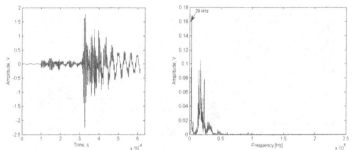

Figure 4. (a) Signal and (b) its amplitude spectrum with the highest energy generated during the sudden failure of the BFS/OPC sample cured for 28 days in a three-point bending test.

The strains measured for both samples as load is increased are shown in Figure 5.

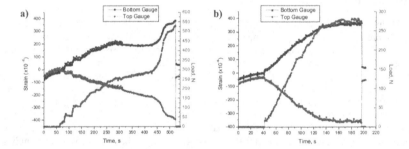

Figure 5. Strains measured for (a) the OPC and (b) the BFS/OPC samples cured for 28 days in a three-point bending test.

For the OPC and BFS/OPC samples negative strain (due to compression) was measured by the strain gauges mounted on the top surface of the prisms (see Figure 1b). The strain gauges on the bottom measured strains with positive sign caused by tension. The maximum strains measured were approximately plus and minus 400 microstrains ($\times 10^{-6}$) before the sudden failure and show symmetry as a positive and negative value at any given load.

The fracture surfaces of the samples (Figure 6) reveal brittle-like failure also reported by Chotard et al. [6] for calcium aluminate cement samples during tensile tests.

Figure 6. Fracture surfaces of (a) the OPC and (b) the BFS/OPC samples cured for 28 days after a three-point bending test.

However, the fracture surfaces of the samples cured for 14 days (Figure 7) reveal propagation of pre-existing microcracks associated with the failure, observed also in some of the calcium alumunate cement samples tested by Chotard et al. [6].

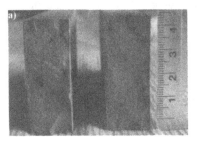 

Figure 7. Fracture surfaces of (a) the OPC and (b) the BFS/OPC samples cured for 14 days after a three-point bending test.

The plots in Figure 8 confirm that the recorded AE from the cementitious samples cured for 14 days was associated with their failure. This occurred at loads of 275 and 220 N for the OPC and BFS/OPC samples respectively. The load of failure of the OPC sample was 1.85 times less than that of the sample cured for 28 days. However, the slower strength development of the BFS/OPC (compared with the OPC) [20] was confirmed by the difference of only 55 N between the loads of failure of the structures cured for 14 and 28 days. Consequently the total ABS energy measured by the WD sensor for the OPC sample was 2.97×10^8 aJ or 6.36 times less than that for the sample cured for 28 days. For the BFS/OPC specimen the total ABS energy was 2.97×10^8 aJ or 2 times less than that recorded for the 28 days old sample. Similar single distinctive, high energy signals associated with the failure of the samples were recorded as that shown in Figure 4.

The strains measured by the gauges attached to the OPC and BFS/OPC samples show again symmetry as positive and negative values at any given load with a maximum of approximately plus and minus 300 microstrains when failure occurred.

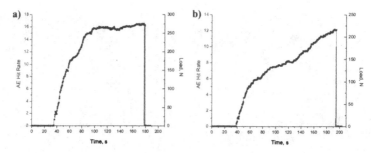

Figure 8. AE hit rate (per second) in relation to the applied load recorded by the WD sensor for (a) the OPC and (b) the BFS/OPC samples cured for 14 days.

AE during compression tests on cementitious samples

During three consecutive loading cycles under compression 522 signals were detected from OPC sample at a rate shown in Figure 9a. 454 (86.97% of the total hits) were recorded during the first cycle (Figure 9a). This AE activity was almost certainly associated with microcrack nucleation or propagation of pre-existing microcracks [8,9] confirmed also by post-experimental visual observation on the sample.

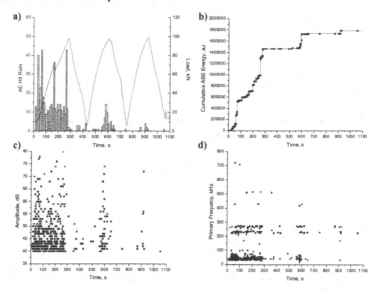

Figure 9. (a) AE hit rate (per second) with variation of the applied load, (b) cumulative ABS energy, (c) amplitudes and (d) primary frequencies of the signals recorded from the OPC sample in compression test.

362

The total ABS energy of the signals was 1.79×10^6 aJ (Figure 9b) as their amplitudes did not exceed 80 dB (or 1 V at the sensor output). The primary frequencies of the recorded waveforms were mainly below 100 kHz (see Figure 9d) but there were also significant numbers with frequencies between 220 and 280 kHz and above 400 kHz (up to 750 kHz).

The number of signals detected from BFS/OPC sample was 1888 or 3.61 times more than these from the OPC. Therefore a larger number of AE sources (pre-existing microcracks) were active within the composite cement specimen compared with the OPC sample [8]. 1648 (87.28% of the total hits) were recorded during the first loading-unloading cycle with only a few signals detected during the next loadings (Figure 10a). This AE response was analogous to Kaiser's effect [3,9] associated with significant decrease or lack of AE before the level of previous loading to be exceeded. The total ABS energy of the signals was 2.62×10^5 aJ (Figure 10b) or 6.83 times less than that for the OPC sample. The amplitudes of the signals were less than 70 dB (Figure 10c) and their primary frequencies were mainly below 100 kHz and between 220 and 280 kHz (Figure 10d).

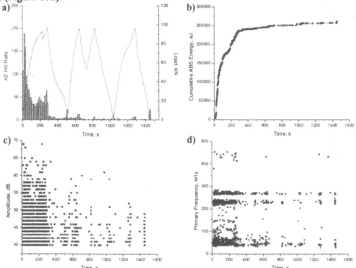

Figure 10. (a) AE hit rate (per second) with variation of the applied load, (b) cumulative ABS energy, (c) amplitudes and (d) primary frequencies of the signals recorded from the BFS/OPC sample under compression.

From the stress-strain linear plots acquired during the second and third loading cycles the elastic modulus E was calculated average 19.59 GPa for the OPC and 13.84 GPa for the BFS/OPC samples. The Poisson ratio ϑ was determined from the longitudinal and transverse strains as 0.261 for the OPC and 0.148 for the composite cement. During the first loading-unloading cycle the stress-strain plots for both samples were non-linear due to heterogeneity of the materials and release of residual stresses (as shown by the recorded AE) induced by the preparation and drying of the samples as well as microcrack propagation under loading [9]. The density ρ of the OPC and BFS/OPC samples was determined as 2091.8 kg/m^3 for the OPC and 2050.2 kg/m^3 for the BFS/OPC samples. The elastic constants and density were used to calculate the velocities of the longitudinal (v_p) and transverse (v_s) waves for the OPC sample as follows [11]:

$$v_p = \sqrt{\frac{E(1-\vartheta)}{\rho(1+\vartheta)(1-2\vartheta)}} = 3388.5 \ m/s \tag{1}$$

$$v_s = \sqrt{\frac{E}{2\rho(1+\vartheta)}} = 1927.01 \ m/s \tag{2}$$

For the BFS/OPC sample the corresponding wave velocities were $v_p = 2667.6 \ m/s$, and $v_s = 1714.7 \ m/s$.

These ultrasonic wave velocities can be applied as a first approximation in a future work including AE source location determined from the time of wave arrival at different sensors attached to real-size cementitious wasteforms.

CONCLUSIONS

The three-point bending tests on OPC and 7:3 BFS/OPC samples revealed that their sudden failure generates AE distinguished by single, high energy signals and a total ABS energy dependent on the curing age associated with the strength developed. The results from the compression tests showed that the AE associated with microcracking of the samples (before failure) is characterised by hundreds and thousands of signals with amplitudes up to 80 dB and primary frequencies mainly between 20 and 280 kHz. In terms of AE the structures showed Kaiser's effect which combined with the AE results reported in [3-5] and in the present work can be beneficial for quantitative and qualitative AE data interpretation over long periods of monitoring of real cementitious wasteforms with encapsulated radioactive metals such as Al or Magnox.

REFERENCES

[1] M. I. Ojovan and W. E. Lee, An introduction to nuclear waste immobilization, Elsevier, Amsterdam, 2005.
[2] A. Setiadi, N.B. Milestone, J. Hill and M. Hayes, Adv. Appl. Ceram. 105 (4), 191-196 (2006).
[3] L. M. Spasova and M. I. Ojovan, J. Hazard. Mater., 138 (3), 423-432 (2006).
[4] L. M. Spasova, M. I. Ojovan and C. R. Scales, J. Acoustic Emission, 25, 51-68 (2007).
[5] L. M. Spasova and M. I. Ojovan, Mat. Res. Soc. Symp. Proc., 985, Paper N 0985-NN10-03 (2007).
[6] T. J. Chotard, D. Rotureau and A. Smith, J. Eur. Ceram. Soc., 25, 3523-3531(2005).
[7] H. F. W. Taylor, Cement chemistry, Thomas Telford Publishing, London, 1990.
[8] J. P. Gorce and N. B. Milestone, Cem. Concr. Res. 37, 310-318 (2007).
[9] M. T. Tam and C. C. Weng, Cem. Concr. Res. 24(7), 1335-1346 (1994).
[10] E. Landis and S. Shah, J. Eng. Mech. 121(6), 737-743 (1995).
[11] A. Van Hauwaert, J. F. Thimus and F. Delannay, Ultrasonics, 36 209-217 (1998).

Mater. Res. Soc. Symp. Proc. Vol. 1124 © 2009 Materials Research Society   1124-Q10-12

Modeling of pH Elevation Due to the Reaction of Saline Groundwater With Hydrated Ordinary Portland Cement Phases

A.Honda[1] , K.Masuda[1], H.Nakanishi[1], H.Fujita[2], and K.Negishi[2]
[1]Japan Atomic Energy Agency, 4-33, Muramatsu, Tokai-mura, Naka-gun, Ibaraki, 319-1194, Japan
[2]Taiheiyo Consultant Co., Ltd., 2-4-2, Osaku, Sakura-shi, Chiba, 285-8655, Japan

ABSTRACT

Thermodynamic calculations of the reaction between hydrated OPC phases and saline groundwater indicate an elevated pH > 13, which is not associated with the well known initial release of the alkalis. Instead, the pH elevation is attributed to the generation of OH^- accompanying the precipitation of Friedel's salt ($Ca_3Al_2O_6 \cdot CaCl_2 \cdot 10H_2O$; AFm-$Cl_2$) from the reaction of portlandite ($Ca(OH)_2$; CH) and hydrogarnet ($Ca_3Al_2O_6 \cdot 6H_2O$; C_3AH_6) with chloride ions from the saline groundwater. If such a reaction mechanism were to occur in the context of the geological disposal of radioactive wastes, the impact of a hyper-alkaline plume on other barrier components, such as the host rock or bentonite buffer, could be significant. Experimental investigations were therefore conducted using only CH and C_3AH_6 to represent hydrated OPC and NaCl solution to represent a saline groundwater. The pH elevation was confirmed and showed good agreement with thermodynamic calculations. The experiments were repeated using hardened OPC paste to confirm this reaction mechanism in the presence of other hydrated OPC phases. In this case, however, the pH elevation was not as high as expected. This deviation can be explained by the residual aluminum, after being partially consumed by ettringite ($Ca_6Al_2(OH)_{12}(SO_4)_3 \cdot 32H_2O$; AFt), monosulfate ($Ca_4Al_2O_6SO_4 \cdot 12H_2O$) and/or monocarbonate ($Ca_4Al_2O_6CO_3 \cdot 11H_2O$; AFm-$CO_3$) phases, not being wholly assigned to C_3AH_6. A better agreement between the thermodynamic calculations and the experimentally measured results can be made assuming a fraction of aluminum is incorporated into the calcium silicate hydrate (C-S-H) gel phase.

INTRODUCTION

Current proposals for the geological disposal of TRU waste in Japan involve the construction of a purpose-built repository based on the combined performance of natural and engineered barriers [1]. Using cementitious materials in the engineered barrier, however, will almost certainly produce a hyper-alkaline plume as they are subjected to leaching by invading groundwaters. A hyper-alkaline plume will have detrimental effects on the performance of the host rock and bentonite backfill, which are known to be thermodynamically unstable at high pH [2].

The pH and composition of hydrated OPC during leaching in pure water can be subdivided into three regions (Region I, Region II and Region III) [3]. Region I is characterized by a high pH (>13), which is attributed to the release of the completely soluble alkalis (Na and K). Region II has a pH ≈ 12.5, which is controlled by equilibrium with CH. Following the loss of CH, the pH of Region III is largely controlled by the incongruent dissolution of C-S-H gel.

It is possible that the groundwater at the repository site will be saline [4] and so the pore water composition of hydrated OPC was estimated for the saline groundwater case [5]. The estimation showed that the time period during which pH exceeds 13 in the saline groundwater case is about 4 times longer than for freshwater type groundwater. The result suggests the condition of pH > 13 continues even after Region I in the saline groundwater case and is attributed to the generation of OH^- accompanying the formation of AFm-Cl_2 from the reaction of C_3AH_6, CH and chloride ions :

$$CH + C_3AH_6 + 4H_2O + 2Cl^- \rightarrow AFm\text{-}Cl_2 + 2OH^- \tag{1}$$

The calculation of Glasser et al. [6] also suggested pH elevation by the immersion of C_3AH_6 in NaCl solution.

To verify the pH elevation according to the reaction in Equation 1, immersion experiments were conducted to investigate the reaction of CH and C_3AH_6 with chloride ions in solution. The experimental results were interpreted with the assistance of thermodynamic modeling to confirm the pH elevation mechanism. Similar experiments were then conducted using hardened OPC paste and the results interpreted with the assistance of thermodynamic modeling to confirm if the same reaction mechanism occurs in the presence of other hydrated OPC phases.

EXPERIMENT

Batch immersion of mixture of CH and C_3AH_6 in NaCl solution

C_3AH_6 was synthesized according to Breval [7] and CH obtained as a commercial reagent (KANTO CHEMICAL CO.,INC.). All experiments were conducted in a glove box under an Ar atmosphere to minimize carbonation effects. A mixture of CH and C_3AH_6 (1:1 mole ratio) was placed in polypropylene bottles containing 0.6 mol dm^{-3} NaCl solution at a liquid/solid (L/S) ratio of 4.4 (equivalent to 0.5 mol of each solid for 1 kg of solution). The same solid phases were also immersed in deionized water as a blank for the measurement of pH. The temperature was maintained at 293 ± 6 K and the pH was periodically measured up to 360 days. After this time, solid phases were characterized by X-Ray diffraction and a wet chemical analysis to determine composition, according to Japan Industrial Standard (JIS R 5202), and the liquid phases were analyzed for composition using a combination of ICP-AES and IC.

Batch immersion of hardened OPC in NaCl solution

OPC clinker was hydrated with deionized water at a water/cement (W/C) ratio of 0.5 and cured for 5 years in a sealed container. The hardened OPC paste was then ground into a powder (< 150 μm). To minimize pH elevation due to alkali release, the ground OPC was rinsed by immersion in deionized water at L/S = 10 for a week prior to the immersion tests. The powdered hardened OPC paste was then immersed in 0.6 mol dm^{-3} NaCl aqueous solution at L/S = 2.5 for the same temperature and time as described above. The same hardened OPC paste was also immersed in deionized water as a blank. Solid phases were again characterized by X-Ray diffraction and a wet chemical analysis and the liquid phases were again analyzed using a combination of ICP-AES and IC. Additionally, Differential scanning calorimetry (DSC) was used to quantify AFm and the Al content of C-S-H gel was measured by Electron Probe Micro Analysis (EPMA).

DISCUSSION

Batch immersion of mixture of CH and C₃AH₆ in NaCl solution

The pH of the 0.6 mol dm⁻³ NaCl solution increased and stabilized at a value of 13.5 after 180 days, whereas the blank test showed an almost constant value around 12.6 throughout the immersion period (Figure 1).

An X-Ray diffraction pattern of the solid phase after 360 days immersed in the 0.6 mol dm⁻³ NaCl solution confirmed the precipitation of AFm-Cl₂ (Figure 2).

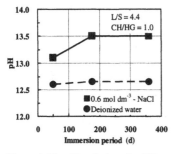 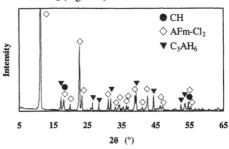

Figure 1. The pH evolution of a CH and C₃AH₆ mixture immersed in 0.6 mol dm⁻³ NaCl solution and deionized water.

Figure 2. X ray diffraction pattern of the solid phase after C₃AH₆ and CH were immersed in 0.6 mol dm⁻³ NaCl solution for 360 days.

The solution composition was measured and predicted using the geochemical computer code PHREEQC [8] with the JNC-TDB.TRU [9] database (Table I).

Table I . The measured and predicted composition of the 0.6 mol dm⁻³ NaCl solution in which C₃AH₆ and CH were immersed after 360 days. Element concentrations are in mol dm⁻³.

Items	Measured Values	Calculated Values
pH	13.5	13.5
Ca	5.3×10^{-4}	6.7×10^{-4}
Al	9.1×10^{-4}	1.9×10^{-3}
Na	6.1×10^{-1}	6.0×10^{-1}
Cl	8.4×10^{-2}	5.9×10^{-2}

The measured decrease of Cl concentration could be calculated and was consistent with the diffraction pattern showing the precipitation of AFm-Cl₂. The calculated pH values were in good agreement with the experimental results and the concentrations of the elements were within the range of ±50 %. On the basis of Table I, the experimental and predicted mole ratio of the removed Cl⁻ and the released OH⁻ (almost corresponding to pH 13.5) for 1 dm³, assuming all of the Cl⁻ removed from solution was consumed by the formation of AFm-Cl₂ and the activity coefficient of OH⁻ as 0.62 (= the predicted value in this simulation) were 1.02 and 1.06 respectively. The ratio of the removed Cl⁻ and the released OH⁻ for 1 dm³ solution determined by Equation 1 is 1.0, which is very close to both the experimental and predicted mole ratio. The occurrence of the pH elevation to 13.5 by the reaction between CH, C_3AH_6 and NaCl solution to form AFm-Cl_2 and OH⁻ has therefore been confirmed.

Batch immersion of hardened OPC in NaCl solution

The pH of the 0.6 mol dm^{-3} NaCl solution stabilized at a value of 12.8 after 70 days, whereas the pH of the blank test shows a slightly lower value around 12.6 throughout the immersion period (Figure 3).

Figure 4 shows the X-Ray diffraction pattern of initial hardened OPC paste (Initial-OPC-Solid), the hardened OPC paste immersed for 360 days in the deionized water (OPC-Solid-DW) and that immersed for 360 days in the 0.6 mol dm^{-3} NaCl solution (OPC-Solid-SW).

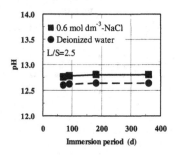

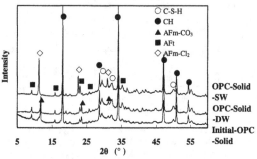

Figure 3. The pH evolution of 0.6 mol dm^{-3} NaCl solution and deionized water in which hardened OPC pastes were immersed.

Figure 4. X ray diffraction pattern of initial hardened OPC paste (Initial-OPC-Solid), the hardened OPC pastes immersed for 360 days in the deionized water (OPC-Solid-DW) and that immersed for 360 days in the 0.6 mol dm^{-3} NaCl solution (OPC-Solid-SW).

CH, C-S-H gel and AFt were identified, but not C_3AH_6. The X-Ray diffraction pattern of the OPC-Solid-SW shows that AFm-Cl_2 was the only AFm phase formed. In both Initial-OPC-Solid and OPC-Solid-DW, only AFm-CO_3 was formed. The AFm phases in the previous three solid phases were also quantified using DSC. The content of AFm-Cl_2 in the OPC-Solid-SW was 8 wt %, while the content of the AFm-CO_3 in the Initial-OPC-Solid and that in the OPC-Solid-DW were 8 wt % and 6 wt %, respectively.

The liquid phase compositions for OPC-Liquid-SW and OPC-Liquid-DW are shown in Table II. The Cl concentration in the OPC-Liquid-SW decreased to 62 % of the initial value. The decrease was consistent with the X-Ray diffraction pattern showing the formation of AFm-Cl_2. The concentration of Na in OPC-Liquid-SW also decreased to 75 % of the initial value, which is possibly due to Na sorption onto OPC hydrates such as C-S-H gel.

PHREEQC [8] was used with the JNC-TDB.TRU [9] database for geochemical calculations. Hydrated OPC was modeled using the reference composition of OPC [5] on the basis of existing studies [10, 11]. A normative hydrated phase assemblage was determined from the reference OPC composition (Table III) by the following assumptions:

Table II. The pH and concentration of major elements (mol dm^{-3}) in the 0.6 mol dm^{-3} NaCl solution (SW) and deionized water (DW) in which hardened OPC pastes were immersed for 360days.

	pH	Ca[aq]	Si[aq]	Al[aq]	Na[aq]	Cl[aq]
OPC-Liquid-SW	12.8	9.00E-03	7.00E-06	<1.00E-06	4.49E-01	3.73E-01
OPC-Liquid-DW	12.6	2.34E-02	6.00E-06	<1.00E-06	2.30E-04	<1.00E-06

Table III. Reference oxide composition of OPC. [5] [wt %]

SiO$_2$	Al$_2$O$_3$	Fe$_2$O$_3$	CaO	MgO	SO$_3$	Na$_2$O	K$_2$O
21.3	5.31	2.57	64.8	1.95	1.94	0.24	0.56

- OPC is fully hydrated.
- All sulfate is located in AFt: Ca and Al forming AFt are adjusted.
- All Fe forms Fe-Hydrogarnet (Ca$_3$Fe$_2$O$_6$·6H$_2$O; C$_3$FH$_6$): Ca is adjusted.
- Remaining Al is used to form C$_3$AH$_6$: Ca forming C$_3$AH$_6$ is adjusted.
- Remaining Ca and Si forms C-S-H gel. If the Ca/Si ratio is greater than 1.8, CH is formed.
- All of Mg is assigned to brucite (Mg(OH)$_2$; MH).

The calculated normative mineralogical composition of the initial hydrated OPC defined as case A is shown in Table IV. As previously described in the experimental procedure, NaOH and KOH were removed from the hydrated OPC prior to the experiment and were therefore neglected.

Table IV. Calculated normative mineralogical composition of initial OPC hydrated phases for case A, B and C. Units in mol kg^{-1} hydrated OPC paste without NaOH and KOH.

	CH	C-S-H gel (C$_{1.8}$SH$_{1.9}$)	C$_3$AH$_6$	C$_3$FH$_6$	MH	AFt	AFm-CO$_3$
CASE A	2.407	5.309	0.367	0.134	0.403	0.067	-
CASE B	2.209	5.216	0.204	0.132	0.396	0.066	0.156
CASE C	2.796	5.360	0.039	0.135	0.407	0.068	0.157

Cement chemistry shorthand nomenclature: C: CaO, S: SiO$_2$, H: H$_2$O, M: MgO.

The model of Atkinson et al. [12] was used to describe the incongruent dissolution of C-S-H gel. In order to maintain thermodynamic consistency, the log K values of C-S-H gel phases were recalculated using log K values of end members in the JNC-TDB.TRU database. In the recalculation, the log K values of C-S-H gel phases were given in 0.1 increments of Ca/Si ratio from 0.1 to 1.8. The applicability of the recalculated log K values for C-S-H gel phases was confirmed by comparing predicted values with the experimental data [10] as shown in Figure 5. The calculated results are shown with experimental data in Table V.

The calculated pH did not show a good agreement with the measured pH nor did the concentrations of the major elements in the case of 0.6 mol dm^{-3} NaCl solution. The difference of pH and element concentrations between the experimental and calculated results can be attributed to Al not being wholly present as C$_3$AH$_6$. Although the modeled initial solid phases were in good agreement with experimental data in the case of deionized water, the results suggest that it is not sufficient to account for saline groundwater.

Another initial solid setting for geochemical calculation was, therefore, attempted. AFm-CO$_3$ was observed and quantified as 8 wt % of the Initial-OPC-Solid. In order to improve the initial solid model, the AFm-CO$_3$ was taken into account, thereby reducing the amount of Al present

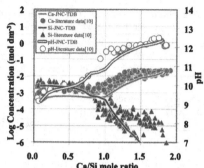

Figure 5. Experimental pH and calcium and silicon concentrations (symbols) in solution compared with calculated values (solid lines) using the C-S-H gel model described by Atkinson et al [12].

Table V. The pH and aqueous concentration of major elements(mol dm^{-3}) calculated using the initial hydrated OPC phases in which Al is assigned to AFt and C_3AH_6, compared to measured values in the OPC immersion tests.

Case		pH	Ca[aq]	Si[aq]	Al[aq]	Na[aq]	Cl[aq]
Deionized water case	Calculated Values [AFt + C$_3$AH$_6$]	12.5	2.00E-02	4.05E-07	1.39E-06	---	---
	(Measured Values)	(12.6)	(2.34E-02)	(6.00E-06)	(<1.00E-06)	(2.30E-04)	(<1.00E-06)
0.6mol dm^{-3} NaCl solution case	Calculated Values [AFt + C$_3$AH$_6$]	13.1	2.38E-03	8.91E-07	1.17E-06	6.00E-01	2.55E-01
	(Measured Values)	(12.8)	(9.00E-03)	(7.00E-06)	(<1.00E-06)	(4.49E-01)	(3.73E-01)

as C_3AH_6. The amended composition of Initial-OPC-Solid defined as case B is shown in Table IV. Despite lacking consistency with the JNC-TDB.TRU database, the log K value of the AFm-CO$_3$ from Zhang et al. [13] was used. The calculated solution composition using the amended Initial-OPC-Solid did not show much improvement with the measured values (Table VI).

Table VI. The pH and aqueous concentration of major elements(mol dm^{-3}) calculated using the initial hydrated OPC phases in which Al is assigned to AFt, AFm-CO$_3$ and C_3AH_6, compared to measured values in the OPC immersion tests in 0.6 mol dm^{-3} NaCl. Units of aqueous species are in mol dm^{-3}.

Case	pH	Ca[aq]	Si[aq]	Al[aq]	Na[aq]	Cl[aq]
Calculated Values [AFt + AFm-CO$_3$ + C$_3$AH$_6$]	13.1	2.49E-03	8.60E-07	1.09E-06	6.00E-01	2.62E-01
(Measured Values)	(12.8)	(9.00E-03)	(7.00E-06)	(<1.00E-06)	(4.49E-01)	(3.73E-01)

Though C-S-H gel was modeled as a binary solid solution between CH, C-S-H gel and amorphous silica (SiO$_{2(am)}$) it is known that C-S-H gel can incorporate Al into its structure [14]. The Al content of C-S-H gel in the Initial-OPC-Solid was measured by EPMA analysis. Mole ratios of Si/Ca and Al/Ca are plotted with data from Richardson and Groves [14] (Figure 6), which shows good agreement between the two datasets. The variation of Al content in unit weight of C-S-H gel prior to immersion in the Initial-OPC-Solid was determined for changing L/S (Figure 7), by geochemical calculation using the relationship between Si/Ca and Al/Ca given by Richardson and Groves [14].

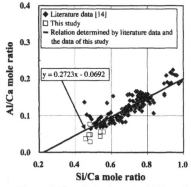

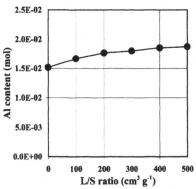

Figure 6. The relation between Al/Ca and Si/Ca mole ratios of the C-S-H gel in the Initial-OPC-Solid with the data reported by [14].

Figure 7. Predicted variation of Al content per unit weight of C-S-H gel prior to immersion in the Initial-OPC-Solid during incongruent dissolution.

The calculation suggested Al was not released from C-S-H accompanied with dissolution because the Al content increased with an increase in C-S-H gel dissolution, i.e. as Ca/Si ratio decreased. Using the recalculated composition of the Initial-OPC-Solid, considering Al incorporation into C-S-H, defined as case C shown in Table IV, the composition of the equilibrated solution was calculated as shown in Table VII.

Table VII. The pH and aqueous concentration of major elements(mol dm^{-3}) calculated using the initial hydrated OPC phases in which Al is assigned to AFt, AFm-CO_3, C-S-H gel, and C_3AH_6, compared to measured values in the OPC immersion tests in 0.6 mol dm^{-3} NaCl.

Case	pH	Ca[aq]	Si[aq]	Al[aq]	Na[aq]	Cl[aq]
Calculated Values (AFt + AFm-CO_3 + C_3AH_6 +CSH gel)	12.8	6.17E-03	3.50E-07	2.09E-07	6.00E-01	4.12E-01
(Measured Values)	(12.8)	(9.00E-03)	(7.00E-06)	(<1.00E-06)	(4.49E-01)	(3.73E-01)

Finally, the calculated solution composition shows a better agreement with the experimental results, except for Si concentration. At high Ca/Si ratios, measured Si concentrations are very low, ranging from 10^{-4} to 10^{-6} mol dm^{-3}, and are typically underestimated with currently available C-S-H models. This discrepancy is therefore considered beyond the scope of the current paper.

CONCLUSIONS

The reaction of CH and C_3AH_6 in saline water results in a pH increase to 13.5 by the precipitation of AFm-Cl_2 and the release of OH^-. This was confirmed experimentally and supported by geochemical calculations.

Further experiments were conducted using hardened OPC paste to confirm this reaction mechanism in the presence of other hydrated OPC phases. In this case, the pH elevation was not as high as expected. This deviation can be explained by the residual aluminum, after being consumed as AFt, AFm and AFm-CO$_3$, not being wholly assigned to C$_3$AH$_6$. A better agreement between the experimentally measured results and thermodynamic calculations can be made assuming a fraction of aluminum is incorporated into the C-S-H gel phase.

The experimental and calculated results demonstrate the importance of Al assignment in the initial OPC hydrated phases in saline groundwaters.

ACKNOWLEDGMENTS

The authors would like to thank Dr. Colin Walker for valuable discussions.

REFERENCES

1. Japan Atomic Energy Agency and The Federation of Electric Power Companies of Japan, *Second Progress Report on Research and Development for TRU Waste Disposal in Japan*, JAEA-Review 2007-010, FEPC TRU-TR-2007-01 (2007).
2. I.Mackinley, R.Alexander, W.Kickmaire and F.Neall, *Proceeding of the International Workshop on Bentonite-Cement Interaction in Repository Environments*, NUMO-TR-04-05, p.A3-17 (2004).
3. M.H. Bradbury and F.A. Sarott, *Sorption Databases for the Cementitious Near-Field of a L/ILW Repository for Performance Assessment*, NAGRA TECHNICAL REPORT 93-08, p.21_(1994).
4. M. Yui, H. Sasamoto and R. Arthur, *Geochem. Journal* **38**, 33 (2004).
5. Japan Nuclear Cycle development Institute and The Federation of Electric Power Companies of Japan, *Second Progress Report on Research and Development for TRU Waste Disposal in Japan - Supporting Report 2-*, JNC TY1450 2005-001(2), FEPC TRU-TR-2005-04 (2005).
6. F.P. Glasser, M. Tyrer, K. Quillin, D. Ross, J. Pedersen, K. Goldthorpe, D. Bennett, M. Atkins, *The Chemistry of Blended Cements and Backfills Intended for Use in Radioactive Waste Disposal*, R&D Technical Report P98, U.K. Environment Agency, p.61 (1998).
7. E. Breval, *Cement, Concrete Res.***6**, 129 (1976).
8. D.L. Parkhurst and C.A.J. Appelo, *User' s guide to PHREEQC (version 2)-A computer program for speciation batch-reaction, one-dimensional transport, and inverse geochemical calculations*, U. S. Geological Survey Water-Resources Investigations Report 99-4259 (1999).
9. R. Arthur, H. Sasamoto, C. Oda, A. Honda, M. Shibata, Y. Yoshida and M.Yui, *Development of Thermodynamic Database for Hyperalkaline, Argillaceous System*, JNC TN8400 2005-010 (2005).
10. U.R. Berner, *Waste Mgt.***12**, 201 (1992).
11. F.B. Neall, *Modelling of the near-field chemistry of the SMA repository at the Wellenberg Site*, PSI Bericht Nr. 94-18, p.14 (Paul Scherrer Institut, Switzerland, 1994).
12. A. Atkinson, J.A. Hearne and C.F. Knights, *Aqueous chemistry and thermodynamic modeling of CaO-SiO$_2$-H$_2$O gels*, AERE R-12548 (United Kingdom Atomic Energy Authority, 1987).
13. F. Zhan, Z. Zhou and Z. Lou, Solubility Product and Stability of Ettringite, *Proc. 7th Int. Symp. Chem. Cem.*, Vol II, Paris, pp.88-93 (1980).
14. I.G. Richardson, G.W. Groves, *J. Mat. Sci.* **27**, 6204 (1992).

Mater. Res. Soc. Symp. Proc. Vol. 1124 © 2009 Materials Research Society   1124-Q10-13

Cement Based Encapsulation Experiments of Low-Radioactive Phosphate Effluent

Atsushi Sugaya, Kenichi Horiguchi, Kenji Tanaka and Shigeru Akutsu
Tokai Reprocessing Research and Development Centre, Japan Atomic Energy Agency (JAEA),
4-33 Muramatsu, Tokai-Mura, Naka-Gun, Ibaraki, 319-1194, Japan

ABSTRACT

The Tokai Reprocessing Plant produces a phosphate effluent consisting of NaH_2PO_4 at an approximate concentration of 440 $g\ell^{-1}$, with only very minor quantities of other species. JAEA intends to encapsulate this waste in cement. Results are presented from non-active trials that demonstrate that it is possible to encapsulate this waste using ground granulated blastfurnace slag or Super-Cement. It was found that to achieve a stable wasteform it was necessary to pre-treat the phosphate effluent with lime before encapsulation. The optimum pre-treatment and cement formulations were investigated. Compressive strengths of grouted waste simulant exceeded 10 MPa after 28 days curing and a waste loading exceeding 13 wt% was achieved.

INTRODUCTION

The operation of a nuclear fuel reprocessing plant generates a number of different types of effluent treatment wastes. It is necessary for all these wastes to be disposed of in a safe, sustainable and economic manner. In Japan these wastes must be disposed of on land, and not at sea, using either deep or shallow burial depending on the radioactive inventory of each waste type. Following the strategy adopted widely by a number of countries, effluent wastes from JAEA nuclear fuel reprocessing operations will be immobilised in solid matrices that convert the waste into a safer and more convenient form for transportation, storage, and ultimate disposal. Until 1997 low level radioactive liquid wastes such as solvent washing effluent, radionuclide analysis waste, off-gas scrubber wastes, and decontamination wastes were concentrated by evaporation and then bituminisation. However, this process was discontinued in 1997 after a fire in the bituminisation facility. A cementation facility is now being designed with construction starting during 2012 and operation from 2014.

Here we focus on research into the encapsulation of effluent waste containing phosphate salts that is generated during the solvent extraction processes at the Tokai Reprocessing Plant. This phosphate waste consists principally of an aqueous solution of monosodium phosphate of typical concentration 440 $g\ell^{-1}$, with additional minor species.

The phosphate effluent has a pH of ~ 4. Therefore, it is necessary to 'pre-treat' the waste before mixing to increase the pH to make it compatible with the cement powder. An option within the process is also available to evaporate the waste to reduce the water content of the phosphate effluent. For this option to be used the phosphate waste must first be treated to increase the pH to ~ 7 to prevent corrosion and damage of the evaporator.

The work presented here gives the results of an investigation into the pre-treatment of the phosphate waste and its subsequent encapsulation in cement using Ground Granulated Blastfurnace Slag (GGBS) or 'Super Cement' (SC). Super Cement is the commercial name of a proprietary formulation blended cement manufactured by JGC Corporation (Japan) and is an alkali activated slag cement with various minor additives to enhance process and product performance.

PRE-TREATMENT TRIALS

The pre-treatment process investigated involved two steps: first, pre-treatment with sodium hydroxide to raise the pH of the waste to permit the use of the evaporation option; and secondly, pre-treatment with lime (calcium hydroxide). Note that if the evaporation option is not used then the sodium hydroxide step can be omitted.

The reaction with sodium hydroxide is a simple acid-base reaction:

$$NaH_2PO_4 + NaOH \rightarrow Na_2HPO_4 + H_2O.$$

A 1:1 neutralisation with NaOH produces hydrated Na_2HPO_4 crystals. Generally these have a low melting point. The principal product is $Na_2HPO_4.12H_2O$ with a melting point of 34.6°C, but $Na_2HPO_4.7H_2O$ with melting point 48.1°C may also form.[1] At higher temperatures the hydrogen phosphate species are soluble, therefore pre-treatment with sodium hydroxide is suitable for raising the waste pH for evaporation. However, if the waste cools below the melting points of hydrated disodium phosphate then, at the high concentration of hydrogen phosphate in the waste, solidification of the waste occurs owing the hydrated crystals being capable of incorporating all the free water originally present in solution. This has the potential to cause problems during encapsulation because water is needed for the cement hydration reaction. Moreover, the low melting point for the crystals, which releases water, is anticipated to affect the stability of the cement encapsulated waste because of dimensional instability and the potential mobilization of activity whenever the melting/solidification point is crossed.

To remove the difficulty of hydrated disodium phosphate crystals forming in the phosphate waste, treatment with lime to produce a precipitate of calcium phosphate has been investigated. It is possible that several types of calcium phosphate may be precipitated (see Table I). The most thermodynamically stable precipitation product is hydroxyapatite, but the precise composition of the precipitate depends upon the reaction conditions used and generally amorphous material is precipitated first that may subsequently convert to a crystalline phase.[2]

Table I: Types of calcium phosphate

Calcium phosphate	Composition	Ca/P ratio
Dicalcium phosphate	$CaHPO_4.2H_2O$	1
Octacalcium phosphate	$Ca_4H(PO4)_3.3H_2O$	1.33
Tricalcium phosphate	$Ca_3(PO_4)_2$	1.5
Hydroxyapatite	$Ca_{10}(PO_4)_6(OH)_2$	1.67
Amorphous calcium phosphate	-	~1.5

Common features of all calcium phosphates are their low solubilities and high melting points. This means that lime pre-treatment cannot be used to raise the pH of the effluent before evaporation because calcium phosphate would precipitate and be deposited inside the evaporator. However, calcium phosphates are stable for encapsulation in cement.

Experimental

Monosodium phosphate dihydrate ($NaH_2PO_4.2H_2O$) and sodium hydroxide (NaOH) were obtained from Fisher Scientific UK, and lime ($Ca(OH)_2$) was sourced from Tarmac Buxton Lime (UK) in the form of standard grade Limbux lime (>97% $Ca(OH)_2$).

Experiments were performed by preparing a solution of sodium hydroxide of the appropriate concentration for the trial. The temperature of this sodium hydroxide solution was raised to ~50°C and monosodium phosphate dihydrate was added with continuous stirring. To simulate the temperature of the waste following evaporation, the solution temperature was increased to 60°C, and then the required quantity of lime was added under continuous stirring. Stirring was maintained until the temperature of the slurry fell to ~35°C. After cooling, any water lost by evaporation (determined by the change in mass) was replaced.

For trials using only lime pre-treatment, 79.4 g monosodium phosphate dihydrate per 100 g water was used for the waste simulant. In trials with both sodium hydroxide and lime pre-treatment 101.3 g monosodium phosphate dihydrate per 100 g water was used to simulate a higher phosphate concentration for evaporated waste.

Two scales of mix were performed: firstly ~0.4 ℓ trials performed by hand mixing, and secondly ~2 ℓ scale trials using a 5 ℓ Hobart planetary mixer. A comparison between the two scales of working revealed no significant difference in the results and therefore differentiation between the two sets of results has not been highlighted.

Results and discussion

The pre-treatment formulation is expressed as the molar ratio of sodium hydroxide to monosodium phosphate (in the untreated waste) and the molar ratio of lime to monosodium phosphate (in the untreated waste). The range of pre-treatment formulations investigated is shown in Figure 1(a).

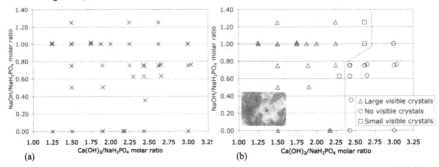

Figure 1: (a) Pre-treatment formulations performed given as molar ratios of NaOH and Ca(OH)$_2$ to the NaH$_2$PO$_4$ component of the waste simulant. (b) As for (a) but with the formation of needle crystals shown as a function of pre-treatment formulation and the dashed line indicates approximately the boundary for crystal formation.

A key observation from the trials was that if insufficient lime was added during pre-treatment then small needle shaped crystals formed that were visible under an optical microscope (typical length 0.25-0.5 mm). In these cases the viscosity of the simulant was initially very high, although it was found to decrease significantly during the following weeks of storage; this behaviour was undesirable owing to the initial high viscosity that may prevent mixing of the treated waste. However, if visible crystals were not formed then the viscosity remained unchanged over time. The precise identity of the crystals formed was not determined in these trials, although the low melting point of these needle crystals ~ 30°C and observed removal of free water when they form is consistent with Na$_2$HPO$_4$.12H$_2$O crystallisation.

The minimum lime addition needed to prevent the formation of crystals was a lime to monosodium phosphate molar ratio of ~2.4, with a sodium hydroxide to monosodium phosphate molar ratio of <0.8 (Figure 1(b)). To raise the pH of the waste to ~7, necessary for use of the evaporator, a sodium hydroxide to monosodium phosphate ratio of 0.75 was sufficient. The level of lime addition also affected the rheology of the phosphate waste following pre-treatment, and a minimum viscosity of the pre-treated waste coincides with a $Ca(OH)_2/NaH_2PO_4$ molar ratio of ~2.4 that was just sufficient to prevent the formation of crystals in the pre-treated waste. In contrast, below a $NaOH/NaH_2PO_4$ molar ratio of 0.8 the level of sodium hydroxide used did not have a significant effect on the viscosity of the mixture. It was also found that if the sample was cooled too rapidly then incomplete reaction occurred with the lime, but maintaining the samples at a high temperature for several days did not appear to change the appearance of crystals for $Ca(OH)_2/NaH_2PO_4$ molar ratios of <2.4. Future work will investigate further the importance of temperature control on the pre-treatment process.

ENCAPSULATION TRIALS

An investigation of the encapsulation of the pre-treated waste was conducted using the identified optimum pre-treatment formulation and either GGBS and Super Cement.

Experimental

Pre-treatment of the waste simulant for trials was performed at ~2 ℓ scale using a 5 ℓ Hobart planetary mixer. For encapsulation trials using GGBS the pre-treatment was a $NaOH/NaH_2PO_4$ molar ratio of 0.75 and a $Ca(OH)_2/NaH_2PO_4$ molar ratio of 2.43. For the trials using Super Cement the pre-treatment step using sodium hydroxide was omitted because it caused the setting time for the cement to be very short; this is discussed later.

GGBS was obtained from Civil Marine (SiO_2 36.41%, Al_2O_3 12.82%, Fe_2O_3 0.61%, CaO 41.25%, MgO 7.56%, MnO 0.23%, Mn_2O_3 0.26%, TiO_2 0.47%, Na_2O 0.17%, K_2O 0.31%, and low fineness of 293 m^2kg^{-1}) and Super-Cement was obtained from the JGC Corporation (SiO_2 36.9%, Al_2O_3 10.6%, Fe_2O_3 0.4%, CaO 42.6%, MgO 4.3%, Na_2O 0.3%, K_2O 0.2%, and high fineness of over 800 m^2kg^{-1}). Mixing of the cement was performed using a 5 ℓ Hobart planetary mixer. Slow mixing (speed 1) was used for 4 minutes mixing, which included addition of the cement powder during the first ~2 minutes. The mixing speed was then increased to speed 2 for a further 10 minute mixing time. Samples were cured at a humidity >90% and a temperature of 20±1°C for up to 28 days.

In order to quantify the availability of water in the pre-treated waste a nominal water to cement ratio (w/c) was defined by assuming that during pre-treatment each $H_2PO_4^-$ ion was neutralised to produce two H_2O molecules. The waste loading (L) was defined as the mass of NaH_2PO_4 from the untreated waste simulant that was immobilized in the cement, given as a weight percentage of the immobilised waste.

Results and discussion

The range of cement formulations investigated using GGBS is shown in Figure 2(a), and the increase in compressive strength of the cemented effluent simulant over time is shown in Figure 2(b). In all cases setting of the grout occurred within 24 hours, without bleed water being present at this time. It was found that the compressive strength of the cemented waste is largely determined by the w/c ratio, whilst the waste loading has a much smaller, secondary effect over the ranges investigated. Unfortunately, the waste loading has a strong effect on the viscosity of the grout, which increases significantly as the waste loading is increased for a fixed w/c ratio, and this is the limiting factor for the waste loading. Using the GGBS cement it was possible to exceed a compressive strength of 10 MPa after 28 days with waste loadings in the range 12-14.4 wt%, w/c 0.58-0.75.

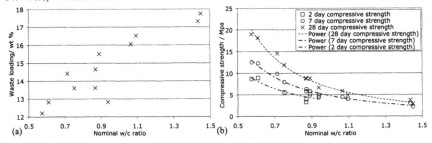

Figure 2: (a) Waste loading and nominal w/c ratio for the cementation trials performed using GGBS. (b) Compressive strength after 2, 7 and 28 days curing as a function of the nominal w/c ratio for the cementation trials using GGBS. The dashed lines are power law fits of the data points to guide the eye.

For trials investigating encapsulation of waste using Super Cement the range of formulations tested is shown in Figure 3(a), with the corresponding compressive strength of the samples shown in Figure 3(b). All trials set within 24 hours with no bleed produced. Similar to the GGBS trials, the compressive strength of the grouted waste mainly depends on the w/c ratio, rather than the waste loading. A compressive strength of 10 MPa after 28 days was achieved up to w/c=1.65 with waste loadings exceeding 13 wt%.

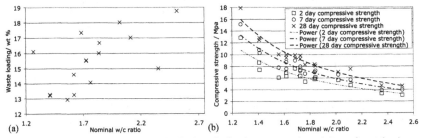

Figure 3: (a) Waste loading and nominal w/c ratio for the cementation trials performed using Super Cement. (b) Compressive strength after 2, 7 and 28 days curing as a function of the nominal w/c ratio for the cementation trials using Super Cement. The dashed lines are power law fits of the data points to guide the eye.

The JAEA acceptance criterion for compressive strength is 10 MPa after 28 days curing, which gives a significant margin over the waste transportation requirements of 8 MPa. Both GGBS and Super Cement encapsulated wastes can achieve this for loadings up to 13 wt% and future work will now investigate the longer-term stability of the encapsulated waste.

As expected, the viscosity of the grout increases with increasing waste loading, but decreases with increasing w/c ratio. Acceptable mixing to give homogeneous grouts was achieved in all cases reported here, although for three mixes with high waste loadings (L=16.1%, w/c=1.27; L=17.4, w/c=1.69; and L=18.0%, w/c=2.02) the grout was very viscous and could not be poured. In comparison with the GGBS mixes it was notable that the minimum w/c ratio that could be mixed was significantly higher for Super Cement, probably due to its higher fineness.

As noted above, Super Cement could not be used to encapsulate phosphate waste that had been pre-treated with sodium hydroxide because the setting time for the grout was very short compared with the GGBS trials. This is because the fineness of Super Cement is higher than the GGBS powder, which causes it to be more reactive. Removing the pre-treatment with sodium hydroxide increases the setting time of the grout by reducing the alkalinity of the mix, and reduces the activation of the slag. In both the GGBS and Super Cement trials the highest waste loading that could be achieved with acceptable mixing properties was 18.8 wt% at w/c=2.5. Producing a grout with this waste loading does not require evaporation of the original waste effluent (440 $g\ell^{-1}$ monosodium phosphate). Consequently, at present the pre-treatment with sodium hydroxide and evaporation of the phosphate waste is not needed for encapsulation of this waste in cement, but this was not known during the investigation of the pre-treatment process.

CONCLUSIONS

Pre-treatment of phosphate effluent using lime to precipitate calcium phosphate can be used to stabilize phosphate waste for encapsulation in cement. Encapsulation was successfully performed with both GGBS and Super Cement, with the encapsulated waste exceeding the JAEA requirement of 10 MPa after 28 days curing at waste loadings of up to 14.4 wt% for GGBS and 16.1 wt% for Super-Cement. Future work will investigate further the importance of temperature control during pre-treatment, longer term stability of the encapsulated waste, and will perform pre-treatment and encapsulation at full 200 ℓ scale to assess scaling-up the processes.

ACKNOWLEDGMENTS

The experimental work was carried out during a secondment of A. Suguya at the UK National Nuclear Laboratory. R. M. Orr and I. H. Godfrey of the UK National Nuclear Laboratory are acknowledged for their support in the preparation of this contribution.

REFERENCES

1. Knovel Critical Tables (2nd Edition), Knovel, (2003).
2. E. D. Eanes: Amorphous calcium phosphate: Thermodynamic and kinetic considerations, in *Calcium Phosphates in Biological and Industrial Systems*, Ed. Z. Amjad, (1997).
3. Y.SAITOH *et al.*: *Development of new treatment process for low level liquid waste at Tokai Reprocessing Plant, Proceedings of international symposium on radiation safety management, Daejeon, Korea, P8*, (2007).
4. A.SUGYA *et al.*: *2008 Autumn Meeting of Atomic Energy Society of Japan, Kochi, Japan, O24*, (2008).

Mater. Res. Soc. Symp. Proc. Vol. 1124 © 2009 Materials Research Society   1124-Q10-15

Diffusion Behavior of Organic Carbon and Iodine in Low-Heat Portland Cement Containing Fly Ash

Taiji Chida and Daisuke Sugiyama
Central Research Institute of Electric Power Industry, 2-11-1, Iwado-Kita, Komae, Tokyo 201-8511, Japan

ABSTRACT

The diffusion of radionuclides in cementitious materials used as an engineered barrier is an important parameter in the performance assessment of the sub-surface repository system used for low-level radioactive waste disposal in Japan. In particular, organic carbon-14 and iodine-129 would provide large contributions to the dose evaluation, because of their low ability to be adsorbed on cementitious materials. In this study, the diffusion of acetate and iodide in hardened cement pastes was examined by through-diffusion experiments. Low-heat Portland cement containing 30 wt% fly ash (FAC), which is a candidate cement material for the construction of the sub-surface repository, was prepared for the diffusion experiments. The effective diffusion coefficients, D_e, of the trace ions for hardened FAC cement pastes were estimated to be on the order of 10^{-13} m^2 s^{-1} at the beginning of the diffusion experiments. Then, the rate of diffusion of the trace ions decreased over the experimental period of 1-15 months. This is probably due to the change in the microstructure of the FAC as the result of a pozzolanic reaction. After a few months, the values of D_e were estimated to be on the order of 10^{-14} m^2 s^{-1}. These results suggest that an engineered barrier made of FAC can act as an effective barrier inhibiting the diffusion of trace ions such as organic carbon and iodine.

INTRODUCTION

In the sub-surface repository system for the disposal of low-level radioactive waste in Japan, cementitious materials can be used as an engineered barrier to limit the migration of radionuclides [1]. In particular, organic carbon-14 [2] and iodine-129 would provide large contributions to the dose evaluation, because they are sorbed less strongly on cementitious materials [3]. Low-heat Portland cement containing 30 wt% fly ash (FAC) is a candidate cement material for the construction of the sub-surface repository because of its closely packed structure and its resistance to cracking. Therefore, the diffusion of radionuclides in cementitious materials is a very important parameter in performance assessment when considering the release of these radionuclides from waste and their migration in a cementitious repository environment. To clarify the barrier properties of FAC materials, in this study we examined the diffusion coefficients for organic carbon and iodine in FAC.

EXPERIMENT

Materials

Solid disk-shaped samples of FAC were prepared for the diffusion experiments. As a comparison, ordinary Portland cement (OPC), which is the most commonly used cementitious material, was also used in the experiments. The cements were mixed with deionized water in water/cement clinkers (w/c) with different w/c mixing ratios of 0.35 and 0.70, to produce different diffusivities in the solid. To prevent bleeding (the separation of water at the top of the paste) at the high w/c ratio of 0.70, the cement pastes were repeatedly mixed before setting. The cement pastes were allowed to set for 24 hours at 30°C, and then were cured in water for 91 days at 50°C. The chemical compositions and porosities of the hardened FAC and OPC pastes used in this study are shown in Table I. The porosities of the hardened cement pastes were measured by mercury intrusion porosimetry. The hardened FAC and OPC were cut into samples of 50 mm diameter and 5 mm thickness using a diamond cutter and fixed to sample holders with epoxy resin. Cement-equilibrated solutions were prepared by placing particles of crushed hardened FAC or OPC paste (< 250 μm) in contact with deionized water at a solid/liquid ratio of 1/100 (g cm^{-3}). After 28 days of equilibration, each solution was filtered through a membrane filter with 0.45 μm pore size.

Through-diffusion experiment

The diffusion of acetate, which is a typical chemical form of organic carbon, and iodide in hardened FAC and OPC paste was examined by through-diffusion experiments. Figure 1 shows an illustration of the experimental setup. The hardened cement paste sample saturated by the cement-equilibrated solution forms a partition between two liquid-filled compartments (made of acrylic). One compartment contains the tracer species at a known concentration, and the diffusion of the trace ions through the cement sample is monitored by measuring the increase in concentration in the other compartment as a function of time. For the high-concentration side, the initial tracer concentration was set to about 0.1 mol dm^{-3} for iodide (in the form of potassium iodide) or acetate (in the form of sodium acetate). The concentration of the iodide was measured by inductively coupled plasma atomic emission spectrometry (ICP-AES). The concentration of acetate was measured as the organic carbon concentration by a total organic carbon analyzer. During the experiments, the concentration of acetate or iodide in the high-concentration side was fixed at ± 5%, and the pH was also fixed at 12.2 ± 0.2. Both the preparation of the solutions and the through-diffusion experiments were carried out in an argon-filled glove box.

Table I Chemical compositions and porosities of OPC and FAC samples.

Cement	w/c ratio	Chemical component [%]								Porosity [%]
		SiO_2	Al_2O_3	Fe_2O_3	CaO	MgO	SO_3	Na_2O	K_2O	
OPC35	0.35	17.02	4.53	2.49	51.22	0.90	1.57	0.18	0.24	16
OPC70	0.70	16.85	4.49	2.47	50.72	0.89	1.55	0.18	0.24	48
FAC35	0.35	20.84	3.66	2.24	38.21	0.68	1.41	0.17	0.28	32
FAC70	0.70	20.63	3.47	2.10	34.42	0.62	1.32	0.16	0.24	52

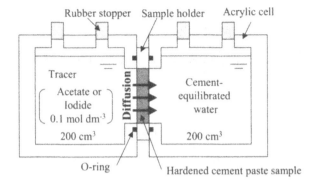

Figure 1 Experimental apparatus used for the through-diffusion experiment.

Determination of effective diffusion coefficient

The change in concentration C (g m^{-3}) at a point x (m) in a one-dimensional system is given by Fick's second law,

$$\alpha \frac{\partial C}{\partial t} = D_e \frac{\partial^2 C}{\partial x_2} , \tag{1}$$

$$\alpha = \varepsilon_{tot} + \rho R_d , \tag{2}$$

where D_e is the effective diffusion coefficient (m^2 s^{-1}), α is the capacity factor, t is the time (s), ε_{tot} is the total pore volume, ρ is the density of the matrix (g m^{-3}) and R_d (m^3 g^{-1}) is the distribution ratio defined as the ratio of concentrations in the adsorbed and liquid phases [4].

For the case of diffusion through a porous plate of thickness L (m), the inlet concentration C_1 at $x = 0$ is constant, and the outlet concentration C_2 at $x = L$ is assumed to be $C_2 = 0$ because C_2 is sufficiently smaller than C_1 in this system. We also assumed the following conditions:

Initial condition: $C(x,0) = 0, \quad 0 \leq x \leq L$ (3)

Boundary condition: $C(0,t) = C_1, \quad C(L,t) = C_2 = 0 \ (\ll C_1).$ (4)

Then, the concentration C can be written as

$$\frac{C(x,t)}{C_1} = 1 - \frac{x}{L} - \frac{2}{\pi} \sum_{n-1}^{\infty} \frac{1}{n} \sin \frac{n\pi x}{L} \exp\left(-\frac{D_e n^2 \pi^2 t}{L^2 \alpha}\right) . \tag{5}$$

Assuming that the cross-sectional area of the porous plate is A (m^2) and the volume of the outlet side is V (m^3), the total amount of the diffusing substance $Q \ (= C_2(t) \cdot V)$ is

$$\frac{Q}{LC_1A} = \frac{D_e t}{L^2} - \frac{\alpha}{6} - \frac{2\alpha}{\pi^2}\sum_{n=1}^{\infty}\frac{(-1)^n}{n^2} - \exp\left(-\frac{D_e n^2\pi^2 t}{L^2\alpha}\right). \tag{6}$$

At the steady state after a long time (t approaches infinity), Equation (6) can be written as

$$\frac{Q}{LC_1A} = \frac{D_e t}{L^2} - \frac{\alpha}{6} . \tag{7}$$

When Equation (7) is rewritten as a function of time, we obtain

$$C_2(t) = \frac{AC_1}{VL}D_e t - \frac{ALC_1}{6V}\alpha . \tag{8}$$

Therefore, the effective diffusion coefficient, D_e, is obtained from the rate of change of tracer concentration in the low-concentration side [5].

RESULTS AND DISCUSSION

Figures 2a and 2b show the results of the through-diffusion experiments for iodide. As shown in Figure 2, after an initial period of 400 hours, the rate of change of iodide concentration for FAC35 is much lower than that for OPC35 (Figure 2a), and the diffusion rate of iodide decreased over the experimental period (Figure 2b). Similar tendencies were observed in the case of acetate. The effective diffusion coefficients, D_e, estimated from the through-diffusion experiments are shown in Table II. The ranges of D_e in Table II show the variation of D_e estimated from several experimental results. For the FAC samples, the values of D_e are estimated separately for the early stage and the late stage. Figure 3 shows the dependence of D_e on the porosity (The "+" symbols in Figure 3 denote results from Kaneko et al. [6].). The values of D_e for the FAC samples in Figure 3 are those for the early stage given in Table II. For both FAC and

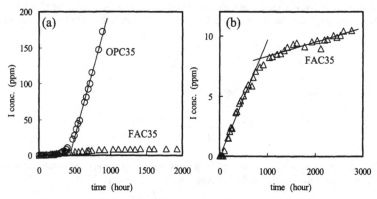

Figure 2 Diffusion of iodide through the hardened cement pastes.

Table II Effective diffusion coefficients.

Cement	Effective diffusion coefficient [$m^2\,s^{-1}$]	
	Acetate	iodide
OPC35	$3.4 - 3.6 \times 10^{-12}$	$3.5 - 4.3 \times 10^{-12}$
OPC70	$6.7 - 7.1 \times 10^{-11}$	$-$
FAC35	Early-stage $0.99 - 2.1 \times 10^{-13}$	Early-stage 1.2×10^{-13}
	Late-stage $4.7 - 9.4 \times 10^{-14}$	Late-stage $1.5 - 5.0 \times 10^{-14}$
FAC70	Early-stage $3.5 - 3.8 \times 10^{-13}$	Early-stage $4.1 - 6.3 \times 10^{-12}$
	$-$	Late-stage $2.3 - 3.0 \times 10^{-12}$

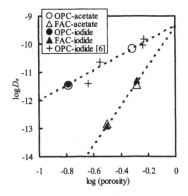

Figure 3 Porosity dependence of the effective diffusion coefficients.
(The values of D_e for the FAC are those for the early stage given in Table II)

OPC, there is a correlation between the porosity and D_e as shown by the dotted lines. Furthermore, D_e for iodide and acetate are nearly equal for each porosity and type of cement. When FAC is compared with OPC, the values of D_e in FAC are one to two orders of magnitude smaller than those in OPC for both acetate and iodide.

We now focus on the decrease in the diffusion rates for the FAC samples as shown in Figure 2 and Table II. This behavior can be ascribed to the change in the physical properties of the FAC sample in which trace ions diffuse, because the experimental conditions, i.e., the concentration of the high-concentration side and the boundary condition (Eq. (4)), did not change during the experiments. A possible reason for the decrease in the diffusion rate is that the pozzolanic reaction induced by fly ash may affect the pore structure in the solid-cement matrix. Generally, the reaction in which pozzolanic materials react with portlandite to form insoluble silicate compounds requires a comparatively long time [7, 8]. The microstructure of the FAC solid matrix is expected to have altered during the through-diffusion experiments because of the progress of the pozzolanic reaction. Figure 4 shows the pore size distributions and porosities of the FAC35 samples determined by mercury intrusion porosimetry. Figure 4a shows the result for the FAC35 sample before the through-diffusion experiment, and Figure 4b shows the result 6 months after the beginning of the experiment. With time, both the pore size and the porosity

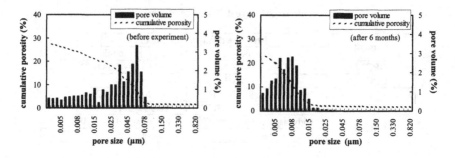

Figure 4 Pore size distributions for FAC35.

decrease. It can be considered that these changes in the FAC samples contribute to the decrease in the diffusion coefficients of the trace ions.

CONCLUSIONS

In this study, the diffusion behavior of organic carbon and iodine in FAC and OPC was examined by through-diffusion experiments. The effective diffusion coefficients of trace ions for FAC were one to two orders of magnitude smaller than those for OPC. Furthermore, the diffusion rates of trace ions for FAC decreased with time because the pore size and the porosity decreased, probably due to a pozzolanic reaction. These results suggest that an engineered barrier made of FAC can act as an effective barrier inhibiting the diffusion of trace ions such as organic carbon and iodine, and good barrier performance can be expected in the long term after the closure of the repository.

REFERENCES

1. K. Kato, T. Waki, N. Saito, F. Ono, T. Ohma and M. Ozaki, *J. Nucl. Fuel Cycle Environ.* **13**, No. 1, 49-64 (2006) (*in Japanese*).
2. T. Yamaguchi, S. Tanuma, I. Yasutomi, T. Nakayama, H. Tanabe, K. Katsurai, W. Kawamura, K. Maeda, H. Kitao and H. Saigusa, *A Study on Chemical Forms and Migration Behavior of Radionuclides in Hull Waste* Proceedings of ICEM'99, edited by ASME (1999) (CD-ROM).
3. J. Matsumoto and T. Banba, *Sorption Behavior of Organic Carbon-14 onto Cementitious Materials* Proceedings of ICEM'99, edited by ASME (1999) (CD-ROM).
4. K. Skagius and I. Neretnieks, *Water Resour. Res.* **22**(5), 389-398 (1986).
5. D. A. Lever, M. H. Bradbury and S. J. Hemingway, *J. Hydrol.* **80**, 45-76 (1985)
6. M. Kaneko, M. Toyohara, K. Nakata, N. Mitsutsuka and K. Haga, *Study on Diffusion Coefficients of some Elements for Altered Cement* Abstracts book of 2001 Fall Meeting of the Atomic Energy Society of Japan, **III**, 887(O9) (2001) (*in Japanese*).
7. P. K. Mehta, *Cement Concr. Res.* **11**, 507-518 (1981).
8. K. Wang, T. Noguchi and F. Tomosawa, *Cement Concr. Res.* **34**, 2269-2276 (2004).

Geological Disposal and
Waste Treatment

Mater. Res. Soc. Symp. Proc. Vol. 1124 © 2009 Materials Research Society   1124-Q07-04

Uranium-Series Disequilibrium Studies in Bedrock for the Safety Case of Deep Geological Disposal of Spent Fuel

Kari Rasilainen[1], Juhani Suksi[2], Petteri Pitkänen[1], Nuria Marcos[3], and Torbjörn Carlsson[1]
[1]VTT, Technical Research Centre of Finland, P.O. Box 1 000, FI-02044 VTT, Finland
[2]University of Helsinki, P.O. Box 55, FI-00014 University of Helsinki, Finland
[3]Saanio&Riekkola, Laulukuja 4, FI-00420 Helsinki, Finland

ABSTRACT

The possibility of glacial meltwaters disturbing the stable conditions around a nuclear waste repository can be studied by the USD method. The method is based on the perception of uranium isotope mobilisation relative to ^{230}Th occurring due to past meltwater intrusions. Uranium mobilisation occurs when redox potential increases. In shallow bedrock redox-potential may change under normal recharge. Similar changes probably occur when meltwater penetrates a deeper reducing aquifer. The role of redox conditions in U mobilisation appears dominant, as direct α recoil induced signals seem too weak to be observed for rock and water samples. The USD method can be used to provide support to the Finnish safety case in the area of complementary safety arguments and scenario formulations for transport calculations.

INTRODUCTION

The classical application of U-series disequilibrium (USD) studies has been ^{230}Th/^{234}U dating in Quaternary geology and palaeoclimatology. USD dating is based on the mobility of ^{234}U and the immobility of ^{230}Th. It takes around 350 ka for the parent-daughter pair to reach radioactive equilibrium and thus the method covers time scales including the two last European glaciations Saalian (~ 350 – 130 ka BP) and Weichselian (~115 – 10 ka BP). U-series elements form part of the spent nuclear fuel inventory and this has inspired nuclear waste disposal related applications [1,2,3]. The potential of USD in indicating glacial effects has been studied in [4,5].

A role considered for the host bedrock in nuclear waste disposal is to provide a chemically stable environment for the repository. Dilute oxygen-rich glacial meltwaters, driven by high hydraulic gradients, could introduce a redox perturbation in the bedrock. If low ionic strength meltwater could reach the repository, it could possibly chemically erode the bentonite buffer and subsequently increase canister corrosion [6]. Light oxygen isotope compositions observed in deep groundwaters in Fennoscandia [7,8,9] are generally interpreted as indicating penetration of glacial meltwaters, but whether the deeply penetrating water has been oxidising is still unclear.

In this paper we discuss possible contributions of USD observations to the safety case of spent nuclear fuel repositories in the Fennoscandian Shield, considered to be subject to future glaciations. Examples are given of natural analogue (NA) studies in which USD has clarified the understanding of the behaviour of the bedrock-groundwater system and radionuclide migration in situ.

SCIENTIFIC BASIS OF USD STUDIES

The geochemistry of U and Th is well known, thus providing a scientifically sound basis for the method. U occurs in all geological environments in two relatively stable oxidation states, less soluble U(IV) and soluble U(VI) (e.g., see the NEA data base for U [10, 11]). The nuclear geochemistry of U is well known, too. Natural U consists of three long-lived isotopes: primordial ^{238}U and ^{235}U, and the decay product ^{234}U. A "hot atom" chemistry model has been developed for ^{234}U [13,14,15,16] predicting that in-growing ^{234}U tends toward the more mobile ^{234}U(VI), enabling a valence contrast between ^{234}U and ^{238}U if ^{238}U was originally U(IV). The model assumes that a recoiling ^{234}Th atom "pushes" oxygen atoms it meets to the end of its trajectory and thus the ^{234}U will be generated in a environment with increased redox potential (Figure 1). Recoil-induced mineral lattice damages further contribute to ^{234}U mobilisation.

Evidence of the hot atom model is higher ^{234}U/^{238}U activity ratio in the U(VI) fraction in rock which has been systematically observed when U oxidation states have been studied [17].

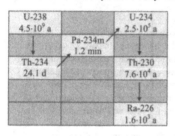

Figure 1. The part of the ^{238}U decay chain used in repository-related USD studies [4].

U isotope fractionation — preferential ^{234}U release due to valence contrast — necessitates that U has first existed as U(IV) for a relatively long time. When U is precipitated or adsorbed as U(IV), both ^{238}U and ^{234}U are immobilised at the same time and at the same valence state (Figure 2). After that, chain decay starts to replace the original immobilised ^{234}U by in-grown ^{234}U and to generate ^{230}Th. It takes about 300 ka of a closed chain decay to generate a ^{234}U(VI) inventory equal to the original ^{234}U(IV) inventory. After a period of around 2 Ma all the original ^{234}U(IV) has been replaced by ingrown ^{234}U (part of which will be U(VI)), with the ^{238}U inventory remaining in the reduced U(IV) form.

Decay chain modelling (Figure 1) is based on the inherent tendency of a radioactive decay chain to evolve towards radioactive equilibrium at a rate set by the decay constants of the chain members. Mass flows of individual nuclides are the unknown parameters. Supporting mass balance calculations are, therefore, done (i) to derive the mass flow scenarios used in USD modelling and (ii) to test the obtained modelling results. Mass balance calculations can, for example, be based on simple diffusion-controlled mass flows [18].

USD METHODOLOGY

Below, the basic components of the USD methodology are given; the overall pros and cons of the methodology have been summarised in [4].

The working hypothesis of consecutive U release processes along a groundwater flow route includes three steps: (1) oxidative congruent release in the beginning of the route at depths close to the ground surface, (2) selective ^{234}U release further along the route at the redox front, and (3) no release at distances past the redox front. This sequence is due to the fact that dissolved oxygen in groundwater will gradually be consumed as the water flows along the route. In rock samples, steps 1 and 2 create ^{230}Th/^{234}U>1 disequilibrium and step 3 ^{230}Th/^{234}U=1 equilibrium. The release hypothesis could be tested by sampling at all three parts of a route.

Sampling strategy aims to obtain representative samples. In fractured crystalline aquifers, three types of samples must be considered: (i) groundwater, (ii) fracture surface with secondary minerals, and (iii) the rock matrix adjacent to fractures. Groundwaters may have travelled long distances and via different fractures, so their U isotopic compositions reflect the flow history rather than local conditions [19]. Fracture coatings better represent the sampling location, but as fractures may have different hydraulic properties, comparable sampling is complicated. The rock matrix adjacent to the fracture face can be considered as the most representative local sample material for studying past changes. The rock fabric includes pores and fine fissures extending through the rock structure [20], which provide a route for redox perturbation to propagate from a fracture into the rock matrix to mobilise U.

In igneous rocks U occurs [21]: (1) in U minerals, (2) in crystallographic sites or structural defects of major rock forming minerals and minor accessory minerals, (3) in cation exchange positions, (4) adsorbed and precipitated on mineral surfaces, (5) dissolved in intergranular fluids, and (6) dissolved in fluid inclusions (Figure 2). It is mainly modes 3-5 (collectively called labile U) that are involved in U mobilisation [22] and thus in USD studies. The rest is called refractory U. Labile U can be U(IV), U(VI), or a mixture, and refractory U occurs predominantly as U(IV). For U(IV) mode to develop significant valence contrast it should be older than 300 ka.

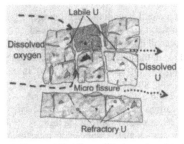

Figure 2. Schematic distribution of U in low-porosity igneous rocks. There are two main phases. Labile U occurs in rock pores easily accessible to fluids. Refractory U occurs chiefly as various uraniferous inclusions (e.g. zircon, apatite) and is less accessible to fluids.

U modes may display both ^{230}Th/^{234}U disequilibrium and equilibrium. When the sample material is dissolved completely, all U is taken into the solution and the disequilibrium signal in labile U may be diluted, that is, disequilibrium is modified towards equilibrium:

$$\frac{^{230}Th}{^{234}U} = \sum_i w_i \left(\frac{^{230}Th}{^{234}U} \right)_i , \; (\Sigma w_i = 1) \tag{1}$$

where w_i is the inventory fraction of mode i of U occurrence.

Dating the diluted ^{230}Th/^{234}U and ^{234}U/^{238}U disequilibria will yield an upper limit for the age of the event that caused the disequilibrium. More accurate dating can be approached, for example, by separating labile U selectively or studying it directly using LA-ICP-MS.

USD modelling aims to quantify the mass flows between the groundwater, fracture coating, and rock matrix samples. The following coupled equations state that the mass flow from the U source (here fracture coating material [23]) will enter the surrounding water phase:

$$\frac{dC_{i,f}}{dt} = -\lambda_i C_{i,f} + \lambda_{i-1} C_{i-1,f} - \frac{J_i}{V_{fc}}, \; i=1, \, ..., \, n; \; C_{i,f}(t=0)=C_{i,f}^0 \tag{2}$$

$$\frac{dC_{i,W}}{dt} = -\lambda_i C_{i,W} + \lambda_{i-1} C_{i-1,W} + \frac{J_i}{V_b} + \frac{q}{V_b} C_{i,W}^0 - \frac{q}{V_b} C_{i,W} , \; i=1, \, ... \, , \, n; \; C_{i,W}(t=0)=C_{i,W}^0 \tag{3}$$

where $C_{i,f}$ = concentration of nuclide i in fracture coating (atoms/m^3), $C_{i,W}$ = concentration of nuclide i in fracture water (atoms/m^3), t = time (s), λ_i = radioactive decay constant of nuclide i (1/s), J_i = transfer rate of nuclide i into water via α recoil (atoms/s), q = flow rate of groundwater in half-fracture (m^3/s), V_b = volume of half–fracture (m^3), and V_{fc} = volume of fracture coating (m^3). The term q/V_b (1/s) represents water turnover in the water conducting fracture.

Despite its sound basis, the current USD methodology appears to face its technical limits. The uncertainties involved cannot be removed or circumvented, and thus the interpretations of USD as a stand-alone methodology cannot be improved. An improved USD methodology could include additional coupled hydro-chemical (HC) modelling. This would enable us to model kinetically the oxygen consuming reactions in and around a water conducting fracture, which are now simply assumed to be instantaneous. USD modelling can provide the necessary mass flows to HC modelling. With a more detailed modelling tool, more detailed geochemical hypotheses can be tested.

ON THE LINKAGE BETWEEN USD AND GROUNDWATER CONDITIONS

^{234}U can be released from a U source either via direct α recoil (within the recoil range around the source) or via chemical release. An important observation supporting the role of chemical release in U isotope fractionation is that direct α recoil does not appear to be important. In rock samples the α recoil induced signals are diluted [23] in the overall U mass of the samples, which are normally ground before USD analysis (recoil range 20 nm vs. grain size 1 mm). In groundwater samples the recoiling ^{234}U atoms entering flowing groundwater are diluted in the ^{234}U mass flow conveyed by the water [23], cf. Equations 2 and 3. Groundwater samples are taken from locations of good yield, which may indicate locations of relatively high groundwater flow. High flow routes are relevant, as meltwater induced redox changes spread via these.

We have several observations of episodic U release from rock matrix, suggesting more than one meltwater pulses from successive deglaciations. Release rates were bounded by (i) the measured ^{230}Th/^{234}U activity ratios and (ii) by maximum mass flows matrix diffusion can convey in rock matrix per meltwater intrusion episode [18].

So far, the best analysed Finnish samples from the Palmottu natural analogue site have been taken from shallow depths, in the oxidising zone above the redox front. At this depth glacial meltwater disturbances are obvious if there is hydraulic connection. Due to the shallow sampling, the working hypothesis for U release has not been tested systematically.

Palmottu is an exceptional site, as it is close to the ice margin formation at Salpausselkä formed when the retreat of the last ice sheet paused for around 1 ka, thus providing huge amounts of meltwater. Therefore, meltwater signals can be studied expressly at Palmottu.

Driven by high hydraulic gradients, the glacial meltwater induced disturbance (generally oxidising conditions) may have propagated to a distance far beyond the range of modern recharge. Therefore, a U release signal (^{230}Th/^{234}U>1) found in the rock matrix further from a fracture indicates strong and long-lasting redox changes in the past.

USD has clarified the behaviour of bedrock-groundwater systems in NA studies, mainly due to its dating property. As process analogues, the occurrence of in situ matrix diffusion of U in crystalline bedrock was shown in [24], and new information on reversible/irreversible in situ U fixation in minerals was obtained in [25,26]. As material analogues, the longevity of a Cu deposit in crystalline bedrock at Hyrkkölä was studied in [26], and that of secondary U minerals co-existing with UO$_2$ in [27]. USD has not yet been used as a standard method in site investigations, as we have been developing the methodology concerning sampling, experimental, and interpretation methods using the observations at Palmottu.

USD UNDERSTANDING CAN SUPPORT A SAFETY CASE

A safety case is the documentation that the applicant organisation presents to competent authorities to support its licence application to site, construct, operate or close a nuclear waste disposal facility. It is the synthesis of evidence, analyses and arguments that quantify and substantiate the safety of the planned facility. It includes a quantitative safety assessment, but also complementary considerations, such as natural analogues, heuristic conclusions, and other reasoning independent of the safety assessment (Figure 3).

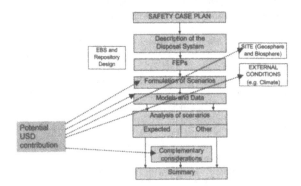

Figure 3. Possible USD contribution to the current Finnish safety case plan (modified from [28]).

The safety case also covers the assessment of remaining uncertainties and open issues, their potential effects, and ways to resolve them. Developing a safety case is a step-wise process, as individual reports will be updated at intervals, becoming more detailed as the nuclear waste management programme proceeds from siting to construction.

As USD studies lead to improved understanding of the bedrock-groundwater system under redox perturbation, it appears that safety case components dealing with the site, processes, and evolution of the site and repository, as well as radionuclide transport, could benefit from USD input. The USD contribution could be channelled in complementary considerations and scenario formulations for transport calculations (Figure 3).

CONCLUSIONS

The geochemistry of U is known relatively well and the radioactive behaviour of decay chains is known very well, forming a good basis for the USD method. The pros and cons of the method are well-known. Changes in groundwater redox conditions appear to dominate U mobilisation and U isotope fractionation, as direct α recoil induced signals appear too weak to be detected in water and rock samples. The USD contribution to safety case could be channelled in complementary considerations and scenario formulations for transport calculations.

ACKNOWLEDGEMENTS

Conducted under the Finnish Research Programme on Nuclear Waste Management (KYT).

REFERENCES

[1] W. Murphy and D.A. Pickett, Mater. Res. Soc. Symp. Proc. **713**, pp. 867-874 (2002).

[2] D.A. Pickett and W. Murphy, Mater. Res. Soc. Symp. Proc. **824**, pp. 555-560 (2004).

[3] W.M. Murphy, Mater. Res. Soc. Symp. Proc. **824**, pp. 533-541 (2004).

[4] K. Rasilainen, J. Suksi, T. Ruskeeniemi and P. Pitkänen, Mater. Res. Soc. Symp. Proc. **807**, pp. 565-570 (2004).

[5] E.-L. Tullborg, J.A.T. Smellie and A.B. MacKenzie, Mater. Res. Soc. Symp. Proc. **807**, pp. 571-576 (2004).

[6] SKB, TR 06-09 (SKB, Stockholm, 2006).

[7] R. Blomqvist and S.K. Frape, SKI Report 97:13 (A5-A6) (SKI, Stockholm, 1997).

[8] SKB, SKB R-05-17 (ISSN 1402-3091) (SKB, Stockholm, 2005).

[9] SKB, SKB R-06-70 (ISSN 1402-3091) (SKB, Stockholm, 2006).

[10] H. Wanner and I. Forest, *Chemical Thermodynamics 1. Chemical thermodynamics of uranium.* (Elsevier Science Publishing Company, Inc., New York, 1992) p. 715.

[11] I. Grenthe, I. Puigdomenech, M.C.A. Sandino and M.H. Rand, *Chemical Thermodynamics of Uranium* (Appendix D in Chemical Thermodynamics of Americium. Chemical Thermodynamics 2), OECD Nuclear Energy Agency, (Elsevier, Amsterdam, 1995), pp. 347-374.

[12] D. Langmuir, *Geochim. Cosmochim. Acta* **42**, 547 (1978).

[13] K. Rössler in *Gmelin Handbook of Inorganic Chemistry, 8th Edition, Uranium* (Supplement Volume A6), (Springer Verlag, Berlin, 1983), pp. 135-164.

[14] J.-C. Petit, Y. Langevin and J.-C. Dran, *Bull. Minéral* **108**, 745 (1985).

[15] E. Ordonez-Regil, J.J. Schleiffer and J.P. Adloff, *Radiochim. Acta* **47**, 177 (1989).

[16] J.P. Adloff and K. Rössler, *Radiochim. Acta* **52/53**, 269 (1991).

[17] J. Suksi and K. Rasilainen, Mater. Res. Soc. Symp. Proc. **663**, pp. 961-969 (2001).

[18] K. Rasilainen, J. Suksi, T. Ruskeeniemi, P. Pitkänen and A. Poteri, *J. Contam. Hydrol.* **61**, 235 (2003).

[19] J. Suksi, K. Rasilainen and P. Pitkänen, *Phys. Chem. Earth* **31**, 556 (2006).

[20] M. Siitari-Kauppi, Report Series in Radiochemistry 17/2002 (Doctoral Dissertation), (University of Helsinki, Helsinki, 2002).

[21] G. Neuerburg, U.S. Geol. Surv. Prof. Pap. 300, (1956) pp.55-64.

[22] J.S. Stuckless and C. Pires Ferreira, IAEA-SM-208/17, (1976) pp. 717-729.

[23] K. Rasilainen, H. Nordman, J. Suksi and N. Marcos, Mater. Res. Soc. Symp. Proc. **932**, pp. 1041-1048 (2006).

[24] K. Rasilainen, VTT Publications 331 (Doctoral Dissertation) (VTT, Espoo, 1997).

[25] J. Suksi, Report Series in Radiochemistry 15/2001 (Doctoral Dissertation) (University of Helsinki, Helsinki, 2001).

[26] N. Marcos, Acta Polytechnica Scandinavica Ci 124 (Doctoral Dissertation)(Helsinki University of Technology, Espoo, 2002).

[27] R.J. Finch, J. Suksi, K. Rasilainen and R.C. Ewing, Mater. Res. Soc. Symp. Proc. **412**, pp. 823-830 (1996).

[28] Posiva, Posiva Report 2008-05 (Posiva, Eurajoki, 2008).

Mater. Res. Soc. Symp. Proc. Vol. 1124 © 2009 Materials Research Society   1124-Q10-02

Uranium Mineralogy at the Askola Ore Deposit, Southern Finland

Mira Markovaara-Koivisto[1], David Read[2], Antero Lindberg[3], Marja Siitari-Kauppi[4] and Nuria Marcos[5]
[1] Helsinki University of Technology, PO Box 6200, FIN-02015 TKK, Finland
[2] Enterpris, The Old Library, Lower Shott, Leatherhead Road, Great Bookham, Surrey, UK
[3] Geological Survey of Finland, Betonimiehenkuja 4, FIN-02150 ESPOO, Finland
[4] University of Helsinki, PO Box 55, FIN-00014 HELSINKI UNIVERSITY, Finland
[5] Saanio & Riekkola Oy, Laulukuja 4 FIN-00420 HELSINKI, Finland

ABSTRACT

Understanding uranium retention processes is essential when assessing the safety of a spent nuclear fuel repository as uranium forms over 95% of the spent fuel. It is crucial to establish whether migration mainly occurs continuously through slow diffusion, as usually postulated, or if groundwater-mediated transport occurs episodically, for example, during periods of intense, glacial recharge. This study provides insights into the principal retention mechanisms of natural uranium in granitic bedrock at Askola in Southern Finland.

The morphology and chemical composition of uranium minerals, found on the surfaces of water-conducting fractures ('open fractures') as well as in the matrix adjacent to fractures, were investigated in polished rock slabs and thin sections optically, by scanning electron microscopy (SEM) and by energy and wavelength dispersive spectrometers (EDS and WDS). In addition, the radiogenic isotopes were analyzed *in situ* by inductively coupled plasma mass spectrometry coupled to a high-performance Nd:YAG deep UV (213 nm) laser ablation system.

The predominant uranium ore mineral at Askola was identified as uraninite in studies carried out from 1963-1979, occurring as primary grains in the rock matrix. In the present study, abundant secondary uranium phases were found in micro-fractures, typically associated with goethite, pyrite and chalcopyrite. Chemical analyses indicate an assemblage of uranium (VI) silicates and hydroxides; phosphate is also present. Uranium (VI) hydroxides were found to fill micro-fractures propagating from open fractures to depths of several millimetres. If released into the bedrock, U would be expected to precipitate in such micro-fractures, preferentially in association with the various iron minerals present.

INTRODUCTION

It is planned to dispose of spent nuclear fuel from Finnish power plants deep in the bedrock at Olkiluoto (Western Finland). A comprehensive assessment of the future performance of the repository is required. Early failure of the engineered barrier system, in particular defects in the copper–iron canisters containing the fuel could lead to interactions with the local groundwater. On the longer term, glacial melt waters, which may intrude into the bedrock due to the high hydrostatic pressure under an ice sheet, have to be taken into account [1]. These conditions may lead to uranium migration and retention in the far field, i.e. the bedrock surrounding the repository and the local environment.

A variety of secondary uranium minerals have been identified at ore deposits in Finland [2, 3, 4, 5] and Sweden [6]. The topic is currently of interest from the perspective of both spent nuclear fuel disposal and the mining of uranium ores. It is essential to clarify which secondary minerals are likely to precipitate under specific geochemical conditions. This information can then be used to predict uranium migration behaviour and to evaluate the capacity of the bedrock for uranium retention.

In this study, secondary uranium precipitates were investigated in samples from the Askola ore deposit, located 50 km north-east of Helsinki (Figure 1). As with other sites studied in the region [7], isotopic fractionation can be used to elucidate the conditions under which they were formed and the timescales over which migration occurred.

Figure 1. Askola uranium deposit is located ~50 km north-east of Helsinki.

EXPERIMENTAL

Samples

The samples were taken from two drill cores; DH303 at lengths of 15.55 m and 19.25 m and DH311 at 21.45 m, corresponding to vertical depths of 10.49 m, 12.98 m and 15.59 m, respectively (Figure 2).

Samples adjacent to open fractures were selected by α monitoring. Yellow crystals of uranophane can be seen with the naked eye. The samples comprise quartz (~40 % in DH303 and 20 % in DH311), K-feldspar (~20 % and 60 %, respectively), plagioclase and biotite; plagioclase is almost totally sericitized in DH303 samples (~20-25%). Fe-oxyhydroxide staining is common and apatite is present in sample DH303 at 19.25 m.

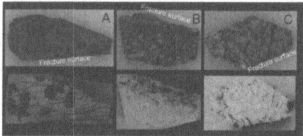

Figure 2. Photograph of A) drill core sample DH303 15.55 m, B) DH303 19.25 m, D) DH311 21.45 m and corresponding autoradiographs. Dark areas correspond to PMMA porosity and radioactive mineral phases. Width of samples: A 4.9 cm, B 4.1 cm and C 4.3 cm.

Autoradiography

The samples were impregnated with ^{14}C labelled polymethylmethacrylate (PMMA) to reveal the pore structure and porosity of the granites. The method [8, 9] involves impregnating centimetre-scale rock cores with ^{14}C labelled methylmethacrylate (^{14}C-MMA) in vacuum. This is followed by irradiation polymerization using a ^{60}Co-source, sample partitioning and autoradiography. In the case of samples containing loosely bound mineral phases or clay, the PMMA prevents the samples from crumbling and the water-soluble secondary uranium phases from dissolving during preparation. The distribution of radioactive minerals and pores in the selected rock slab samples were visualized on roentgen film (Figure 2). The sawn and slightly polished rock surfaces were exposed on Kodak X-omat MA roentgen film for 7-14 days. The film darkens in areas which are porous such that inter- and intra-granular fissures are clearly visible. Mineral phases containing natural radioactivity also cause film darkening but the U phases can be distinguished from the porous network in the autoradiograph images by the halos they produce around the darkened features.

Microscopy

The mineralogical composition of the samples is evident from thin sections under the polarizing microscope. Smaller scale observations were made on polished sections of the PMMA impregnated rock samples using a Hitachi FE-SEM S-4800 field emission scanning electron microscope. The elemental composition of mineral phases was determined semi-quantitatively using a INCA350 EDS system with a UTW Si(Li) detector operated in the backscattered electron mode (BSE). The operating conditions were 20 kV acceleration voltage and a 10 μA beam current. Wavelength Dispersive (WDS) analyses were performed with a Cameca SX100-type microprobe with 15 keV acceleration voltage and 10-25 nA beam current.

Laser ablation mass-spectrometry

Uranium minerals with radius more than 25 μm were chosen for LA-ICP-MS analyses using an Agilent 7500ce, quadrupole ICP-MS with an Octopole Reaction System coupled to a high-performance Nd:YAG deep UV (213 nm) laser ablation system. The latter provides flat craters with a diameter from 10 to 100 μm and high absorption for the analyses. The ablation frequency was 15 Hz, laser beam size from 20 to 55 μm, laser efficiency 65 % and ablation time 10 sec. The laser was focused manually. The isotopic mass ratio ^{234}U/^{238}U was measured to calculate corresponding activity ratio. No internal or external calibration was used.

RESULTS AND DISCUSSION

Emphasis in earlier studies at Askola [2] was placed on the primary uranium phases in order to determine ore potential. In the present study, samples were taken close to the surface, as the aim was to investigate secondary uranium phases formed by precipitation in a relatively open oxidized system.

Samples from core DH303 are altered and highly porous. Only a few radioactive minerals were detected in the sample from 15.55 m, mostly next to open fractures. Uranium content is high (UO_3 ca. 55 wt.%) due to co-precipitation at the rims of goethite nodules (Figure 3); levels decrease towards the interior. Formation of the nodules may well be microbially mediated. Similar goethite nodules are found deeper in the matrix, but U rich rims do not occur.

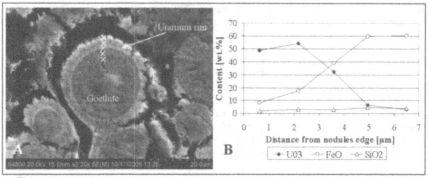

Figure 3. A) BSE image showing goethite nodules in sample DH303 15.55 m. Uranium is enriched (ca. 55 wt.%) at the edge of the nodules shown as a white rim. **B)** Uranium and iron content as wt.% in a line scan crossing the rim and the inner parts of the nodule as marked in **A**.

At 19.25 m autoradiography shows clearly that uranium is enriched in a 1 cm wide zone extending from an open fracture, indicating widespread penetration by oxidising waters. Uranium is associated with pyrite and chalcopyrite (Figure 4 A) in the form of a silicate-phosphate(s) (Table I). The uranium content (UO_3) is greater than 40 wt.%. Silica (SiO_2) content varies from 10 to 20 wt.% and calcium content is around 2% indicating uranophane as the dominant silicate phase. Phosphorus (P_2O_5) content varies from 4 to 6 wt.%.

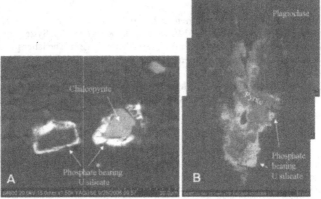

Figure 4. A) BSE image of a phosphate-bearing uranium silicate surrounding an euhedral chalcopyrite grain in sample DH303 19.25 m. **B)** BSE image of a phosphate-bearing uranium silicate surrounding a weathered pyrite grain in sample DH311 21.45 m.

Table I. Elemental analysis with EDAX of the uranium precipitates associated with chalcopyrite and pyrite seen in Figure 4.

Elements as oxides	Na_2O	Al_2O_3	SiO_2	P_2O_5	K_2O	CaO	UO_3
Figure A	-	2.69	21.32	3.69	-	1.67	64.75
Figure B	0.42	2.34	13.26	5.89	0.60	3.37	45.04
Inferred mineral	Uranophane $Ca(UO_2)_2SiO_3(OH)_2 \cdot 5(H_2O)$ with possible apatite inclusions						

The sample from core DH311 is significantly less altered and grain boundary porosity dominates. At 21.45 m the uranium phases occurs associated with pyrite and filling micro-fractures. In addition to uranophane (Figure 4 B), uranium occurs as (hydr)oxides (Figure 5).

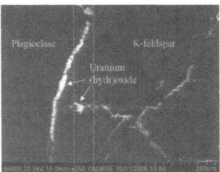

Figure 5. BSE image of uranium (hydr)oxide filling a micro-fracture in sample DH311 21.45 m.

Part of the 20 mm micro-fracture which cuts the whole of sample DH311 21.45 m is filled with uranium (hydr)oxide (Figure 5). Silica (SiO_2) and calcium (CaO) contents are ca. 4 wt.% and the uranium (UO_3) content more than 80 wt.%.

The isotopic mass ratio $^{234}U/^{238}U$ was measured in the secondary uranium silicates from DH311 21.45 m (e.g. Figure 4 B) to establish whether it represents direct precipitation from groundwater or an earlier hydrothermal event. The LA technique allows *in situ* measurements from precipitates a few micrometers in size. Mass ratios were converted to activity ratios (ARs); information needed for later age determination. Average $^{234}U/^{238}U$ activity ratios at three measurement positions were 1.4, 1.7 and 1.9 confirming that the uranium precipitated from groundwater [10], in the geologically recent past.

CONCLUSIONS

At Askola, uranium is associated with goethite nodules in open fractures, with pyrite and chalcopyrite grains in the matrix, possibly acting as local reducing agents, and in micro-fractures. The precipitates were observed to a maximum depth of one centimetre from the water-conducting zones. Fracture fillings were found to be principally uranophane and uranium (hydr)oxides with some U-bearing phosphate also present. Activity ratios of the uranophane indicate relatively recent precipitation from groundwater. The formation of secondary uranium minerals in oxidizing conditions is the primary retardation mechanism for uranium released into groundwater.

ACKNOWLEDGMENTS

Financial support for this study was provided by the Finnish Research Programme on Nuclear Waste Management (KYT).

REFERENCES

1. D. Read, S. Black, T. Buckby, K-H. Hellmuth, N. Marcos and M. Siitari-Kauppi. Global and Planetary Change **60** (3-4), 235-249 (2007).
2. H. Appelqvist. Report No. M 60/3022/-82/1, Geological Survey of Finland, 1982.
3. R. Blomqvist, T. Ruskeeniemi, J. Kaija, L. Ahonen, M. Paananen, J. Smellie, B. Grundfelt, K. Pedersen, J. Bruno, L. Pérez del Villar, E. Cera, K. Rasilainen, P. Pitkänen, J. Suksi, J. Casanova, D. Read and S. Frape. CEC Report No. EUR 19611, 2000.
4. N. Marcos and L. Ahonen. Report No. 99-23, Posiva, 1999.
5. M. Markovaara-Koivisto, N. Marcos, D. Read, A. Lindberg, M. Siitari-Kauppi and K. Loukola-Ruskeeniemi in *Scientific Basis for Nuclear Waste Management XXXI*, edited by W.E. Lee, J.W. Roberts, N.C. Hyatt and R.W. Grimes, (Mater. Res. Soc. Symp. Proc. **1107**, Sheffield, UK, 2008), pp. 637-644.
6. R. Löfvendahl and E. Holm. Lithos **14**, 189-201 (1981).
7. H. Appelqvist. Report No. M 19/3022/-72/1/10, Geological Survey of Finland, 1972.
8. K-H. Hellmuth, M. Siitari-Kauppi and A. Lindberg. Journal of Contaminant Hydrology **13**, 403-418 (1993).
9. M. Siitari-Kauppi, E.S. Flitsiyan, P. Klobes, K. Meyer, K-H. Hellmuth in *Scientific Basis for Nuclear Waste Management XXI*, edited by McKinley I.G., McCombie C. (Mater. Res. Soc. Symp. Proc. **506**, Boston, MA, 1998), pp. 671-678.
10. D. Porcelli and R. Swarzenski, in *Uranium-series Geochemistry*, edited by B. Bourdon, G.M. Henderson, C.C. Lundstrom and S.P. Turner (Mineralogical Society of America, Washington, 2003), p. 656.

Mater. Res. Soc. Symp. Proc. Vol. 1124 © 2009 Materials Research Society   1124-Q08-06

Secondary Minerals and Ambient Fluid Flow at Yucca Mountain

William M. Murphy
Geological and Environmental Sciences, California State University, Chico, CA 95929-0205
USA

ABSTRACT

Secondary calcite and opal deposits in fractures and cavities of the unsaturated zone at Yucca Mountain have been interpreted to mark infiltrating groundwater flow paths under natural conditions. An evaluation of reports of the distribution of these deposits and their textural features suggests that the mechanism of their precipitation is evaporation, which may be consequence of gas flow that is warming, not downward groundwater percolation. Evaporation provides the chemical potential for mineral precipitation, is consistent with coexisting calcite and opal, and helps explain the heterogeneity in unsaturated zone water chemistry.

INTRODUCTION

Yucca Mountain, Nevada is the proposed permanent geologic repository site for high level nuclear waste in the United States. The proposed waste emplacement horizon is in fractured, welded, rhyolite tuff (Topopah Spring Tuff) approximately 300 m below the ground surface and 300 m above the groundwater table. Studies of the flow of water from the ground surface through the thick sequence of partially water saturated fractured rock at Yucca Mountain have revolutionized the science of unsaturated groundwater flow (see review in [1]). The gas phase in the unsaturated zone also plays an important role in transporting and buffering reactive volatile components including H_2O, O_2, and CO_2 under ambient and perturbed (e.g., exploration and repository) conditions. Understanding the mechanisms of groundwater and gas flow is fundamental to evaluating the capacity of the proposed geologic repository to isolate radioactive wastes.

Secondary deposits of calcite and opal occur in certain fractures and lithophysal cavities at depth in the unsaturated zone at Yucca Mountain. Petrographic textures, chemistry, isotopy, and fluid inclusions of these deposits have been studied in detail (e.g., [2,3,4] and others cited below). Secondary mineral deposits are interpreted commonly as markers of groundwater flow paths, particularly in the fields of metamorphic petrology and hydrothermal ore deposition (e.g., [5]), in saturated and unsaturated limestone caves (e.g., [6,7]), and specifically for the unsaturated zone at Yucca Mountain (e.g., [8,9]). Secondary mineral precipitation in the unsaturated zone at Yucca Mountain records the occurrence of water, which is probably pervasive on surfaces in the natural unsaturated system because of flow and because of interactions between hygroscopic mineral surfaces and a vapor phase saturated with water. However, relations are incompletely understood between fluid (water and gas) flow and secondary calcite and opal deposits. The objective of this study is to explore these relations and in particular to examine the hypothesis that secondary calcite and opal deposits in the unsaturated zone at Yucca Mountain are produced in areas of warming gas flow in the ambient system and are not necessarily markers or accurate measures of downward percolating water flow in fractures.

SITE CHARACTERIZATION DATA

Most fractures in the welded Topopah Spring Tuff at Yucca Mountain are free of macroscopic deposits of secondary minerals (e.g., [3]), and the fractional volumes of secondary calcite and opal are small. In the Exploratory Studies Facility (ESF) less than ten percent of the abundant fractures and lithophysal cavities contain secondary mineralization [8]. Secondary deposits of calcite and/or opal show considerable heterogeneity in occurrences, textures, and continuity of deposition. However, early stage calcite is commonly capped by silica and later calcite is commonly precipitated with opal (e.g., 4,8]). Precipitates commonly occur at the base of lithophysal cavities and on the footwall of dipping fractures reflecting effects of gravity on the distribution of water in the unsaturated system [8], and secondary minerals are generally absent from the matrix of the welded Topopah Spring Tuff [B.D. Marshall, written communication].

Late stage calcite intergrown with opal has precipitated at Yucca Mountain for the last 1.9 to 2.9 million years [4]. Microstratigraphic dating of particular opal deposits from the ESF shows nearly continuous deposition at a steady rate over at least the last one million years [10]. The rate of precipitation is extremely slow, less than one nm/y [10]. Petrographic textures indicate that opal and calcite were commonly deposited together. Whelan et al. [8] refer to textural relationships among calcite and opal as "intergrowths" consistent with "concurrent growth mechanisms" and "intimately intergrown ... consistent with either coprecipitation of calcite and opal, or replacement of calcite by opal." "In many deposits, the calcite-opal interfaces are irregular and may reflect competitive growth textures formed by concurrent deposition" [8]. Textures of calcite and opal in relation to the tuff substrate also provide compelling evidence that a gas phase was present at the time of their formation (e.g., [3,8]), and fluid inclusions in secondary minerals precipitated during the last five million years attest to low temperatures (e.g., [4,8]).

Ambient unsaturated zone groundwater chemistry at Yucca Mountain is notably heterogeneous exhibiting differences between pore water and perched water (e.g., [11,12]). Nevertheless, the unsaturated zone groundwaters are all dilute, oxidizing, and close to chemical equilibrium with both calcite and opal (e.g., [12,13]). Aqueous calcium concentrations tend generally to decrease with depth in the unsaturated zone, which tends to compensate for the decrease in calcite solubility with increasing temperature. Thermodynamic interpretations of unsaturated zone groundwater chemistry show no systematic relationship between depth and the chemical saturation state of pore water with respect to calcite [12]. Corrected to achieve equilibrium aqueous speciation, charge balance, and equilibrium with gas phase CO_2, the pH of unsaturated zone groundwaters is almost uniform near 8.1 [12].

The formation gas phase in the unsaturated zone at Yucca Mountain is atmospheric air, effectively at equilibrium with dilute liquid water and enriched in CO_2 (e.g., [11,14]). The CO_2 content of the ambient formation gas at depth in Yucca Mountain is remarkably constant and homogeneous, but it is easily contaminated with atmospheric gas or drilling fluids introduced in boreholes or drifts.

Models using measured hydraulic, thermal, and geochemical properties of the system indicate that fluid flow in the welded tuffs is dominated by flow in the fracture network. Net infiltration has large lateral and temporal variations. Modern average net infiltration and percolation of water though the unsaturated zone is estimated to be 3.6 mm/y for the larger infiltration domain of Yucca Mountain, and it is estimated to be 4.7 mm/y for the proposed emplacement area domain [15]. Past and future net infiltration is estimated to be greater because of wetter climates. Relatively little research has been devoted to natural gas flow at depth in the mountain. Deep gas flow is inferred to be minimal based particularly on depleted [14]CO_2 at depth (e.g., [11,16]). Yang et al. [11] conclude that gravity flow of water is faster than diffusive transport of gas. Thorstenson et al. [14] conclude that overlying rocks are a barrier to gas flow in the Topopah Spring Tuff and that barometric effects at depth are "extremely small." Finsterle et al. [17] presented models suggesting that gas flow in faults could affect temperature profiles and complicate estimates of water percolation fluxes from temperature data. Bodvarsson et al. [18] presented models for water percolation using borehole temperature profiles and showed that gas fluxes have negligible effects for deep boreholes.

MECHANISMS OF CALCITE AND OPAL PRECIPITATION

The ambient unsaturated zone aqueous geochemical system at depth in Yucca Mountain is close to equilibrium with calcite, opal, and a gas phase saturated with nearly pure water at a nearly fixed CO_2 pressure (e.g., [12]). A change to this equilibrium system is required to provide the chemical potential for precipitation of secondary calcite and opal, and several mechanisms are possible (Table I). Equilibrium calcite precipitation due to retrograde solubility along descending, warming groundwater flow paths has been shown in reactive transport models to be a possible mechanism (e.g., [19,20]). These models have not included possible effects of gas flow. A key constraint on the mechanism of secondary mineral precipitation is that warming water alone cannot account for opal precipitation, because of its prograde solubility, or for simultaneous calcite and opal precipitation, which are common and steady occurrences.

Of the mechanisms listed in Table I, evaporation of water is consistent with simultaneous calcite and opal precipitation and the textural evidence for the presence of a gas phase at the time of precipitation (e.g., [3,8]). Evaporation requires the gas phase to be less than saturated with liquid water. Although vigorous subsurface gas circulation near the ground surface causes evaporation [21], persistent local gradients in the vapor pressure of H_2O (or CO_2) are unlikely at depth in Yucca Mountain under ambient conditions because of the omnipresent liquid water, homogenization of the gas phase by diffusion, and the extremely slow rate of change in the natural geochemical system. Whelan et al. [8] conclude that "Although mineral formation is driven by evaporation and gas loss, calcite $\delta^{18}O$ values in the TSw show no clear effects of ^{18}O-enrichment due to evaporation. Evaporative ^{18}O-enrichment is, however, likely to be minimal if the processes of water migration, gas removal, and mineral deposition are slow." Descending

Table I. Possible Mechanisms for Calcite and Opal Precipitation at Yucca Mountain

---

Decreasing calcite solubility with increasing temperature (retrograde solubility)
Decreasing opal solubility with decreasing temperature (prograde solubility)
Calcite precipitation with increasing pH or Ca^{2+}: $HCO_3^- + Ca^{2+} + OH^- \rightarrow CaCO_3 + H_2O$
Calcite precipitation with CO_2 degassing: $2\ HCO_3^- + Ca^{2+} \rightarrow CaCO_3 + CO_2(g) + H_2O$
Incongruent dissolution of matrix silicates (e.g., albite $\rightarrow$ smectite + amorphous silica):
$\quad$ $2.5\ NaAlSi_3O_8 + 2\ H^+ \rightarrow Na_{0.5}Al_{2.5}Si_{3.5}O_{10}(OH)_2 \bullet H_2O + 4\ SiO_2(am) + 2\ Na^+ + H_2O$
Equilibration of fluids with heterogeneous pCO_2 and pH_2O in fractures and matrix [9]
Elevated water vapor pressure at convex surfaces [8]
Evaporation of water

---

water in fractures is an unlikely mechanism for evaporation. This percolation will encounter and extract heat from rocks, matrix water, and formation gas that are at a higher temperature and higher water vapor pressure, and descending water will tend to displace gas upward. Ascending, cooling gas flow in the geothermal gradient will cause condensation, dilution of the liquid phase, and dissolution of secondary minerals at equilibrium with formation water.

A plausible hypothesis is that evaporation as a mechanism for secondary mineral precipitation results from warming (e.g., descending) gas flow. The mechanism of natural gas flow is uncertain, but it may be due to minor convective or barometric forces. Calcite and opal fracture mineralization in the unsaturated zone at Yucca Mountain may represent paths of slow and episodic flow of warming gas over geologic times and on small space scales. This model for the origin and significance of secondary opal and calcite in the unsaturated zone at Yucca Mountain differs from the common idea that the secondary mineralization marks paths of downward percolating groundwater flow.

Evaporation is also a likely mechanism contributing to heterogeneity in unsaturated groundwater chemistry. Evaporation occurs mainly near the ground surface, but there is also geochemical evidence for its effects at depth. Contrasting fracture water and pore (matrix) water chemistry data are available from tunnels and boreholes at Rainier Mesa [22], which is north of Yucca Mountain where net infiltration rates are greater. Pore waters at Rainier Mesa are enriched in nonvolatile anions, chloride and sulfate, and they are relatively depleted in bicarbonate (which can be volatilized by $HCO_3^- + H^+ \rightarrow CO_2 + H_2O$) relative to fracture water. Apparently pore waters have been affected by evaporation to a greater extent than active fracture waters. A similar pattern exists at Yucca Mountain where pore waters collected above the Topopah Spring Tuff are generally lower in total dissolved carbonate than perched waters collected below the Topopah Spring Tuff, which tend to resemble chemically the saturated zone water [12]. Nevertheless, the unsaturated zone groundwaters at Yucca Mountain are all close to equilibrium with calcite [12,13].

CONCLUSIONS: SOME CURIOUS FEATURES OF YUCCA MOUNTAIN

Evaporation will occur in the unsaturated zone due to warming gas flow. Models and data for the properties of the media indicate that flow occurs predominantly in fractures. Fractures with secondary mineralization produced under natural conditions do not necessarily correspond to paths of downward percolating groundwater flow. Liquid water will be drawn from the matrix by water pressure gradients to fracture surfaces where evaporation occurs and secondary minerals are deposited. Reequilibration of the matrix system can result in the greater effects of evaporation on matrix water than on fracture water, which is episodically refreshed by percolation.

In the thermal regime of the nuclear waste repository, the heat and fluid flow systems will be changed dramatically, but a similar mechanism of warming gas flow in fractures and flow of water to fracture surfaces due to the water pressure gradient should lead to secondary mineral precipitation on fracture surfaces. However, as the system approaches dryness in the repository near field, the last residual water is likely to be retained in the finest pores of the matrix, which are likely sites for ultimate precipitation of carbonate, sulfate, chloride, or nitrate salts [23].

Secondary deposits of calcite do not mark zones of modern water flow in the tuffs of the groundwater saturated zone in the vicinity of Yucca Mountain. Water extracted from saturated zone boreholes is consistently undersaturated in calcite [16,24], and zones of water production from boreholes in the saturated zone correspond to intervals that lack calcite cement in fractures [16]. Calcite deposits in fractures in the saturated zone tuffaceous aquifer mark rocks where reduced fracture permeability isolates the rocks from relatively high flux groundwater flow in unmineralized fracture networks [16].

Relations between unsaturated groundwater flow paths and fracture-fill minerals can be assessed further by examining data on the fracture-fill characteristics for sites of bomb-pulse [36]Cl occurrences, sites of observed unsaturated zone fracture flow in the south ramp of the ESF or elsewhere, and relations between heat flow and fracture mineralization. Characterization of secondary minerals in fractures that showed flowing water in the ESF south ramp in 2005 is poorly documented. With present or future closure and sealing of the ESF and the East-West Cross Drift at Yucca Mountain it seems possible that rehydration could occur and that seepage into the tunnels would recommence. If so, its characterization and corresponding studies of fracture mineralization would provide valuable information on relations between flow and mineralization and on seepage phenomena in general. Mineralogical characterization of flowing fractures at depth in Rainier Mesa provides analog data (e.g., [22]), and additional study could be valuable if these sites were accessible.

Unresolved questions surround the bomb-pulse [36]Cl problem at Yucca Mountain. However, reported data for [36]Cl/Cl from the ESF (e.g., [25,26]) have been interpreted to indicate that that fast flowing groundwater containing a bomb-pulse [36]Cl contribution occurs at about twenty-three percent of the ESF sites initially examined for [36]Cl/Cl at the level of the proposed waste emplacement horizon [27]. The fraction of fractures that carry fast groundwater flow exceeds the fraction of fractures that show evidence for secondary mineralization. Levy et al. [26] report that roughly half of the places in the ESF where high [36]Cl/Cl ratios were observed reveal no macroscopic calcite. It seems likely that a large fraction of percolating water in the deep unsaturated zone at Yucca Mountain occurs along unmineralized fracture pathways. Putative one-to-one correlations between fracture mineralization and groundwater flow paths are apparently contradicted by the site characterization data.

Data on heat flow at Yucca Mountain [28] show systematic variations on a scale of hundreds of meters. Heat flow heterogeneity could promote warming, descending, convective gas flow at depth in the unsaturated zone at Yucca Mountain. Relations between mineral deposits and heat flow could aid interpretations of the mechanisms of fracture mineralization.

The nearly continuous precipitation of opal in a small fraction of unsaturated zone fractures at Yucca Mountain over a geologic time scale is an extraordinary and enigmatic observation. What makes these sites special? These precipitates indicate the persistence of water. However, fluid flow paths in fractures would be likely to shift over geologic time at Yucca Mountain, particularly in consideration of the active tectonic setting, frequent earthquakes, and variations in climate. Again the one-to-one correspondence is questionable between paths of descending fracture flow and geologically persistent, heterogeneously distributed sites of fracture mineralization. Relations between fracture mineralization and paths of gas flow remain hypothetical, but warming gas flow provides a plausible mechanism for evaporation and for secondary precipitation of calcite and opal.

ACKNOWLEDGMENTS

This paper presents my independent observations. Conversations with David Diodato helped me formulate some ideas, but he is not accountable for anything presented here. Constructive comments on the text were provided by Lauren Browning and Bruce Kirstein. An authoritative and constructive review provided by Brian Marshall contributed to the clarity and precision of the presentation. Encouragement and consideration by David Pickett, a symposium organizer and editor, merit my thanks. I gratefully acknowledge the US Nuclear Waste Technical Review Board for providing support for me to participate in the MRS Symposium on the Scientific Basis for Nuclear Waste Management.

REFERENCES

1. A.L. Flint, L.E. Flint, G.S. Bodvarsson, E.M. Kwicklis, and J.T. Fabryka-Martin, Jour. Hydrology **247**, 1 (2001).
2. S.S. Levy, Proc. 2nd Internat. Conf. High Level Rad. Waste Man., (Amer. Nucl. Soc., 1991) pp. 447-485.
3. J.B. Paces, L.A. Neymark, B.D. Marshall, J.F. Whelan, and Z.E. Peterman, *Ages and Origins of Calcite and Opal in the Exploratory Studies Facility Tunnel, Yucca Mountain, Nevada*, USGS Water-Resources Invest. Rep. 01-4049 (2001).
4. N.S.F. Wilson, J.S. Cline, and Y.V. Amelin, Geochim. Cosmochim. Acta **67**, 1145 (2003).
5. F.S. Spear, *Metamorphic Phase Equilibria and Pressure-Temperature-Time Paths*, Mineralogical Soc. Amer. (1993).
6. I.J. Winograd, T.B. Coplen, J.M. Landwehr, A.C. Riggs, K.R. Ludwig, B.J. Szabo, P.T. Kolesar, and K.M. Revesz, Science **258**, 255 (1992).
7. A. Baker, D. Genty, W. Dreybrodt, W.L. Barnes, N.J. Mockler, and J. Grapes, Geochim. Cosmochim. Acta **62**, 393 (1998).
8. J.F. Whelan, J.B. Paces, and Z.E. Peterman, App. Geochem. **17**, 735 (2002).
9. B.D. Marshall, L.A. Neymark, and Z.E. Peterman, Jour. Contamin. Hydro. **62-63**, 237 (2003).
10. J.B. Paces, L.A. Neymark, J.L. Wooden, and H.M. Persing, Geochim. Cosmochim. Acta **68**, 1591 (2004).
11. I.C. Yang, G.W. Rattray, and P. Yu, *Interpretation of Chemical and Isotopic Data from Boreholes in the Unsaturated Zone at Yucca Mountain, Nevada*, USGS Water-Resources Invest. Rept. 96-4058 (1996).
12. L. Browning, W.M. Murphy, B.W. Leslie, and W.L. Dam in *Scientific Basis for Nuclear Waste Management XXIII*, edited by R.W. Smith and D.W. Shoesmith (Mater. Res. Soc. Symp. Proc. **608**, Warrendale, PA, 2000) pp. 237-242.
13. J.A. Apps in *The Site Scale Unsaturated Zone Model of Yucca Mountain, Nevada, for the Viability Assessment* edited by G.S. Bodvarsson, T.M. Bandurraga, and Y.S. Wu, LBNL-40376 (1997) pp. 1-30, Chap. 9.
14. D.C. Thorstenson, E.P. Weeks, H. Haas, E. Busenberg, L.N. Plummer, and C. Peters, Water Resources Res. **34**, 1507 (1998).
15. Bectel SAIC Corporation, *Simulation of Net Infiltration for Modern and Potential Future Climates*, ANL-NBS-HS-000032 Rev 00 ICN02 (2001).

16. W.M. Murphy in *Scientific Basis for Nuclear Waste Management XVIII,* edited by T. Murakami and R.C. Ewing (Mater. Res. Soc. Symp. Proc. **353**, Warrendale, PA, 1995) pp. 419-426.

17. S. Finsterle, T.M. Bandurraga, C. Doughty, and G.S. Bodvarsson in *Development and Calibration of the Three-Dimensional Site-Scale Unsaturated-Zone Model of Yucca Mountain, Nevada,* edited by G.S. Bodvarsson and T.M. Bandurraga, LBNL-39315 (1996) pp. 447-479.

18. G.S. Bodvarsson, E. Kwicklis, C. Shan, and Y.S. Wu, Jour. Contamin. Hydro. **62-63**, 3 (2003).

19. L. Browning, W.M. Murphy, C. Manepally, and R. Fedors, Comput. Geosci. **29**, 247 (2003).

20. T. Xu, E. Sonnenthal, and B. Bodvarsson, Jour. Contamin. Hydro. **64**, 113 (2003).

21. E.P. Weeks in *Flow and Transport Through Unsaturated Fractured Rock,* edited by D.D. Evans, T.J. Nicholson, and T.C. Rasmussen (Geophysical Monograph 42, second edition, Amer. Geophys. Union, 2001) pp. 53-59.

22. A.F. White, H.C. Claassen, and L.V. Benson, *The Effect of Dissolution of Volcanic Glass on the Water Chemistry in a Tuffaceous Aquifer, Rainier Mesa, Nevada,* USGS Water Supply Pap. 1535-Q (1980).

23. W.M. Murphy in *Uncertainty Underground,* edited by A.F. Macfarlane and R.C. Ewing (MIT Press, 2006) pp. 271-283.

24. J.F. Kerrisk, *Groundwater Chemistry at Yucca Mountain, Nevada, and Vicinity,* LANL LA-10929-MS (1987).

25. J.T. Fabryka-Martin, A.V. Wolfsberg, P.R. Dixon, S.S. Levy, B. Liu, and H.J. Turin, *Summary Report of Chlorine-36 Studies: Sampling, Analysis, and Simulation of Chlorine-36 in the Exploratory Studies Facility,* LANL LA-13352-MS (1997).

26. S. Levy, S. Chipera, G. WoldeGabriel, J. Fabryka-Martin, J. Roach, and D. Sweetkind in *Faults and Subsurface Fluid Flow in the Shallow Crust,* edited by W.C. Haneberg, P.S. Mozley, J.C. Moor, and L.B. Goodwin, Amer. Geoph. Union Geophys. Monograph **113** (1999) pp. 159-184.

27. W.M. Murphy in *Scientific Basis for Nuclear Waste Mangement XXI,* edited by I.G. McKinley and C. McCombie (Mater. Res. Soc. Symp. Proc. **506,** Warrendale, PA, 1998) pp. 407-414.

28. J.H. Sass, A.H. Lachenbruch, W.W. Dudley, S.S. Priest, and R.J. Munroe, *Temperature, Thermal Conductivity and Heat Flow Near Yucca Mountain, Nevada: Some Tectonic and Hydrologic Implications,* USGS Open File Rep. 87-649 (1988).

Mater. Res. Soc. Symp. Proc. Vol. 1124 © 2009 Materials Research Society 1124-Q07-17

A systematic approach to evaluate the importance of concerns
affecting the geological disposal of radioactive wastes

Takao Ohi* Manabu Inagaki*, Makoto Kawamura*, Takeshi Ebashi*
*JAEA, Tokai-mura, Ibaraki-Ken, Japan.

ABSTRACT

A systematic approach was developed to represent the relative importance of concerns which are the phenomena affecting the safety of a geological disposal system for radioactive wastes. This approach is composed of a Total Assessment Work Frame (TAWF) and a "Redundant Approach". TAWF is a unified structure for all scenarios and is able to describe the change of conditions of the fields relating to the concerns. "Redundant Approach" is a sensitivity analysis method to analyze the importance of the concerns based on a successful condition, which is composed of the combination of parameter ranges that do not exceed target dose values in safety assessment, and which is regarded as a practical indicator for the examination of relative importance of the concerns. In this paper, an example using the findings of the Japanese H12 safety assessment report and the existing sensitivity analysis was represented to illustrate the usefulness of this approach. By using this approach, the propagation of the influence of the concerns on the safety was represented as a chain. Also, relative importance of the concerns was examined based on this propagation and the successful condition for the parameter representing the safety functions. Furthermore, throughout these examinations, significance and contribution of the relevant research item in the safety assessment were represented. Future investigations based on this approach will help to conduct the relevant research effectively and improve the reliability of the safety assessment of geological disposal.

INTRODUCTION

Concerns relating to the safety assessment for geological disposal have been identified and have been the subject of extensive research [1, 2, 3 and 4]. Being able to demonstrate the relative importance of the concerns systematically is a critical issue to prioritize and undertake relevant research more effectively as well as to improve the reliability of safety assessment.

In the implementation of geological disposal, there is a growing need to identify the relative importance of various activities included in the projects such as site investigation, research/development and assessment, which must be included in the overall program in some reasonable and efficient manner. Multiple criteria are needed to judge the importance of each activity from the point of view, such as safety, feasibility, cost and reliability of the geological disposal system. A comprehensive examination that balances these different criteria is required in order to truly understand the importance of a given activity. Therefore a method, such as Multi - Attribute Utility Analysis (MAUA) similar to that used in the YMP [5] was applied to identify the importance of various activities.

In this study, focusing on the safety among the several criteria and the sufficiency of relevant data and/or information, a systematic approach has been developed to assess the relative importance of concerns affecting the safety of the geological disposal system. This paper demonstrates this approach and presents an example application based on the results of the H12 report [3] and existing sensitivity analysis [6].

METHODOLOGY

In order to identify the relative importance of the concerns on the safety of geological disposal system, the following two procedures were adopted
- Organization of the information based on TAWF which is a unified structure for all scenarios and is able to describe the change of conditions of the fields relating to the safety assessment.
- Examination of the relative importance based on a successful condition [7] obtained from sensitivity analysis for a multivariable system under the prerequisite arranged by using TAWF.

Organization of the information based on TAWF

TAWF is a structure to organize the results of FEP analysis and to describe the information on several concerns in a unified way. It is composed of common survey items and of multistep fields for typifying (refers to figure 1). According to this structure, a change of the conditions of fields affected by the concerns are organized and typified effectively, and are described as prerequisites of the safety assessment in as quantitatively as possible.

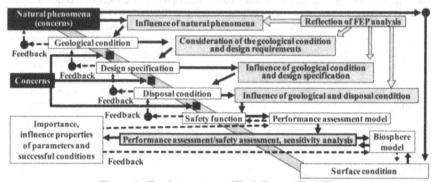

Figure 1. Total Assessment Work Frame (TAWF)

In this TAWF, several fields are considered in order to assess the change of conditions and information. The use of several fields allows us to examine similarities through multiple stages and to typify the information effectively. "Natural phenomena", "Surface condition", "Geological condition", "Design specification", "Disposal condition" and "Safety function" are set as fields in this study. Each field is divided into "Vitrified waste", "Over pack", "Buffer material", "Disposal facilities (inc. Excavated Damaged Zone (EDZ))" and "Host rock" as areas in which the change of the condition and/or the information is examined. Also, based on the existing scenario analysis methodology [8, 9], the organization of the changes of the conditions and/or the information in each field and each area is classified in factor of T [Thermal], H [Hydrological], M [Mechanical], C [Chemical] and G [Geometrical]. In the organization in each area, discussion points and representative parameters with respect to discussion points are extracted by THMCG. The following information is organized as quantitatively as possible.
- The handling method and values of representative parameters in the reference case
- The handling method of the concerns affecting the representative parameters
- The values/the variation of the representative parameters affected by concerns

In the extraction of the concerns, in addition to the existing realized concerns such as Natural phenomena, Gas effects, Colloid effects, Microbe effects and Radiation effects, the following cases are also recognized as concerns,: the case that the assumptions and the prerequisites which are considered in the setting of representative parameter in the existing reference case do not hold true, and the case of the model uncertainties.

By organizing the relationship of the representative parameters and related information throughout all fields, the propagation of influence of concerns is represented as a chain of the representative parameters and the relevant information, which eventually connect to safety assessment parameters. Also, the influencing factors which are key parameters locating among the chain of representative parameters relating to concerns are identified. And the typifying of the influence of concerns is attempted. The TAWF could contribute to the improvement of comprehensiveness and rationality of the organization and the typifying of information relating to concerns. The information relating to the variation of representative parameters including safety assessment parameters are arranged as prerequisites for the assessment of concerns.

Examination of the relative importance based on the successful condition

In the sensitivity analysis, the values or the variation ranges of all safety assessment parameters are set based on the prerequisites. The sensitivity analysis with multi-parameter variations is then conducted under these conditions. As a result, the parameters which have a large impact on dose are identified and the characteristic of the influence of the parameter on dose (influence property) is understood. The successful conditions [7], which are composed of the combination of parameter ranges that do not exceed target dose values in safety assessment, are extracted in each prerequisite based on the large impact parameters and those influence properties. The concept of a successful condition is similar to savior condition of SPARC [10].

In the case that the prerequisites and relationship of the chain are given quantitatively, the comparison among the successful conditions, the influence properties and the value and/or variation range of the safety assessment parameters give the information with respect to the extent of the safety margin or the conformity to the safety of the safety assessment parameter included in the representative parameter. Such information allows us to discuss the relative importance of the concerns from the view point of their impact on safety.

On the other hand, only qualitative information on the relationship of the chain of the representative parameters might be available in the current status because there is very little quantitative information on the influence of the concerns or the disposal site has not yet been determined. Furthermore, it is expected that the value or the variation range estimated under such conditions should include large uncertainties. In this case, the examination of the importance of concerns would be conducted based on the following procedure:

- Conservative setting of the prerequisites for the concern based on the latest data available and the expert judgment
- Extraction of the successful conditions and the influence properties based on those prerequisites and
- Examination of the extent or perspective of satisfaction of successful condition based on the impact of the influencing factors on the safety assessment parameters extracted as a successful condition, the impact of the concerns on the influencing factors and the influence properties of the safety assessment parameters, either semi-quantitatively or qualitatively

The conservative successful condition given by conservative setting of prerequisites is the conservative threshold value with respect to the safety as the results of bounding analysis in the safety assessment. Under no sufficient quantitative information on the influence of the concerns, the examination of perspective of satisfying this conservative successful condition and the examination of sufficiency of relevant data and/or information give the minimum information on the relative importance of the concerns and prioritize additional research and data acquisition. This approach based on this sensitivity analysis is named a "Redundant Approach".

RESULTS

The application of this systematic approach to represent the relative importance of the concerns was attempted using the findings of the H12 report [3] and of an existing sensitivity analysis [6]. In this application, based on the TAWF, the handling method and values of representative parameters in the reference case, the handling method of the concerns affecting the representative parameters and the values/the variation of the representative parameters affected by concerns were rearranged. Based on such information, the relationship of those parameters between the fields were organized and represented as a chain.

In figure 2, the example of the chain of the discussion points including those parameters was shown using the safety function of vitrified waste as a starting point. The volcanoes/magma activity was selected as an example of the concerns. Figure 2 shows that geological conditions such as Hydrology of host rock, Temperature of host rock and Chemical property of host rock and groundwater are affected by volcanoes/magma activity. Then, temperature of vitrified glass and composition of water were identified as the influencing factors on the safety assessment parameters, such as glass dissolution rate.

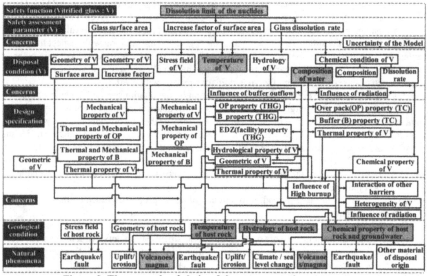

Figure 2. The example of chain of the representative parameters

It is difficult to specify and to organize the quantitative information on the influences of volcanoes/magma activity on the geological conditions owing to the uncertainty caused by long temporal scales and having no specific characterization site. Therefore, a "Redundant Approach" was applied. In order to understand the safety margin of the engineered barrier system under these kind of uncertain geological conditions, *Ebashi et.al* [6] found the successful conditions on the glass dissolution rate, buffer thickness, solubility, De and Kd in engineered barrier system for 10μSv/y, 300μSv/y of target dose with extremely conservative geological condition which prevents the natural barrier system from functioning properly. In this examination, such extremely conservative geological condition was assigned safety assessment parameters as the conservative prerequisite caused by the influence of the volcanoes/magma activity. The successful conditions of Cs-135 among the existing findings [6] were referred.

In table I, the variation range of key parameter considered based on the existing analysis [3, 4], the prerequisite (hatched section) set in the analysis of *Ebashi et.al* [6] and the extracted successful condition focused on glass dissolution rate for 10μSv/y, 300μSv/y of target dose are shown. The successful conditions for the glass dissolution rate in table I are not affected by further increase of transmissivity because the existing analysis [6] show that 10^{-6} m^2/s is threshold value of transmissivity affecting the dose. Also, the successful conditions are not affected by further decrease of Kd or buffer thickness and further increase of diffusion coefficient because the release rate from the buffer in the successful condition is equal to the maximum release rate in steady state, which is equal to the glass dissolution rate and, which does not depend on Kd, diffusion coefficient or buffer thickness. Such influence property is important information contributing to the identification of the importance of the concerns [11].

Table I. Prerequisite and successful condition focused on the glass dissolution rate

| | key Parameters | Dimension | Variation range | | Reference value of H12 |
			Minimum	Maximum	
Engineered barrier	Glass surface area	m^2/can	1.7		1.7
	Increase factor of surface area	-	10		10
	Buffer thickness	m	0.1	3	0.7
	Glass dissolution rate	g/m^2/y	3.65×10^{-2}	2.43×10^4	0.365
	Solubility	mol/l	soluble element		soluble element
	Distribution coefficient (Kd) in buffer	m^3/kg	0.001 (Saline type)	0.01 (Fresh type)	0.01 (Fresh type)
	Diffusion coefficient in buffer (De)	m^2/s	3.0×10^{-10} (Saline type)	6×10^{-10} (Fresh type)	6×10^{-10} (Fresh type)
natural barrier	Hydraulic gradient	-	0.001	0.23	0.01
	Average value of Transmissivity	m^2/s	1×10^{-11}	1×10^{-6}	1×10^{-10}
	Matrix diffusion depth	m	0.01	1	0.1
	The proportion of fracture surfaces for matrix	-	0.01	1	0.5
	Porosity	-	0.003	0.5	0.02
	Kd in host rock	m^3/kg	0.001	10	0.05
Successful conditions of Glass dissolution rate(g/m^2/y)			2.5> for 10μSv/y		-
			75> for 300μSv/y		-

The each successful condition for glass dissolution rate is regarded as the practical indicator in the study on the glass dissolution rate. Therefore, the study on the relationship between the change of the glass dissolution rate and the influencing factors, such as increase of temperature of vitrified glass and the change of composition of water is the important research issue for obtaining the perspective of satisfying successful condition, in addition to the model study of glass dissolution. From the view point of the impact of the volcanoes/magma activity, the investigation of the perspective of the occurrence of the volcanoes/magma activity that bring the above increase of temperature of vitrified glass and the change of composition of water are also important research issue as well as the accumulation of the geoscientific knowledge relating to the scale, style, time-dependence and properties of such an occurrence. The organization of such information might allow us to identify the relative importance of concerns and the relevant research item either semi-quantitatively or qualitatively.

CONCLUSIONS

A systematic approach to represent the relative importance of the concerns including potentially detrimental factors to the safety of a geological disposal system for radioactive wastes was developed based on the TAWF and the "Redundant approach". To illustrate the usefulness of this approach, the application of this systematic approach was attempted using the findings of the H12 report and the successful condition focused on the soluble element Cs referred from an existing sensitivity analysis. This approach might allow us to identify the relative importance of concerns and the relevant research item either semi-quantitatively and to demonstrate significance and contribution of the concerns and of the relevant research item in the safety assessment. In order to examine the importance of several concerns from the view point of the safety of geological disposal system, the examination covering all important nuclides on the safety assessment and the examination of the wide range of prerequisites are required. Future investigations based on this approach will help to conduct the relevant research effectively and improve the reliability of the safety assessment of geological disposal.

ACKNOWLEDGMENTS

The authors would like to thank E. K. Webb for precious advice in editing this paper.

REFERENCES

1. Nagra, Technical Report NTB 02-05, (2002).
2. SKB, Technical Report TR-06-09, (2006).
3. JNC, JNC TN1410 2000-004, (2000).
4. JAEA&FEPC, JAEA-Review 2007-010, FEPC TRU-TR2-2007-01, (2007).
5. G. D. LeCain, D. Barr, D. Weaver, R. Snell, S. W. Goodin, F. D. Hansen, Development of the Performance Confirmation Program at Yucca Mountain, Nevada, IHLRWM 2006, Las Vegas, NV, pp.1058-1065 (2006).
6. T. Ebashi, S. Koo, T. Ohi: "Assessment approach for a safety margin of parameters on the high-level radioactive waste disposal", Journal of Nuclear Fuel Cycle and Environment (2008), [in Japanese], [Submitted].
7. T. Ohi, H. Takase, M. Inagaki, K. Oyamada, T. Sone, M. Mihara, T. Ebashi, K. Nakajima: Application of a comprehensive sensitivity analysis method on the safety assessment of TRU waste disposal in JAPAN, Mater. Res. Soc. Symp. Proc.Vol. 985, pp.129-134 (2007).
8. SKI, SKI Report95:26, (1995).
9. M. Kawamura, T. Ohi, H. Makino, K. Umeda, T. Niizato, T. Ishimaru, T. Seo: Study on evaluation method for potential impacts of "natural phenomena" on a HLW disposal system, 2006 EAFORM Conference, pp.350-367 (2006).
10. S. T. Ghosh, Dr.Thesis: Risk-Informing Decisions about High-Level Nuclear Waste Repositories, Massachusetts Institute of Technology, p125, (2004).
11. T. Ohi. K. Nakajima: BOUNDING ANALYSIS FOR SOLUBILITY, Mater. Res. Soc. Symp. Proc.Vol. 465, pp1091-1098 (1997).

Mater. Res. Soc. Symp. Proc. Vol. 1124 © 2009 Materials Research Society 1124-Q07-18

Estimation of Ra Concentration in High-Level Radioactive Waste Disposal System

Yasushi Yoshida[1] and Hideki Yoshikawa[2]
[1]Inspection Development Corporation, 4-33, Muramatsu, Tokai-mura, Naka-gun, Ibaraki 319-1112, Japan.
[2]Japan Atomic Energy Agency, 4-33, Muramatsu, Tokai-mura, Naka-gun, Ibaraki 319-1194, Japan

ABSTRACT

Concentrations of Ra in groundwater are thought to be affected by substitution reactions with alkaline earth elements contained in a mineral, because the concentration of Ra is less than the solubility of known Ra containing phases. The substitution reaction was simulated using a partition coefficient. Calcite ubiquitously exists in geological system and easy to access to react with Ra. Therefore, calcite is thought to be an important mineral for the substitution reaction. However, previous identification of reactive layers of calcite or reversibility of the substitution reaction has not been confirmed. A re-distribution experiment was therefore undertaken and it was found that an estimated 21 ± 13 layers near the surface were reactive and within them substitution equilibrium was achieved. Using these results, a model to estimate Ra concentration was established and adopted to analyze Ra migration. The effects of substitution of Ra and Ca were reasonably simulated.

INTRODUCTION

Performance assessment of a geological disposal system typically includes an analysis of migration of radionuclides. Concentrations of radionuclides in groundwater are estimated based on solubility of oxides and carbonate containing phases [1-3]. Concentrations of Ra, which is one of the more important radionuclides, are thought to be limited by substitution reaction for alkaline earth elements in minerals because its concentration is less than the solubility of $RaCO_3(cr)$ or $RaSO_4(cr)$ [1]. This effect has been considered with solid solution theory by Berner [2] and with coprecipitation by Azuma et al. [3]. However, applicability of the solid solution equation of Berner [2] was not confirmed experimentally and the over-simplified assumption in Azuma et al. [3] revealed a discrepancy against experimental results.

In the present study, a partition coefficient (D) [4-6] is applied to describe a substitution reaction. It expresses distribution of tracer (Tr) and carrier (Cr) element for coprecipitation and is given by:

$$D = \frac{(Tr/Cr)_{solid}}{(Tr/Cr)_{solution}} \quad (1)$$

where $(Tr/Cr)_{solid}$ and $(Tr/Cr)_{solution}$ are mole ratios of Tr/Cr in the solid and solution, respectively. D is derived from the results of a coprecipitation experiment and is thought to be a reasonable empirical parameter to describe distribution of trace element in coprecipitation reactions. Since alkaline earth elements and counter ligands are dominated by Ca and carbonate, respectively, in the geological environment, calcite $(CaCO_3)$ is a commonly occurring mineral. D of Ra in calcite was determined by Yoshida et al. [7] and can be converted to be a distribution coefficient (K_{cp}) with molar quantities of Ca in the solid phase and solution, given by:

$$K_{cp} = \frac{V_{Ra}}{C_{Ra}} = \frac{V_{Ca}}{C_{Ca}} \times D_{Ra} \quad (2)$$

where $V_{Ra,Ca}$ and $C_{Ra,Ca}$ are molar quantities of Ra and Ca in solid and solution, respectively and D_{Ra} is the partition coefficient of Ra in calcite. K_{cp} can be adopted in migration analysis directly, but values of V_{Ca} and C_{Ca} have to be defined. For migration calculations, C_{Ca} is defined and V_{Ca} can be calculated from the amount of calcite in the system, if Ra diffuses quickly into the calcite matrix. However, the diffusion coefficient of Ra into calcite is estimated to be very slow from the results of diffusion experiments for Sr into calcite [8]. This means that Ra tends to react with Ca in calcite near the mineral surface. Hence, V_{Ca} can

413

be expressed with total molar quantities of Ca in system ($V_{Ca-total}$) and mole ratio of reactive layers as:

$$V_{Ca} = V_{Ca-total} \times (\text{mole ratio of reactive layers}) \quad (3)$$

Therefore, it is necessary to know the number of layers in calcite that are available for substitution reaction. D is generally derived from coprecipitation experiments whose initial solution is oversaturated with respect to the solid phase. This reaction progresses with precipitation dominating. However, substitution reactions between Ra and Ca are thought to occur under calcite equilibrium conditions. The applicability of D to a reversible system has to be confirmed.

In the present study, re-distribution experiments for Ca and alkali earth elements were performed to determine the numbers of layers available for substitution reactions in calcite and confirm the applicability of D for a reversible reaction. With the derived value of K_{cp}, a migration analysis with the effect of substitution in calcite was calculated. As an example, a result of the migration analysis of 4n+2 series radionuclide with K_{cp} was shown with comparison to the results of JNC [9].

EXPERIMENTAL METHODS

Two experiments for re-distribution were carried out for calcite containing trace amounts of Sr and Ba in a solution equilibrated with pure calcite (experiment I) and trace amounts of Sr, Ba and Ra in a solution equilibrated with calcite in the presence of pure calcite (experiment II). For experiment I, calcite containing trace amounts of Sr or Ba was made by the free drift method [7]. The initial solution was equilibrated with pure calcite and filtered with a membrane filter (MILLIPORE, 0.45 μm). The solid was added to the initial solution at a solid/solution ratio of 0.01 g/100 ml, 0.1 g/100 ml and 0.25 g/50 ml. Solutions were shaken by a rotary shaker and after an aging period of 30 days, solutions were sampled and filtrated solids were dissolved in acid. The concentrations of Ca, Sr and Ba in solution and the dissolved solid were measured by ICP-AES (Perkin Elmer, OPTIMA 3300XL, detection limit is 2×10^{-7} molal for Ca , 1×10^{-7} molal for Sr and 7×10^{-8} molal for Ba) or by ICP-MS (Agilent Technologies, Agilent 7500, detection limit is 1×10^{-9} for Sr and 7×10^{-10} molal for Ba). For experiment II, pure calcite was added to the initial solution at a solid/solution ratio of 0.1 g/100 ml, 0.6 g/ 30 ml and 2 g/100 ml. After more than three days on a rotary shaker to reach steady state between solid and solution, a stock solution was added of SrCl$_2$, BaCl$_2$ or acidified ^{226}Ra. Again, solutions were shaken by a rotary shaker and after an aging period of 30 days, the amounts of Ca, Sr, Ba, and Ra in solid and solution were measured. The concentration of ^{226}Ra was measured one month after sampling, to ensure secular equilibrium with daughters had been attained (a closed system assures no loss of Rn), using a liquid scintillation counter (PACKARD, TRI-CARB 2750TR/LL, detection limit 1×10^{-11} molal). In all experiments temperature was maintained at 24 ± 2 °C. The initial solids were grained and surface area of them was determined by the BET method (QUANTASORB QS-18 particle size analyzer). Surface area of some solids reacted with solution was also measured. Samples were frozen and dried by freeze dryer. Experiments with Sr and Ba were carried out in an N$_2$ atmosphere glove box, whereas experiments with Ra were carried out under ambient

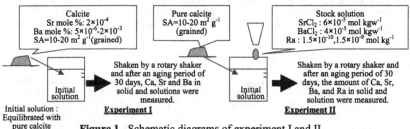

Figure 1 . Schematic diagrams of experiment I and II

conditions. A summary of experimental conditions is shown in tables I and II and figure 1.

Table I. Re-distribution experiment calcite containing Sr and Ba react with solution

Exp. #	Me	Solid liquid ratio	Me/Ca mol initial solid	Surface area, m² g⁻¹
Sr-A1		0.1g/100ml		
Sr-A2				
Sr-A3				19.5±0.5
Sr-A4		0.01g/100ml		
Sr-A5				
Sr-A6	Sr		2.0×10^{-4}	
Sr-A7		0.1g/100ml		
Sr-A8				27.0±0.3
Sr-A9		0.01g/100ml		
Sr-A10				
Sr-A11				
Ba-A12				
Ba-A13				
Ba-A14				
Ba-A15		0.01g/100ml	5.9×10^{-6}	17.9±0.4
Ba-A16				
Ba-A17				
Ba-A18				
Ba-A19		0.009g/100ml		
Ba-A20			4.1×10^{-5}	16.6±0.2
Ba-A21			2.2×10^{-3}	10.6±0.3
Ba-A22				
Ba-A23	Ba	0.01g/100ml		18.7±0.2
Ba-A24			5.3×10^{-6}	13.7±0.3
Ba-A25			4.4×10^{-3}	20.9±0.4
Ba-A26			1.3×10^{-3}	21.20±0.05
Ba-A27			5.9×10^{-4}	17.9±0.4
Ba-A28			4.1×10^{-5}	16.6±0.2
Ba-A29		0.1g/100ml		
Ba-A30			2.2×10^{-3}	18.7±0.2
Ba-A31				
Ba-A32			5.3×10^{-6}	13.7±0.3
Ba-A33		0.25g/50ml	4.4×10^{-3}	20.9±0.4
Ba-A34			1.3×10^{-3}	21.20±0.05

RESULTS AND DISCUSSION

Re-distribution experiment

Surface area of some solids reacted with solution was measured. Surface area was observed to be a 10 to 20 % decrease comparing with those of initial solid. This decrease was gradually achieved through duration period, while substitution reaction was rather fast. Therefore the effect of decrease of surface area is thought to be negligible.

Measured values for experiment I are shown in table III. As a result of replicate measurement, error (2σ) of concentration in solution was estimated to be less than 2%. An error of 2.8% was estimated for the mole ratio of Me/Ca in the solid. Release of Sr and Ba ions is thought to be by substitution reaction between Ca in solution and Sr and Ba in the solid. Since the released amounts of Sr and Ba were not directly compared due to different solid-liquid ratios and surface areas, the extent of release of Sr and Ba was evaluated as numbers of layers which contribute towards the substitution reaction. It can be assumed that the crystal is constituted with unit cell of calcite (rhombohedron structure of a = 6.37×10^{-10} m, α =46.05° and ρ =2.71×10^{6} g m⁻³ [10]) as a cubic sedimentation (figure. 2). With these data, surface area (m² g⁻¹) can be expressed as a function of the number of unit cells from the center to surface of crystal (n). With measured values of surface area, n can be calculated (table IV) [11]. The mole ratio of reactive layers corresponding to the numbers of layers is given by:

$$\text{Mole ratio of reactive layers} = \frac{(2n+1)^3 - (2(n-x)+1)^3}{(2n+1)^3} \quad (4)$$

where x is the number of reactive layers. Since the mole ratio of reactive layers was calculated with initial and final Sr/Ca and Ba/Ca ratio in the solid and n was calculated with a measured value of surface area (table IV), values of x were derived. Derived values of x are 34 ± 17 and 21 ± 13 layers for Sr and Ba, respectively (table III). This indicates that Sr and Ba were released from the inside of the crystal structure as well as the surface and that substitution reactions of Sr occur at greater depths in the crystal than for Ba. Coordination number (CN) for a metal ion in calcite is 6 and for this CN, the radius of Ca, Sr and Ba is

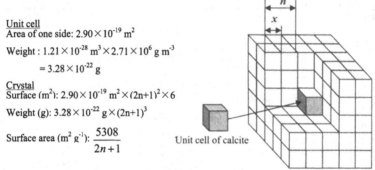

Unit cell
Area of one side: $2.90 \times 10^{-19} \text{ m}^2$

Weight : $1.21 \times 10^{-28} \text{ m}^3 \times 2.71 \times 10^6 \text{ g m}^{-3}$

$= 3.28 \times 10^{-22} \text{ g}$

Crystal
Surface (m^2): $2.90 \times 10^{-19} \text{ m}^2 \times (2n+1)^2 \times 6$

Weight (g): $3.28 \times 10^{-22} \text{ g} \times (2n+1)^3$

Surface area ($\text{m}^2 \text{ g}^{-1}$): $\dfrac{5308}{2n+1}$

Unit cell of calcite

Figure 2. Assumption of crystal structure of calcite. In this figure, the number of unit cells from the center to surface of crystal (n) is 2 and the number of reactive layers (x) is 1

Table II. Re-distribution experiment calcite react with solution containing Sr, Ba and Ra

Exp. #	Me	Solid liquid ratio	Initial tracer concentration	Surface area, $\text{m}^2 \text{ g}^{-1}$
Sr-B1	Sr	0.1g/100ml	5.7×10^{-5}	22.8 ± 0.2
Sr-B2				
Sr-B3				
Sr-B4		0.6/30ml		27.5 ± 0.9
Sr-B5				
Sr-B6				
Ba-B7	Ba	0.1g/100ml	3.5×10^{-5}	22.8 ± 0.2
Ba-B8				
Ba-B9				
Ba-B10		0.6g/30ml		23.4 ± 0.1
Ba-B11				
Ba-B12				
Ra-B13	Ra	2g/100ml	1.5×10^{-9}	11.0 ± 0.3
Ra-B14				
Ra-B15				
Ra-B16			1.5×10^{-10}	
Ra-B 17				
Ra-B 18				
Ra-B19			1.5×10^{-9}	18.2 ± 0.5
Ra-B20				
Ra-B21				

Table III. Results of re-distribution experiment (I) with calcite containing Sr and Ba reacted with solution

Case	Final Ca in solution	Final Me in solution	Me/Ca mol ratio in solid	Layers
Sr-A1	$2.04(\pm0.04)\times10^{-4}$	$6.65(\pm0.13)\times10^{-7}$	$1.04(\pm0.03)\times10^{-4}$	25.8
Sr-A2	$2.04(\pm0.04)\times10^{-4}$	$6.65(\pm0.13)\times10^{-7}$	$1.02(\pm0.03)\times10^{-4}$	26.5
Sr-A3	$1.42(\pm0.03)\times10^{-4}$	$2.13(\pm0.04)\times10^{-7}$	$6.31(\pm0.18)\times10^{-5}$	42.8
Sr-A4	$1.42(\pm0.03)\times10^{-4}$	$2.13(\pm0.04)\times10^{-7}$	$6.16(\pm0.17)\times10^{-5}$	43.5
Sr-A5	$1.43(\pm0.03)\times10^{-4}$	$2.13(\pm0.04)\times10^{-7}$	$6.10(\pm0.17)\times10^{-5}$	43.8
Sr-A6	$1.42(\pm0.03)\times10^{-4}$	$2.13(\pm0.04)\times10^{-7}$	$6.12(\pm0.17)\times10^{-5}$	43.7
Sr-A7	$1.85(\pm0.04)\times10^{-4}$	$1.07(\pm0.02)\times10^{-6}$	$1.14(\pm0.03)\times10^{-4}$	17.5
Sr-A8	$1.32(\pm0.03)\times10^{-4}$	$2.26(\pm0.05)\times10^{-7}$	$6.10(\pm0.17)\times10^{-5}$	32.8
Sr-A9	$1.32(\pm0.03)\times10^{-4}$	$2.26(\pm0.05)\times10^{-7}$	$6.05(\pm0.17)\times10^{-5}$	33.0
Sr-A10	$1.32(\pm0.03)\times10^{-4}$	$2.26(\pm0.05)\times10^{-7}$	$5.90(\pm0.17)\times10^{-5}$	33.5
Sr-A11	$1.30(\pm0.03)\times10^{-4}$	$2.26(\pm0.05)\times10^{-7}$	$6.01(\pm0.17)\times10^{-5}$	33.1
Ba-A12	$1.29(\pm0.03)\times10^{-4}$	$<8.0\times10^{-9}$	$4.14(\pm0.12)\times10^{-6}$	16.4
Ba-A13	$1.27(\pm0.03)\times10^{-4}$	$<8.0\times10^{-9}$	$3.97(\pm0.11)\times10^{-6}$	18.2
Ba-A14	$1.30(\pm0.03)\times10^{-4}$	$<8.0\times10^{-9}$	$3.94(\pm0.11)\times10^{-6}$	18.5
Ba-A15	$1.28(\pm0.03)\times10^{-4}$	$<8.0\times10^{-9}$	$3.50(\pm0.10)\times10^{-6}$	23.6
Ba-A16	$1.26(\pm0.03)\times10^{-4}$	$<8.0\times10^{-9}$	$3.79(\pm0.11)\times10^{-6}$	20.2
Ba-A17	$1.28(\pm0.03)\times10^{-4}$	$<8.0\times10^{-9}$	$4.14(\pm0.12)\times10^{-6}$	16.4
Ba-A18	$1.32(\pm0.03)\times10^{-4}$	$<8.0\times10^{-9}$	$3.72(\pm0.11)\times10^{-6}$	21.0
Ba-A19	$1.31(\pm0.03)\times10^{-4}$	$<8.0\times10^{-9}$	$4.90(\pm0.14)\times10^{-6}$	8.7
Ba-A20	$1.36(\pm0.03)\times10^{-4}$	$2.40(\pm0.05)\times10^{-8}$	$2.29(\pm0.06)\times10^{-5}$	28.4
Ba-A21	$1.25(\pm0.03)\times10^{-4}$	$7.21(\pm0.10)\times10^{-7}$	$1.50(\pm0.04)\times10^{-3}$	30.7
Ba-A22	$1.26(\pm0.03)\times10^{-4}$	$7.29(\pm0.10)\times10^{-7}$	$1.47(\pm0.04)\times10^{-3}$	31.9
Ba-A23	$1.36(\pm0.03)\times10^{-4}$	$1.27(\pm0.03)\times10^{-6}$	$1.03(\pm0.03)\times10^{-3}$	31.9
Ba-A24	$1.34(\pm0.03)\times10^{-4}$	$<8.0\times10^{-9}$	$3.52(\pm0.10)\times10^{-6}$	24.8
Ba-A25	$1.36(\pm0.03)\times10^{-4}$	$2.40(\pm0.05)\times10^{-8}$	$2.33(\pm0.07)\times10^{-5}$	24.5
Ba-A26	$1.33(\pm0.03)\times10^{-4}$	$6.01(\pm0.10)\times10^{-7}$	$6.84(\pm0.19)\times10^{-4}$	23.7
Ba-A27	$1.34(\pm0.03)\times10^{-4}$	$1.60(\pm0.03)\times10^{-8}$	$3.69(\pm0.10)\times10^{-6}$	21.4
Ba-A28	$1.39(\pm0.03)\times10^{-4}$	$1.20(\pm0.02)\times10^{-7}$	$2.68(\pm0.08)\times10^{-5}$	21.3
Ba-A29	$1.38(\pm0.03)\times10^{-4}$	$1.36(\pm0.03)\times10^{-7}$	$2.71(\pm0.08)\times10^{-5}$	20.7
Ba-A30	$1.39(\pm0.03)\times10^{-4}$	$8.09(\pm0.20)\times10^{-6}$	$1.43(\pm0.04)\times10^{-3}$	19.3
Ba-A31	$1.36(\pm0.03)\times10^{-4}$	$8.81(\pm0.20)\times10^{-6}$	$1.32(\pm0.04)\times10^{-3}$	22.3
Ba-A32	$1.56(\pm0.03)\times10^{-4}$	$7.21(\pm0.10)\times10^{-8}$	$4.12(\pm0.12)\times10^{-6}$	15.6
Ba-A33	$1.50(\pm0.03)\times10^{-4}$	$4.49(\pm0.09)\times10^{-7}$	$3.57(\pm0.10)\times10^{-5}$	8.9
Ba-A34	$1.55(\pm0.03)\times10^{-4}$	$1.29(\pm0.03)\times10^{-5}$	$1.04(\pm0.03)\times10^{-3}$	8.5

1.00×10^{-10}, 1.18×10^{-10} and 1.35×10^{-10} m [12], respectively. The deeper penetration of Sr than Ba may result from the similarity of the Ca and Sr ionic radii. The ionic radius of Ra at CN=6 is 1.39×10^{-10} m [13]. Since the ionic radius of Ra is similar to that of Ba, the number of reactive layers for Ra was estimated to be 21 ± 13. The number of reactive layers for substitution reactions was estimated to be 25 by Lorens (1981) [4], in good agreement with that derived in the present study.

Measured values for experiment II are shown in table V. Values of error were the same as those in experiment I except for Ra for which a 10% solution concentration error was estimated. Measured Me/Ca mole ratio in solid (table V) is derived from concentration of dissolved solid which does not consider the reactive layer. Concentration of the trace element in solution tends to decrease due to substitution for Ca in calcite. Here, the basic mechanism of reaction is same as that of experiment I, meaning that crystal layers contributing to the substitution reaction can be assumed to be 34 ± 17 for Sr and 21 ± 13 for Ba (and hence Ra of a similar ionic radius). Then, D can be calculated by eq. (1) from the final concentration of Ca, Sr, Ba and Ra in solution and the mole ratio of reactive layers (table IV) and measured mole ratio of Me (Sr, Ba and Ra) and Ca (table V). Calculated D values are, $D_{Sr} = 1.3(\pm0.3)\times10^{-2}$, $D_{Ba} = 1.0(\pm0.5)\times10^{-2}$ and $D_{Ra} = 5(\pm4)\times10^{-2}$.

Reported D were derived in coprecipitation experiment [4,5 and 7] and reversibility of the substitution reaction in their experiments has not been confirmed before. The values of D calculated in experiment II were derived under reversible system. Therefore, by evaluating the similarity of D values for same element between reported in literature and calculated in the experiment II, a reversibility of the substitution reaction can be confirmed. Reported D is $2.1(\pm0.3)\times10^{-2}$ for D_{Sr} [5], $1.6(\pm1.1)\times10^{-2}$ for D_{Ba} [7] and $1.5(\pm0.6)\times10^{-1}$ for D_{Ra} [7]. Mean values of calculated D is smaller than those of reported D. Smaller values of D mean that amount of incorporated Me in calcite is smaller. Mean values of calculated D for Sr, Ba,

Table IV. Calculated number of layers and mole ratio reactive layers against surface area

Measured surface area, $m^2\ g^{-1}$	The number of layers from center to surface	Mole ratio of reactive layers	
		34 layers	21 layers
10.6	249.9	0.35	0.23
11.0	240.8	0.37	0.24
13.7	193.2	0.44	0.29
16.6	159.4	0.51	0.34
17.9	147.8	0.54	0.37
18.2	145.3	0.55	0.37
18.7	141.4	0.56	0.38
19.5	135.6	0.58	0.40
20.9	126.5	0.61	0.42
21.2	124.7	0.61	0.42
22.8	115.9	0.65	0.45
23.4	112.9	0.66	0.46
27.0	97.8	0.72	0.51
27.5	96.0	0.73	0.52

Table V. Results of re-distribution experiment (II) with calcite reacted with solution containing Sr, Ba and Ra.

Case	Final Ca in solution 10^{-4} mol kgw^{-1}	Final Me in solution	Me/Ca mol ratio in solid	D
Sr-B1	1.58($\pm$0.03)	3.98($\pm$0.08)$\times10^{-5}$	1.84($\pm$0.05)$\times10^{-3}$	1.12($\pm$0.57)$\times10^{-2}$
Sr-B2	1.55($\pm$0.03)	3.94($\pm$0.08)$\times10^{-5}$	1.84($\pm$0.05)$\times10^{-3}$	1.11($\pm$0.57)$\times10^{-2}$
Sr-B3	1.57($\pm$0.03)	3.93($\pm$0.08)$\times10^{-5}$	1.95($\pm$0.06)$\times10^{-3}$	1.20($\pm$0.61)$\times10^{-2}$
Sr-B4	3.87($\pm$0.08)	1.21($\pm$0.02)$\times10^{-5}$	3.55($\pm$0.10)$\times10^{-4}$	1.55($\pm$0.79)$\times10^{-2}$
Sr-B5	3.46($\pm$0.07)	1.20($\pm$0.02)$\times10^{-5}$	3.59($\pm$0.10)$\times10^{-4}$	1.41($\pm$0.72)$\times10^{-2}$
Sr-B6	3.38($\pm$0.07)	1.18($\pm$0.02)$\times10^{-5}$	3.54($\pm$0.10)$\times10^{-4}$	1.37($\pm$0.70)$\times10^{-2}$
Ba-B7	1.52($\pm$0.02)	2.92($\pm$0.01)$\times10^{-5}$	6.59($\pm$0.19)$\times10^{-4}$	7.7($\pm$4.9)$\times10^{-3}$
Ba-B8	1.48($\pm$0.02)	2.88($\pm$0.01)$\times10^{-5}$	6.54($\pm$0.19)$\times10^{-4}$	7.6($\pm$4.8)$\times10^{-3}$
Ba-B9	1.49($\pm$0.02)	2.92($\pm$0.01)$\times10^{-5}$	6.70($\pm$0.19)$\times10^{-4}$	7.7($\pm$4.9)$\times10^{-3}$
Ba-B10	2.92($\pm$0.04)	7.54($\pm$0.02)$\times10^{-6}$	1.42($\pm$0.04)$\times10^{-4}$	1.21($\pm$0.77)$\times10^{-2}$
Ba-B11	2.96($\pm$0.04)	7.52($\pm$0.02)$\times10^{-6}$	1.45($\pm$0.04)$\times10^{-4}$	1.25($\pm$0.80)$\times10^{-2}$
Ba-B12	2.94($\pm$0.04)	7.34($\pm$0.02)$\times10^{-6}$	1.41($\pm$0.04)$\times10^{-4}$	1.24($\pm$0.79)$\times10^{-2}$
Ra-B13		1.2($\pm$0.1)$\times10^{-9}$	1.6($\pm$0.2)$\times10^{-8}$	2.7($\pm$1.7)$\times10^{-2}$
Ra-B14		9.72($\pm$0.97)$\times10^{-10}$	1.6($\pm$0.2)$\times10^{-8}$	3.5($\pm$2.3)$\times10^{-2}$
Ra-B15		1.0($\pm$0.1)$\times10^{-9}$	1.7($\pm$0.2)$\times10^{-8}$	3.4($\pm$2.3)$\times10^{-2}$
Ra-B16		1.0($\pm$0.1)$\times10^{-10}$	1.4($\pm$0.1)$\times10^{-9}$	2.9($\pm$1.9)$\times10^{-2}$
Ra-B17	4.90($\pm$0.10)	6.7($\pm$0.7)$\times10^{-11}$	1.6($\pm$0.2)$\times10^{-9}$	5.0($\pm$3.3)$\times10^{-2}$
Ra-B18		7.1($\pm$0.7)$\times10^{-11}$	1.8($\pm$0.2)$\times10^{-9}$	5.3($\pm$3.5)$\times10^{-2}$
Ra-B19		3.3($\pm$0.3)$\times10^{-10}$	1.8($\pm$0.2)$\times10^{-8}$	7.1($\pm$4.7)$\times10^{-2}$
Ra-B20		3.3($\pm$0.3)$\times10^{-10}$	2.1($\pm$0.2)$\times10^{-8}$	8.3($\pm$5.4)$\times10^{-2}$
Ra-B21		3.5($\pm$0.3)$\times10^{-10}$	1.9($\pm$0.2)$\times10^{-8}$	7.2($\pm$4.8)$\times10^{-2}$

and Ra is 0.6, 0.6 and 0.3 times as large as those in literature, respectively. Although those difference are so large that error of mean values of calculated D for Sr and Ra can not overlap the reported D, calculated D was derived with reactive layer which has large error (34 ± 17 for Sr and 21 ± 13 for Ba and Ra) and as a results each value of calculated D has corresponding range of error (table V). Considering the range of error of them, most values of Sr, all values of Ba and four values for Ra cover the range of reported D. It implies that values of reported D are consistent with those calculated. Therefore, as it indicates that substitution reaction in coprecipitation experiment is reversible, applicability of D to estimate partition of Sr, Ba or Ra and Ca in solid phase and solution in reversible system is confirmed.

Migration analysis

With a value of K_{cp} for Ra in calcite, a migration analysis was performed for the 4n+2 series radionuclides for the reference case in JNC (2000) [9]. K_{cp} can be calculated from eq.(2) knowing values of D_{Ra}, C_{Ca} and V_{Ca}. C_{Ca} was estimated by Oda et al., [14] to be 5.25×10^{-2} mol m^{-3}. V_{Ca} can be estimated from the mole % of 21 layers and the total amount of calcite in the buffer material. The length of a side of a calcite rhombohedron was estimated to vary from 2.2×10^{-5} (measured by Ito et al. [15]) to 7.4×10^{-5}m, the latter value corresponding to the maximum size of crystal [16]. With these values, mole % of Ca of 21 layers was estimated as 0.1 to 0.4 (following the same procedure in the 'Re-distribution experiment'). Calcite in the buffer material was measured by Sasaki et al. [17] to be 2.35 wt. %, which converts to 0.24 moles of calcite in 1 kg of buffer material. V_{Ca} was calculated to be 2×10^{-4} to 1×10^{-3} moles of calcite per kg of bentonite by multiplying the mole % in 21 layers by the mole amount of calcite. Consequently, K_{cp} for Ra is calculated to be 6×10^{-4} to 3×10^{-3} m^3 kg^{-1}. However, the measured value of K_d for Ra in bentonite is 1×10^{-2} m^3 kg^{-1} [18]. K_d is an empirical parameter which shows all distribution reactions, including the effect of K_{cp}. Therefore, $K_d > K_{cp}$ is reasonable. For migration analysis in this study, K_d can be adopted as a parameter which includes the effect of substitution reactions. With this assumption, the migration.analysis was performed and the results are shown in figure 3. The release rate of Ra on the surface of the engineered barrier is calculated to be largest in 4n+2 series radionuclide and that at 100m point in the rock is not notable. Since solubility was defined by the simplified coprecipitation theory [3] for the calculation of JNC[9], the release rate of Ra becomes constant at high Ra concentration on the surface of the engineered barrier. Such a limited concentration from solubility is inappropriate for a substitution reaction. The model in the present study can express the solubility behavior of Ra by a substitution reaction. For a low release rate of Ra at 100m point in the rock, Ra concentration is affected by K_d in both cases. Therefore, almost the same results are shown.

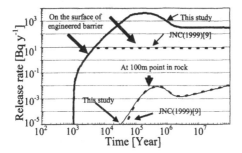

Figure 3. Release rate of Ra on a surface of engineered barrier and at 100 m point in rock.

CONCLUSIONS

The partition coefficient is applied to a model to estimate Ra concentration in groundwater for a substitution reaction with Ca in calcite. Since numbers of reactive layers of calcite had not been identified and reversibility of reaction had not been confirmed, re-distribution experiments were carried out. From the experiments, the number of reactive layers in calcite was determined to be $34(\pm 17)$ for Sr and $21(\pm 13)$ for Ba and Ra, respectively. With the evaluated reactive layers and experimental results, a partition coefficient was calculated, which showed reasonable agreement with values reported in the literature. From this result, the reversibility of substitution within reactive layers was confirmed. A migration analysis in a geological disposal system for 4n+2 series radionuclides was carried out and the effects of substitution of Ra and Ca in calcite were reasonably simulated.

ACKNOWLEDGEMENTS

We are grateful to Mr. T. Nakazawa, Mitsubishi Materials Corporation, for support in the experimental work on Ra. We thank Dr. I. Mckinley, Mckinley consulting, for constructive, critical comments. We also thank Dr. C. Walker for valuable and insightful review. We thank anonymous reviewers for critical comments which helped us very much in revising our manuscript.

REFERENCE

1. U. Berner, PSI Bericht Nr. 95-07, (1995).
2. U. Berner, PSI Bericht Nr. 02-22 (2002).
3. J.Azuma, M.Shibata, M.Yui, T.Shibutani, S.Notoya and Y.Yoshida, JNC TN8400 99-071 (1999),in Japanese.
4. R.B. Lorens, *Geochim. Cosmochim. Acta* **45**, 553-561 (1981).
5. A.J. Tesoriero and J.F. Pankow, *Geochem. Cosmochim. Acta* **60**, 1053-1063 (1996).
6. E. Curti, PSI-Bericht Nr. 97-10 (1997).
7. Y. Yoshida, H. Yoshikawa and T. Nakanishi, Geochem. J. **42**, 295-304 (2008).
8. D.J. Cherniak, Geochim. Cosmochim. Acta **61**, 4173-4179 (1997).
9. Japan Nuclear Cycle Development Institute, JNC TN1410 2000-004 (2000).
10. C. Klein and C. S. Hurlbut,Jr., Manual of mineralogy twenty-first edition,revised, JOHN WILEY&SONS, INC., NEW YORK (1993).
11. Y. Yoshida and H.Yoshikawa, JAEA-Research 2008-015 (2008), in Japanese.
12. R.D. Shannon, *Acta Cryst.* **A32**, 751-767 (1976).
13. D.A. Sverjensky and P.A. Molling, *Nature* **356**, 231-234 (1992).
14. C. Oda, M. Shibata and M. Yui, JNC TN8400 99-078 (1999), in Japanese.
15. M.Ito, M.Okamoto, M.Shibata, Y.Shibata, T.Danbara, K.Suzuki, T.Watanabe, PNC TN8430 93-003 (1993), in Japanese.
16. Japan Nuclear Cycle Development Institute, JNC TN1410 2000-003(2000).
17. T. Murakami and R.C. Ewing in *Experimental Studies on the Interaction of Groundwater with Bentonite*, entitled by Y.Sasaki, M.Shibata, M.Yui and H.Ishikawa, (Mater. Res. Soc. Symp. Proc. **353**, Kyoto, 1994) pp.337-344.
18. Y.Tachi,T.Shibutani,M.Sato and M.Yui, *J. Contam. Hydrol.* **47**, 171-186 (2001).

Mater. Res. Soc. Symp. Proc. Vol. 1124 © 2009 Materials Research Society 1124-Q07-20

Tribochemical treatment for immobilisation of radioactive wastes

Olga G. Batyukhnova
State Unitary Enterprise Scientific and Industrial Association "Radon", Moscow, The 7th
Rostovsky Lane, 2/14, 119121, Moscow, Russia

Michael I. Ojovan
Immobilisation Science Laboratory, Department of Engineering Materials,
University of Sheffield, Mappin Street, Sheffield S1 3JD, UK

ABSTRACT

Tribochemical treatment of waste-additive mixtures was used in an attempt to improve the efficiency of self-sustaining high-temperature synthesis of ceramics for immobilisation of operational nuclear power plant radioactive wastes. Good quality ceramics were produced with significant diminution of volatile nuclide carryovers from 12.5 to 3 wt.% achieved when the tribochemical treatment was used.

INTRODUCTION

Immobilisation of radioactive wastes in glasses and ceramics involves utilisation of high temperatures which lead to volatilisation of radionuclides and generation of secondary wastes which result in a diminishing of overall efficiency. Typical temperature limitations to vitrification of high-level waste from nuclear fuel reprocessing and operational radioactive waste from nuclear power plants are ~1200 °C [1]. The longer high-temperature processing time the higher the amount of secondary radioactive waste, and so the lower the overall efficiency of immobilisation. A potential route to enhance the efficiency of immobilisation is to use mechanochemical (tribochemical) treatment of waste before high-temperature processing [2]. The mechanomechanical treatment of powders is a well known method providing activation and production of many materials at low temperatures [3,4] in both crystalline and amorphous state [5,6]. The tribochemical treatment of raw nuclear waste and waste batches intended for high temperature processing can enable utilisation of lower processing temperatures and decrease processing times. Utilisation of tribochemical treatment at low temperatures diminishes generation of secondary radioactive wastes. These can result in an increased processing efficiency. Herein we describe the results of preliminary investigations to use the tribochemical treatment in a self-sustaining high-temperature synthesis (SHS) of ceramics for immobilisation of operational nuclear power plant (NPP) radioactive wastes. Note that initial studies of waste termochemical processing were carried out at the end of 1980s and were intended to study the feasibility of SHS. A novel, highly efficient, ecologically safe and cheap calcinations method evolved from this was with SHS being supplied by PMF (powder metal fuel) components selectively interacting with nitrogen oxides [7]. This method, however, has not been used on an industrial scale, since the Russian vitrification processes used a one-stage calcinations-melting approach utilizing non-calcined wastes.

EXPERIMENTAL

The immobilisation of simulant operational NPP waste was conducted using SHS reactions in a mixture of pre-calcined and pre-treated waste with Fe_2O_3-Al thermite. This is one of optional variants of self-sustaining immobilisation of powder-like wastes [8]. A schematic of immobilisation route used is shown in Fig. 1.

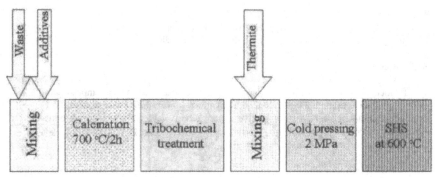

Figure 1. Self-sustaining high temperature synthesis (SHS) immobilisation scheme used in experiments.

For experiments NPP operational radioactive waste simulants were used. The composition of simulant was (wt.%): $NaNO_3$ (70), NaCl (20), CaO (6), Na_2SO_4 (2.4), $MgSO_4$ (0.6), $NaPO_3$ (1). To provide the self-sustaining immobilisation process we have used the standard thermite mixture Fe_2O_3:Al (ASD-6 powder)=3:1. To ensure a stable and controlled self-sustaining immobilisation process and to produce a durable wasteform, additives were required. Several waste immobilising additives (matrix materials) were used: (i) crushed glass of composition (wt.%) SiO_2 (71.6-76.8)-Al_2O_3 (13-16)-R_2O (Na_2O, K_2O) (5.63-7.6)-Fe_2O_3 (0.44-1.44)-CaO (0.8-1.8)-MgO (0.11-0.51), (ii) granite, (iii) silica gel, (iv) quartz sand and (v) clay soil. The clay soil which had the volumetric density $1.9 - 2.4$ g/cm^3 and specific density $2.5 - 2.7$ g/cm^3 and contained up to 30 wt.% clay particles [9], was thoroughly washed and dried at 300 °C during 3 hours before use to eliminate the organic impurities and excess water.

The waste simulant was mixed with additives in various ratios, calcined during 2.5 h at 700 °C and cooled to room temperature. The tribochemical treatment was carried out in the planetary mill M3 with steel balls with diameters 5-7 mm of mass 300 g. Batches of powder mixtures were placed in steel cylindrical drums of volume 1 L, the atmosphere was air. The acceleration at treatment was 45g. In a preliminary analysis of the evolution of powders during tribochemical treatment is was found that the most important changes in the particle median diameter occur within first 10 to 30 min of treatment (Fig. 2). Because of this the tribochemical treatment was carried out for all mixtures during 15-20 min.

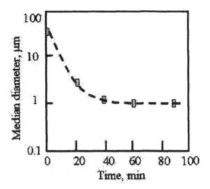

Figure 2: The evolution of median diameter of quartz sand additive with tribochemical treatment time.

The tribochemical treatment was also carried out in a mechanical disintegrator with two counter-rotating rotors with steel beater paddles-fingers. The rotation speed of two rotors was 1000-4500 rot/min. For comparison in a parallel series of experiments the product of calcination was simply crushed in alundum mortar. Then both treated and untreated mixtures were mixed with thermite, cold pressed in pellets at 2MPa, put in pre-weighted alundum crucibles in a preheated resistance furnace (600 °C) and ignited for SHS reactions. The synthesis of ceramics was carried out in air atmosphere. The overall carryover was measured comparing the weights of pellets before and after the SHS immobilisation process. Ceramics obtained were analysed for mechanical parameters such as porosity, mechanical strength and leachability. The porosity was measured by mercury intrusion porosimetry. The normalised leaching rates were measured using the standard IAEA test protocol ISO 6961-1982. As waste simulant was used rather than actual NPP waste the leaching rates were calculated for Na which is an analogue of radioactive ^{134,137}Cs typically present in operational NPP wastes [1]. Phase composition of ceramics obtained was determined on DRON-UM 1 difractometer using CuK_α radiation. Microstructure of samples was studied by means of micro-X-ray spectral analysis (Camebax).

RESULTS

Initially the SHS immobilisation was carried out with the purpose to identify the most perspective mixtures from the point of view of processing efficiency. The overall carryover of batch components was the main parameters in selecting the optional waste-additive mixture. Table 1 shows compositions (wt%) of several mixtures tested without tribochemical treatment.

The optimal results were obtained for mixtures (wt.%): waste (40): thermite (25): sand (25): clay (10). The immobilisation SHS process temperature was 1300-1500 °C at wave velocity 20-30 mm/min. These mixtures were tested for the effect of tribochemical treatment (Table 2).

Table 1. Waste-additive-thermite mixtures tested without tribochemical treatment.

No.	Waste, wt.%	Additive*, wt.%	Thermite, wt.%	Carryover overall, wt.%
1	60	10(i)+20(ii)	10	30
2	60	30(ii)	10	27
3	60	20(i)+10(ii)	10	24
4	60	10(i)+20(ii)	10	25
6	60	30(i)	10	35
7	60	20(i)	20	16
8	60	30(iv)	10	40
9	60	20(iv)	20	30
10	60	15(i)+5(iv)	20	20
11	60	15(i)+10(iv)	15	20
12	60	15(i)+15(iv)	10	20
13	45	33.7(i)+11.3(iii)	10	25
14	40	30(i)+10(iv)	20	23
15	35	26.3(i)+ 8.7(iv)	30	18
16	37.5	28(iv)+9.5(v)	25	12.5
17	40	25(iv)+10(v)	25	15

* The additives (matrix materials) used: (i) crushed glass of composition (wt.%) SiO_2 (71.6-76.8)-Al_2O_3 (13-16)-R_2O (Na_2O, K_2O) (5.63-7.6)-Fe_2O_3 (0.44-1.44)-CaO (0.8-1.8)-MgO (0.11-0.51), (ii) granite, (iii) silica gel, (iv) quartz sand and (v) clay soil .

Table 2. Optimal waste-additive-thermite mixtures tested using tribochemical pre-treatment.

No.	Waste, wt.%	Additive*, wt.%	Thermite, wt.%	Carryover overall, wt.%
1TBC	37.5	28(iv)+9.5(v)	25	5
2TBC	40	25(iv)+10(v)	25	3

* The additives (matrix materials) used: (iv) - quartz sand, (v) - clay soil.

The ceramics produced had colours from light yellow to yellow-brown and demonstrated dense and partly glass-like views of their surfaces (Fig. 3a and 3b). Metallic spheres up to several mm in diameter within the uniform matrix were observed. A study of the sample with scanning electron microscopy with energy dispersive system has shown that it is composed of both glassy and crystalline phases (see [10]). The glassy phase has an uniform texture and composition. The crystalline phases were found to be complex silicates and metallic Fe. Metallic Fe is uniformly distributed in the ceramics. Fe is also present in spheres as isometric particles. Fig. 3c demonstrates this for the accidental case when a large spherical particle of reduced Fe was formed in the ceramics body.

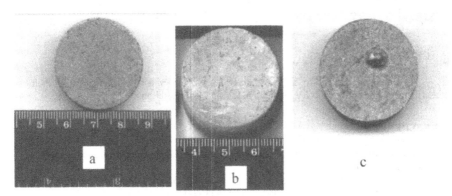

Figure 3. Ceramics produced via SHS of waste simulants, additives and thermite mixtures: a – ceramic TBS1 (Table 2); b – ceramic #7 (Table 1); c – ceramic sample with reduced Fe-inclusion, the composition same as of sample TBS1.

Data from Tables 1 and 2 show that the use of tribochemical treatment diminished the overall carryover from 12.5-15 without treatment to 3 – 5 wt.% when the tribochemical treatment was used. Samples obtained using optimal batches (Table 2) had the porosity 5%. The breaking strength of samples was within 45-50 MPa. The sodium normalised leaching rates on IAEA test protocol ISO 6961-1982 were below 10^{-5} g/cm^2 day on 7th day of tests. One can see that the tribochemical pre-treatment of batch mixture improved the overall efficiency of treatment as less secondary waste is being produced.

DISCUSSION

Mechanical activation of materials was widely used achieving significant improvements of many processes, e.g. by using desintegrators the proportion of cement in the concrete can be reduced maintaining its solidity, etc. [3,4]. The initial powder-like materials and dispersions used are activated by mechanical impacts in the ball mill and disintegrator. Owing to these impacts solid materials are subjected to defect formation in their microstructure and macrostructure this being in addition to a comminution of the grains, thus leading to a change of the grain size spectrum and the specific surface of the particles [11]. The investigations show that the production of particles with a median particle size lower than 10 nm is possible by wet grinding in a stirred media mill [12]. Mechanical activation was used in various applications including intensification of SHS similarly to current work [13]. The SHS is a perspective method for nuclear waste immobilisation particularly for relative small volume waste streams and legacy wastes [8,14].

CONCLUSIONS

The results obtained here show that tribochemical treatment of batches before SHS immobilisation significantly diminishes the overall carryover and hence improves the process efficiency.

ACKNOWLEDGEMENTS

The authors are grateful to A.I. Alexandrov, I.A. Sobolev and T.D. Scherbatova for their assistance in the work and to I.V. Semenova for sample preparation.

REFERENCES

[1] M.I. Ojovan and W.E. Lee, *An Introduction to Nuclear Waste Immobilisation* (Elsevier, Amsterdam, 2005).

[2] O.G. Batyukhnova, A.I. Alexandrov, M.I. Ojovan, I.A. Sobolev. Method of solidification of liquid radioactive waste. Patent of Russia 2009556 (1994).

[3] G. Heinicke. *Tribochemistry*. Akademie-Verlag, Berlin (1984).

[4] E.G. Avvakumov, *Mechanical Methods of the Activation of Chemical Processes*. Nauka, Novosibirsk (1986).

[5] V. Sepelak. Nanocrystalline materials prepared by homogeneous and heterogeneous mechanochemical reactions. *Ann. Chim. Sci. Mat.*, **27**, 6, 61-76 (2002).

[6] C.C. Koch. Amorphization of single composition powders by mechanical milling. *Scripta Materialia*, **34**, 21-27 (1996).

[7] O.G.Batyukhnova, N.S.Makhonin, M.I.Ojovan, I.A.Sobolev, L.S.Popov. Method of calcination of nitrate-conteining radioactive waste. USSR Patent 1604066 (1989).

[8] M.I. Ojovan, W.E. Lee. Self sustaining vitrification for immobilisation of radioactive and toxic waste. *Glass Technology*, **44** (6) 218-224 (2003).

[9] M.I. Ojovan, W.E. Lee, A.S. Barinov, I.V. Startceva, D.H. Bacon, B.P. McGrail, J.D. Vienna. Corrosion of low level vitrified radioactive waste in a loamy soil. *Glass Technol.*, **47** (2), 48-55 (2006).

[10] M.I. Ojovan, G.A. Petrov, S.V. Stefanovsky, B.S. Nikonov. Processing of large scale radwaste-containing blocks using exothermic metallic mixtures. *Mat. Res. Soc. Symp. Proc.*, **556**, 239-245 (1999).

[11] P.Yu. Butyagin. Problems in mechanochemistry and prospects for its development. *Russ. Chem. Rev.* **63**, 965-976 (1994).

[12] S. Mende, F. Stenger, W. Peukert, J. Schwedes. Production of sub-micron particles by wet comminution in stirred media mills. *J. Mater. Sci.*, **39**, 5223 – 5226 (2004).

[13] Yu.S. Nayborodenko, N.G. Kasatsky, E.G. Sergeeva, O. Lepakova. The effect of mechanical activation oh high-temperature synthesis and phase formation low-calorie intermetallic compounds. *Chemistry for Sustainable Development*, **10**, 155-160 (2002).

[14] M.I. Ojovan, W.E. Lee, I.A. Sobolev. S.A. Dmitriev, O.K. Karlina, V.L. Klimov, G.A. Petrov, C.N. Semenov. Thermochemical processing using powder metal fuels of radioactive and hazardous waste. *J. Process Mechanical Engineering*, **218E**, 261-269 (2004).

Mater. Res. Soc. Symp. Proc. Vol. 1124 © 2009 Materials Research Society 1124-Q10-04

Resorcinarenes and Aza-Crowns as New Extractants for the Separation of Technetium-99

Patricia Paviet-Hartmann[1], Jared Horkley[2], Joshua Pak[2], Eric Brown[3], and Terry Todd[4]
[1]University Nevada Las Vegas, Dept. Chemistry, 4505 Maryland Parkway, Las Vegas, NV 89074, USA
[2]Idaho State University, Dept. Chemistry, 921 S. 8th Avenue, Pocatello, ID 83209, USA
[3]Boise State University, Dept. Chemistry and Biochemistry, Boise, ID 83725, USA
[4]Idaho National Laboratory, Fuel Cycle Science and Technology Division, Idaho Falls, ID 83415-6000, USA

ABSTRACT

Among the radionuclides considered for separation within the UREX + concept is Technetium-99, a uranium fission product with a low energy beta emission and a half life of 2.13 x 10^5 years. The fission products, present in the high level waste (HLW) issued from the PUREX process, are mainly responsible for the long term radiotoxicity of this HLW stream. Partitioning and transmutation as a means of reducing the burden on a geological repository requires Tc-99 to be removed from the HLW (partitioning) and then fragmented by fission (transmutation), which allows reduction of the radiotoxicity inventory of the remaining waste. As an example, a single isotopic species (Tc-99) can be transmuted by single neutron capture into the stable noble metal ruthenium (Ru-100). Selective extraction of Tc-99 needs to be investigated. We describe herein the synthesis of new macro-compounds which can be functionalized with oxygen, sulfur, and nitrogen donating functional groups with various cavity sizes and/or bonding modes, and the assessment of their extraction properties towards Tc-99.

INTRODUCTION

Many countries are currently developing electricity generation systems for the near future that are environmentally sound and maximize resources. It is obvious that closed fuel cycle technologies are key to sustainable nuclear energy, waste reduction, and proliferation control. Depending on the country, the spent nuclear fuel is either reprocessed through the PUREX process or will be stored in a geological repository. Furthermore, new separation schemes are developed to complement or ultimately replace the PUREX process, which include the UREX+ process [1]. Either one of these two paths, reprocessing or direct disposal of spent nuclear fuel (SF), has several issues, among them: Technetium-99 (Tc-99). The presence of this undesirable radioelement in the UREX+ raffinate or in the spent fuel, as a result of the thermal neutron fission of uranium-235, makes it a unique challenge to safely dispose of nuclear waste as spent fuel or vitrified high level waste (HLW). Furthermore, the vitrification of Tc-99 in HLW glasses is problematic due to volatility issues [2]. With a half-life of 2.13 × 10^5 yrs, and an affinity to form highly mobile pertechnetate species, TcO$_4^-$, Tc-99 is considered a long-term hazard in nuclear waste disposal. There is a need for the development of new extractant systems, such as systems based on crown ethers (CE) or resorcinarenes, which may serve to selectively extract and separate this long lived radionuclide from HLW. We have started to investigate the use of macrocompounds as a method to selectively separate Tc-99 from real effluents [3,4]. The development of new extractant systems that may selectively extract Tc-99 is presented here along with liquid-liquid extraction results on the separation of pertechnetate from acidic and basic medium.

EXPERIMENTAL

All glassware was oven-dried before each experiment. The anaerobic experiments were placed under N_2 or Argon atmosphere. For water sensitive reactions, the solvents were distilled before use. All chemicals and materials were purchased form scientific catalogs of Aldrich and Acros Organics. Instruments used for the data analysis were NMR VARIAN MERCURY 300, Thermo Nicolet FT-IR NEXUS 600, Shimadzu UV-2101PC, Thermo FINNEGAN GC-MS, Waters Autospec Ultima Magnetic Sector MS, PERKIN ELMER 2400 Series II CHNS/O Analyzer and Perkin Elmer LS 50B fluorescence spectrometer. Ammonium pertechnetate $NH_4^{99}TcO_4$ was purchased from Oak Ridge National Laboratory and purified as described elsewhere [5]. Water was purified to 18 MΩ by a MilliQ system. All other chemicals were used as received except the ones that were synthesized. The liquid-liquid extractions were performed in 15 mL corning tubes; equal volumes of organic and aqueous phases contacted on a vortex, for a specified time at room temperature. The aqueous and organic phases were sampled and measured by (1) liquid scintillation counting and (2) ICP-MS for the analysis of Tc-99 and Re respectively. The distribution coefficient D_A was calculated as the ratio $[A]_{org}/[A]_{aq}$ after reaching the chemical equilibrium.

DISCUSSION

Synthesis

4,4',5,5'-Tetraiododibenzo-24-crown-8 (9) was synthesized according to Scheme 1 [6]. In a round bottom flask equipped with a condenser, Cs_2CO_3 (14.07g, 43.175mmol) was mixed with MeCN (334 ml) and refluxed under N_2. Compound 8 (8.07 g, 8.635 mmol) and diiodocatechol (4) (3.124 g, 8.635 mmol) were dissolved in 50 ml of MeCN and added drop wise via syringe pump over 24 hrs. Reaction was refluxed for an additional two days. The reaction mixture was concentrated and run through a plug of silica gel using EtOAc/MeCN/Hexanes (3:3:4) mixture. White solid was isolated without further purification in 85% (6.99 g). ^{1}H NMR (CDCl$_3$, 300 MHz) δ 7.25 (s, 4H), 4.08 (t, J = 3.6 Hz,8H), 3.88 (d, J = 3.9 Hz, 8H), 3.78 (s, 8H). ^{13}C NMR (CDCl$_3$, 75 MHz) δ 69.8, 71.5, 96.8, 124.1, 149.6. MS (m/z) 952.0 (M+, 2.31), 826.1(M-I, 3.32), 698.1(M-2I-H, 4.20), 572.1(M-3I-H, 23.8). IR (cm^{-1}) 3054, 2986, 2305, 1494, 1232, 896, 705. Anal. Calc. for $C_{24}H_{28}I_4O_8 \cdot CH_3CN$ (992.822): C, 31.44; H, 3.15. Found compound: C, 31.45; H, 3.18.

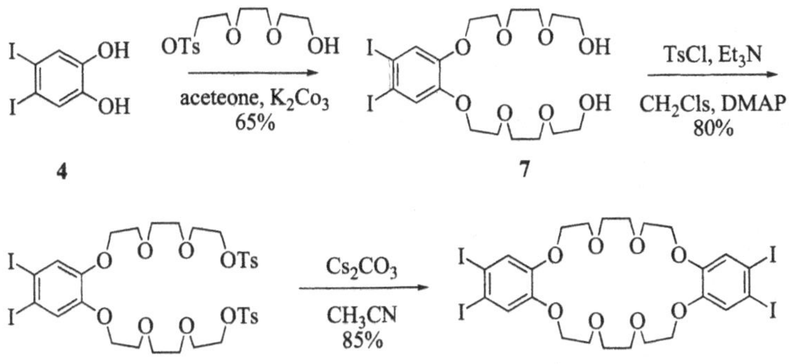

Scheme 1. Preparation of tetraiodoCE (**9**).

The synthesis of Tetrakis(*N*-2-pyridonyl) methylresorcin[4]arene pentyl foot (**12c**) (Scheme 2) was performed by using tetra-bromo resorcin[4]arene (**10**) (100 mg, 0.084 mmol), NaH (60% suspension in mineral oil) (16. mg, 0.40 mmol), and 2(1*H*)-pyridone (40 mg, 0.42 mmol). The reaction mixture was evaporated to dryness and redissolved in EtOH and subsequently ran through a short silica plug with MeOH to remove NaBr. The product was isolated on preparatory TLC (silica, 67% acetone in hexanes) yielding 79 mg (76%) of a white solid. 1H NMR (300 MHz, CDCl$_3$): d (ppm) 7.43 (t, *J*=8.1 Hz, 4H), 7.32 (d, *J*=6.0 Hz, 4H), 7.14 (s, 4H), 6.79 (d, *J*=8.1 Hz, 4H), 6.33 (t, *J*=6.0 Hz, 4H), 5.76 (d, *J*=7.5 Hz, 4H), 4.98 (s, 8H), 4.71 (t, *J*=7.8 Hz, 4H), 4.43 (d, *J*=7.5 Hz, 4H), 2.26-2.14 (m, 8H), 1.47-1.25 (m, 24H), 0.90 (t, *J*=6.8 Hz, 12H). 13C NMR (75 MHz, CDCl$_3$): d (ppm) 162.74, 154.18, 139.41, 138.26, 137.71, 122.11, 120.90, 120.72, 106.29, 99.62, 42.67, 37.07, 32.19, 30.34, 27.74, 22.85, 14.29. HRMS (M++Na+) Calcd: 1267.5983, Found: 1267.6014.

The synthesis of Tetrakis(*N*-2-benzothiazolonyl) methylresorcin[4] arene pentyl foot (**12d**) (Scheme 2) used tetra-bromo resorcin[4]arene (**10**) (1.00 g, 0.841 mmol), NaH (60% in mineral oil) (158 mg, 3.95 mmol), and 2-benzothiazolone (636 mg, 4.21 mmol), and was crystallized from CH$_2$Cl$_2$ and EtOAc yielding 870 mg (70%) of white needles. (ppm) 7.41 (d, 1H NMR (300 MHz, CDCl$_3$): d *J*=7.2 Hz, 4H), 7.26-7.06 (m, 16H), 5.78 (d, *J*=7.5 Hz, 4H), 4.95 (s, 8H), 4.70 (t, *J*=8.0 Hz, 4H), 4.47 (d, *J*=7.5 Hz, 4H), 2.19-2.09 (m, 8H), 1.38-1.22 (m, 24H), 0.85 (t, *J*=6.9 Hz, 12H). 13C NMR (75 MHz, CDCl$_3$): d (ppm) 170.45, 154.13, 138.26, 137.39, 126.53, 123.13, 122.53, 121.04, 120.75, 111.72, 99.86, 37.60, 36.89, 32.06, 30.25, 27.57, 22.83, 14.41, 14.28. HRMS (M++Na+) Calcd: 1491.4866, Found: 1491.4890.

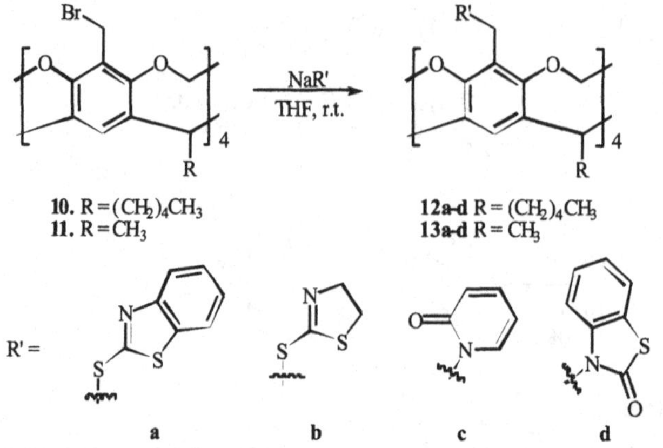

10. R = (CH$_2$)$_4$CH$_3$
11. R = CH$_3$

12a-d R = (CH$_2$)$_4$CH$_3$
13a-d R = CH$_3$

R' =

a   b   c   d

Scheme 2. Preparation of Tetrakis(*N*-2-pyridonyl) methylresorcin[4]arene pentyl foot (**12c**) and Tetrakis(*N*-2-benzothiazolonyl) methylresorcin[4] arene pentyl foot (**12d**).

Extraction of Rhenium with synthesized 4,4',5,5'-Tetraiododibenzo-24-Crown-8-Ether

Extraction of perrhenate from acidic and basic solutions (table I) by 4,4',5,5'-Tetraiododibenzo-24-Crown-8-Ether (10^{-4} M) in cyclohexanone, nitrobenzene, 1:1 nitrobenzene/acetone, 1:1 nitrobenzene/methyl isobutyl ketone, 1:4 benzene/acetone, and toluene was examined to investigate the relationship between diluent properties (such as polarity) and extractability of the crown ether-perrhenate complex. A simple system composed of perrhenate dissolved in an acidic or alkaline medium phase was used. The nature of the diluent was examined and was shown to be the most influential variable in controlling the extraction efficiency, encompassing one order of magnitude in the range of observed extraction yield for ReO$_4^-$. Though some of the diluents tested are not suitable for process applications, this insight provided the means to find a suitable diluent for Tc-99 extraction that would meet process requirements.

Table I. ReO$_4^-$ Extraction Yield Obtained With 4,4',5,5'-Tetraiododibenzo-24-Crown-8 Ether, [ReO$_4^-$] = 10^{-4} M, [Crown Ether] = 10^{-4} M

Diluent	Aqueous Phase	Dielectric Constant	% Extraction
Cyclohexanone	0.1M HNO$_3$	18.2 (20°C)	92
Nitrobenzene	0.1M HNO$_3$	34.8(20 °C)	5.6
Nitrobenzene/Acetone 1:1	0.1M HNO$_3$	34.8(20 °C)/20.7(25°C)	50
Nitrobenzene /methyl isobutyl ketone 1:1	0.1M HNO$_3$	34.8(20 °C)/13.1(20 °C)	21
Benzene/Acetone 4:1	4M NaOH	2.3(20 °C)/20.7(25°C)	42.2
Toluene	4M KOH	2.4(20 °C)/20.7(25°C)	28.5
Toluene	4M NaOH	2.4(20 °C)/20.7(25°C)	26.6

The extraction yields were low with non polar solvent such as toluene. By adding a more polar solvent such as acetone to the mixture benzene/acetone, the extraction yield was increased from 26.6% to 42.2%. Surprisingly, a more polar solvent such as nitrobenzene did not perform as expected with a low extraction yield of 5.6. Results were outstanding for cylcohexanone with an extraction yield of 92%. But cyclohexanone is fairly volatile (BP 155.6° C, flashpoint 46° C), and significantly water soluble (50 grams per liter at 30° C), thus making this diluent less suitable for investigating technetium extraction.

Extraction of Tc-99 with newly synthesized resorcinarenes

Tetrakis(N-2-pyridonyl) methylresorcin[4]arene pentyl foot (12c), and Tetrakis(N-2-benzothiazolonyl) methylresorcin[4] arene pentyl foot (12d) were tested for the selective extraction of Tc-99 from nitric acid solution (0.1M). The results are displayed in table II.

Table II. Distribution Coefficient D for TcO_4^- Extraction With Resorcinarenes (12c) and (12d)

Extractant	[Tc-99]/M	D_{Tc-99}	%E= 100 D/(D+1)
12c*	1.10^{-4}	0.027	2.7
12c*	1.10^{-4}	0.024	2.3
12d**	1.10^{-4}	0.034	3.0
12d**	1.10^{-4}	0.048	4.5

*12c was diluted in a mixture of toluene, 2-propanol, 3 heptanone and methylene chloride
**12d was diluted in a mixture of toluene, methylene chloride and 2-propanol

The low distribution coefficient of Tc-99 obtained with resorcinarenes implies that these types of macrocompounds may not be suitable for the selective extraction of Tc-99. For this reason, we pursue the extraction of Tc-99 with other crown ethers and aza crown ethers.

Extraction of Tc-99 with commercially available crown-ethers and aza crown ethers

Seven crown ethers and aza-crown ethers with ring sizes of 12-crown-4 to 18-crown-6 (10^{-3}M) in methylisobutyl ketone medium were evaluated for efficiency of pertechnetate (620 Bq.mL^{-1}) extraction from nitric acid solution (table III).

Table III. Distribution Coefficient D for TcO_4^- Extraction With Crown-Ethers from 0.1N HNO_3

Crown Ether	Cavity Diameter (Å)	D_{Tc-99}	%E = 100 D/(D+1)
Aza 12 Crown 4	1.2 - 2.5	0.82	45
Aza 15 Crown 5	1.7 - 2.2	0.785	44
Benzo 15 Crown 5	1.7 - 2.2	**0.83**	45
1,2 Diaza 15 Crown 5	1.7 - 2.2	0.75	43
Aza 18 Crown 6	2.6 - 3.2	0.76	43
Diaza 18 Crown 6	2.6 - 3.2	0.78	44
DC18 Crown 6	**2.6 - 3.2**	**2.5**	**72**

DC18C6 was revealed to be the most efficient crown ether to extract Tc-99. It appears that the cyclohexano groups attached to the crown ether tend to increase the distribution coefficient of

Tc-99. The presence of cyclohexano groups in DC18C6 tends to increase the basicity of the ether oxygens and also allow the ring to have more flexibility, thus, cyclohexano crowns tend to have higher binding constants for a given alkali metal ion than their benzo counterparts, which could explain the value of the distribution coefficient $D_{Tc-99} = 2.5$ obtained with DC18C6 in methylisobutyl ketone.

CONCLUSION

To satisfy the needs of new extractant systems we have demonstrated that crown ether (CE) based systems have the potential to serve as selective extractants for the separation of technetium-99, a long lived radionuclide, from HLW, intermediate level nuclear waste (ILW), and low level nuclear waste (LLW) streams. So far, the experimental results have shown that Dicyclohexano-18-Crown-6 (DC18C6) is a promising extractant under our experimental conditions. Studies are in progress to evaluate the extraction of Tc-99 by 4,4',5,5'-Tetraiododibenzo-24-Crown-8 Ether, as well as to examine the stability of these macrocompounds against radiolysis and hydrolysis.

ACKNOWLEDGMENTS

This research work was supported through the INL CAES LDRD program, grant n° 00037/00043028. The authors wish to thank Ian Kihara and Jeff Hess from ISU.

REFERENCES

1. M.C. Thompson, M.A. Norato, G.F. Kessinger, R.A. Pierce, T.S. Rudisill, and J.D. Johnson, in *Demonstration of the UREX solvent extraction process with Dresden reactor fuel solutions*, WSRC-TR-2002-00444 (2002).
2. D.A. McKeown, A.C. Buechele, W.W. Lukens, D.K. Shuh, and I.L. Pegg, in *Tc and Re behavior in borosilicate waste glass Vapor Hydration Test*, edited by D.S. Dunn, C. Poinssot, and B. Begg, (Mat. Res. Soc. Symp. Proc. **985**, Boston, MA, 2007) Scientific Basis for Nuclear Waste Management XXX, pp. 199-204.
3. P. Paviet-Hartmann, in *Solvent extraction of Tc-99 from radioactive intermediate liquid waste by Dibenzo-18-Crown-6*, Int. Conf. Proc. Waste Management (2002), Tucson, AZ, USA, February 24-28.
4. P. Paviet-Hartmann P., and T. Hartmann, in *Separation of long-lived fission products ^{99}Tc and ^{129}I from synthetic effluents by crown ethers*, Int. Conf. Proc. Waste Management (2006), Tucson, AZ, USA, February 26-March 2.
5. R. Kopunec, F. N. Abudeah, and I. Makaiova, *J. Radioanal. Nucl. Chem.*, **208**, 207-228 (1996).
6. P. Paviet-Hartmann, J. Pak, J. Horkley, and E. Wolfrom, in *Selective Extraction of Perrhenate and Pertechnetate by New Macrocyclic Compounds*, Int. Conf. Proc. Atalante (2008): Nuclear Fuel Cycles for a Sustainable Future, Montpelliers, France, Paper P1-19, May 19-23.

Container Corrosion

Mater. Res. Soc. Symp. Proc. Vol. 1124 © 2009 Materials Research Society   1124-Q09-01

Comparison of Alloy 22 Crevice Corrosion Repassivation Potentials From Different Electrochemical Methods

Ricardo M. Carranza
Dept. Materiales - Comisión Nacional de Energía Atómica
Instituto Sabato - Univ. Nac. de San Martín / CNEA
Av. Gral. Paz 1499, San Martín, B1650KNA Buenos Aires, Argentina

ABSTRACT

Alloy 22 is a candidate waste package outer container material for disposal of high-level waste. Alloy 22 has all the required alloying elements for protection against corrosion in a variety of environments. It was designed to withstand the most aggressive industrial applications, including reducing acids such as hydrochloric and oxidizing acids such as nitric. Alloy 22 is practically immune to pitting corrosion. However, it may be susceptible to crevice corrosion in chloride-containing solutions.

There are several electrochemical methods to determine the susceptibility of Alloy 22 to crevice corrosion. Crevice corrosion repassivation potential ($E_{R,CREV}$) is the most accepted electrochemical parameter to determine crevice corrosion susceptibility.

The main purpose of this paper is to compare the most commonly used electrochemical methods and to discuss their suitability for crevice corrosion prediction measurements based on recently published data. In this paper three methods are discussed: (i) the cyclic potentiodynamic polarization (CPP) method ASTM G61–86, (ii) the potentiodynamic polarization plus intermediate potentiostatic hold (PD – Pot –PD) method, and (iii) the Tsujikawa – Hisamatsu Electrochemical (THE) method ASTM G 192-08.

INTRODUCTION

Alloy 22 belongs to the corrosion resistant Ni-Cr-Mo family of nickel based alloys and resist corrosion damages in a variety of environments.[1,2,3] The base metal Ni, and the main alloying elements Cr and Mo provide resistance to environmentally assisted cracking in hot concentrated chloride solutions.[1,4,5] Cr and Mo also provide resistance to localized corrosion such as pitting and crevice corrosion in chloride containing solutions. Some of the Ni-Cr-Mo alloys also contain a small amount of tungsten (W), which may act in a similar way as Mo regarding protection against localized corrosion.[6] Ni-Cr-Mo alloys are practically immune to pitting corrosion in chloride solutions at near boiling temperatures. However, they may be susceptible to crevice corrosion in these conditions.[7]

In localized crevice corrosion the attack progresses in a non-uniform way at preferential sites. Crevice corrosion occurs only when the corrosion potential E_{CORR} is equal to or greater than the repassivation potential for crevice corrosion $E_{R,CREV}$. That is, if $E_{CORR} < E_{R,CREV}$, only general corrosion will occur. $E_{R,CREV}$ can be defined as a the potential above which crevice corrosion of Alloy 22 can take place. The margin of safety against localized corrosion will be given by the value of $\Delta E = E_{R,CREV} - E_{CORR}$. The higher the value of ΔE, the larger the margin of safety for localized corrosion.[7]

ELECTROCHEMICAL TESTS FOR CREVICE CORROSION

Various electrochemical methods are used to assess the susceptibility of Alloy 22 to crevice corrosion (Table I).[1,8-13] Most of them, except immersion tests which are used to find critical crevice temperatures (CCT)[1,8,10], measure the repassivation potential as the key parameter for crevice corrosion susceptibility. For Alloy 22, the most used electrochemical test is the cyclic potentiodynamic polarization (CPP) or ASTM G 61.[8] There are two ways to quantify the repassivation potential from a CPP curve: the potential at which the anodic current in the reverse scan attains the passive current value (E_{R1}, E_{R10}, etc) or the cross-over of the reverse and forward scans (E_{CO}). Other tests are the Tsujikawa - Hisamatsu Electrochemical (THE) method (ASTM G 192)[11], and the potentiostatic (POT) test, which does not have ASTM standard.[9,12] In the THE method the repassivation potential ($E_{R,CREV}$) is the potential at which the anodic current ceases to increase in the potentiostatic decreasing steps. A method implying a potentiostatic hold in the middle of two potentiodynamic runs in opposite direction is also used (PD – Pot – PD method).[13]

TABLE I – Electrochemical Methods to assess the susceptibility of Alloy 22 to crevice corrosion. CCT: Critical Crevice Temperature. CPP: Cyclic Potentiodynamic Polarization. THE: Tsujikawa-Hisamatsu Electrochemical. POT: Potentiostatic. PD – Pot – PD: Mixed Potentiodynamic and Potentiostatic. E_{RCO}, E_{R1}, E_{R2}, E_{R10}, and $E_{R,CREV}$: Repassivation Potentials. Definitions of each one is given while describing the thechniques later on in the paper

METHOD	PARAMETER	ASTM STANDARD
Immersion	CCT	ASTM G 48[1,8,10]
CPP	E_{CO}, E_{R1}, E_{R10}, etc.	ASTM G 61[8]
THE	$E_{R,CREV}$	ASTM G 192[11]
POT	$E_{R,CREV}$	No ASTM[9,12]
PD – POT -PD	E_{R2}, E_{CO}	No ASTM[13]

Alloy 22 was mostly tested to obtain crevice repassivation potentials.[9,12-28] Whenever crevice corrosion readily occurs in an aggressive environment, the values of the repassivation potential for Alloy 22 seem comparable.[12] Under less aggressive conditions, when crevice corrosion is more difficult to initiate, the value of repassivation potential may differ from method to method.[29,30] Chloride is the most aggressive anion regarding localized corrosion at temperatures. Chloride concentration has a significant effect on the resistance of Alloy 22 to crevice corrosion. As the chloride concentration increases, the susceptibility of Alloy 22 to crevice corrosion increases.[12,17,18,20,23,31-35] Inhibition of crevice corrosion by oxyanions such as nitrate, was reported.[30] The value of the ratio R = [Inhibitor]/[Cl⁻] has a strong effect on the susceptibility to crevice corrosion [13,19,21,24-29,31,36]. The higher the R value, the stronger is the inhibition.

Cyclic Potentiodynamic Polarization method

In the CPP curves (ASTM G 61)[8] the potential scan is usually started 0.100 to 0.150 V below E_{CORR} or at the end potential of a 10 minute galvanostatic cathodic treatment of -50

$\mu A.cm^{-2}$,[37] in the anodic direction at a scan rate of 0.167 mV/s. The scan direction is reversed when the current density reached usually 5 mA/cm^2 in the forward scan. Figure 1 shows a typical CPP curve obtained for Alloy 22 using a Prism Crevie Assemby (PCA) specimen (fabricated based on ASTM G 48[8]) immersed in a 1M NaCl solution. The curve presents a passive zone with low current densities in the order of 1 $\mu A.cm^{-2}$. After the passive domain, the increase of current with potential is due to crevice corrosion initiation and, at higher potentials, to transpassive dissolution also. Current hysteresis is always observed between the forward and the reverse scans. The method used to determine the repassivation potential (the cross-over repassivation potential E_{CO}) is also illustrated in Figure 1, as the potential at which reverse and forward current scans intersect each other. Some times E_{R1}, E_{R10}, are used, indicating the potential at which current density reach 1 or 10 $\mu A.cm^{-2}$ in the reverse scan, respectively.

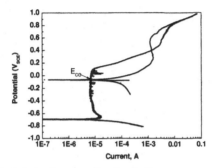

Figure 1: CPP method. Polarization curve E vs log I. E_{CO}: cross-over repassivation potential. Alloy 22 in 1M NaCl solution at 90 °C. Scan rate: 0.167 mV.s^{-1}. Scan reversed at 5 mA.cm^{-2}.

In the CPP method the current reached high anodic values (not less than 5 mA.cm^{-2}) and high anodic potentials (approx. 1V$_{SCE}$) in the full transpassive domain. The whole experiment lasts about 5.3 hours for the system showed in Figure 1.

Figure 2 shows the aspect of a PCA specimen after a CPP test. The attack is mainly shiny crystalline. The grain structure of the alloy is revealed as crystalline planes are attacked in a different degree according to their surface energy. No preferential grain boundary attack is detected. The surface of the specimen that was not covered by the crevice former shows usually signs of heavy transpassive dissolution.

The question that arises with CPP method is whether the fact of maintaining for long times the specimen at potentials in the full transpassive domain modifies or not the repassivation potential E_{CO}. In going to high anodic potentials one is not only allowing crevice corrosion propagation at higher intensity but increasing transpassive dissolution in the whole specimen as well.

Tsujikawa – Hitsamatsu Electrochemical method

In the THE method (ASTM G 192)[11] after 15 minutes at the open circuit potential the specimen is cathodically polarized at E_{CORR} – 0.1V. The potential scan is then started at a scan rate of 0.167 mV/s in the anodic direction until the anodic current density reaches a value of 30 μA (PCA specimen area aprox. 14 cm^2 and current density aprox. 2 μA.cm^{-2}). After this potentiodynamic polarization, the polarization control is changed to galvanostatic mode and kept constant for two hours at 30 μA. After that, two hour potentiostatic holds are applied decreasing the potential in 10 mV steps until the anodic current density cease to increase as a function of time. Figure 3 shows a typical result for the application of the THE method to an Alloy 22 PCA specimen immersed in a 1M NaCl solution. The results are represented as the typical E and I time transients.

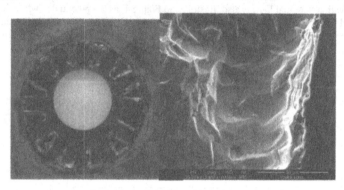

Figure 2: Photograph and SEM images of Alloy 22 PCA specimen after CPP test in 1M NaCl at 90°C.

During the galvanostatic step the potential decreases due to crevice corrosion propagation to a potential accordingly to the crevice corrosion rate corresponding to the applied anodic current. In the same figure, the method used to determine the repassivation potential $E_{R,CREV}$ is illustrated. $E_{R,CREV}$ corresponds to the potential at which the current decreases continuously in time in the potentiostatic step part of the THE method.

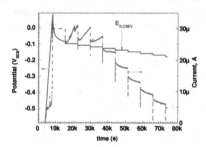

Figure 3: THE method. Potential and current time-ransients. $E_{R,CREV}$: repassivation potential. Alloy 22 in 1M NaCl solution at 90 °C. Scan rate: 0.167 mV.s^{-1} + 2 h hold at 30 μA + 2 h potentiostatic steps -10 mV apart from each other.

In the THE method the anodic current reaches very low anodic values (but not lower than 2 $\mu A.cm^{-2}$) and low anodic potentials (approx. 0.1 V_{SCE} in the system corresponding to Figure 3) below the transpassive domain. In Figure 3 the whole experiment lasted about 14 hours but the duration depends on the area exposed to crevice attack during the galvanostatic step. The higher this area in the initial attack is, the lower the end potential of the galvanostatic step will be and the lower the duration of the test accordingly.

Figure 4 shows the aspect of a PCA specimen after a THE test. The morphology of the attack is similar to that found after the CPP test but much less pronounced under the crevice former teeth and sometimes only small edge pits are found under the crevice formers. The surface of the specimen that was not covered by the crevice former does not show signs of transpassive dissolution.

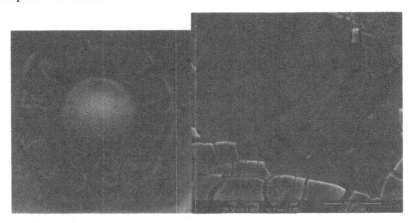

Figure 4: Photograph and SEM images of Alloy 22 PCA specimen after THE test in 1M NaCl at 90°C.

The problem posed to the CPP method regarding transpassive dissolution is avoided in the THE method. Nevertheless, is a high time consuming method.

Potentiodynamic polarization plus intermediate potentiostatic hold method

In the mixed PD – POT – PD method the potential scan was started at a potential slightly lower to E_{CORR}[13] or at the end potential of a 10 minute galvanostatic cathodic treatment of -50 $\mu A.cm^{-2}$[38], in the anodic direction at a scan rate of 0.167 mV/s. At an anodic potential lower than the transpassive potential the potential is hold at that value during a certain time (several hours[13] or two hours[38]). The potentiodynamic scan was then reversed at a scan rate of 0.0167 $mV.s^{-1}$[13] or 0.100 $mV.s^{-1}$[38] until the current changes from anodic to cathodic.

Figure 5 shows the results for the application of the mixed PD – Pot – PD method to an Alloy 22 PCA specimen immersed in 1M NaCl. Results are represented as a polarization curve E vs log I. In the figure, the method used to determine the repassivation potential E_{CO} is illustrated. Sometimes E_{R2} values are used, indicating the potential at which current density reach 2 μA.cm^{-2} in the reverse scan.

In the mixed PD – Pot – PD method with a potential hold at 0.3 V_{SCE} (E_{hold}) the anodic current and potential values lay between those obtained with the other two methods. With this technique the sample was maintained for more than 2h at potentials significantly lower than those of the CPP method. In the PD – Pot – PD method and in the conditions indicated in Figure 5 the whole experiment lasts about 6 hours.

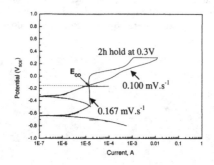

Figure 5: PD – Pot – PD method. Polarization curve E *vs* log I. E_{CO}: cross-over repassivation potential. Alloy 22 in 1M NaCl solution at 90 °C. Forward scan rate: 0.167 mV.s^{-1} + 2h potentiostatic hold at 0.300 V_{SCE} + reverse scan rate 0.100 mV.s^{-1}

Figure 6 shows the aspect of a PCA specimen after a PD – Pot – PD test.

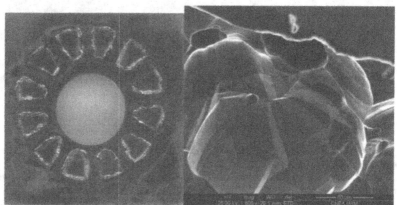

Figure 6: Photograph and SEM images of Alloy 22 PCA specimen after PD-Pot-PD test in 1M NaCl at 90°C.

The morphology of the attack is similar to that found after the CPP and THE tests. The attack is less pronounced under the crevice former teeth than that obtained after the CPP test

(Figure 2) and much pronounced than after the THE test (Figure 4). The surface of the specimen which was not covered by the crevice former does not show signs of high transpassive dissolution.

The problem posed to the CPP method regarding transpassive dissolution is avoided in the PD – Pot – PD method, and this method is less time consuming than the THE method. Besides, the PD – Pot – PD method has not been deeply analyzed and can be modified in order to reduce time consumption without losing prediction capabilities.

Carranza et al[38] measured E_{CO} using the PD – Pot –PD method varying E_{hold}. They tested Alloy 22 in 1M NaCl solution at 90°C using PCA specimens. Figure 7 shows the variation of the repassivation potential (E_{CO}) as a function of E_{hold} obtained by the authors for this highly aggressive solution. The full line corresponds to the linear fit of the data considering only $E_{CO} \geq$ 0.000 V_{SCE}. It can be observed that E_{CO} does not varied significantly with E_{hold} or even diminishes slightly with E_{hold} for $E_{hold} > E_{CO} + 0.1$ V, meaning that an increase in the severity of the crevice attack may produce a slight increase of the repassivation potential. A systematic study of this technique is needed in order to select the best parameters for the test. These parameters are the reverse scan rate and E_{hold}. A carefully selection of these parameters would render conservative repassivation potentials in short time experiments.

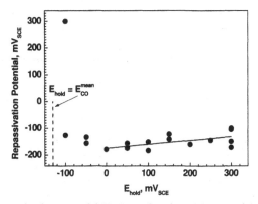

Figure 7: Repassivation potencial (E_{CO}) as a function of the potential applied during the potentiostatic step in PD – Pot – PD tests (E_{hold})[38]. Alloy 22 in 1M NaCl solution at 90 °C. Solid line: linear fit of the data considering only those for which $E_{hold} \geq 0.000$ V_{SCE}.

The main differences of the three electrochemical methods analyzed in this paper are summarized in Table II.

EFFECT OF TORQUE AND CREVICE FORMER MATERIAL

Before comparing the repassivation potentials of the three methods, considerations must be taken regarding the crevicing conditions. Most of the crevice corrosion studies for Alloy 22 use artificially creviced specimens. The crevicing mechanism commonly used is based on ASTM G

48[8] which contained 24 artificially creviced spots or teeth formed by PTFE (polytetrafluoroethylene) washers or PTFE-wrapped ceramic washers. These specimens include multiple crevice assemblies (MCA)[13-19,33,39] and prism crevice assemblies (PCA)[12,31,37,40,41,42] Different amounts of torque were applied to the washers to obtain a tight crevice. However, the depth and the gap of the obtained crevice have never been quantified.[29] It has been argued that PTFE-wrapped ceramic washers are more demanding than PTFE washers.[12,29,31] Pressure applied to the latter can relax in time as PTFE is a polymeric material.[31] Another type of specimens which contains an unintentional artificial crevice formed by a PTFE compression gasket is a variation of the ASTM G 5 specimen usually referred to as prismatic.[20,43-45] This type of PTFE compression gasket is not considered demanding enough to study crevice corrosion of Alloy22.[45]

TABLE II – Elapsed time, and upper E and i values for the electrochemical methods. Values for PCA Alloy 22 specimens in 1M NaCl at 90°C. CPP: Cyclic Potentiodynamic Polarization. THE: Tsujikawa-Hisamatsu Electrochemical. PD – Pot – PD: Mixed Potentiodynamic and Potentiostatic. E_{hold}: potential of the potentiostatic step of PD – POT – PD method

METHOD	TESTING TIME h	UPPER E VALUE V_{SCE}	UPPER i VALUE A.cm^{-2}
CPP	5 to 6	~ 1	~ 5 m
THE	14 to 24	~ 0.1	2 μ to 20 μ
PD – POT -PD	2 to 6	$0 \le E_{hold} \le 0.3$	20 μ to 0.4 m

The crevice depth (x) and the crevice gap (g) define the crevice geometry and the ratios x^2/g and x/g are the parameters most commonly used as scaling factors for correlations in the study of geometrical aspects of crevice corrosion.[46-50] Therefore, the material of the crevice former and the torque applied to it, may be of fundamental significance in crevice corrosion susceptibility. Factors affecting the crevice geometry, and the surface roughness and mechanical properties of the crevice former could affect the severity of crevice corrosion.

Shan and Payer[51] tested Alloy 22 with PTFE, PCTFE (Kel-F™), ceramic, and PTFE tape covered ceramic crevice formers in Alloy 22. They found that PTFE tape covered ceramic was the most active crevice former. Ceramic crevice formers not covered with PTFE tape caused no crevice corrosion. In a recent publication, Shan and Payer[52] studied the crevice corrosion of Alloy 22 in 4M NaCl solutions at 100 °C. They found that surface finish of the crevice former and metal specimen affected crevice corrosion. With 1200 grit surface finish, the crevice corrosion of Alloy 22 initiated immediately after the anodic potential was applied, while when the surface finish of Alloy 22 crevice former/specimen was rougher than 600 grit no crevice corrosion was observed. Smoother surfaces created tighter crevices (smaller gap) and correspondingly more severe crevice corrosion.

Carranza et al[38] found that the torque applied to the PTFE tape covered ceramic crevice former was not crucial for repassivation potential determinations provided that torque values higher than 2 N.m are applied. It seems that PTFE tape act as a surface smoother, making PTFE tape covered ceramic one of the best crevice former. The severity of the crevice must be related to the volume of solution allowed within it. The higher the surface roughness, of both the crevice former and the metal, the higher the volume of stagnant solution inside the crevice and

442

the more difficult will be to maintain critic conditions for crevice corrosion initiation and propagation. All the other material tested as crevice formers, PTFE, Metal, ceramic, PCTFE, are less demanding and could render higher $E_{R;CREV}$ and even prevent crevice corrosion to form depending on the surface finish of both metal and crevice former.

Therefore, in order to compare repassivation potentials of different electrochemical techniques, care must be taken so that similar crevicing conditions are used for all the methods besides similar environmental conditions like temperature and solution composition.

REPASSIVATION POTENTIALS

Carranza et al[38] measured repassivation potentials obtained for the three electrochemical methods using Alloy 22 PCA specimens with a PTFE tape wrapped ceramic crevice formers at different applied torque values. They used 1M NaCl at 90°C as an aggressive solution and 0.15M $NaNO_3$ + 1M NaCl at 90°C as a less aggressive solution (Figure 8).

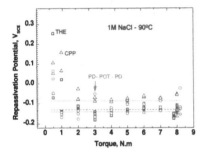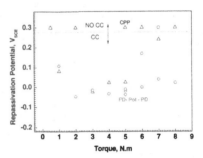

Figure 8 Variation of the repassivation potential with applied torque for Alloy 22 at 90°C. Left: 1M NaCl solution. Right: 0.15M $NaNO_3$ + 1M NaCl solution. Using different electrochemical techniques: Δ CPP, • THE and O PD – POT – PD.[38]

They found that repassivation potentials are practically identical for THE and PD – POT –PD methods in the aggressive 1M NaCl solution and when torque values are higher than 2N.m. These are then considered highly aggressive conditions. E_{CO} values obtained by CPP technique render slightly higher values. A higher intensity of the crevice corrosion attack and/or upper potentials in the transpassive domain may justify these slightly higher values.

For weakly aggressive solutions containing an inhibitor like nitrate ions it seems that CPP tests are not severe enough to produce crevice corrosion attack under the crevice formers. PD – POT – PD method is more effective in producing crevice attack in the same experimental conditions. Systematic research is needed in order to establish the best parameter of this technique to produce conservative and reproducible repassivation potential values. It is expected that THE method render more reproducible repassivation potentials for these weakly aggressive solutions.

SUMMARY

The susceptibility of Alloy 22 to crevice corrosion can be evaluated through the value of the repassivation potential. Existing electrochemical methods seem to give similar results on repassivation potentials when highly aggressive solutions are used. For aggressive environements like 1M NaCl solutions the three methods are suitable for crevice corrosion predicitons.

When less aggressive conditions are present, i.e. when an inhibitor is added or when the torque is released, the CPP method is not demanding enough to produce crevice corrosion. It is not yet clear enough if this is due to the short period of time that the sample is subjected to potentials higher than the repassivation potential, or to the fact that transpassivity may stifle crevice attack. The mixed PD – Pot – PD method seems to be more demanding for less aggressive environments than the CPP method, and it could be still improved by modifying E_{hold}, potential reverse scan rate, and waiting time at E_{hold}. It is expected that THE method be also demanding for less aggressive solutions but it is highly time-consuming.

PTFE tape covered ceramic crevice formers demonstrated to be able to produce severe crevices if torque values higher than 1 - 2 N.m are applied, depending on the aggressiveness of the environment. PTFE tape act as a surface smoother diminishing the volume of stagnant solution inside the crevice.

ACKNOWLEDGMENTS

R. M. Carranza acknowledges financial support from the ANPCyT of the Ministerio de Educación, Ciencia y Tecnología from Argentina.

REFERENCES

1. R.B. Rebak in Corrosion and Environmental Degradation, Volume II, p. 69 (Wiley-VCH, 2000: Weinheim, Germany).
2. ASTM International, Annual Book of ASTM Standards, Volume 02.04 "Non-Ferrous Metals" Standard B 575 (West Conshohocken, PA: ASTM International, 2002).
3. R.B. Rebak and J.H. Payer, "Passive Corrosion Behavior of Alloy 22," International High-Level Radioactive Waste Management Conference, Las Vegas, NV April 30 to May 04, 2006 (American Nuclear Society 2006: La Grange Park, IL).
4. R.B. Rebak and P. Crook, Transportation, Storage and Disposal of Radioactive Materials, PVP-Vol. 483, p. 131 (American Society of Mechanical Engineers, 2004: New York, NY).
5. R.B. Rebak and P. Crook, *Advanced Materials and Processes*, February 2000.
6. R.B. Rebak and P. Crook, Proceeding of the Symposium Critical Factors in Localized Corrosion III, PV 98-17, p. 289 (The Electrochemical Society, 1998: Pennington, NJ).
7. R.B. Rebak and J.C. Estill, Review of corrosion modes for Alloy 22 regarding lifetime expectancy of nuclear waste containers, Mat. Res. Soc. Symp. Proc. Vol. 757-II4.1 (MRS, 2003: Warrendale, PA), pp. 713–721.
8. Annual Book of ASTM Standards, Volume 03.02 Wear and Erosion; Metal Corrosion, (West Conshohocken, PA: ASTM International, 2005).

9. K.J. Evans, L.L. Wong and R.B. Rebak, Transportation, Storage and Disposal of Radioactive Materials, PVP-Vol. 483, p. 137 (American Society of Mechanical Engineers, 2004: New York, NY).

10. D.C. Agarwal, *Uhlig's Corrosion Handbook*, p. 831 (J.Wiley & Sons, Inc.,2000: New York).

11. Annual Book of ASTM Standards, Volume 03.02 Wear and Erosion; Metal Corrosion, (West Conshohocken, PA: ASTM International, 2008).

12. K.J. Evans, A. Yilmaz, S.D. Day, L.L. Wong, J.C. Estill and R.B. Rebak, JOM, January 2005, (TMS, 2005. Warrendale, PA) pp 56-61.

13. D.S. Dunn, Y.-M. Pan, K. Chiang, L. Yang, G.A. Cragnolino and X. He, JOM, January 2005, (TMS 2005: Warrendale, PA), pp 49-55.

14. K.J. Evans, R.B. Rebak, J. ASTM International Proc. Symp, on Advances in Electrochemical Technology for Corrosion Monitoring and Measurement, Vol. 4, N° 9, 2007.

15. S. Tsujikawa and Y. Hisamatsu, Corr. Eng. (Japan), 29, 37 (1980).

16. K.A. Gruss, G.A. Cragnolino, D.S. Dunn and N. Sridhar; Paper 98149 Corrosion/1998, (Houston, TX: NACE International 1998).

17. B.A. Kehler, G.O. Ilevbare and J.C. Scully, Corrosion, Vol. 57, N° 12, p. 1042 (2001).

18. K.J. Evans, S.D. Day, G.O. Ilevbare, M.T. Whalen, K.J. King, G.A. Hust, L.L. Wong, J.C. Estill and R.B. Rebak, Transportation, Storage and Disposal of Radioactive Materials, PVP-Vol. 467, p. 55 (ASME, 2003: New York, NY).

19. D.S. Dunn, L. Yang, C. Wu and G.A. Cragnolino, Mat. Res. Soc. Symp. Proc. Vol. 824 (MRS, 2004: Warrendale, PA).

20. K.J. Evans, and R.B. Rebak, PV 2002-13, p.344 (The Electrochemical Society 2002, Pennington, NJ)

21. K.J. Evans and R.B. Rebak, Repassivation Potential of Alloy 22 in Chloride plus Nitrate Solutions using the Potentiodynamic-Galvanostatic-Potentisotatic Method, in Scientific Basis for Nuclear Waste Management XXX, Vol. 0985-NN03-13 (Materials Research Society 2007: Warrendale, PA).

22. R.B. Rebak, K.J. Evans and G. O. Ilevbare, Crevice repassivation Potentials of Alloy 22 in Simulated Concentrated Waters, paper 07584, Corrosion/07, (Houston, TX: NACE International 2007).

23. D.S. Dunn, L. Yang, Y.-M. Pan and G.A. Cragnolino, Paper 03697, Corrosion/03 (Houston, TX: NACE International 2003).

24. D. S. Dunn and C. S. Brossia, Paper 02548, Corrosion/02 (Houston, TX: NACE International 2002).

25. G.O. Ilevbare, K.J. King, S.R. Gordon, H.A. Elayat, G.E. Gdowski and T.S.E. Gdowski, J. Electrochem. Soc. 152/12 (2005) B547-B554.

26. J.H. Lee, T. Summers and R.B. Rebak, Paper 04692, Corrosion/04 (Houston, TX: NACE International 2004).

27. G.A. Cragnolino, D.S. Dunn and Y.-M. Pan "Localized Corrosion Susceptibility of Alloy 22 as a Waste Package Container Material," Scientific Basis for Nuclear Waste Management XXV, Vol. 713, p. 53 (Materials Research Society 2002: Warrendale, PA).

28. G.O. Ilevbare, Corrosion, 62, 340 (2006).

29. R.B. Rebak, Paper 05610 Corrosion/05, (Houston, TX: NACE International 2005).

30. R.B. Rebak, Mechanisms of Inhibition of Crevice Corrosion in Alloy 22, in Scientific Basis for Nuclear Waste Management XXX, Vol. 0985-NN08-04 (Materials Research Society 2007: Warrendale, PA).

31. R.M. Carranza, M.A. Rodríguez and R.B. Rebak, Corrosion 63/05 (2007) 480-490.
32. C.S. Brossia, L. Browning, D.S. Dunn, O.C. Moghissi, O. Pensado and L. Yang "Effect of Environment on the Corrosion of Waste Package and Drip Shield Materials," Publication of the Center for Nuclear Waste Regulatory Analyses (CNWRA 2001-03), September 2001.
33. D.S. Dunn, Y.-M. Pan, L. Yang, G.A. Cragnolino, Corrosion, Vol.61 N° 11 p. 1078 (2005).
34. D.S. Dunn, Y.-M. Pan, L. Yang and G.A. Cragnolino, Corrosion, 62, 3-12 (2006).
35. R.M. Carranza, M.A. Rodríguez, and R.B. Rebak, Anodic and Cathodic Behavior of Mill Annealed and Topologically Closed Packed Alloy 22 in Chloride Solutions, paper 08579, Corrosion/08, (Houston, TX: NACE International 2008).
36. G.O. Ilevbare, R.A. Etien, J.C. Estill, G.A. Hust, A. Yilmaz, M.L. Stuart, and R.B. Rebak, "Anodic Behavior of Alloy 22 in High Nitrate Brines at Temperatures Higher than 100°C", Paper 93423 in the Proceedings of PVP2006-ICPVT-11, 2006 ASME Pressure Vessels and Piping Division Conference, July 23-27, 2006, Vancouver, BC, Canada.
37. R.M. Carranza, C.M. Giordano, M.A. Rodríguez, and R.B. Rebak. paper 08578, Corrosion/08, (Houston, TX: NACE International 2008).
38. R.M. Carranza, M.A. Rodríguez, C.M. Giordano, R.B. Rebak, paper 09427, accepted for Corrosion/09, (Houston, TX: NACE International 2009).
39. V. Jain, D.S. Dunn, N. Sridhar and L.Yang, Paper 03690, Corrosion/03, (Houston, TX: NACE International 2003).
40. R.M. Carranza, M.A. Rodríguez and R.B. Rebak, Paper 06626, Corrosion/06 (Houston, TX: NACE International 2006).
41. A. Yilmaz, P. Pasupathi and R.B. Rebak, Paper PVP2005-71174, ASME Pressure Vessels and Piping Conference, 17-21 July 2005, Denver, CO (ASME, 2005: New York, NY).
42. R.M. Carranza, C.M. Giordano, M.A. Rodríguez, R.B. Rebak, paper 06627, Corrosion/06, (Houston, TX: NACE International 2006).
43. M.A. Rodríguez, R.M. Carranza and R.B. Rebak, Met. Trans. A Vol. 36A-No.5, 2005, pp. 1179-1185.
44. R.B. Rebak, T.S.E. Summers, T. Lian, R.M. Carranza, J.R. Dillman, T. Corbin and P. Crook, Paper 02542, Corrosion/02, (Houston, TX: NACE International 2002).
45. M. A. Rodríguez, Ms. Sc. Thesis, University of General San Martín, Argentina (2003).
46. P. Combrade in Corrosion Mechanisms in Theory and Practice, p. 349, Second Edition, edited by P. Marcus (Marcel Dekker 2002, New York, NY).
47. H.W. Pickering, J. Electrochem. Soc. 150 (5) pp. K1-K13, 2003.
48. J. S. Lee, M.L. Reed and R.G. Kelly, J. Electrochem. Soc. 151 (7) pp. B423-B433, 2004.
49. M. Vankeerberghen, Corrosion Vol. 60 N° 8, p. 707 (2004).
50. Z. Szklarska-Smialowska, Pitting and Crevice Corrosion of Metals, Chap. 19, (Houston, TX: NACE International 2005), pp. 459-497.
51. X. Shan, J.H. Payer, paper 07582, Corrosion/07, (Houston, TX: NACE International 2007).
52. Xi Shan and Joe H. Payer, Effect of crevice former on the evolution of crevice damage, paper 08575, Corrosion/08, (Houston, TX: NACE International 2008).

Mater. Res. Soc. Symp. Proc. Vol. 1124 © 2009 Materials Research Society   1124-Q09-02

A Comparison of the Corrosion Resistance of Iron-Based Amorphous Metals and Austenitic Alloys in Synthetic Brines at Elevated Temperature

Joseph Farmer[1]

[1]Lawrence Livermore National Laboratory, Livermore, CA 94550, USA

ABSTRACT

Several hard, corrosion-resistant and neutron-absorbing iron-based amorphous alloys have now been developed that can be applied as thermal spray coatings. These new alloys include relatively high concentrations of Cr, Mo, and W for enhanced corrosion resistance, and substantial B to enable both glass formation and neutron absorption. The corrosion resistances of these novel alloys have been compared to that of several austenitic alloys in a broad range of synthetic brines, with and without nitrate inhibitor, at elevated temperature. Linear polarization and electrochemical impedance spectroscopy have been used for *in situ* measurement of corrosion rates for prolonged periods of time, while scanning electron microscopy (SEM) and energy dispersive analysis of X-rays (EDAX) have been used for *ex situ* characterization of samples at the end of tests. The application of these new coatings for the protection of spent nuclear fuel storage systems, equipment in nuclear service, steel-reinforced concrete will be discussed.

INTRODUCTION

The outstanding corrosion resistance that may be possible with amorphous metals was recognized several years ago [1-3]. Compositions of several Fe-based amorphous metals were published, including several with very good corrosion resistance. Examples included: thermally sprayed coatings of Fe-10Cr-10-Mo-(C,B), bulk Fe-Cr-Mo-C-B, and Fe-Cr-Mo-C-B-P [4-6]. The corrosion resistance of an Fe-based amorphous alloy with yttrium (Y), $Fe_{48}Mo_{14}Cr_{15}Y_2C_{15}B_6$ was also been established [7-9]. Yttrium was added to this alloy to lower the critical cooling rate. Several nickel-based amorphous metals were developed that exhibit exceptional corrosion performance in acids, but are not considered in this study, which focuses on Fe-based amorphous metals. Thermal spray coatings of crystalline nickel-based alloy coatings have been deposited with thermal spray technology, but appear to have less corrosion resistance than comparable nickel-based amorphous metals [10]. Two Fe-based amorphous alloys have been found with exceptional corrosion resistance compared to other such Fe-based amorphous alloys, and can be applied as a protective thermal spray coating: SAM2X5 ($Fe_{49.7}Cr_{17.7}Mn_{1.9}Mo_{7.4}W_{1.6}B_{15.2}C_{3.8}Si_{2.4}$)

447

and SAM1651 ($Fe_{48}Mo_{14}Cr_{15}Y_2C_{15}B_6$). These materials incorporate chromium (Cr), molybdenum (Mo) and tungsten (W) for enhanced corrosion resistance, boron (B) to enable glass formation and neutron absorption, and yttrium (Y) for lower critical cooling rates [11-17]. SAM2X5 appears have significantly better corrosion resistant than its Mo-deficient parent alloy ($Fe_{52.3}Cr_{19}Mn_2Mo_{2.5}W_{1.7}B_{16}C_4Si_{2.5}$).

EXPERIMENTAL

Thermal spray coatings

The coatings discussed here were made with the high-velocity oxy-fuel (HVOF) process, which involves a combustion flame, and is characterized by gas and particle velocities that are three to four times the speed of sound (mach 3 to 4). This process is ideal for depositing metal and cermet coatings, which have typical bond strengths of 5,000 to 10,000 pounds per square inch (5-10 ksi), porosities of less than one percent (< 1%) and extreme hardness. The cooling rate that can be achieved in a typical thermal spray process such as HVOF are on the order of ten thousand Kelvin per second (10^4 K/s), and is high enough to enable many alloy compositions to be deposited above their respective critical cooling rate, thereby maintaining the vitreous state. However, the range of amorphous metal compositions that can be processed with HVOF is more restricted than those that can be produced with melt spinning, due to the differences in achievable cooling rates. Both kerosene and hydrogen have been investigated as fuels in the HVOF process used to deposit SAM2X5 and SAM1651.

X-ray diffraction

X-ray diffraction (XRD) measurements of SAM2X5 and SAM1651 gas atomized powders and HVOF coatings has been performed, verifying that both the feed powder, as well as the final thermal spray coatings were indeed amorphous. In general, a broad halo is observed at $2\theta \sim 44°$ which indicates that these coatings was predominately amorphous [21-22]. In cases where very small sharp peaks are observed with SAM2X5, they are generally attributed to the presence of minor crystalline phases including Cr_2B, WC, $M_{23}C_6$ and bcc ferrite, which are known to have a detrimental effect on corrosion performance. These potentially deleterious precipitates deplete the amorphous matrix of those alloying elements, such as chromium, responsible for enhanced passivity. Coatings with less residual crystalline phase have been observed.

Salt fog testing

Several reference samples and amorphous-metal coatings have been made and subjected to salt fog testing. Salt fog tests were conducted according to the standard General Motors (GM) salt fog test, identified as GM9540P. Figure 1 shows the condition of several samples after testing: (a) 1018 carbon steel reference specimens; (b) an HVOF SAM2X5 coating on Type 316L stainless steel, (c) an HVOF SAM2X5 coating on Ni-based Alloy C-22 and (d) an HVOF

SAM2X5 coating on half-scale spent nuclear fuel (SNF) container made of Type 316L stainless steel, all after 8 full cycles in GM salt fog test. Clearly, the thermal-spray coatings of SAM2X5 have good resistance to corrosive attack in such environments. Similar testing was done with SAM1651 with similar positive results.

a)   b)   c)   d)

Figure 1. Photographs of (a) 1018 Steel, (b) SAM2X5 on 316L, (c) SAM2X5 on C-22 and (c) SAM2X5 on prototype spent nuclear fuel container.

Immersion testing

Linear polarization, weight-loss and dimensional-change measurements have been made to determine the rates of corrosion of SAM2X5 and SAM1651 HVOF coatings in a wide variety of synthetic brines including but not limited to: natural seawater at 90°C; 3.5-molal NaCl solution at 30°C and 90°C; 3.5-molal NaCl and 0.525-molal KNO₃ solution at 90°C; SDW (simulated dilute water) at 90°C; SCW (simulated concentrated water) at 90°C; and SAW (simulated acidic water) at 90°C. Data for testing in heated seawater and sodium chloride brines, with and without nitrate inhibitor are reported here.

In regard to the SAM2X5 coating, the worst corrosive attack occurred in 3.5m NaCl without nitrate inhibitor at 90°C. The corrosive attack of coated rods and plates is shown in Figures 2 and 3, respectively. The addition of nitrate was observed to be very effective in mitigating corrosive attack of the SAM2X5 coating in these near-boiling brines.

The optimized SAM1651 coating experienced less corrosion in these aggressive environments than the SAM2X5, as shown in Figures 4 and 5. The corrosion performance demonstrated by this amorphous metal coating appears to be outstanding. As shown in Figure 6, in the case of near-boiling seawater, virtually all corrosion spots could be prevented through the use of an inorganic sealant with the thermal spray coating. The demonstrated performance is comparable to, or better than that of Fe- and Ni-based austenitic steels.

Relative corrosion rates based upon linear polarization measurements of SAM2X5 and SAM1651 are respectively: (a) 5-7 and 4-5 μm/yr in natural seawater at 90°C; (b) 71-116 and 4-5 μm/yr in 3.5-molal NaCl solution at 90°C; and (c) 2 and 6 μm/yr in 3.5-molal NaCl with 0.525-molal KNO₃ solution at 90°C. The corrosion rate for SAM1651 in the 3.5-molal NaCl solution was far less than that for SAM2X5, which is consistent with the discussion of Figures 2 through 6. Figure 7 shows that corrosion, when it does occur, does not penetrate to the substrate.

Figure 2. Examples of corrosive attack of optimized HVOF SAM2X5 coatings on rods used for linear polarization and electrochemical impedance spectroscopy after several months in (a) natural seawater 90°C, (b) 3.5m NaCl at 90°C, and (c) 3.5m NaCl + 0.525m KNO$_3$ 90°C.

Figure 3. Examples of corrosive attack of optimized HVOF SAM2X5 coatings on flat plates after several months in (a) natural seawater at 90°C, (b) 3.5m NaCl at 90°C, and (c) 3.5m NaCl + 0.525m KNO$_3$ 90°C. Near boiling 3.5 m NaCl proved to be the most aggressive aqueous environment of those tested.

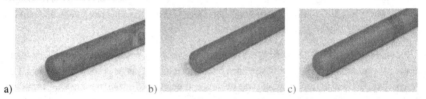

Figure 4. Examples of corrosive attack of optimized HVOF SAM1651 coatings on rods used for linear polarization and electrochemical impedance spectroscopy after several months in (a) natural seawater 90°C, (b) 3.5m NaCl at 90°C, and (c) 3.5m NaCl + 0.525m KNO$_3$ 90°C.

Figure 5. Examples of corrosive attack of optimized HVOF SAM1651 coatings on flat plates after several months in (a) natural seawater at 90°C, (b) 3.5m NaCl at 90°C, and (c) 3.5m NaCl + 0.525m KNO$_3$ 90°C. Near boiling seawater proved to be the most aggressive aqueous environment of those tested.

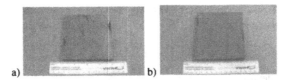

Figure 6. Benefits of using inorganic sealant with optimized HVOF SAM1651 coatings on flat plates after several months in natural seawater at 90°C (a) without sealant and (b) with sealant.

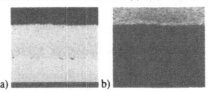

Figure 7. Un-Optimized SAM1651 Coating After Approximately Four Months in 90°C Natural Seawater: (a) Back-Scattered Electron Image – BEI and (b) Secondary Electron Image – SEI.

CONCLUSIONS

Early Fe-based amorphous metal coatings had very poor corrosion resistance and failed salt-fog tests. The HPCRM Program has developed new Fe-based amorphous-metal alloys with good corrosion resistance, high hardness, and exceptional absorption cross-sections for thermal neutrons. More than forty high-performance Fe-based amorphous alloys were systematically designed and synthesized. Cr, Mo and W were added to enhance corrosion resistance; Y was added to lower the critical cooling rate; and B was added to render the alloy amorphous and to enhance capture thermal neutrons. Enriched boron could be used for the further enhancement of the absorption of thermal neutrons. The most profound observations are that the optimized SAM1651 HVOF coating is far more corrosion resistant than the SAM2X5 coating in the most aggressive environment, 3.5 molal NaCl solution without nitrate inhibitor, and that the nitrate anion suppresses corrosive attack of the SAM2X5 coating in concentrated chloride solutions.

ACKNOWLEDGEMENTS

This work was performed under the auspices of the U.S. Department of Energy by Lawrence Livermore National Laboratory under Contract DE-AC52-07NA27344. Work was co-sponsored by the Office of Civilian and Radioactive Waste Management (OCRWM) of the United States Department of Energy (DOE), and the Defense Science Office (DSO) of the Defense Advanced Research Projects Agency (DARPA). The guidance of Jeffrey Walker at DOE OCRWM and Leo Christodoulou at DARPA DSO is gratefully acknowledged. The production of gas atomized powders by The NanoSteel Company and Carpenter Powder Products, and the production of coatings from these powders by Plasma Technology Incorporated and Caterpillar are gratefully acknowledged. Laboratory work by S. Daniel Day is gratefully acknowledged.

REFERENCES

1. M. Telford, Materials Today, 3 (2004) 36-43.

2. N. Sorensen, R. Diegle, Corrosion, Metals Handbook, 9th Ed., Vol. 13, J. R. Davis, J. D. Destefani, Eds., ASME, 1987, pp. 864-870

3. D. Polk, B. Giessen, Metallic Glasses, J. Gilman, H. Leamy, Eds., ASME, 1978, pp. 2-35

4. K. Kishitake, H. Era, F. Otsubo, J. Thermal Spray Tech. 5, 2 (1996) 145-153.

5. S. Pang, T. Zhang, K. Asami, A. Inoue, Materials Transactions 43, 8 (2002) 2137-2142.

6. S. Pang, T. Zhang, K. Asami, A. Inoue, Acta Materialia 50 (2002) 489-497.

7. F. Guo, S. Poon, G. Shiflet, Metallic Applied Physics Letters, 83, 13 (2003) 2575-2577.

8. Z. Lu, C. Liu, W. Porter, Metallic Applied Physics Letters, 83, 13 (2003) 2581-2583.

9. V. Ponnambalam, S. Poon, G. Shiflet, JMR 19, 5 (2004) 1320.

10. D. Chidambaram, C. Clayton, M. Dorfman, Surf. Coatings Tech. 176 (2004) 307-317.

11. J. Farmer et al., JMR 22, 8 (2007) 2297-2311.

12. J. Farmer et al., Critical Factors in Localized Corrosion 5, N. Missert, Ed., Electrochem. Soc. Trans. 3 (2006).

13. J. Farmer et al., Sci. Basis Nucl. Waste Mgmt. XXX, MRS Symp. Series, Vol. 985, 2006.

14. J. Farmer et al., PVP2006-ICPVT11-93835, ASME, New York, NY, 2006

15. J. Farmer et al. PVP2005-71664, ASME, New York, NY, 2005.

16. D. Branagan, U.S. Pat. Appl. No. 20040250929, Filed Dec. 16, 2004.

17. D. Branagan, U.S. Pat. Appl. No. 20040253381, Filed Dec. 16, 2004.

18. T. Lian et al., Sci. Basis Nucl. Waste Mgmt. XXX, MRS Symp. Series, Vol. 985, 2006.

19. J-S. Choi et al., Sci. Basis Nucl. Waste Mgmt. XXX, MRS Symp. Series, Vol. 985, 2006.

20. J. C. Farmer, J. Nucl. Tech. 161, 2 (2008) 169-189.

21. C. K. Saw, Nanotechnologies for the Life Science, Challa Kumar, Ed., Wiley-VCH Verlag GmbH and Company, KGaA, Weinheim, 2006.

22. C. K. Saw, R. B. Schwarz, J. Less-Common Metals, 140 (1988) 385-393.

Mater. Res. Soc. Symp. Proc. Vol. 1124 © 2009 Materials Research Society   1124-Q09-03

Studies of the Effect of Chloride and Nitrate on the Propagation of Localized Corrosion of Alloy 22

Fraser King,[1] Jingli Luo,[2] Licai Mao,[2] Mick Apted,[3] John H. Kessler,[4] and Andrew Sowder[4]

[1]Integrity Corrosion Consulting Ltd, Nanaimo, BC, Canada V9T 1K2
[2]Dept. Materials and Chem. Eng., University of Alberta, Edmonton, AB, Canada T6G 2R3
[3]Monitor Scientific, Denver, CO, USA 80235
[4]Electric Power Research Institute, Charlotte, NC, USA 28262

ABSTRACT

Alloy 22 is susceptible to localized corrosion in concentrated chloride solutions at elevated temperature. A number of oxyanions, most notably nitrate, sulfate, and carbonate, inhibit the aggressiveness of the chloride ion.

The results of a preliminary experimental program to study the effects of chloride and nitrate ions on the propagation and stifling of the crevice corrosion of Alloy 22 are described. A coupled-electrode technique is being used, in which a large planar cathode is electrochemically coupled to a smaller creviced anode electrode.

The crevice corrosion of Alloy 22 is difficult to initiate, even in concentrated chloride solutions at elevated temperature and with the imposition of a constant current to force initiation. In the presence of nitrate, rapid stifling occurred following the period of imposed galvanostatic initiation. Only limited initiation (as evidenced by the value of the potential during the initiation phase and the absence of extensive localized attack) occurred in a solution with a $[NO_3^-]:[Cl^-]$ ratio of 0.1.

The current results indicate that, as well as inhibiting the initiation of crevice corrosion, nitrate ions also inhibit the propagation of localized attack, although the mechanism is currently uncertain.

INTRODUCTION

Crevice corrosion of Alloy 22 waste packages in the Yucca Mountain repository may occur during the retrograde cooling phase of the thermal pulse, but only if certain conditions are met [1]. First the drip shields must have failed, permitting seepage drips to contact the waste package surface. Second, there must exist on the waste package surface some type of occluded region that promotes the spatial separation of anodic and cathodic processes. Third, the combination of temperature and the composition of the seepage drips must be such that the material is susceptible to the initiation of localized corrosion. In this last regard, chloride ions are known to promote the initiation of localized corrosion, whereas a number of oxyanions, most notably nitrate, sulfate, and carbonate, inhibit the process.

Based on the difference between the corrosion (E_{CORR}) and repassivation (E_{RCREV}) potentials as a measure of the susceptibility to localized corrosion, nitrate inhibits the aggressiveness of chloride ions [2]. Although nitrate ions may ennoble E_{CORR} they shift the value of E_{RCREV} to more positive potentials to a greater extent, so that the potential difference ($E_{RCREV}-E_{CORR}$) increases with increasing nitrate:chloride ratio. As a consequence, a nitrate:chloride ratio of 0.1 is considered sufficient to inhibit the initiation of localized corrosion of Alloy 22 for a wide range of chloride concentrations and temperatures.

If crevice corrosion does initiate, it is by no means certain that the crevice will continue to propagate. Studies in nitrate-free chloride solutions have shown that there is a tendency for the rate of crevice propagation to slow with time, in a process commonly referred to as stifling [3,4]. The mechanism of this stifling process is not known, but is of interest since rapid crevice propagation, were it to occur, could lead to relatively early waste package failure. Furthermore, since nitrate is known to have an effect on E_{RCREV}, which, despite its use as a measure of crevice corrosion <u>initiation</u>, is actually the potential at which an actively <u>propagating</u> crevice re-passivates, it might be expected that nitrate would also affect the stifling process.

The aims of the present work, therefore, are to investigate further the mechanism of stifling and, in particular, to determine the role of nitrate in inhibiting the propagation of the crevice corrosion of Alloy 22

EXPERIMENTAL

The specimens were made from 3/16"-thick Alloy 22 plate material (Allegheny Ludlum Corp.). The chemical composition is listed in Table I. Specimens were polished to 600 grit.

Figure 1 shows a schematic of the specimen configuration, which was chosen to avoid localized attack at the crevices formed by the nuts and washers by raising these locations above the level of the solution. Both a PTFE-metal and a metal-metal crevice were formed by placing an Alloy 22 block (2.2 cm × 1.7 cm × 0.5 cm) between the PTFE crevice former and the second, bent Alloy 22 specimen. The Alloy 22 block sample was small enough to fit in the Auger spectrometer chamber for surface analysis and had the second advantage of providing both a metal-metal and a metal-PTFE crevice in the same experiment.

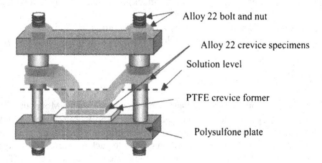

Alloy 22 bolt and nut

Alloy 22 crevice specimens

Solution level

PTFE crevice former

Polysulfone plate

Figure 1: Schematic Illustration of Creviced Electrode Geometry.

Figure 2 shows the electrochemical test cell used to study the crevice corrosion of Alloy 22 at a temperature of 120°C. The counter and planar electrodes were prepared from the same Alloy 22 material as the working (creviced) electrodes. The creviced working electrode was galvanically coupled to the counter electrode through a zero resistance ammeter and the coupled current and, with the use of a custom-made internal Ag/AgCl/KCl reference electrode, the potentials of the creviced and planar electrodes were measured as a function of time. All potentials have been converted to SHE (standard hydrogen electrode), unless otherwise specified.

Table I. Elemental Composition of the Alloy 22 Plate.

Element	Ni	Cr	Mo	W	Fe	Co	V	Mn	Cu	Si	P	C	S
Weight %	56.53	21.23	13.43	2.64	4.51	0.904	0.170	0.154	0.065	0.072	0.015	0.0073	0.0002
Atomic %	58.99	25.01	8.57	0.88	4.95	0.940	0.204	0.172	0.063	0.160	0.030	0.037	0.0004

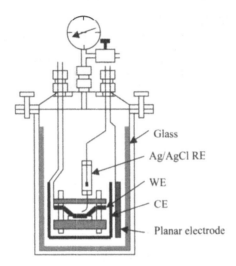

Figure 2: Schematic of Glass-lined Autoclave and the Experimental Arrangement. RE, WE, and CE stand for reference electrode, working electrode, and counter electrode, respectively.

Tests were performed in aerated 5 $mol \cdot dm^{-3}$ NaCl solution, with additions of 0, 0.1 $mol \cdot dm^{-3}$, or 0.5 $mol \cdot dm^{-3}$ NaNO$_3$ (Table II), providing [NO$_3^-$]:[Cl$^-$] ratios of 0, 0.02 and 0.1, respectively. A second set of tests were conducted in CaCl$_2$ + NaNO$_3$ mixtures, but the results are not discussed in detail here.

Crevice corrosion was initiated by applying a constant anodic current density of 10 $\mu A \cdot cm^{-2}$ to the working electrode for a set period, or until such time that initiation was thought to have taken place based on the behavior of the creviced potential (see below). Following initiation, the initiating current was switched off and the propagation process followed by measuring the coupled current and the potentials of the creviced and planar electrodes.

After each experiment, the creviced specimen was carefully rinsed with water, opened and examined for indications of localized attack. In some cases, the ratio of nitrate to chloride and the distribution of alloying elements in the creviced region was measured using X-ray photoelectron spectroscopy (XPS) and Auger electron spectroscopy (AES), respectively.

RESULTS

Crevice corrosion tests

Crevice corrosion could be electrochemically initiated in 5 mol·dm^{-3} NaCl solutions, both with and without added NO$_3$. Figure 3 shows the variation of the potential of the creviced samples during an 11-hour initiation period during which a constant current of 10 µA·cm^{-2} was applied (based on the geometric area of the creviced region).. Crevice initiation, as characterized by an initial positive shift in potential followed by a decrease to more-active values, was observed in the absence of NO$_3$ (specimen #15) and for a [NO$_3$]:[Cl] ratio of 0.02 (specimen #11). The initiation time (as measured by time at which the potential drops to active values) was shorter, and the degree of initiation (as indicated by the absolute value of the potential post-initiation) was greater, in the absence of NO$_3$. Visible corrosion was observed following these two tests but, interestingly, crevice corrosion initiated at the metal-metal crevice in the absence of NO$_3$ (specimen #15) but at the metal-PTFE crevice in the presence of NO$_3$ (specimen #11). Generally, it is considered that a metal-PTFE crevice is required to initiate localized corrosion of Alloy 22, although recent evidence suggests that a PTFE tape covered crevice former creates the most severe conditions [5,6].

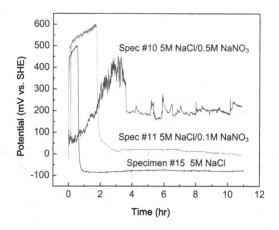

Figure 3: Time Dependence of the Potentials of the Creviced Electrode During the Galvanostatic Initiation Stage in 5 mol·dm^{-3} NaCl + NaNO$_3$ Solutions at 120°C.

Initiation was more difficult to achieve in the 5 mol·dm^{-3} NaCl solution at a [NO$_3$]:[Cl] ratio of 0.1 (specimen #10, Figure 3). An active potential shift characteristic of crevice initiation was observed after 3.8 hours, but the potential did not drop to the more-active values seen for specimens #15 and #10, but instead remained relatively noble and constant at ~0.2 V$_{SHE}$. The coupled current after the 11-hour initiation period was small (0.5 µA), suggesting minimal crevice propagation. In order to induce greater crevice corrosion on Specimen #10, a second period of galvanostatic initiation was imposed for a further 24 hours with a higher applied current density of 26 $\mu A \cdot cm^{-2}$ (Figure 4). Although the potential once again dropped after ~2.5 hours of galvanostatic polarization, the more-active potential (of ~0.1 V_{SHE}) was not maintained and the potential shifted to more passive values after ~15 hours. No evidence for crevice corrosion was observed visually on either face of the specimen at the end of the test.

The extent of crevice propagation of those crevices that initiated in the nitrate-free solution and at a $[NO_3^-]:[Cl^-]$ ratio of 0.02 was followed by measuring the time dependence of the coupled current and potential. Figure 5 shows the time dependence of the coupled potential and current and of the potential of the planar specimen following initiation of crevice corrosion in 5 $mol \cdot dm^{-3}$ NaCl solution (specimen #15). Over the 11-day period the coupled current remained relatively constant at between 2 μA and 5 μA, with no indication of stifling. However, over the same period, the potential of the creviced sample drifted to more-passive values suggesting that the crevice would passivate eventually. During the course of the measurements, the difference between the creviced and planar electrode potentials decreased from about 200 mV initially to 75 mV after 11 days, consistent with a slow re-passivation process. Grain boundary pitting was observed on the corroded specimen following the test, although the overall metal loss was minimal.

Following initiation of crevice corrosion at a $[NO_3^-]:[Cl^-]$ ratio of 0.02 (specimen #11) there was an immediate positive shift in the coupled potential followed by a period of relatively constant potential of 0.04-0.05 V_{SHE} (Figure 6). During the same period, the coupled current was $\ll 1$ μA, indicating minimal propagation.

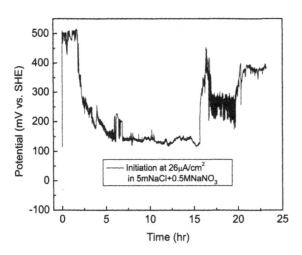

Figure 4: Time Dependence of the Potential of the Creviced Electrode During the Galvanostatic Initiation Stage in 5 $mol \cdot dm^{-3}$ NaCl + 0.5 $mol \cdot dm^{-3}$ NaNO₃ Solutions at a Current Density of 26 $\mu A \cdot cm^{-2}$ at 120°C.

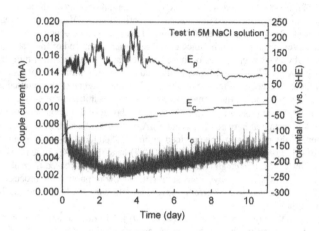

Figure 5: Time Dependence of the Coupled Current and Potential and of the Planar Potential Following Initiation of Crevice Corrosion of Alloy 22 in 5 mol·dm^{-3} NaCl at 120°C.

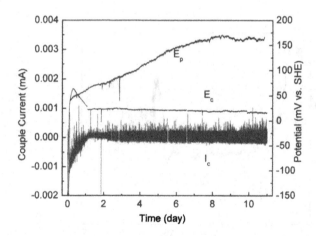

Figure 6: Time Dependence of the Coupled Current and Potential and of the Planar Potential Following Initiation of Crevice Corrosion of Alloy 22 in 5 mol·dm^{-3} NaCl + 0.1 mol·dm^{-3} NaNO₃ at 120°C.

Surface analysis

Surface analysis was performed on a number of samples to investigate the degree of selective dissolution within the crevice and to determine whether there was any evidence the enrichment of nitrate in the crevice solution.

Auger electron spectroscopy was used to map the elemental distribution on the creviced surface of specimen #11 (5 mol·dm^{-3} NaCl + 0.1 mol·dm^{-3} NaNO$_3$, [NO$_3^-$]:[Cl$^-$] ratio of 0.02). Two areas were investigated, one near the edge and another in the center of the visibly corroded area. There was significant enrichment of Mo with respect to Ni in the nine specific locations investigated within the two areas, although greater enrichment was observed in the center of the visibly corroded area. The mean Mo:Ni atomic ratio for the nine locations was 1.6 (0.66 for four locations around the edge and 2.3 for the five locations in the center), compared with an atomic ratio of 0.12 in the as-received material. Molybdenum was also enriched with respect to Cr, especially in the center of the corroded area. The mean Mo:Cr atomic ratio was 0.36 on the edge and 0.48 in the center of the corroded area, compared with a ratio of 0.34 in the as-received material. The Mo enrichment was more pronounced at grain boundaries in the middle of the corroded area, with a Mo:Cr ratio of 0.63 compared with slight depletion (Mo:Cr = 0.28) in the grain body.

Photoelectron X-ray spectroscopy (XPS) was used to determine the NO$_3^-$:Cl$^-$ ratio on different areas of the creviced sample. Table II lists the results of analyses on three of the specimens. There is no evidence that NO$_3^-$ preferentially migrates into the crevice, although the analyses described here are a relatively simple attempt at determining what the crevice chemistry was during the test, and are subject to changes in composition during the cool-down step and sampling procedure.

SUMMARY AND CONCLUSIONS

The crevice corrosion of Alloy 22 is difficult to initiate, even in concentrated chloride solutions at elevated temperature and with the imposition of a constant current to force initiation. However, initiation was observed in 5 mol·dm^{-3} NaCl solutions, both with and without added NO$_3^-$. In the presence of nitrate, rapid stifling occurred following the period of imposed galvanostatic initiation. Only limited initiation (as evidenced by the value of the potential during the initiation phase and the absence of extensive localized attack) occurred in a solution with a [NO$_3^-$]:[Cl$^-$] ratio of 0.1.

No evidence was found for the enhanced transport of nitrate into the crevice region, which has been proposed as a possible explanation for the inhibitive effect of NO$_3^-$. Significant enrichment of Mo (with respect to both Ni and Cr) was observed in the creviced region, especially at grain boundaries.

The current results indicate that, as well as inhibiting the initiation of crevice corrosion, nitrate ions also inhibit the propagation of localized attack, although the mechanism by which they do so is not understood at present.

ACKNOWLEDGMENTS

This work was funded by the Electric Power Research Institute, Charlotte, NC.

461

REFERENCES
1. DOE. 2008. Yucca Mountain Repository License Application. Safety Analysis Report, U.S. Department of Energy, Office of Crystalline Radioactive Waste management, DOE/RW-0573, Rev. 0, June 2008.
2. Sandia 2007a. General and Localized Corrosion of Waste Package Outer Barrier. ANL-EBS-MD-000003 REV 03C, May 2007.
3. K. G. Mon, G.M. Gordon, and R.B. Rebak. 2005. "Stifling of crevice corrosion in Alloy 22." Proc. 12th Int. Conf. Environmental Degradation of Materials in Nuclear Power System – Water Reactors, T.R. Allen, P.J. King, and L. Nelson (eds.), The Minerals, Metals & Materials Society (Warrendale, PA), p. 1431-1438.
4. X. He and D.S. Dunn. 2007. Crevice corrosion penetration rates of Alloy 22 in chloride-containing waters. Corrosion 63, 145-158.
5. X. Shan and J. Payer. 2008. Effect of crevice former on the evolution of crevice damage. Proc. CORROSION/2008, NACE International (Houston, TX), paper 08575.
6. R.M. Carranza. 2008. The crevice corrosion of Alloy 22 in the Yucca Mountain nuclear waste repository. J. of Metals 60, 58-65.

Table II. Results of Post-test XPS Analysis of the NO$_3^-$:Cl$^-$ Ratio in the Creviced Area.

Solution composition	Location of the analysis	Measured [NO$_3^-$]:[Cl-] on surface of creviced region	[NO$_3^-$]:[Cl-] in bulk solution
5 M CaCl$_2$ + 1M NaNO$_3$	Corroded area	0.163	0.1
5 M CaCl$_2$ + 0.5 M NaNO$_3$	Corroded area	0.0835	0.05
5 M NaCl + 0.1M NaNO$_3$	Corroded area	N not detected	0.02
5 M NaCl + 0.1M NaNO$_3$	Un-corroded area	N not detected	0.02

Mater. Res. Soc. Symp. Proc. Vol. 1124 © 2009 Materials Research Society   1124-Q09-04

Corrosion Behavior of Weld Zone of Carbon Steel Overpack for HLW Geological Disposal

Yutaka Yokoyama,[1, 3] Hiroyuki Mitsui,[1, 3] Rieko Takahashi,[1] Hidekazu Asano,[1]
Naoki Taniguchi,[2] Morimasa Naito[2]
[1]Radioactive Waste Management Funding and Research Center (RWMC), 1-15-7, Tsukishima,
Chuo-ku, Tokyo, Japan
[2]Japan Atomic Energy Agency (JAEA), 4-33, Muramatsu, Tokai-mura, Ibaraki, Japan
[3]Mitsubishi Heavy Industries, Ltd., 2-1-1, Shinhama, Arai-cho, Takasago, Hyogo, Japan

ABSTRACT

The corrosion behavior of the weld zone of carbon steel overpack, a candidate container for geological disposal of high-level radioactive waste (HLW), was studied using electrochemical measurement and slow strain rate testing (SSRT) technique, and immersion tests in aerobic/anaerobic conditions. Several welding processes were applied to low carbon steel, and corrosion behavior of the weld zone, which consists of weld metal and heat-affected zone (HAZ), was compared with that of the base metal. The integrity of weld zone of overpack under disposal environment is also discussed.

INTRODUCTION

In Japan, carbon steel is one of the candidate materials for the disposal container (overpack) for HLW. Overpack seals vitrified waste and is currently required to isolate it from contact with groundwater for at least 1,000 years under the present concept for geological disposal in Japan. Its integrity for 1,000 years is evaluated in consideration of mechanical strength and corrosion property. The corrosion behaviors of the base metal have been evaluated along with the corrosion scenarios of carbon steel overpack presented by JNC [1, 2]. However, the corrosion behavior of the weld zone of overpack under repository conditions has not been investigated enough to assess its corrosion lifetime along with the corrosion scenario of a carbon steel overpack. In this study, to evaluate the corrosion behavior of the weld zone produced by TIG (GTAW), MAG (GMAW) and EBW (electron beam welding), which have been validated for applicability to remote welding overpack [3], the corrosion modes which were the main concerns in the weld zone were extracted from the corrosion scenario of base metal, and four tests were conducted: passivation behavior, corrosion property under aerobic conditions, susceptibility to stress corrosion cracking (SCC), and corrosion property under anaerobic conditions.

EXPERIMENT

Material preparation

The test specimens were cut from the weld zone and base metal of carbon steel test pieces (SF340A; carbon steel forgings for general use in JIS standards (JIS G3201)) that model the full-scale lid structure of overpack and welded by TIG, MAG, and EBW in a circumferential direction. The details of the welding processes and chemical compositions of the specimens are summarized in Table 1 and Table 2.

Table 1. Welding parameter

	TIG (GTAW)	MAG (GMAW)	EBW
Groove depth [mm]	190	190	100
Welding position	Horizontal	Flat	Horizontal
Shield gas	Ar	Ar-20% CO_2	-
Welding current [A]*[1]	100–300	350	0.6 *[2]
Voltage [V]	11.0–13.5 *[3]	34–36 *[3]	100 000 *[4]
Travel speed [cm/min]	5.7–11.7	28	25
Layer	38	20	1
Number of weld passes	54	40	1
Energy input [kJ/cm]	23–90	51–54	144
Welding time [h]	24.5	2.4	0.13

*1: Base current, *2: Beam current, *3: Arc voltage, *4: Accelerating voltage

Table 2. Chemical Composition (wt%)

		C	Si	Mn	P	S	Cu	Cr	Mo
TIG	Weld metal	0.11	0.67	1.29	0.009	0.011	0.25	0.02	< 0.01
	Base metal*[1]	0.15	0.19	0.36	0.006	0.002	0.01	0.05	0.01
	Filler metal	0.10	0.73	1.40	0.011	0.014	0.24	0.03	-
MAG	Weld metal	0.082	0.58	1.07	0.010	0.014	0.20	0.03	<0.01
	Base metal*[2]	0.11	0.25	0.65	0.007	0.002	0.05	0.04	0.01
	Filler metal	0.10	0.76	1.37	0.010	0.014	0.24	-	-
EBW	Weld metal	0.11	0.25	0.70	0.011	0.003	0.05	0.11	0.02
	Base metal*[2]	0.12	0.25	0.65	0.012	0.004	0.05	0.11	0.02

*1: 910 °C × 2 h air-cooling, *2: 900 °C × 10.5 h air-cooling+ 600 °C × 5 h furnace-cooling

Investigation of passivation behavior

To investigate the passivation behavior of the carbon steel overpack in a disposal environment it is important to evaluate its corrosion scenarios and to assess its corrosion lifetime. According to the investigation of the base metal, it is assumed that the carbon steel (base metal) is not passivated under the disposal environment, and general corrosion occurs [1]. In this study, the passivation behavior in a weld zone was investigated by an anodic polarization experiment. As test solutions, 0.01–0.1 M carbonate-bicarbonate solution (carbonate ions are known to act as passivators for carbon steel in groundwater; pH was varied within pH 8.5–10 by adjusting the ratio of carbonate-bicarbonate in the solution), synthetic seawater (ASTM D1141: simulating saline groundwater) and synthetic freshwater (2.5mM NaCl-2.5mM $NaHCO_3$: simulating fresh groundwater) were used. Polarization curves were also measured in buffer material (compacted bentonite 70% and silica sand 30% mixture) permeated with test solutions. The test temperature was 80 °C and the potential sweeping rate was 20mV/min. For the ohmic potential drops compensation, the ohmic resistances between the working and reference electrodes were measured by the AC impedance method.

Investigation of stress corrosion cracking susceptibility

Some environments which induce SCC on carbon steel are thought to contain caustic alkali, nitrate, phosphate, $CO-CO_2-H_2O$, high temperature water and carbonate [4]. It is now recognized that there are two forms of SCC due to carbonate. One is intergranular SCC and is usually called "high pH SCC". The other, near-neutral pH SCC, which is engendered by dilute ground water with a relatively low pH of around 6.5, has a transgranular SCC, and is designated" low pH SCC". The disposal tunnels may need support in the soft rock system. If concrete is used as a support material, the pH of the groundwater passing through the tunnel may rise [1]. It was reported that susceptibility to SCC ("high pH SCC") becomes high after receiving a certain amount heat treatment [5]. In this study, therefore, the SCC ("high pH SCC") susceptibility and SCC initiation critical conditions of a weld zone were investigated. Uniaxial tensile specimens were machined from the base metal, HAZ, and the weld metal. SSRT was conducted in 0.2–1.5 M of sodium bicarbonate-sodium carbonate solution (pH 9.2–9.5) at 80 °C under potentiostatic control. The application potential of the SCC test was determined corresponding to the active-passive transition zone of anodic polarization curves. Specimens were strained at a constant strain rate of 8.3×10^{-7} s^{-1}.

Investigation of corrosion behavior under aerobic/anaerobic conditions

In this study, the corrosion behavior of the weld zone under aerobic conditions was investigated by immersion tests. Synthetic seawater (ASTM D1141) and synthetic freshwater (2.5mM NaCl-2.5mM NaHCO$_3$) were used as test solutions. In order to evaluate the influence of the buffer material, immersion tests were also performed under solution conditions in which buffer material (bentonite) was added (solid/liquid ratio: 1g/1mL). To simulate the aerobic conditions at the early stage of the post-closure repository phase, air was supplied to the test vessels, and the test temperatures were kept at 80 °C for 90 days and 1 year.

Corrosion behavior of the weld zone under anaerobic conditions was also investigated by immersion tests. The immersion tests were also performed in buffer material (a compacted mixture of 70% bentonite and 30% silica sand) permeated with test solutions. Synthetic seawater and synthetic freshwater were used as test solutions. To simulate the anaerobic conditions after oxygen was consumed, the immersion was conducted in a glovebox (<0.1ppm O$_2$) replaced with N$_2$ gas. Test temperatures were kept at 80 °C for 90 days and 1 year.

Investigation of the hydrogen absorption and assessment of hydrogen embrittlement

Some of the hydrogen produced by corrosion reaction under anaerobic conditions is absorbed into the carbon steel. It is known that hydrogen embrittlement occurs when the concentration of absorbed hydrogen exceeds a given value depending on the yield strength of the steel [1]. Therefore, to investigate the hydrogen absorption behavior of the weld zone, an immersion test under anaerobic conditions was carried out. The specimen was dehydrogenated in advance. The immersion test conditions were the same as those in the investigation of corrosion behavior under anaerobic conditions. The concentration of diffusible hydrogen which contributes to hydrogen embrittlement was analyzed by thermal desorption spectroscopy (TDS) method (heating rate of 12 °C / min, Ar carrier gas:1 L/min). In this study, the hydrogen emitted up to 250 °C was classified as diffusible hydrogen because the hydrogen desorption peak in low temperature finished up to 250 °C under these conditions [6]. In order to investigate the quantity and properties of the trapping sites for diffusible hydrogen in the specimens, Specimens dehydrogenated by heat-treatment (at either 30 °C x 500 h or 200 °C x 2 h) were hydrogen-charged by a cathode charge under constant conditions (30 °C, 50 A/m^2, 24 h, in an H$_2$SO$_4$ + 0.09%NH$_4$SCN solution), and the hydrogen trapped in the specimen was analyzed by TDS.

RESULTS AND DISCUSSION

Passivation behavior of the weld zone in a disposal environment

The carbon steel was easy to be passivated, so that the carbonate-bicarbonate concentration or pH was high (Table 3). In the 0.01 M carbonate-bicarbonate solution (pH 10), synthetic seawater, and synthetic freshwater, neither the base metal nor the weld zone were passivated. In buffer material, the acidity at which neither the base metal nor the weld zone was passivated went up to pH 11.5. These results suggest that the passivation of carbon steel is hindered in a buffer material. Under some conditions in which a base metal was passivated, the depassivation of the weld metal of TIG and MAG was observed. In an actual repository condition, carbonate concentration is less than 0.1 M, and the overpack is contact with compacted bentonite [1]. Therefore, it appears unlikely that both the base metal and the weld zone would be passivated in a repository condition.

Table 3   Results of investigation of passivation behavior

Test environment	TIG			MAG			EBW		
	BM	HAZ	WM	BM	HAZ	WM	BM	HAZ	WM
0.01M carbonate solution - pH 8.5	○	○	○	○	○	○	○	○	○
0.01M carbonate solution - pH 10.0	●	●	△	●	●	△	●	●	●
0.1M carbonate solution - pH 8.5	●	●	△	●	●	△	●	●	●
0.1M carbonate solution - pH 10.0	●	●	●	●	●	●	●	●	●
Synthetic seawater	○	○	○	○	○	○	○	○	○
Synthetic freshwater	○	○	○	○	○	○	○	○	○
0.1M carbonate solution - pH 8.5 + Buffer material	○	○	○	○	○	○	○	○	○
0.1M carbonate solution - pH 10.0 + Buffer material	○	○	○	○	○	○	○	○	○
0.1M carbonate solution - pH 11.5 + Buffer material	○	○	○	○	○	○	○	○	○
0.1M carbonate solution - pH 13 + Buffer material	●	●	△	●	●	△	●	●	●
Synthetic seawater + Buffer material	○	○	○	○	○	○	○	○	○
Synthetic freshwater + Buffer material	○	○	○	○	○	○	○	○	○

BM: Base metal, HAZ: Heat affected zone, WM: Weld metal,
●: Passivation, ○: Non-passivation (general corrosion), △: Depassivation in high potential

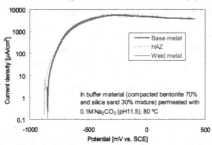

Fig. 1 Anodic polarization curves in buffer material (TIG specimens)

SCC susceptibility of the weld zone in a disposal environment

Both the base metal and the weld zone were not susceptible to SCC at 0.2 M or less of carbonate-bicarbonate concentration. At concentrations over 0.5 M, SCC occurred in both the base metal and the weld zone. In order to compare the SCC susceptibility of the base metal and the weld zone, the area ratio of SCC fracture and the crack growth rate of the base metal and weld zone were compared. The results show that the weld zone was less susceptible to SCC rather than the base metal, irrespective of welding method (Fig. 2). As a result of metallographic observation, the SCC morphology seen in the base metal was in intergranular mode (Fig. 3 (a)). On the other hand, the weld metal microstructure was so fine and complicated that it is difficult to definitively characterize the SCC morphology, although the crack paths appeared to be either transgranular or a mixture of intergranular and transgranular (Fig. 3 (c)). It was assumed that crack propagation in the weld zone was suppressed by the fine-grained microstructure and/or distribution of carbon around the crack path. Since the SCC susceptibility of the weld zone is less than that of the base metal, SCC due to carbonate is not likely to be initiated in the weld zone under the repository conditions like in base metal [1]. The susceptibility to SCC under near-neutral pH condition (near-neutral pH SCC) and the influence of buffer material (bentonite) on the susceptibility to SCC are subjects for further study.

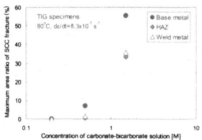

Fig. 2 Relationship maximum area ratio of SCC fracture and concentration of carbonate solution (TIG)

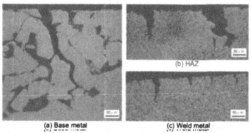

Fig. 3 Cross-sectional micrograph of crack near fracture surface after SSRT (TIG, 1.5M carbonate-bicarbonate solution)

Corrosion behavior of the weld zone under aerobic/anaerobic conditions

Under aerobic conditions, the weight loss in synthetic seawater and synthetic freshwater was almost equal, but the localization of corrosion was high in synthetic freshwater. In synthetic seawater with buffer material, weight loss was smaller than the weight loss in synthetic seawater without buffer material. It was suggested that the supply of oxygen might have been restricted, due to the presence of buffer material. It deserves to be mentioned that preferential corrosion was observed in the weld metal of TIG and MAG (Fig. 4). This preferential corrosion occurred along welding layers or entire parts of weld metals. On the other hand, in EBW specimens, the preferential corrosion of weld metal was not observed and the corrosion depth of the weld zone was less than or equal to the base metal. As a result of detailed investigation, it was assumed that the causes of preferential corrosion were chemical composition changes in the weld metal due to use of filler wire during TIG and MAG processes, and microstructure changes at the weld zone due to the thermal process induced by welding. The concentrations of Si, Mn, and S in the weld metal of TIG and MAG were higher than in the base metal. Si, Mn, and S are known as elements which have detrimental influences in the corrosion resistance of carbon steel welds [7, 8]. On the other hand, there is no significant difference between the chemical composition of the base metal and the weld metal of EBW. On the other hand, the weld metals of TIG and MAG are multi-layer structures that consist of as-welded and reheated areas. The preferential corrosion occurred in the as-welded area that mainly consists of acicular ferrite and grain boundary ferrite. Thus, it was assumed that this preferential corrosion occurred mainly due to the influence of their metallurgical factors, such as chemical composition and/or microstructure of the weld metals, which are affected by the filler wire and the welding heat cycle. Therefore, improvements to the filler wire and welding heat cycle might be needed to control the preferential corrosion.

On the other hand, under anaerobic conditions, the average corrosion depth was noticeably small compared with that under aerobic condition, and the corrosion rate decreased temporally. Furthermore, the preferential corrosion on the weld metal of TIG and MAG was restricted.

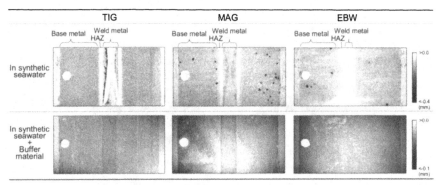

Fig. 4 The distribution of corrosion depth of specimens after immersion test under aerobic conditions (90 days)

Assessment of corrosion lifetime of the weld zone

The corrosion allowance of carbon steel overpack is assumed to be 40 mm, obtained by rounding up the greatest estimate of 31.8 mm for the corrosion depth of carbon steel (base metal) over a period of 1,000 years [1]. This estimate for the corrosion depth consists of 11.8 mm for the maximum corrosion depth due to oxygen at the early stage of the repository phase and 20 mm for the maximum corrosion depth due to water under anaerobic condition.

Since localization of corrosion was observed under aerobic conditions, a conservative evaluation based on estimating the maximum corrosion depth by extreme value statistical analysis was conducted. The estimate of the maximum corrosion due to oxygen at the early stage of the repository phase was calculated by following equation (1) showing the relation between the average corrosion depth and the maximum corrosion [1]:

$$P = Xm + 7.5\, Xm^{0.5} \qquad (1)$$

In this equation, P is the upper limit (mm) of estimated value of maximum corrosion depth, and Xm is the average corrosion depth (mm). The estimate of the maximum corrosion depth of the weld zone was less than P calculated by the average corrosion depth (Fig. 5). In this study, the maximum corrosion depth of the weld zone under aerobic condition did not exceed the present estimate in a base metal.

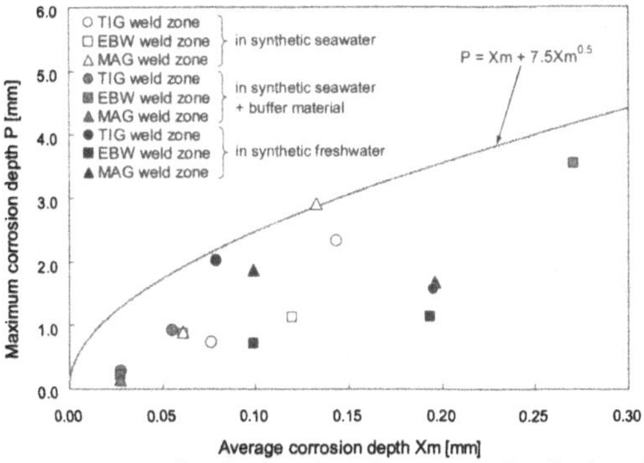

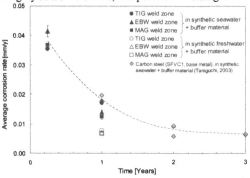

Fig. 5 Relation between average corrosion depth and maximum corrosion depth under aerobic conditions

The corrosion allowance for the anaerobic period was estimated based on the corrosion depth after 1000 years calculated by corrosion rate under anaerobic conditions. Considering the tendency of the corrosion rate to decrease with time, as observed under aerobic conditions, it is expected that the corrosion rate of the weld zone would also decrease to less than 10 μm/year— the value used to determine the present corrosion allowance—after several years (Fig. 6).

Thus, the present corrosion allowance based on evaluation of a base metal must be sufficient to also maintain the integrity of the weld zone, irrespective of welding method.

Fig. 6 Time dependence of average corrosion rate under anaerobic conditions

Hydrogen absorption behavior of the weld zone

In synthetic seawater, significant high absorption of diffusible hydrogen was observed in the weld metal of TIG in early stages of immersion (Fig. 7). In other specimens, the concentration of diffusible hydrogen decreased after immersion. The investigation of diffusible hydrogen trapping sites shows that the most hydrogen was trapped in weld metal of TIG (Fig. 8). Since the trapped hydrogen decreased in specimens heat-treated at 200 °C x 2 h before hydrogen charging, it was assumed that the trapping effect of the many lattice defects, such as the atomic vacancy which was extinguished at 200 °C [9], in the weld metal of TIG upon hydrogen diffusion was the main cause of the high concentration of hydrogen in the weld metal of TIG. A detailed investigation of the hydrogen embrittlement susceptibility of TIG is needed.

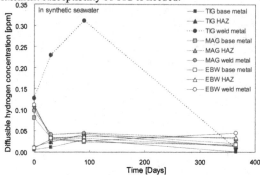

Fig. 7 The concentration of absorbed diffusible hydrogen in synthetic seawater

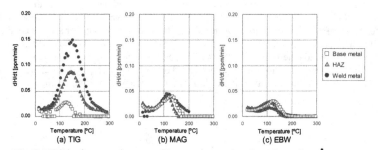

Fig. 8 TDS hydrogen release profiles continuous heating of specimen[*]
hydrogen-charged under constant condition[**]

      [*] Dehydrogenated by heat-treatment (30 °C x 500 h) in advance
[**] Hydrogen charge condition: $50A/m^2$, 24 h, in an H_2SO_4 + 0.09%NH_4SCN solution, 30 °C

CONCLUSIONS

In this study, to evaluate the corrosion property of the weld zone of carbon steel overpack, several tests were examined. As a result, it appears that both the base metal and weld zone would not be passivated and susceptible to SCC in the disposal environment. Although the preferential corrosion observed in the weld metal of TIG and MAG under aerobic/anaerobic conditions, the maximum corrosion depth of the weld zone does not exceed the present estimate for a base metal. In conclusion, the present corrosion allowance based on evaluation of a base metal must be sufficient to also maintain the integrity of the weld zone, irrespective of welding method.

ACKNOWLEDGMENTS

The authors would like to extend their appreciation to Dr. T. Haruna of Kansai university and Dr. H. Inoue of Osaka Pref. University for their helpful suggestions in connection with this work. This investigation constitutes part of the results obtained in the program entitled, 'Investigation for characteristics of engineered barrier material' carried out under grant from the Ministry of Economy, Trade and Industry.

REFERENCES

1. *H12: Project to Establish the Scientific and Technical Basis for HLW Disposal in Japan, Project Overview Report, Second Progress Report on Research and Development for the Geological Disposal of HLW in Japan*, JNC TN1410 2000-003 (JNC, Ibaraki, Japan, 2000).
2. *Development of repository concepts for volunteer siting environments*, NUMO-TR-04-01, Nuclear Waste Management Organization of Japan (NUMO, Tokyo, Japan, 2004).
3. H. Asano and M. Aritomi, *J.Nucl.Sci.Technol.*, **42** (5), 470 (2005).
4. J. A. Beavers, N. G. Thompson, R. N. Parkins, *Stress Corrosion Cracking of Low-Strength Carbon Steels in Candidate High-Level Waste Repository Environment*, NUREG/CR-3861 (1987).
5. R. Shu, T. Shibata, T. Haruna, *Proc. Japan Conf. Mat. Env.*, **44**, 429, (1997).
6. *Investigation on the performance of engineered barrier system considering the weld joint of overpack materials*, RWMC H17 Report (RWMC, Tokyo, Japan, 2006).
7. E. Raesaenen and K.Relander, *Scand. J. Metall.*, **7** (1), 11-17 (1978).
8. L. Malik, L. D. Parkinson,V. Mitrovic-Scepanovic,R. J. Brigham, *Proc. Int. Conf. Offshore Mech. Arct. Eng*, **9** (13), 639–646 (1990).
9. K.Takai, H.Shoda, H.Suzuki, and M.Nagumo, *Acta Materialia*, **56**, 5158–5167 (2008).

General Corrosion Behavior of N06022 in Super Concentrated Brines at Temperatures Higher than 100°C

Raul B. Rebak
GE Global Research, Schenectady, NY 12309, U.S.A.

ABSTRACT

Alloy 22 (N06022) is a highly corrosion resistant nickel based alloy. Extensive research has been conducted in the last eight years on the corrosion behavior of Alloy 22, mainly regarding its resistance to localized corrosion. Less attention has been paid to the general corrosion resistance in highly concentrated brines that may result from the deliquescence of salts contained in dust. Salts such as mixtures of NaCl, KCl, $CaCl_2$, $NaNO_3$, and KNO_3 may deliquesce at temperatures above 100°C through absorption of moisture from the air. Electrochemical tests were used to assess the general corrosion behavior of Alloy 22 in brines with chloride and nitrate concentrations ranging from 8 molal to 100 molal in the temperature range 100 to 160°C. The effect of mixed anions and cations was also studied. Results show that, even for short-term immersion periods, the corrosion rate of Alloy 22 in high temperatures super concentrated brines is generally below 10 μm/year.

INTRODUCTION

Alloy 22 is a nickel base alloy designed to be resistant to all forms of corrosion. Alloy 22 contains approximately 56% nickel (Ni), 22% chromium (Cr), 13% molybdenum (Mo), 3% tungsten (W) and 3% iron (Fe) (ASTM B 575) [1]. Because of its high level of Cr, Alloy 22 remains passive in most industrial environments and therefore has an exceptionally low general corrosion rate [2-6]. The combined presence of Cr, Mo and W imparts Alloy 22 with high resistance to localized corrosion such as pitting corrosion and stress corrosion cracking even in hot concentrated chloride (Cl^-) solutions [7-12]. It has been reported that Alloy 22 may suffer localized corrosion such as crevice corrosion when it is anodically polarized in chloride-containing solutions [8-10,13-15]. It is also known that the presence of nitrate (NO_3^-) in the solution minimizes or eliminates the susceptibility of Alloy 22 to crevice corrosion [8-10,16-23]. The value of the ratio $[NO_3^-]/[Cl^-]$ has a strong effect of the susceptibility of Alloy 22 to crevice corrosion [10,16-25]. The higher the nitrate to chloride ratio, the stronger is the inhibition by nitrate. This ratio may depend on other experimental variables such as total concentration of chloride or temperature. Other anions in solution, such as sulfate, fluoride and carbonate / bicarbonate were also reported to inhibit crevice corrosion in Alloy 22 [10,19-20,26-28].

The general corrosion rate of Alloy 22 in has been measured using weight loss immersion tests in several types of concentrated electrolyte solutions at 60°C and 90°C [29-30]. The general corrosion rate after 5 years of immersion was below 0.03 μm/year [29-30]. Mon et al. published a model for the prediction of the general corrosion behavior of Alloy 22 using both immersion weight loss data and short-term electrochemical polarization resistance data [31]. Mon et al. calculated a maximum corrosion rate of 7.4 μm/year at 200°C [31]. Immersion tests have been performed using creviced and non-creviced Alloy 22 specimens in highly concentrated NaCl-

$NaNO_3$-KNO_3 brines at temperatures up to 180°C and immersion times of up to 80 days [32]. Salt concentrations ranged from ratios of nitrate over chloride up to 20 (in molal units). Most of the reported corrosion rates values were below 2 μm/year; however a few reported values were between 4 and 10 μm/year [32]. The corrosion rate in the vapor phase was lower than in the liquid phase [32].

In other studies, polarization resistance electrochemical tests were used to assess the general corrosion behavior of Alloy 22 in concentrated brines at temperatures higher than 100°C [33-35]. After more than 500 days of immersion in aerated 5 molar $CaCl_2$ solutions at 120°C, the corrosion rate was below 1 μm/year [33]. Ilevbare et al. reported that the general corrosion rate in deaerated highly concentrated brines of 18 m $CaCl_2$ + 9 m $Ca(NO_3)_2$ at 160°C and 12 m $CaCl_2$ + 6 m $Ca(NO_3)_2$ at 130°C was below approximately 1 μm/year after 24 hour immersion in the brines [34]. Rodriguez et al. used the polarization resistance method to evaluate the corrosion rate of Alloy 22 as a function of the immersion time for up to 600 days of testing in 18 m $CaCl_2$ + 9 m $Ca(NO_3)_2$ (R = 0.5) and in 18 m $CaCl_2$ + 0.9 m $Ca(NO_3)_2$ (R = 0.05) solutions at 155°C [35]. They reported that the corrosion rate was below 1 μm/year for the R = 0.5 solution and slightly higher in the R = 0.05 solution. In the R = 0.5 solution, the corrosion potential (E_{corr}) was consistently higher than +500 mV SSC for the entire testing period but in the R = 0.05 solution the E_{corr} varied widely between +100 mV and +500 mV for the same testing time [35].

The objective of the current work was to evaluate the general corrosion behavior of Alloy 22 in highly concentrated brines containing both Ca and Na + K salts at temperatures higher than 100°C using the polarization resistance technique. References 34 and 36 reported repassivation potentials for the same specimens and therefore the complete list of the specimens and brines used for the current results were listed before [34,36].

EXPERIMENTAL TECHNIQUE

Alloy 22 specimens were prepared from 1-inch thick plate. The specimens were creviced using a ceramic washer and PTFE tape (ASTM standard G 192). There were two types of specimens, multiple crevice assemblies (MCA), [34] and prism crevice assemblies (PCA) [36-37]. All the tested specimens had a finished grinding of abrasive paper number 600 and were degreased in acetone and treated ultrasonically for 5 minutes in de-ionized (DI) water 1 hour prior to testing. All the specimens were in the as-welded (ASW) condition. The weld was produced with matching filler metal using Gas Tungsten Arc Welding (GTAW). The welded specimens were not all weld metal but contained a weld seam which varied in width from approximately 8 to 15 mm.

Electrochemical tests were performed in several electrolyte solutions (Table I). The complete lists of solutions are given in References 34 and 36. There are two main types of data sets, $CaCl_2$ + $Ca(NO_3)_2$ solutions [34] and NaCl + KCl + $NaNO_3$ + KNO_3 solutions. [36] The tested electrolytes included pure chloride (R = 0), pure nitrate (R = ∞) and mixtures of chloride plus nitrate solutions (0.05 ≤ R ≤ 100) (Table I). The temperature range was between 100°C and 160°C. Nitrogen (N_2) was purged through the solution at a flow rate of 100cc/min for 24 hours while the corrosion potential (E_{corr}) was monitored. Nitrogen bubbling was continued throughout all the electrochemical tests. The electrochemical tests were conducted in a one-liter, three-electrode, borosilicate glass flask (ASTM G 5) [37]. A water-cooled condenser combined with a water trap was used to avoid evaporation of the solution and to prevent the ingress of air (oxygen). All the tests were carried out at ambient pressure. The reference electrode was saturated silver chloride (SSC), which at ambient temperature has a potential of 199 mV more positive than the standard hydrogen electrode (SHE). The reference electrode was connected to the solution through a water-jacketed Luggin probe so that the electrode was maintained at near ambient temperature. The counter electrode was a flag (36 cm²) of platinum foil spot-welded to a platinum wire. All the potentials in this paper are reported in the SSC scale.

Table I. Solutions used for testing

Type of Electrolyte Solutions	Ranges of composition/temperature studied
$CaCl_2$ + $Ca(NO_3)_2$	10 m $CaCl_2$, 12 m $CaCl_2$ and 18 m $CaCl_2$ Added nitrate to obtain R = 0, 0.15 and 0.5 Temperatures studied, 100, 110, 120, 130, 150, and 160°C
NaCl + KCl + $NaNO_3$ + KNO_3	From pure NaCl + KCl solutions (R = 0) to pure $NaNO_3$ + KNO_3 solutions (R = ∞) Maximum chloride concentration = 8 m Maximum nitrate concentration = 100 m Temperatures studied, 110, 125, 140, and 150°C
$CaCl_2$ + $Ca(NO_3)_2$ and $CaCl_2$ + $NaNO_3$ + KNO_3	From pure $CaCl_2$ solutions (R = 0) to added nitrate to obtain R values of 0.5, 1, 2.5, 5, and 10 Temperatures studied, 110 and 125°C

The corrosion rate can be calculated using electrochemical methods such as the polarization resistance (PR) described in ASTM G 59 and G 102 [37]. An initial potential of 20 mV below the corrosion potential (E_{corr}) is ramped to a final potential of 20 mV above E_{corr} at a rate of 0.167 mV/s. The Polarization Resistance (R_p) is defined as the slope of the potential (E) vs. current density (i) at i = 0. The corrosion current density, i_{corr}, is related to the polarization resistance by the Stern –Geary coefficient B

$$i_{corr} = 10^6 \cdot \frac{B}{R_p} \tag{1}$$

where

$$B = \frac{b_a \cdot b_c}{2.303(b_a + b_c)} \tag{2}$$

where b_a and b_c are the anodic and cathodic Tafel slopes

$$i_{corr} = \frac{10^6}{R_p} \times \frac{b_a \cdot b_c}{2.303(b_a + b_c)} \tag{3}$$

The corrosion rate can then be calculated using the Faraday equation

$$CR(\mu m / yr) = k \frac{i_{corr}}{d} EW \tag{4}$$

Where k is a conversion factor (3.27 x 10⁶ μm·g·A⁻¹·cm⁻¹·yr⁻¹), i_{corr} is the corrosion current density in A/cm² (calculated from the measurements of the resistance to polarization, R_p), EW is the equivalent weight, and d is the density of Alloy 22 (8.69 g/cm³). Assuming an equivalent dissolution of the major alloying elements as Ni^{2+}, Cr^{3+}, Mo^{6+}, Fe^{2+}, and W^{6+}, the EW for Alloy 22 is 23.28 (ASTM G 102) [37]. The corrosion rate reported here was calculated by fitting the experimental data in the potential range ±10 mV with respect to the open circuit potential.

RESULTS AND DISCUSSION

Figure 1 shows the evolution of the corrosion potential (E_{corr}) of Alloy 22 for the 24-hr immersion period in Ca based brines at 130°C [34]. The potential was measured every 3 seconds. The E_{corr} in Figure 1 is for short-term under deaerated conditions and it is not intended to represent the behavior of Alloy 22 for long-term exposure in the aerated brines. It is evident from Figure 1 that as the amount of nitrate in the solution increased, the corrosion potential also increased. For the R = 0 and R = 0.05 solutions, E_{corr} did not vary during the testing period; however, for the R = 0.15 and R = 0.5 solutions the E_{corr} increased as the immersion time increased. It also seems apparent in Figure 1 that the noise in the value of E_{corr} increased as the amount of nitrate in the solution increased. When the ratio R increased from 0 to 0.5, the 24-hr E_{corr} changed approximately 400 mV (from -280 mV SSC to +120 mV SSC).

Figure 2 shows that the corrosion rate calculated from the polarization resistance tests decreased as E_{corr} increased. That is, the highest corrosion rates corresponded to the pure chloride solution (R = 0) when the E_{corr} was the lowest (near -300 mV SSC). The corrosion rates in Figure 2 are for comparative purposes only and are not intended to represent the long-term corrosion rate behavior of Alloy 22 in aerated super concentrated brines.

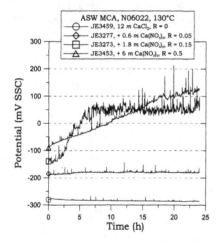

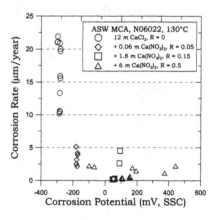

Figure 1. Corrosion potential as a function of the immersion time in Ca brines at 130°C

Figure 2. Corrosion rate calculated using the polarization resistance method ASTM G 59

Figures 3 and 4 show the E_{corr} of Alloy 22 as a function of the immersion time in Na + K brines, at 110°C and at 140-150°C, respectively. Figure 3 shows that as the ratio R increased from 0 to ∞ the short-term E_{corr} increased approximately 200 mV (from -550 mV SSC to -350

476

SSC). When R increased from 0 to 0.5, E_{corr} changed only approximately 50 mV (from -550 mV SSC to -500 mV SSC). Even though the temperatures in Figures 1 and 3 are different, it is apparent that the E_{corr} in Na + K salts did not increase as much as the E_{corr} in the Ca salts for the same change in the R ratio from 0 to 0.5 (changed 50 mV for the Na + K salts vs. 400 mV for the Ca salts). A similar effect of Ca cations promoting higher E_{corr} values was reported before by Lian et al. [38]. Figure 4 shows the E_{corr} evolution as a function of the immersion time for Alloy 22 in Na + K brines at 140°C and 150°C. Figure 4 does not include brines having an R value lower than 25 since such brines do not exist in liquid form at 140°C or 150°C; that is, nitrate is needed in the solution to prevent the brines to dry out [39]. Figure 4 shows that at each temperature the E_{corr} is higher for higher values or R; however the difference in potentials was not large (less than 100 mV).

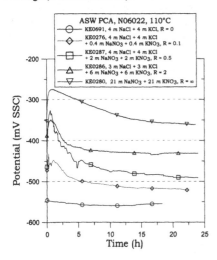

Figure 3. E_{corr} as a function of the immersion time in Na + K brines at 110°C

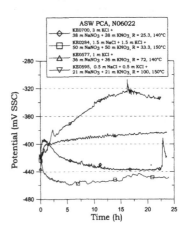

Figure 4. E_{corr} as a function of the immersion time in Na + K brines at 140°C and 150°C

Figure 5 shows the corrosion rate as a function of the corrosion potential after 24-hr immersion in concentrated brines. Similarly as reported before in Figure 2, as the E_{corr} increased the corrosion rate decreased. The corrosion rates in Figure 5 are for comparative purposes only and are not intended to represent the long-term corrosion behavior of Alloy 22 in similarly aerated brines.

Figures 3 and 4 show that the E_{corr} of Alloy 22 remained low even for the highest amounts of nitrate in the solution. The aqueous solutions in Figures 3 and 4 are based on Na and K, and therefore should be or near neutral pH since Na and K do not hydrolyze. Nitrate is not oxidizing at near neutral pH, therefore the E_{corr} remains low. Gray et al. showed that in presence of chloride ions nitrate starts to become oxidizing only a pH below 2 [40]. They also reported that the passive film is thicker in presence of nitrate than in presence of chloride or sulfate solutions of the same pH [40]. Similarly, it was found at near neutral pH at 100°C, the presence of nitrate in solution produces a thicker passive film on Alloy 22 [41].

Figures 2 and 5 show that the maximum short-term corrosion rates of Alloy 22 in concentrated brines of Ca or Na + K salts at T > 100°C are less than 30 µm/year. The corrosion rate values in Figures 2 and 5 are for fully immersed conditions, which are not representative of an exposure to hygroscopic salts, where the amount of developed brines would be extremely small [42]. In Figures 2 and 5, the higher corrosion rates always corresponded to brines containing only chloride. In general, when the R ratio increased above 0.5 the corrosion rates were below 2 µm/year, especially for the Ca based salts. Yang et al. reported that the corrosion rate of Alloy 22 measured by mass loss after 32-80 days immersion in Na + K brines at 150-180°C was generally below 2 µm/year [32]. For only four (4) specimens Yang et al reported corrosion rate values between 2 and 10 µm/year [32]. Current and literature data show that Alloy 22 is highly resistant to general corrosion, even under full immersion in super concentrated high temperature brines.

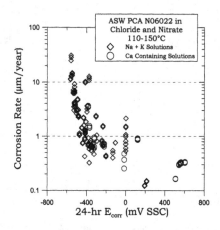

Figure 5. Corrosion rate calculated using the polarization resistance method ASTM G 59 for Alloy 22 immersed in brines from 110°C to 150°C

SUMMARY

1. Polarization resistance tests were used to evaluate the relative corrosion rate of Alloy 22 in concentrated brines in the temperature range 100-160°C
2. As the corrosion potential (E_{corr}) increased the corrosion rate decreased
3. In general, since electrolytes with higher R ratios would give higher values of E_{corr}, the corrosion rate also decreased as the ratio R increased (higher amount of nitrate).
4. Even for highly concentrated brines at temperatures in the order of 150°C, the corrosion rate after 24-hr immersion was generally 10 µm/year and lower.

REFERENCES

1 ASTM International, Standard B575, Vol. **02.04** (ASTM, 2002: West Conshohocken, PA).

2 Haynes International, "Hastelloy C-22 Alloy", Brochure H-2019E (Haynes International, 1997: Kokomo, IN).

3 R. B. Rebak in Corrosion and Environmental Degradation, Volume II, p. 69, Wiley-VCH, Weinheim, Germany (2000).

4 R. B. Rebak and P. Crook, "Nickel Alloys for Corrosive Environments," *Advanced Mater. & Proc.*, 157, 37, 2000.

5 R. B. Rebak and P. Crook, "Influence of the Environment on the General Corrosion Rate of Alloy 22," PVP-Vol. 483 pp. 131-136 (ASME, 2004: New York, NY).

6 R. B. Rebak and Joe H. Payer, "Passive Corrosion Behavior of Alloy 22," ANS Conf. International High Level Radioactive Waste Management, Las Vegas 30Apr-04May 2006.

7 R. B. Rebak and P. Crook, "Improved Pitting and Crevice Corrosion Resistance of Nickel and Cobalt Based Alloys," ECPV 98-17, pp. 289-302 (The Electrochemical Society, 1999: Pennington, NJ).

8 B. A. Kehler, G. O. Ilevbare and J. R. Scully, *Corrosion*, **57**, 1042 (2001).

9 K. J. Evans and R. B. Rebak, "Passivity of Alloy 22 in Concentrated Electrolytes, Effect of Temperature and Solution Composition," PV2002-13, p. 344-354 (The Electrochemical Society, 2002: Pennington, NJ).

10 A. K. Mishra and G. S. Frankel, "Crevice Corrosion Repassivation of Alloy 22 in Aggressive Environments," *Corrosion*, **64**, No. 11, 836-844 (2008).

11 Y-M. Pan, D. S. Dunn and G. A. Cragnolino in Environmentally Assisted Cracking: Predictive Methods for Risk Assessment and Evaluation of Materials, Equipment and Structures, STP 1401, pp. 273-288 (West Conshohocken, PA: ASTM 2000).

12 R. B. Rebak, "Environmentally Assisted Cracking of Nickel Alloys – a Review," in Environment-Induced Cracking of Materials, Vol. 1, p. 435-446 (Elsevier, Amsterdam, 2008).

13 C. S. Brossia, L. Browning, D. S. Dunn, O. C. Moghissi, O. Pensado and L. Yang, "Effect of Environment on the Corrosion of Waste Package and Drip Shield Materials," Publication of the Center for Nuclear Waste Regulatory Analyses (CNWRA 2001-03), September 2001.

14 D. S. Dunn, L. Yang, Y.-M. Pan and G. A. Cragnolino, "Localized Corrosion Susceptibility of Alloy 22," Paper 03697 (NACE International, 2003: Houston, TX).

15 K. J. Evans, A. Yilmaz, S. D. Day, L. L. Wong, J. C. Estill and R. B. Rebak, "Comparison of Electrochemical Methods to Determine Crevice Corrosion Repassivation Potential of Alloy 22 in Chloride Solutions," JOM, p. 56, January 2005.

16 G. A. Cragnolino, D. S. Dunn and Y.-M. Pan, "Localized Corrosion Susceptibility of Alloy 22 as a Waste Package Container Material," Scientific Basis for Nuclear Waste Management XXV, Vol. **713** (Materials Research Society 2002: Warrendale, PA).

17 D. S. Dunn and C. S. Brossia, "Assessment of Passive and Localized Corrosion Processes for Alloy 22 as a High-Level Nuclear Waste Container Material," Paper 02548 (NACE International, 2002: Houston, TX).

18 J. H. Lee, T. Summers and R. B. Rebak, "A Performance Assessment Model for Localized Corrosion Susceptibility of Alloy 22 in Chloride Containing Brines for High Level Nuclear Waste Disposal Container," Paper 04692 (NACE International, 2004: Houston, TX).

19 D. S. Dunn, L. Yang, C. Wu and G. A. Cragnolino, Material Research Society Symposium, Spring 2004, San Francisco, Proc. Vol. 824 (MRS, 2004: Warrendale, PA).

20 D. S. Dunn, Y.-M. Pan, L. Yang and G. A Cragnolino and X. He, "Localized Corrosion Resistance and Mechanical Properties of Alloy 22 Waste Package Outer Containers" JOM, Jan. 2005, pp 49-55.

21 R. B. Rebak, "Factors Affecting the Crevice Corrosion Susceptibility of Alloy 22," Paper 05610, Corrosion/2005 (NACE International, 2005: Houston, TX).

22 R. M. Carranza, "The crevice corrosion of alloy 22 in the Yucca Mountain nuclear waste repository," JOM, **60**, No. 1, 58-65 (2008).

23 G. O. Ilevbare, K. J. King, S. R. Gordon, H. A. Elayat, G. E. Gdowski and T. S. E. Gdowski, *Journal of The Electrochemical Society*, **152**, 12, B547-B554, 2005.

24 D. S. Dunn, Y.-M. Pan, L. Yang and G. A. Cragnolino, *Corrosion*, **61**, 1078-1085 (2005).

25 D. S. Dunn, Y.-M. Pan, L. Yang and G. A. Cragnolino, *Corrosion*, **62**, 3 (2006).

26 G. O. Ilevbare, *Corrosion*, **62**, 340 (2006).

27 R. M. Carranza, M. A. Rodriguez and R. B. Rebak, "Effect of fluoride ions on crevice corrosion and passive behavior of alloy 22 in hot chloride solutions," *Corrosion*, **63**, No. 5, 480-490 (2007).

28 R. B. Rebak, "Mechanisms of Inhibition of Crevice Corrosion in Alloy 22," in Scientific Basis for Nuclear Waste Management XXX, Vol. **985**, p. 261-268 (MRS, 2007: Warrendale, PA).

29 L. L. Wong, D. V. Fix, J. C. Estill, R. D. McCright, and R. B. Rebak, "Characterization of the Corrosion Behavior of Alloy 22 after Five Years Immersion in Multi-ionic Solutions," Materials Research Society Symposium - Proceedings, Vol. **757**, 2003, p 735-741

30 R. B. Rebak and R. D. McCright, "Corrosion of Containment Materials for Radioactive-Waste Isolation," ASM Handbook, Vol. 13C, p. 421-437 (ASM International, Metals Park, OH 2006).

31 K. G. Mon, B. E. Bullard, F. Hua, and G. C. De, "Modeling Alloy 22 General Corrosion in the Yucca Mountain Nuclear Waste Repository," JOM, Vol. **60**, No. 1, p. 52-57 (2008).

32 L. Yang, D. S. Dunn, X. He, V. Jain, G. A. Cragnolino, and Y.-M. Pan, "Corrosion Behavior of Alloy 22 in Concentrated Nitrate and Chloride Salt Environments at Elevated Temperatures," NACE International, Corrosion/2007 conference, Paper 07580 (Houston, TX 2007).

33 J. C. Estill, G. A. Hust, K. J. Evans, M. L. Stuart, and R. B. Rebak, "Long-term corrosion behavior of alloy 22 in 5 M $CaCl_2$ at 120°C," in Proceedings of 2006 ASME Pressure Vessels and Piping Division Conference - PVP2006/ICPVT-11, Vol. 7, p. 581-589, 2006

34 G. O. Ilevbare, J. C. Estill, A. Yilmaz, R. A. Etien, G. A. Hust, M. L. Stuart, and R. B. Rebak, "Anodic Behavior of Alloy 22 in High Nitrate Brines at Temperatures Higher than 100°C," 2006 ASME Pressure Vessels and Piping Division Conference - PVP2006/ICPVT-11, 7, p. 591-600, 2006

35 M. A. Rodriguez, M. L. Stuart and R. B. Rebak, "Long-Term Electrochemical Behavior of Creviced and non-Creviced Alloy 22 in $CaCl_2$ + $Ca(NO_3)_2$ Brines," NACE International, Corrosion/2007 conference, Paper 07577 (Houston, TX 2007).

36 T. Lian, G. E. Gdowski, P. D. Hailey, and R. B. Rebak, "Crevice Repassivation Potential of Alloy 22 in High-Nitrate Dust Deliquescence Type Environments" in ASME Pressure Vessels and Piping Conference - Operations, Applications and Components, Vol. 7, p 413-423 (2008).

37 ASTM International, Volume 03.02 "Wear and Erosion; Metal Corrosion" (ASTM International, 2003: West Conshohocken, PA).

38 T. Lian, G. E. Gdowski, P. D. Hailey, and R. B. Rebak, "Crevice corrosion resistance of Alloy 22 in high-nitrate, high-temperature dust deliquescence environments," *Corrosion*, **64**, n 7, July, 2008, p 613-623

39 R. B. Rebak, "Newer Alloy 22 Data and Their Relevance to High-Temperature Localized Corrosion," presented the Nuclear Waste Technical Review Board Workshop on Localized Corrosion, 25-26 September 2006, Las Vegas, Nevada (http://www.nwtrb.gov/meetings/overheads.html).

40 J. J. Gray; J. R. Hayes, G. E. Gdowski, B. E. Viani; and C. A. Orme, "Influence of solution pH, anion concentration, and temperature on the corrosion properties of alloy 22," *Journal of the Electrochemical Society*, **153**, n 3, 2006, p B61-B67

41 P. Pharkya, S. Xi, and J. H. Payer, "The effect of anions on the passive film properties and localized corrosion behavior of alloy C-22," in ECS Transactions, v 3, n 31 - Critical Factors in Localized Corrosion 5: A Symposium in Honor of Dr. Hugh S. Isaacs, 2007, p 473-484

42 C. R. Bryan, "Evolution of Waste Package Environments in a Repository at Yucca Mountain," presented the Nuclear Waste Technical Review Board Workshop on Localized Corrosion, 25-26 September 2006, Las Vegas, Nevada (http://www.nwtrb.gov/meetings/overheads.html).

Mater. Res. Soc. Symp. Proc. Vol. 1124 © 2009 Materials Research Society   1124-Q09-07

Comparison of wet air and water radiolysis effects on oxidized Zircaloy-4 corrosion

C. Guipponi[1], N. Millard-Pinard[1], N. Bérerd[1,2], E. Serris[3], M. Pijolat[3], V. Peres[3], V. Wasselin-Trupin[4], L. Pinard[5] , L. Roux[4], P. C. Leverd[4]

[1] Université de Lyon, Université Claude Bernard Lyon 1, CNRS UMR 5822, Institut de Physique Nucléaire de Lyon, 4 rue Enrico Fermi, 69 622 Villeurbanne Cedex, France
[2] Université de Lyon, Université Claude Bernard Lyon 1, IUT A, département chimie, 43 Bd du 11 novembre 1918, F-69622 Villeurbanne Cedex, France
[3] Ecole Nationale Supérieure des Mines de Saint Etienne, Centre SPIN, CNRS UMR5148, 158 cours Fauriel, 42 033 Saint Etienne Cedex, France
[4] Institut de Radioprotection et de Sûreté Nucléaire, BP. 17, 92 262 Fontenay aux Roses, France
[5] Laboratoire des Matériaux Avancés, Université Lyon 1, CNRS/IN2P3, Bât. Virgo, 22 bd Niels Bohr - 69622 Villeurbanne Cedex - France

ABSTRACT

After spent nuclear fuel processing, the Zircaloy-4 cladding tubes are compacted in a wafer form, placed into a steel container, and then into a concrete over-pack in order to dispose of in a geological repository. These wastes are mostly composed of activated oxidized metal pieces which also contain traces of fission, activation products and actinides. In the repository, they are exposed to radioactivity in presence of resaturation water which is a water equilibrated with the geological repository. The water radiolysis may accelerate the oxidised metal corrosion processes. Waste degradation is mainly due to corrosion process. A fundamental study was designed to assess the effects of vapour and liquid water radiolysis on Zircaloy-4 corrosion. Various water vapour pressures have been studied with proton irradiation and an aqueous solution simulating resaturated waste water was used with gamma irradiation. The results obtained on samples irradiated or not have been compared to determine the effect of water radiolysis on the stability of the oxide layer. Two phenomena had been observed: an enhancement of the adsorption of H_2O molecules and a tin surface enrichment.

INTRODUCTION

Assessment of the long term disposal properties of nuclear waste requires an understanding of the chemical interaction of the wastes with the surrounding environment. One of the key topics is the understanding and the modeling of the waste package degradation of under repository conditions. During their irradiation in light water the Zircaloy-4 (a zirconium alloy) cladding tubes are activated and oxidised. After processing, these cladding tubes are compacted in a wafer form and placed in a steel container. It is planned that the container be emplaced into a concrete over-pack before being disposed of in a vault of a deep geological disposal. These wastes are mostly composed of activated oxidized metal pieces which also contain activation products, traces of fission products and actinides. In the disposal, the radioactive waste are exposed to humid air in a first phase and to water after the resaturation phase that are likely, due to radiolytic processes, to accelerate the oxidised metal corrosion processes and the release of radionuclides. In this context the IPNL (Institut de Physique Nucléaire de Lyon), the ENSMSE (École Nationale Supérieure des Mines de Saint-Étienne) and IRSN (the French Institute of

Radioprotection and Nuclear Safety), have developed an experimental program to study the degradation process of the nuclear waste packages.

The oxidation of Zircaloy-4 has been extensively studied [2-4]. After a pseudo-parabolic first period, a kinetic transition occurs when the thickness of the oxide layer exceeds a threshold value, which corresponds to an increase in the oxidation rate associated with the appearance of cracks in the zirconia layer. To understand the corrosion processes during the disposal, the effects of water radiolysis on a porous oxide, which simulates the zirconia layer formed during reactor operations have been investigated. Several parameters are studied such as irradiations types, irradiation conditions and medium composition (different water vapour partial pressures, model solution simulating the clay porewater). The radiolytic species concentrations in water depend on irradiations types [5]. So we have chosen to irradiate with charged particles and electromagnetic irradiation. The characterization of the oxide layer surface of samples exposed to water with or without irradiation gives information on modifications induced by water radiolysis.

EXPERIMENTAL

Sample preparation

Samples were 420 µm thick Zircaloy-4 foils. The major alloying element is tin, with a content equal to 1.4 weigth percent. The samples were oxidised at 550°C in a mixed O_2/He atmosphere (flow rate: 2 L.h^{-1}) using a thermobalance (Setaram SETSYS) at the ENSMSE [6]. Oxygen partial pressure is fixed equal to 200 hPa. The oxidation takes 52 hours in order to reach the kinetic transition and obtain a scale of about 6 µm thick. This oxide contains several cracks which are parallel and perpendicular to the metal oxide interface [7]. These samples are called "reference".

Proton irradiations

The irradiation experiments are performed using a 1.5 MeV energy proton beam at the 4 MV Van de Graaff accelerator of the IPNL. The proton beam was extracted from the beam line vacuum to the sample cell through a 10 micrometer Havar (Co /Fe alloy) window [8]. Under normal operation conditions, the diameter of the beam spot is about 1 µm. In order to irradiate the gaseous atmosphere using a homogeneous flux on a well delimited surface (6 x 6 mm^2 in our experimental conditions), a sweeping electrostatic device of the beam was set up irradiation over the whole zirconia surface.

In the cell, synthetic air containing water vapour at a partial pressure fixed using a thermoregulated bath, circulates at 3.5 L.h^{-1} [9]. The distance between the Havar window and the sample surface was equal to 8 mm. Measurements were performed at two water partial pressures, 6 and 50 hPa, during 12 hours. According to the SRIM (the Stopping and Range of Ions in Matter) [10] calculation, the initial proton energy was chosen in order to get the Bragg peak as close as possible to the surface of the sample. The energy is mainly spent in the gaseous phase. Samples obtained after proton irradiations in wet air are called "proton radiolysis"

<u>Gamma irradiation</u>

The gamma irradiation experiments were performed at the IRMA ("IRradiation des MAtériaux") facility of IRSN at Saclay. The samples were placed into 18 mL PYREX tubes filled with a model aqueous solution. This solution simulated the water in contact with the waste and contained carbonate (2.3×10^{-3} mol.L^{-1}), sodium (3.1×10^{-2} mol.L^{-1}), calcium (1.2×10^{-2} mol.L^{-1}) and chloride (2.9×10^{-2} mol.L^{-1}) ions to simulate the resaturated waste water [11]. The tubes are arranged all around a ^{60}Co source at the same distance in order to receive the same dose rate of 3.5 kGy.h^{-1}. The applied dose was 33, 170 and 334 kGy. According waste specifications it was determined that these experiments simulate from 6 000 to 40 000 years after reprocessing. The irradiated samples are called "gamma radiolysis".
Samples with the same operating conditions but without any protons or gamma irradiation are called "blank".

<u>Analytical procedure</u>

For proton irradiation experiments, the same sample is analysed at three different stages. Firstly after oxidation (reference), secondly after wet air contact (blank) and finally after exposition to radiolysed wet air (radiolysis).
Nuclear Backscattering Spectrometry (NBS) was used to monitor oxygen concentration. The extreme surface was characterized by X-ray Photoelectron Spectroscopy (XPS) and Scanning Electron Microscopy (SEM).
For the study regarding gamma irradiations, the reference samples, blanks and radiolysis state were characterised by the same analysis techniques.

RESULTS

<u>Water and hydroxyl species behaviour</u>

The NBS analyse were performed at 172° detection angle and using 7.5 MeV α particles. Figure 1 shows the raw spectra obtained after contact between the sample and the wet air or water. At this energy the $^{16}O(\alpha,\alpha)^{16}O$ nuclear reaction occurs and so the oxygen relative measurements are 120 times more precise than the classical Rutherford backscattering [12]. The spectrum can be split in four components, the oxygen relative signal (between 600 and 2800 keV), the zirconium on the zirconia relative signal (between 4700 and 6300 keV), the zirconium substrate relative signal (between 2800 and 3700 keV) and the zirconium metal/oxide relative signal. It was noticed that the contact with wet air with or without irradiation increased the relative sample oxygen amount but the interface and the zirconium relative signal were not modified. It can be concluded that no oxide loss nor oxidation occurred during the experiment. The oxygen relative signal evolution explication may be due to new oxygen content species on the sample. To identify these species, XPS analysis was realized.
Figure 2a) presents the O 1s signal. In the range 530 – 535 eV, according to binding energy tables and literature [13] this signal corresponds to O^{2-} in ZrO_2. After contact with wet air, the peak shape changed and the maximum of the peak shifted towards a higher binding energy. The oxygen peak seems to be a sum of several contributions. It is mean that oxygen exists under several chemical states at the oxide surface. Centered at 532 eV, this large peak can

be O^{2-}, OH⁻ and H_2O species. After contact with irradiated wet air the relative proportion between the O^{2-} oxide group and the OH⁻/H_2O groups changed. From the position of the maximum of the peak, it can be seen that the wet air enhances the presence of water molecules on the zirconia surface.

Figure 2b) is the oxygen relative peak obtained for various gamma irradiations doses. There is no peak modification for the experiment at 33 kGy. This suggests that no new species were adsorbed after the radiolysis experiments. However, at 170 kGy, the signal exhibited at least two peaks, one with the same binding energy than O^{2-}, the other corresponding to H_2O molecules.

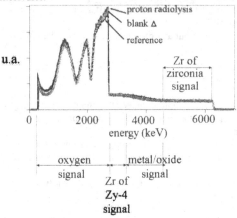

Figure 1. Comparison of NBS spectra obtained before and after water vapour contact (P = 6 hPa) : reference, blank, proton radiolysis.

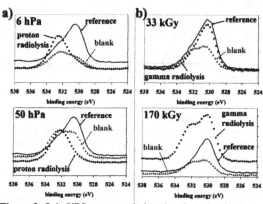

Figure 2. O 1s XPS oxygen relative signal obtained a) before and after contact with vapour contact irradiated and no irradiated by protons and b) before and after contact with water radiolysis by gamma.

Surface tin concentration enhancement

The detection limit of XPS is only about 1 atomic percent, that is, the peaks relative to tin are practically not observable for the reference samples. After the radiolysis experiments, for some conditions, an enhacement of Sn 3d peaks (at 487.4 and 496.1 eV [14]) has been noticed. Figure 3a) shows the Sn 3d region for reference, blank and protons radiolysed samples at 6 and 50 hPa. Only for 50 hPa, a surface enrichment was clearly evidenced. Figure 3b) shows the results obtained with gamma irradiation experiments at 33, 170 and 334 kGy. For each sample, the Sn 3d can be clearly observed compared to the reference spectra. Using the element cross section values according to Scofield [15], the Sn/Zr atomic ratio was found to be equal to 0.4 in the proton radiolysis experiment at 50 hPa, instead of 0.01 for reference ZrO_2. In the case of gamma irradiation, this ratio was about 0.04.

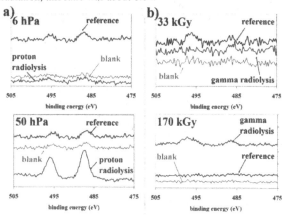

Figure 3. Sn 3d XPS signal obtained a) before and after contact with vapour contact irradiated and no irradiated by protons and b) before and after contact with water radiolysis by gamma.

DISCUSSION

The wet air proton irradiation experiments show that two phenomena occur at the surface of zirconia : water adsorption and tin enhancement. As it could be expected, water adsorption is observed after contact with wet air. The wet air irradiation seems to increase this adsorption. In fact, the relative O^{2-} and H_2O amounts are modified. The radiolysis of water molecules near to the surface of zirconia induces the formation of radicals which are much more reactive than H_2O. The appearance of a large amount of adsorbed water molecules is thus a consequence of reactions between the radicals and the ZrO_2 surface. The NBS analysis shows that in this surface oxygen enhancement is located not only at the extreme surface but also in the cracks of the porous oxide.

The second phenomenon is tin surface enrichment for both proton and gamma irradiations. As previously mentioned we suggest that the radiolysis induces formation of species which interact with the surface to form new bonds. Radicals such as $HZrO_3^{\cdot}$ could be formed, leading to the formation of $HZrO_3^-$ species in the case of gamma radiolysis, according to the Pourbaix diagram [16]. These radicals could be transferred to the gaseous atmosphere in the case of proton radiolysis. In both cases, tin enrichment would be the result of zirconium loss from the surface due to the formation of much radical. The water adsorption and the tin concentration appear be two independent phenomena because the O 1s XPS signal does not systematically change after the radiolysis experiments (see Figure 2b).

CONCLUSION

The irradiation of water molecules by proton or gamma near to the surface of oxidised Zircaloy-4 leads to two phenomena: adsorption enhancement of H_2O molecules and tin surface enrichment. The radical species due to the radiolysis of water are responsible for these modifications. However, more studies are necessary to determine the mechanisms describing the interactions between the radicals and the ZrO_2 surface as well as the role of $HZrO_3^-$.

ACKNOWLEDGEMENTS

The ion beam analysis and the proton irradiation experiments were performed at the Van de Graaff accelerator of the IPNL, Villeurbanne, France. We thank A. Gardon and the accelerator office for their help. We thank also P. Passet for his help in the XPS measurements at the ENSMSE, France and Richard Drogo for the sample oxidation.

REFERENCES

[1] CEA, *Synthèse des connaissances sur le comportement à long terme des bétons, applications aux colis cimentés*, CEA report (2004).
[2] X. Iltis, *J. Nucl. Mater.*, **224**, 109, (1995).
[3] B. Cox, *J. Nucl. Mater.* , **336**, 331, (2005).
[4] K. Poulard, *Nucl. Instrum. Methods Phys. Res., Sect. B*, **181**, 640, (2001).
[5] V. Wasselin-Trupin, PhD thesis, Paris Sud University University, 2000.
[6] P. Barbéris, *J. Nucl. Mater*, **226**, 34, (1995).
[7] B. Cox, *J. Nucl. Mater*, **148**, 332, (1987).
[8] C. Pichon, *Nucl. Instrum. Methods Phys. Res., Sect. B*, **240**, 589, (2005).
[9] V. Wasselin-Trupin, presented at the 2008 IHLRWM, Las Vegas, 2008.
[10] J. F. Ziegler, *The Stopping and Range of Ions in Solids*, edited by Pergamon Press (1985).
[11] ANDRA, *Dossier Argile*, ANDRA Report, (2005).
[12] J. A. Leavitt, *Nucl. Instrum. Methods Phys. Res., Sect. B* , **44**, 269, (1990).
[13] P. Keller, *Corros. Sci.*, **46**, 1939, (2004).
[14] C.T. Liu, *Corros. Sci.*, **49**, 2198, (2007).
[15] J. H. Scofield, *J. Electron. Spectrosc. Relat. Phenom.*, **8**, 129, (1976).
[16] M. Pourbaix, *Atlas of Electrochimical Equilibria*, edited by Pergamon (London, 1966).

Mater. Res. Soc. Symp. Proc. Vol. 1124 © 2009 Materials Research Society   1124-Q09-08

Effect of fluoride ions on passivity and chloride-induced crevice corrosion of Alloy 22

Martín A. Rodríguez[1], Ricardo M. Carranza[1] and Raúl B. Rebak[2]
[1] Depto. Materiales - Comisión Nacional de Energía Atómica
Instituto Sabato - Univ. Nac. de San Martín / CNEA
Av. Gral. Paz 1499, San Martín, B1650KNA Buenos Aires, Argentina
[2] GE Global Research
One Research Circle, CEB2505, Schenectady, NY 12309, USA.

ABSTRACT

Electrochemical tests were carried out in Alloy 22 artificially-creviced specimens in order to assess the effect of fluoride ions on passivity and chloride-induced crevice corrosion. Amperometries at a constant potential of 0.0 V_{ECS} for 24 hours were performed in 0.1M NaCl solutions with different concentrations of fluorides, at 90°C. Fluoride to chloride ratios ranged from 0.1 to 10. An initial decreased of current was observed in the first stage of the test. It was associated with passive film formation and general passive dissolution. The second stage showed an increase of current due to crevice corrosion onset. A third stage showing a decrease of current was only observed in tests with high fluoride to chloride ratios. It was attributed to repassivation of the alloy after some crevice corrosion development. Increasing fluoride to chloride ratio caused a higher current in the first stage (passive film formation and general passive dissolution) and the mitigation of crevice corrosion (current decrease in the second stage). Complete inhibition of crevice corrosion by fluoride ions was not observed in any tested case. Fluoride ions were deleterious for general passive corrosion but they were able to mitigate chloride-induced crevice corrosion for high fluoride to chloride ratios.

INTRODUCTION

Alloy 22 (N06022) is one of the most versatile members of the Ni-Cr-Mo family and was designed to withstand the most aggressive industrial applications in both reducing and oxidizing conditions [1-3]. It is due to its balanced content of Chromium (22%Cr), Molybdenum (13%Mo) and Tungsten (3%W). Alloy 22 has shown excellent resistance to pitting corrosion, crevice corrosion and environmentally assisted cracking, in hot concentrated chloride solutions [1-3]. It has been selected for the fabrication of the corrosion-resistant outer shell of the high-level nuclear waste container for the proposed Yucca Mountain repository [4]. Alloy 22 might suffer crevice corrosion under certain aggressive conditions. It is assumed that localized corrosion will only occur when the corrosion potential (E_{CORR}) is equal or greater than the repassivation potential for crevice corrosion (E_R) [4,5]. Thus, if $E_{CORR} < E_R$, only uniform passive corrosion is expected and the alloy would be free from crevice corrosion attack. E_R of Alloy 22 has been determined in different experimental conditions and by different methodologies [6-9]. Chloride ion is well known as a deleterious species causing crevice corrosion of Alloy 22 [4-9]. A recent work indicates fluoride ion could mitigate and even inhibit chloride-induced crevice corrosion of Alloy 22, in spite of increasing its uniform corrosion rate [9].

The aim of the present work was to study the effect of fluoride ion on chloride-induced crevice corrosion of Alloy 22, by amperometries at a constant potential of 0.0 V$_{ECS}$ for 24 hours, varying fluoride to chloride molar ratio.

EXPERIMENT

Alloy 22 (N06022) specimens were prepared from wrought mill annealed plate stock. The chemical composition of the alloy in weight percent was 59.56% Ni, 20.38% Cr, 13.82% Mo, 2.64% W, 2.85% Fe, 0.17% V, 0.16% Mn, 0.008% P, 0.0002% S, 0.05% Si, and 0.005% C (Heat 059902LL1). Prism Crevice Assemblies (PCA) specimens were used (ASTM G 192) [10]. They were fabricated based on ASTM G 48 which contained 24 artificially creviced spots formed by a ceramic washer (crevice former) wrapped with a PTFE tape [11]. The applied torque was 7.92 N-m (70 in-lb). The tested surface area was approximately 14 cm². The specimens had a finished grinding of abrasive paper number 600 and were degreased in acetone and washed in distilled water. Polishing was performed 1 hour prior to testing. Electrochemical tests were conducted in a one-liter, three-electrode vessel (ASTM G 5) [12]. A water-cooled condenser was used to avoid evaporation of the solution. The temperature of the solution was controlled by immersing the cell in a thermostatized bath, which was kept at a constant temperature of 90°C. The reference electrode was a saturated calomel electrode (SCE), which has a potential of 0.242 V more positive than the standard hydrogen electrode (SHE). The reference electrode was connected to the solution through a water-cooled Luggin probe. The counter electrode consisted in a platinum foil (total area approximately 50 cm²) spot-welded to a platinum wire. All the potentials in this paper are reported in the SCE scale. All the tests were carried out in pH 6, chloride plus fluoride solutions. Small volumes of HF were added to solutions to set pH 6. Nitrogen (N$_2$) was purged through the solution 1 hour prior to testing and was continued throughout the entire test.

RESULTS AND DISCUSSION

Electrochemical tests consisted in the application of a constant potential of 0.0 V$_{SCE}$ for 24 hours to Alloy 22 PCA specimens, in chloride plus fluoride solutions, after remaining 15 minutes at the open circuit potential. The applied potential was in the range of Alloy 22 passivity for all the tested conditions. Current density evolution in time for these tests is shown in Figure 1. Two or three distinctive domains (or stages) can be distinguished according to the solution composition. In the first domain the current density decreased as the time increased. This is attributed to passivation (*i.e.* passive film build-up). Current density was higher for increasing fluoride concentration in this domain. This indicated fluoride ion increased passive dissolution rate. In the second domain the current density increased as the time increased. This was due to the onset of crevice corrosion. Incubation time for the initiation of crevice corrosion was not affected by fluoride concentration (Figure 1). Crevice corrosion initiation occurred after approximately 1 hour of immersion in all the tests. Current density was higher for decreasing fluoride concentration in this domain. This indicated fluoride ion mitigated crevice corrosion propagation. A third domain could be detected only in some solutions with a fluoride to chloride molar ratio equal or higher than two ([F⁻]/[Cl⁻] ≥ 2). The current density decreased as the time increased. It can be attributed to stifling of crevice growth and eventually to repassivation of crevice corrosion. Yilmaz *et al.* [13] state the activity in the corroded area decreases and the specimen regains its passivity in the third domain.

Figure 2 shows images of Alloy 22 PCA specimens after 24 hours of polarization in chloride plus fluoride solutions. Crevice corrosion attack occurred under the crevice formers. The extent of crevice corrosion attack was reduced as fluoride concentration of the solution increased. An important amount of corrosion products were found over the specimen tested in 0.1M NaCl. Very little corrosion products were observed in specimens tested in the other solutions.

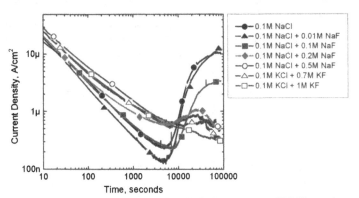

Figure 1. Amperometries at a constant potential of 0.0 V_{ECS} for Alloy 22 PCA specimens in chloride plus fluoride solutions, at pH 6 and 90°C.

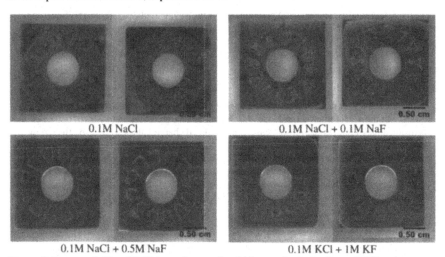

Figure 2. Images of Alloy 22 PCA specimens after 24 hours of polarization at 0.0 V_{ECS} in chloride plus fluoride solutions, at pH 6 and 90°C.

Figure 3 shows the cross-over repassivation potential (E_CO) as a function of fluoride to chloride molar ratio [F⁻]/[Cl⁻] for Alloy 22, in the studied conditions [9]. Crevice corrosion attack was expected for [F⁻]/[Cl⁻] ≤ 2, where the applied potential was higher than E_CO. Nevertheless, localized attack was observed, to different extent, for all the cases. Crevice corrosion was observed even for [F⁻]/[Cl⁻] = 10, where only the first domain was distinguished (Fig.1). There is a difference between present results, obtained by amperometry, and reported results, obtained by potentiodynamic polarization [9]. A recent work indicates that cyclic potentiodynamic polarization would not be a convenient technique for determining repassivation potentials in weakly aggressive media [14]. This is attributed to the short period of time that the specimen is subjected to potentials higher than the repassivation potential, or to the fact that transpassivity may stifle crevice attack. Tsujikawa-Hisamatsu Electrochemical method could be a good option for determining repassivation potentials in these conditions [10].

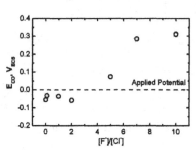 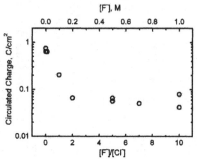

Figure 3. Cross-over repassivation potential (E_CO) as a function of fluoride to chloride molar ratio [F⁻]/[Cl⁻] for Alloy 22 PCA specimens in chloride plus fluoride solutions, at pH 6 and 90°C. [9] Applied potential of 0.0 V_ECS is shown.

Figure 4. Circulated charge as a function of fluoride to chloride molar ratio [F⁻]/[Cl⁻] in amperometries at 0.0 V_ECS for 24 hours, for Alloy 22 PCA specimens in chloride plus fluoride solutions, at pH 6 and 90°C.

Figure 4 shows the total circulated charge in 24 hours of polarization as a function of [F⁻]/[Cl⁻] in the amperometries of Figure 1. Circulated charge is related to penetration of crevice corrosion attack [15]. In the present case, the circulated charge values were also affected by general passive dissolution. An inhibitory effect of fluoride ion on chloride-induced crevice corrosion is evident from Figure 4, specially for [F⁻]/[Cl⁻] ≥ 2. Fluoride ion was able to mitigate but not fully inhibit crevice corrosion. Crevice corrosion attack reduced its rate and also stifled in the presence of increasing concentrations of fluoride. Crevice corrosion repassivation was thought to occur after some propagation for [F⁻]/[Cl⁻] ≥ 2.

Crevice corrosion initiation mechanism

The proposed mechanisms for explaining crevice corrosion initiation in stainless alloys can be classified as follows [16].
1) Passive dissolution

2) Thiosulfate entrapment
3) IR drop and
4) Metastable pitting stabilization

Passive dissolution models are based on the assumption that localized acidification in the crevice occurs by cation hydrolysis, at a rate determined by passive current density. In the present work, fluoride caused an increase of passive current density, and at the same time, inhibited crevice corrosion. The same effect is reported for some organic acids [17]. Passive dissolution mechanism is not consistent with these observations.

MnS inclusions are not found in Alloy 22. Consequently, thiosulfate entrapment mechanism cannot be considered for this alloy.

The theory of IR drop by itself cannot explain crevice corrosion of Alloy 22 in neutral chloride solutions. Localized acidification is needed, as a depassivation pH of 1.7 is reported for Alloy 22 in the studied conditions [15]. Significant IR drop and localized chemistry changes in crevices were observed only after the onset of crevice corrosion [20].

Metastable pitting is a stochastic process whose frequency depends of environmental conditions [1818]. Stable crevice corrosion is reported to occur at the sites where a row of metastable pits form at a critical distance from the crevice mouth [19]. This mechanism is consistent with the observations of the present work. Fluoride ion caused an increase of general passive dissolution rate, but its effect on metastable pitting events rates is unknown.

CONCLUSIONS

Electrochemical tests carried out in artificially creviced Alloy 22 specimens, in fluoride plus chloride solutions at 90°C, indicated the presence of two or three stages (or domains).

- First stage: passive film formation and general passive dissolution. Current density decreased in time.
- Second stage: crevice corrosion propagation onset. Current density increased in time.
- Third stage: repassivation after some crevice corrosion development. Current density decreased in time. It was only observed in tests with high fluoride to chloride ratios ($[F^-]/[Cl^-] \geq 2$).

Fluoride ions were deleterious for general passive corrosion, but they were able to mitigate chloride-induced crevice corrosion for high fluoride to chloride ratios. Complete inhibition of crevice corrosion was not observed in any case, but alloy repassivation was reached for the highest fluoride to chloride ratios.

Considering the proposed mechanisms for explaining crevice corrosion initiation in stainless alloys, passive dissolution cannot explain the observed inhibitory effect of fluoride, as this ion increased passive current density. Metastable pitting is a crevice corrosion initiation mechanism consistent with observations in the present work.

ACKNOWLEDGMENTS

R. M. Carranza acknowledges financial support from the Agencia Nacional de Promoción Científica y Tecnológica of the Ministerio de Educación, Ciencia y Tecnología from Argentina.

491

REFERENCES

1. R. B. Rebak, "Corrosion of Non-Ferrous Alloys", *Corrosion and Environmental Degradation*, Vol. II, ed. M. Schutze (Wiley-VCH, Weinheim, Germany, 2000) pp. 71-111.
2. A. I. Asphahani, *The Arabian Journal of Science and Engineering*, **14**, 317-335 (1989).
3. P. E. Manning, J. D. Smith, and J. L. Nickerson, *Materials Performance*, 67-73, (June 1988).
4. G. M. Gordon, *Corrosion* **58**, 811-825 (2002).
5. R. B. Rebak, Paper 05610, Corrosion/2005, NACE Intl. Houston, TX (2005).
6. B. A. Kehler, G. O. Ilevbare and J. C. Scully, *Corrosion* **57**, 1042 (2001).
7. D. S. Dunn, Y.-M. Pan, K. Chiang, L. Yang, G. A. Cragnolino and X. He, *Journal of Metals*, pp. 49-55 (January 2005).
8. K. J. Evans, A. Yilmaz, S. D. Day, L. L. Wong, J. C. Estill and R. B. Rebak, *Journal of Metals*, pp. 56-61 (January 2005).
9. R. M. Carranza, M. A. Rodríguez and R. B. Rebak, *Corrosion* **63**, 480-490 (2007).
10. ASTM G192-08, Annual Book of ASTM Standards, vol. 03.02 (West Conshohocken, PA: ASTM Intl. 2008).
11. ASTM G48-03, Annual Book of ASTM Standards, Vol. 03.02 (West Conshohocken, PA: ASTM Intl. 2003), pp. 191-201.
12. ASTM G5-94, Annual Book of ASTM Standards, Vol. 03.02 (West Conshohocken, PA: ASTM International, 2004), pp. 53-64.
13. A. Yilmaz, P. Pasupathi and R. B. Rebak, Paper PVP2005-71174, ASME Pressure Vessels and Piping Conference, 17-21 July 2005, Denver, CO (2005).
14. R. M. Carranza, C. M. Giordano, M. A. Rodríguez and R. B. Rebak, Paper to be presented at Corrosion/2009, NACE Intl. Houston, TX (2009).
15. R. M. Carranza, M. A. Rodríguez and R. B. Rebak, Paper 07581, Corrosion/2007, NACE Intl. Houston, TX (2007).
16. N. J. Laycock, J. Steward and R. C. Newman, *Corrosion Science* **39**, pp. 1791-1809 (1997).
17. R. M. Carranza, C. M. Giordano, M. A. Rodríguez and R. B. Rebak, Paper 08578, Corrosion/2008, NACE Intl. Houston, TX (2008).
18. L. Stockert and H. Bönhi, *Materials Science Forum* **44-45**, pp. 313-328 (1989).
19. B. A. Kehler and J. R. Scully, *Corrosion* **61**, 665-684 (2005).
20. N. Sridhar and D. S. Dunn, *Corrosion* **50**, pp. 857-872 (1994).

Mater. Res. Soc. Symp. Proc. Vol. 1124 © 2009 Materials Research Society   1124-Q09-09

Robustness of Passive Films in High Temperature Brines

Joe H. Payer and Pallavi Pharkya

Department of Materials Science and Engineering
Case Western Reserve University
Cleveland, OH

ABSTRACT

A robust passive film provides high corrosion resistance. The ability of the passive film to resist breakdown and to reform after damage is important for the long term performance and reliability of engineering structures. This paper discusses the robustness of passive films as a function of chemical, electrochemical and mechanical stresses. Factors affecting the robustness of passive films are illustrated as a function of corrosion resistance of the alloy and corrosivity of the environment. Alloys covered a range in corrosion resistance from highest resistance alloys [alloy 22 and Fe-based bulk metallic glass, SAM1651], an intermediate alloy [AL6XN] and lowest alloy [316L SS]. Repassivation behavior after mechanical damage was examined.

INTRODUCTION

The primary objective of this work was to determine the robustness of passive films as a function of chemical, electrochemical and mechanical stresses. The ability of the passive film to resist breakdown and to reform after damage is important for the long term performance and reliability of engineering structures. That is, the durability or persistence of the passive film after disruption is important for engineering applications in corrosive environments. Factors affecting the robustness of passive films are illustrated as a function of corrosion resistance of the alloy and corrosivity of the environment.

The environment of interest is oxidizing, high temperature, concentrated chloride brines. These brines are highly aggressive environments, and passive alloys and metals, e.g. Ni-Cr-Mo alloys and Ti, are materials of choice. The passive alloys can have high corrosion resistance but may be susceptible to localized corrosion under aggressive conditions [1,2]. Two highly corrosion resistant alloys [alloy 22 and SAM1651] are compared with an alloy of intermediate corrosion resistant [AL6XN] and an alloy of lower corrosion resistance [316L SS]. Alloy 22 is a crystalline, nickel-chromium-molybdenum alloy, and SAM1651 is a corrosion resistant, iron-based Bulk Metallic Glass (amorphous alloy) containing chromium, molybdenum, boron and yttrium.

Bulk metallic glasses (amorphous metals), such as SAM1651, are an emerging class of alloys which hold high interest due to their ability to be designed for special properties. SAM1651 is an example of bulk metallic glasses that have been fabricated at low cooling rates on the order of 80 K/s using a copper mould casting method [14]. The nominal atomic percent composition of this alloy is 48% Fe, 15% Cr, 14% Mo, 6% B and 2% Y coupled with 15% C. [12,13]. Chromium and molybdenum provide high corrosion resistance while boron enables glass formation and neutron absorption and yttrium lowers critical cooling rates [14]. Because of this elemental composition, unlike many Fe-based amorphous systems, it has shown superior corrosion resistance compared to nickel-based alloy 22 in some highly aggressive environments, including concentrated calcium chloride brines at elevated temperatures [14,15].

Chromium and molybdenum (tungsten) concentrations in the alloys are primary contributors to corrosion resistance. Nominal atomic concentrations for the alloys are

> Alloy 22: $Cr_{25}-Mo_9-W_1$
> SAM1651: $Cr_{15}-Mo_{14}$
> AL6XN: $Cr_{23}-Mo_4$
> 316L SS: $Cr_{18}-Mo_2$

The aggressiveness of the test conditions (chloride, temperature and oxidizing potential) was adjusted to match the corrosion resistance of the alloys.

A robust passive film provides high corrosion resistance, and passive films control the corrosion resistance of stainless steels and Ni-Cr-Mo alloys. The robustness of passive film depends upon both the alloy and the environment. The literature is rich with work that examines these effects; a sampling of references includes [3,4,5,6,7,8,9,10,11]. Robustness increases with increasing corrosion resistance of the alloy and decreases with increasing corrosivity of the environment. Figure 1 illustrates a range of behaviors from highly robust passivity to no passivity, i.e. active corrosion behavior. From (a) to (d), the cyclic polarization diagrams show passive behavior with no hysteresis, with smaller hysteresis, with larger hysteresis and no passivity, respectively. The range of behaviors can be observed for a given alloy in different environments or a given environment with different alloys.

EXPERIMENTAL METHODS

Tests were conducted on SAM1651, alloy 22 (N06022), AL6XN (N08367), and 316L SS (UNS S31600), the latter two are iron based alloys with intermediate and low corrosion resistances, respectively. SAM1651 rods were prepared by drop casting into chilled copper mould at Oak Ridge National Laboratory (ORNL).The rods of the rest of the alloys were procured from commercial vendors. Rods of alloy 22 were received from Haynes International Inc., of AL6XN from Rolled steels and of 316L SS from Southern Steels. Table I shows the nominal compositions (at % and wt %) of SAM1651 and the mill analysis compositions of alloy 22, AL6XN and 316L SS.

The CPP specimen of SAM1651 was a 0.42 cm diameter rod mounted in an epoxy (Dexter Hysol, resin EE4183, hardener HD3561) such that only the circular face of one side of the sample was exposed to the electrolyte. The CPP specimens of alloy 22, AL6XN, and 316L SS were cylinders of 0.625 cm diameter and 0.5 cm length, drilled and tapped, and assembled with a PTFE compression fitting. The samples for mechanical scribing tests were rods partially covered with double-walled, heat shrink, polytetrafluoroethylene (PTFE) tube, and approximately 15 mm of the rods were exposed.

Tests were carried out in two environments: (a) 3.5 wt% saline brine or 0.6M NaCl, 80°C, and (b) 1M HCl, 80°C. The saline and hydrochloric acid solutions were prepared with American Chemical Society (ACS) certified grade chemicals using a reagent grade deionized water (resistivity = 18 MΩ.cm).

Table I. Nominal alloy compositions for Alloy 22, SAM1651, AL6XN, 316L SS

Alloys	Ni	Fe	Cr	Mo	Co	B	W	Y	C	Mn
SAM1651, at%	0.00	48.00	15.00	14.00	0.00	6.00	0.00	2.00	15.00	0.00
SAM1651, wt%	0.00	51.30	14.90	25.70	0.00	1.20	0.00	3.40	3.00	0.00
Alloy 22, at%	59.60	4.20	24.70	8.40	2.20	0.00	0.90	0.00	0.00	0.00
Alloy22, wt%	57.20	3.80	21.00	13.10	2.10	0.00	2.80	0.00	0.00	0.00
AL6XN, at%	23.30	49.79	22.49	3.76	0.21	0.00	0.00	0.00	0.08	0.00
AL6XN, wt%	23.80	48.37	20.29	6.24	0.21	0.00	0.04	0.00	0.02	0.40
316L SS, at%	9.67	69.06	18.15	1.22	0.00	0.00	0.00	0.00	0.09	1.80
316L SS, wt %	10.20	69.07	16.86	2.09	0.00	0.00	0.00	0.00	0.02	1.77

Electrochemical tests were conducted in standard three electrode glass test cells and the details are described elsewhere. Cyclic potentiodynamic polarization (CPP) tests were conducted to determine the electrochemical parameters, namely, open circuit potential (E_{ocp}), breakdown potential (E_{bd}), repassivation potential (E_{rp}) and the passive current density (i_{pass}). Here, breakdown potential is the potential at which the polarization current increases sharply from the passive current range on the forward polarization scan and does not necessarily imply onset of localized corrosion. The tests were conducted with Solartron 1280B and the data collection and analysis were performed using Corrware and Corrview softwares (Scribner Associates, Inc.).

Controlled rotation scratch-repassivation tests were conducted by coupling the sample with a modulated speed rotator AFMSRX from Pine Instrument Company and the details of the experimental set-up are described elsewhere. The cell was equipped with a diamond tip studded scriber which could be adjusted for vertical alignment in z-direction. The diamond tip had the width of 450 μm and the length of 250 μm. The test protocol was to measure open circuit potential for 1 hr. This was followed by ramping up the potential to a pre-selected potential. The sample rotation was also initiated at this stage at the pre-selected rotation speed of 100 rpm. The sample was held at the constant potential for 15 minutes to allow the formation of a stable passive film. The scribing of the sample was initiated at the 15th minute with a diamond tip scriber for the duration of 30 seconds. The current-time was measured for 5 minutes to examine the repassivation behavior. The subsequent scratches under the identical condition were made at the 20th and the 25th minute at different locations. A pristine sample was used for each potential. Table II provides the matrix for performing the controlled rotation scratch-repassivation tests on the four alloys in 0.6M NaCl (open to air) at 80°C.

RESULTS

Cyclic potentiodynamic polarization behavior

The Cyclic Potentiodynamic Polarization (CPP) curves of SAM1651, alloy 22, AL6XN and 316L SS are shown in Figure 2. The data for SAM1651, alloy 22 and AL6XN are in 0.6M NaCl, 80°C. A more benign chloride concentration and temperature (0.1M NaCl, 60°C) were chosen for 316L SS to get some window of passivity at a higher temperature. Both SAM1651 and alloy 22 showed high anodic breakdown potentials and repassivated without hysteresis. The AL6XN showed a large hysteresis before repassivating at a potential below the open circuit potential. The large hysteresis loop indicates that AL6XN is susceptibility to localized corrosion in this environment. In 0.1M NaCl, 60°C, the 316L SS showed a hysteresis loop on the reverse scan. At the potential of -0.104 V_{SCE}, the current changed polarity to cathodic current.

Table II. Matrix of potentials for performing controlled rotation scratch-repassivation tests on alloy 22, SAM1651, AL6XN, in 0.6M NaCl, 80°C and 316L SS in 0.1M NaCl, 60°C

MATERIAL	ENVIRONMENT	E1 (V_{SCE})	E2 (V_{SCE})	E3 (V_{SCE})	E4 (V_{SCE})	ROTATION SPEED (RPM)
Alloy 22	0.6M NaCl-80°C	+0.010	+0.250	~	+0.650	100
SAM1651	0.6M NaCl-80°C	+0.010	~	+0.650	+1.000	100
AL6XN	0.6M NaCl-80°C	+0.010	+0.250	+0.450	~	100
316L SS	0.1M NaCl-60°C	+0.010	~	~	+0.100	100

On comparing the four alloys, the open circuit potential did not show much variation but the breakdown and the repassivation potentials varied significantly. No localized corrosion was observed on either SAM1651 or alloy 22 after the polarization tests. The breakdown potential for SAM1651 was 300 mV more positive than that for alloy 22 and 900 mV more anodic than 316L SS which was tested at a much lower temperature and in a less concentrated solution. The breakdown potential of AL6XN was similar to that of SAM1651 but, unlike SAM1651, it did not repassivate after breakdown. The AL6XN samples sustained deep crevices along the circumference of the specimen holder. The repassivation potential of SAM1651 was approximately 300 mV more positive compared to that of alloy 22 and 1025 mV more positive as compared to AL6XN.

The results showed that SAM1651 was highly resistant to localized corrosion in 0.6M NaCl, 80°C. In the same environment, alloy 22 also showed high corrosion resistance but the resistance was lower compared to SAM1651. AL6XN, on the other hand, had a broad passive range on the forward scan but it was highly prone to localized corrosion as indicated by the reverse scan. Among the four alloys tested, 316L SS, even when exposed to a more benign environment and lower temperature, showed the smallest window of passivity and the least resistance to localized corrosion.

CPP plots for SAM1651, alloy 22, AL6XN and 316L SS in 1M HCl are shown in Figure 3. All the alloys were tested at 80°C except 316L SS which was tested at room temperature. The forward scan of SAM1651 showed a continuous increase in current with increase in potential. The reverse scan retraced the forward scan, but the typical breakdown and the repassivation potentials for passive metals could not be determined. This behavior, which is not typical of either fully passive or active behavior, was also observed for SAM1651 after exposures to concentrated HCl [16]. The behavior was referred to as "pseudo-passivity", and it was accompanied by the formation of a shiny, black film on the metal surface.

For alloy 22, the breakdown and the repassivation potentials were around +0.800 V_{SCE}. The alloy did not show hysteresis on the reverse scan. For AL6XN, the breakdown potential was around +0.800 V_{SCE}. While the breakdown potential of AL6XN was comparable to that of alloy 22, after the passive film brokedown, it repassivated below its open circuit potential with a large hysteresis. The 316L SS in 1M HCl at room temperature was spontaneously active and did not show any passivity, and the sample developed a black coating on its surface.

Comparing the performance of the three alloys in 1M HCl, 80°C, alloy 22 had the best corrosion resistance. The passivity window (E_{bd}-E_{ocp}), an indication of superior corrosion resistance was the largest and the reverse scan hysteresis, which is a measure of the susceptibility to crevice corrosion, was negligible for alloy 22. A large passivity window for AL6XN indicated good corrosion resistance on the forward scan, but the large hysteresis on the reverse scan once the passive film broke down, indicated susceptibility to localized corrosion. The SAM1651, while not exhibiting classic passive behavior, was not fully active in 1M HCl, 80°C. The 316L SS, which had shown small passivity region in neutral brine at elevated temperature, i.e. 0.1M NaCl, 60°C, (Figure 2) did not show any passivity in 1M HCl, 26°C (Figure 3).

These results clearly illustrate the range of robustness of passive films that are observed. The behavior shifts from fully passive to active behavior for a given alloy due to changes in environment and in a given environment due to changes in alloy corrosion resistance.

Repassivation behavior and reformation kinetics of alloys as a function of potential

Controlled rotation scratch-repassivation tests were run to examine the repassivation behavior after mechanical damage. The current-time transients at constant potentials were measured throughout the test. With the sample at the pre-selected potential, rotation was initiated at 100 rpm. The sample was held at the constant potential for 15 minutes, and scribing of the sample was initiated at the 15th minute with a diamond tip scriber for the duration of 30 seconds. The current-time was measured for 5 minutes to examine the repassivation behavior. Subsequent scratches at the same constant potential were made at the 20th and the 25th minute at different locations on the sample. The matrix of potentials for performing controlled rotation scratch-repassivation tests on alloy 22, SAM1651, AL6XN, in 0.6M NaCl, 80°C and 316L SS in 0.1M NaCl, 60°C is presented in Table II. Representative results are presented here to illustrate the robustness of the passive films after mechanical damage. More aggressive conditions are represented by more oxidizing (more positive) potentials, higher temperature and more concentrated chloride solution.

The CPP curve and current-time transients at two potentials are shown for SAM1651 and alloy 22 in Figures 4 and 5, respectively. The first potential shown is in the passive range for the alloy, and the second potential is more positive than the potential where polarization current increases spontaneously in forward scan of the polarization curve. For mechanical damage imposed when the alloys were in the passive range, the current-time transients for SAM1651 and alloy 22 show that current after mechanical damage decayed to the same values as before scribing, indicating that the passive films on these alloys reformed at all the potentials within the passive range. This was the case for the tests at +0.650 V_{SCE} for SAM1651 (Figure 4b) and +0.450 V_{SCE} for alloy 22 (Figure 5b) and for all the lower potentials tested for each alloy. At the potentials above the breakdown potentials, i.e. +1.000 V_{SCE} for SAM1651, shown in Figure 4-c and +0.650 V_{SCE} for alloy 22, shown in Figure 5-c, the current rose above the passive current levels for both alloys during the holding period and before any mechanical damage.

The repassivation behavior for AL6XN and 316SS are shown in Figure 6 along with their CPP curves. The AL6XN and 316L SS were tested in 0.6M NaCl, 80°C and 0.1M NaCl, 60°C, respectively. The passive current on AL6XN at +0.250 V_{SCE} returned to the same level after scribing. However, the current rose to higher values when +0.450 V_{SCE} was applied indicating the breakdown of the passive film. Examination of the specimen indicated that the increase of polarization current was due to initiation of crevice corrosion underneath the specimen holder

497

and not the result of mechanical damage. The passive film on 316L SS reformed at +0.010 V_{SCE} after all the three scribing events.

While each of the alloys repassivated after mechanical damage imposed in their passive region, the magnitude of the peak current during scribing was proportional to the corrosion resistance of the alloy in the environment. The results show that 316L SS underwent greater amount of anodic dissolution once the passive film was damaged as compared to the other alloys. This occurred even though the 316L SS was exposed to a more benign solution and lower temperature.

DISCUSSION

A robust passive film provides high corrosion resistance, and the properties of the passive film depend upon both the alloy and the environment. A range of behaviors was observed for a given alloy or a given environment. The passive films on alloy 22 and SAM1651 are more durable than the passive films on AL6XN and 316L SS. The passive films remain protective under more aggressive conditions, and they repassivate more readily after chemical/electrochemical induced breakdown.

The higher concentration of Cr, Mo and W in the bulk composition of the alloy is a primary contributor to the greater durability of alloy 22 and SAM1651 compared to AL6XN and 316L SS. The observations here are consistent with expectations based upon two empirically based parameters for pitting resistance: PREN -*Pitting Resistance Equivalent Number* and PRE - *Alternative Pitting Resistance Equivalent Number*. These numbers were calculated based on the following equations and alloy compositions from Table I, and are shown below:

$$PREN = \text{wt\% } Cr + 3.3 \text{ wt\% } Mo + 16 \text{ wt\% } N$$

$$PRE = \text{wt\% } Cr + 3.3 \text{ (wt\% } Mo + \text{wt\% } W) + 30 \text{ wt\% } N$$

Alloy	PREN	PRE
SAM 1651	100	100
Alloy 22	64	74
AL6XN	41	41
316L SS	24	24

It is important to note that the parameters are being applied here to alloy compositions beyond the range upon which they are based, i.e. compositions not within the data base for the empirical relationships. However with that caveat in mind, higher PRE/PREN numbers correspond here to higher expected pitting resistance, and the rank order based on PRE/PREN numbers agrees with the measured corrosion behavior. While the rankings were as expected, the large separation in PRE/PREN numbers between SAM1651 and alloy 22 suggested greater difference in corrosion behavior than was observed, i.e. both alloys has high corrosion resistance.

The full range of passive behaviors shown in Figure 1 was observed in this work: (a) broad passive region with no hysteresis loop, (b) passive region with smaller hysteresis loop, (c) passive region with larger hysteresis loop, (d) no passivity, completely active behavior. The shift in behavior was seen for both corrosivity of the environment and for the corrosion resistance of the alloy. Based on a broad range of behaviors reported in the literature for stainless steels and nickel based alloys, the expected corrosive environment and alloy trends for durability and robustness of passive films are indicated in Table III. More aggressive environmental conditions result from increased chloride concentration, increased acidity, increased temperature and more oxidizing (more positive) potentials. Increased nitrate concentration in chloride brines decreases the aggressiveness of the environment. Increasing Cr, Mo and W in the alloy increases corrosion resistance. As corrosivity increases, the passive range narrows, passive current increases and ultimately active dissolution (no passivity) is observed.

Table III - Environment and Alloy Trends in Durability of Passivity

Factor	Shift in Behavior	Effect
Increasing Chloride Ion	→	Detrimental
Increasing Acidity	→	Detrimental
Increasing Temperature	→	Detrimental
Increasing Potential	→	Detrimental
Increasing Nitrate Ion	←	Beneficial
Increasing Cr-Mo-W	←	Beneficial

An objective of this work was to determine the effect of transient, mechanical damage events on repassivation behavior of the alloys. The passive film was removed by scribing a rotating cylinder, and corrosion of the freshly exposed metal surface was measured. When scribed at potentials within the passive range, the passive film reformed on all four alloys after each scribing event caused mechanical damage. The passive film on alloy 22 and SAM1651 reformed with lower corrosion current spikes at higher anodic potentials compared to AL6XN and 316L SS.

It is significant to note that even when the passive film was mechanically damaged at potentials above the repassivation potentials of AL6XN and 316L SS, the passive film still reformed rapidly. This repassivation may seem counter intuitive because of the expectation of an unstable film and onset of crevice corrosion for these conditions. However, in the bulk solution and with a rotating specimen, critical crevice chemistry for the onset of crevice corrosion does not form and persist at the mechanically damaged area, and the passive film can spontaneously reform.

SUMMARY

This paper discusses the robustness of passive films as a function of chemical, electrochemical and mechanical stresses. Results are presented for chemical, electrochemical and mechanical durability as a function of alloy composition and corrosivity of environment. A range of behaviors was found with more robust passive films for more corrosion resist alloys and for less corrosive environments.

In 0.6M NaCl at 80°C, the trend in chemical/electrochemical durability was

SAM1651>Alloy 22>>AL6XN>>316L SS

In 1M HCl at 80°C, the trend in chemical/electrochemical durability was

Alloy 22>SAM1651 >>AL6XN>>316L SS

After mechanical damage at potentials within the passive range, the passive film for all four alloys reformed rapidly. At potentials above the breakdown potentials, the polarization current rose above the passive current levels during the holding period and before any mechanical damage.

ACKNOWLEDGEMENTS

Support by the Science & Technology Program of the Office of the Chief Scientist (OCS), Office of Civilian Radioactive Waste Management (OCRWM), U.S. Department of Energy (DOE), is gratefully acknowledged. The work was performed under the Corrosion and Materials Performance Cooperative, DOE Cooperative Agreement Number: DE-FC28-04RW12252. The views, opinions, findings, and conclusions or recommendations of authors expressed herein do not necessarily state or reflect those of the DOE/OCRWM/OCS.

The authors wish to acknowledge the support of the Defense Science Office (DSO) of the Defense Advanced Research Projects Agency (DARPA). The work was part of a DOE/DARPA program for the development of the corrosion resistance of iron-based amorphous metal coatings, under the direction of Dr. J.C. Farmer at the Lawrence Livermore National Laboratory. Preparation the SAM 1651 cast rods by Oak Ridge National Lab, courtesy of William Peter and Craig Blue, is acknowledged.

REFERENCES

1. J. H. Payer, *The proposed Yucca mountain repository from a corrosion perspective,* in *Materials Research Society Symposium on Nuclear Waste XXIX* 2005, Ghent, Belgium.
2. L. G. McMillion, D. A. Jones, A. Sun and D. D. MacDonald, Metallurgical and Materials Transactions A, **36A**: 1129 (2005).
3. A. C. Lloyd, J. J. Noël, S. McIntyre, D. W. Shoesmith, Electrochimica Acta, **49** (17-18): 3015 (2004).
4. A. C. Lloyd, D. W. Shoesmith, N. S. McIntyre, and J. J. Noël Journal of the Electrochemical Society,. **150**(4): B120 (2003).
5. S. D. Day, K. J. Evans, and G. O. Ilevbare, Electrochemical Society Proceedings, **2002-24**: 534 (2002).
6. R. C. Newman, Corrosion Science, **25**(5): 331 (1985).
7. G. A. Cragnolino, D.S. Dunn, and Y.-M. Pan, in *Scientific Basis for Nuclear Waste Management XXV,* edited by B. P. McGrail and G. A. Cragnolino (Mater. Res. Soc. Symp. Proc. **713**, Warrendale, PA, 2002), pp. 53-60.
8. B. A. Kehler, G. O. Ilevbare, and J. R. Scully, Corrosion Science, **43** (6): 1042 (2001).

9. K. J. Evans, L. L. Wong, and R. B. Rebak, in *Transportation, Storage, and Disposal of Radioactive Materials*, (Pressure Vessels and Piping Conf Proc, **483**: pp. 137-149 2004).

10. K. J. Evans and R. B. Rebak, Electrochemical Society Proceedings, **2002-13**: pp. 344-354 (2002).

11. T. Cohen-Hyams and T. M. Devine, *In Situ Investigation of the Passive Films of Alloy C22 in 1M NaCl and 1M HCl. presented at MS&T'07,Detroit, MI, 2007.*

12. J. R. Scully, A. Lucente, in ASM Handbook **13B**, *Corrosion of Amorphous Metals*, edited by Stephen D. Cramer and Bernard S. Covino Jr., (ASM International, Materials Park, OH, 2006), p. 476.

13. J. R. Scully, A. Gebert., and J. H. Payer, J. Mater. Res. **22**: 302 (2007).

14. J. C. Farmer and J. J. Haslam, S. D. Day, T. Lian, C. K. Saw, P. D. Hailey, J. Choi, N. Yang, C. A. Blue, W. H. Peter, J. H. Payer, D. J. Branagan, ECS Trans. **3** (31), 485 (2007).

15. X. Shan, H. M. Ha, and J. H. Payer. *The Fate of Corrosion Products after Crevice Corrosion of Alloy C-22,* Submitted to the Journal of the Electrical Society, Nov 2008.

16. H. M. Ha, J. R. Miller, and J. H. Payer, *Devitrification of Fe-Based Amorphous Metal SAM 1651 and the Thermal Effect on Corrosion Behavior,* Submitted to Journal of the Electrochemical Society, Nov. 2008.

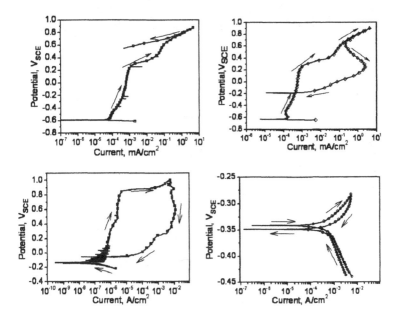

Figure 1. Illustration of the range of passive behaviors that can be observed based on corrosivity of the environment or the corrosion resistance of the alloy.

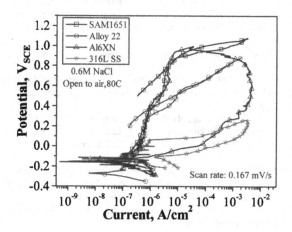

Figure 2. Cyclic potentiodynamic polarization curves for alloy 22, SAM1651 and AL6XN in near-neutral 0.6M NaCl, 80°C and for 316L SS in 0.1M NaCl, 60°C. The solutions were open to air

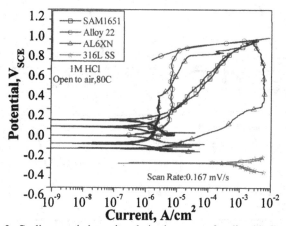

Figure 3. Cyclic potentiodynamic polarization curves for alloy 22, SAM1651, AL6XN in 1M HCl, 80°C. 316L SS was tested at room temperature. The solution was pH 1 and open to air

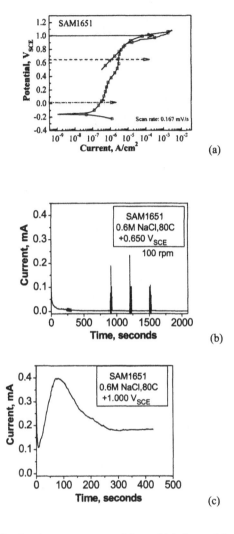

Figure 4. (a) CPP plot showing constant potentials at which the scratch-repassivation tests were conducted for SAM1651 in 0.6M NaCl at 80°C ((b) i-t curve at +0.650 V_{SCE} and (c) i-t curve at +1.000 V_{SCE}

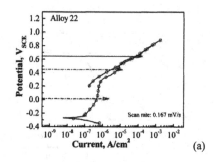

(a)

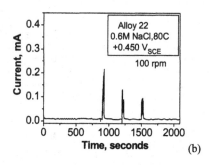

(b)

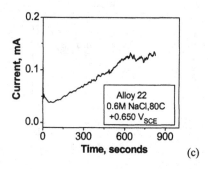

(c)

Figure 5. (a) CPP plot showing constant potentials at which the scratch-repassivation tests were conducted on alloy 22 in 0.6M NaCl at 80°C,(b) i-t curve at +0.450 V_{SCE} and (c) i-t curve at +0.650 V_{SCE}

504

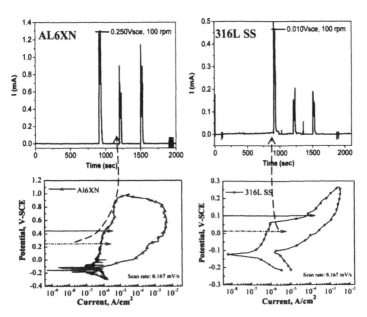

Figure 6. CPP plots and current-time transients for constant potential scratch-repassivation for AL6XN in 0.6M NaCl, 80°C at +0.250 V_{SCE} and 316L SS in 0.1M NaCl, 60°C at +0.010 V_{SCE}. The location of constant potential tests w.r.t. the polarization curves are shown.

505

Migration and Colloids

Mater. Res. Soc. Symp. Proc. Vol. 1124 © 2009 Materials Research Society   1124-Q11-01

The Formation and Modeling of Colloids From the Corrosion of Nuclear Waste Forms

Edgar C. Buck and Rick S. Wittman
Pacific Northwest National Laboratory, 902 Battelle Blvd., Richland, WA 99352, U.S.A.

ABSTRACT

This paper describes a model for determining the stability and associated radionuclide concentrations of colloids that might be present in the nuclear waste package environment from degradation of the nuclear waste forms. The model simplifies radionuclide–colloid behavior by assuming that all colloids can be defined as either smectite clay, a mixed actinide-bearing rare earth-zirconium oxide, iron oxyhydroxide (ferrihydrite) $\{FeOOH\}$, or uranophane $\{Ca(UO_2)_2(SiO_3OH)_2(H_2O)_5\}$. However, for the purposes of predictive stability modeling, the colloids are conceptually represented as montmorillonite, ZrO_2, hematite, and meta-autunite, respectively. The model uses theoretical calculations and laboratory data to determine the stability of modeled colloids with ionic strength and pH. The true nature of colloid composition and heterogeneity, generation, and flocculation will be extremely complex, involving the formation of numerous types of phases, often depending on the composition of the various waste forms and waste package materials. This model strives to capture the uncertainty of the real system using conservatively bound theoretical models. Two of the four representative colloids designed to capture the behavior of the borosilicate glass and spent fuel derived colloids are described in detail here.

INTRODUCTION

Investigations of the degradation of high level borosilicate indicated that colloids form through secondary mineralization of radionuclide-bearing phases within the weathered corrosion rind and associated with colloid-sized clay phases, and later spall from the surface. Spallation of clay fragments, termed primary colloids, and radionuclide-bearing phases from these alteration layers formed on the surfaces may also result in colloid generation. Experiments have demonstrated that colloids can be formed in both static and dynamic borosilicate glass tests. It was also observed that as the ionic strength increased, colloid concentration generally decreased, and ultimately a threshold value was reached above which the colloids were not observed or were observed in low quantities [1]. However, similar clear evidence for the occurrence of colloids from corroded spent nuclear fuel is lacking. Several experiments have failed to indicate significant colloid concentrations; whereas, other tests have shown the presence of colloids.

The rationale for assuming that a small colloid mass fraction may travel un-retarded through the natural barriers is that there have been observations of small mass fractions of plutonium migrating over 1.3 km within 30 years on silica-based colloids at the Nevada Test Site (NTS) [2] and about 4 km in 55 years associated with iron oxide (ferrihydrite) colloids at the Mayak facility in Russia [3]. At the Mayak site, plutonium was shown by micro-analysis to be associated only with iron oxy-hydroxides; whereas, as at the NTS, plutonium was associated with silicate minerals. In these cases, the transport of the plutonium colloid depended on the mobility of another species. More recently, Soderholm and co-workers have shown the formation of nano-particles of plutonium oxide that they claim could be mobile based on their surface charge [4].

If we examine a cross-section through the corrosion rind of a borosilicate glass and a spent nuclear fuel there are several similarities that suggest colloids should be generated from both types of materials. Tests have demonstrated that corrosion results in the accumulation of insoluble elements, including; U, Zr, Th, and rare earth elements, at the corrosion front in borosilicate glass [5,6] and spent fuel [7]. Images of the corroded waste forms represent snap-shots in time at specific locations that may not necessarily represent the overall sample conditions; however, in the case of borosilicate glass, the phases observed within the corrosion rind, typically smectite clay, have been found suspended in the leachate [8,9]. These particles of clay have entrained within them radionuclide-bearing particles. Hence, the mobility of the entrained radionuclide is dependent on the clay particle colloidal chemistry. The same strong relationship between colloids and surface weathering, however, cannot be claimed for spent fuel corrosion tests [1]. Yet, if similar actinide-bearing colloids formed from spent fuel degradation, it is reasonable to assume that they would be chemically similar to the phases found within the corrosion rind.

Direct use of laboratory corrosion tests to extract colloid concentrations is fraught with problems because the tests were designed to accelerate reactions and this might have exacerbated colloid generation while the materials used in the corrosion tests may reduce colloid concentrations, through filtration and physical barriers. A better approach to modeling colloid concentrations is to assume that the types of phases observed in the corrosion rinds could become colloidal. By estimating the colloidal stability of these types of phases, we can provide a stronger scientific basis for their occurrence than a relatively short-term experiment.

Finn and co-workers [10] found actinide-bearing particles in the leachates from intermittent drip spent nuclear fuel tests at early testing times; although these virtually disappeared at longer testing times. In spent fuel corrosion experiments conducted by Wilson [11], significant concentrations of plutonium colloids were observed; however, where cladding was present, plutonium colloid concentrations were reduced by three to four orders of magnitude. These observations were supported by the intermittent drip spent nuclear fuel corrosion tests [1]. Colloidal phases were detected in the leachate of the intermittent drip spent fuel tests at early times [10]; but, at later times, colloid release ceased [1]. In corrosion tests on spent nuclear fuel under immersion conditions in deionized water, plutonium, americium, and neptunium have been found associated with the U(VI) phases [12,13].

THEORY

The Derjaguin, Landau, Verwey and Overbeek (DLVO) theory is the principal method for examining the stability of a colloidal suspension in terms of the repulsive and attractive forces. The theory may have limitations as it does not account for hydrophobic interactions which may be very important for the types of particles and systems under consideration in this model. Hydration forces and geometric effects are accounted for through an approximation. In this model, the DLVO theory has been used to look at the fundamental constraints on particle stability. In DLVO theory the total potential energy $U_T(h)$ between two colloid particles of radius a, with their surfaces separated by distance h, includes the sum of both the electrostatic $U_E(h)$ and van der Waals $U_V(h)$ contributions. The colloid stability can be expressed through the stability ratio, W, which is related to the difference in the rate constants of rapid (van der Waals dominated attraction) and slow coagulations [14].

The prediction is based on calculating the average effect of electrostatic repulsion between charged colloid particle surfaces to inhibit the frequency of particle collisions required for coagulation. This report documents the parameters, form of inter-particle potentials and the surface charge assumptions that were used to develop the stability boundaries for the four colloid types. For each colloid a reference basis was used to establish the required inputs:

- Form of the electrostatic potential.
- Surface potential dependence on pH and ionic strength.
- Form of van der Waals potential.
- Value for Hamaker constant, A_H, and, in some cases, specification of an interaction depth.
- Particle size consistent with colloid form and reference source.

In all four cases (see Table I), the form of the electrostatic potential was expressed according to

$$U_E(h) = 2\pi a \, \varepsilon_r \, \varepsilon_0 \, \zeta_{eff}^2 \, \ln(1 + e^{-h/r_D}) \tag{1}$$

where ζ_{eff} is the effective surface potential, h is the particle separation distance, and r_D the Debye length. Also, ε_0 and ε_r are the free-space permittivity and dielectric constant of water respectively. The Debye length squared can be expressed as:

$$r_D^2 = \frac{\varepsilon_r \, \varepsilon_0 \, k_B T}{2 \, e^2 \, N_A I} \tag{2}$$

where k_B is the Boltzmann constant, T is absolute temperature, e is the charge of the electron, N_A is Avogadro's number and I is the ionic strength in moles per volume units consistent with length scale units of the expression. The fundamental physical constants were obtained from Lide [15][1].

In each of the cases the van der Waals potential is expressed according to:

$$U_V(h) = \frac{-A_H}{12h} \frac{a}{(1 + \frac{h}{\delta})(1 + \frac{h}{2\delta})} \tag{3}$$

where δ is the plate thickness. The electrostatic (Equation 1) and van der Waals (Equation 3) potentials are derived by applying the Derjaguin approximation to the interaction energies per unit area given in Equation 3 of Zheng et al. [16]. For the van der Waals potential, this expression retains the dependence on both separation distance and surface thickness. Both dependences are needed for montmorillonite and meta-autunite of Table I. The stability factor relating the inhibited collision rate to the uninhibited rate can then be expressed as a ratio $W = W_T/W_V$, where the numerator

---

[1] The gas constant, $R = k_B \times N_A$ and $\varepsilon = \varepsilon_0 \varepsilon_r$, ρL (where $L \equiv N_A$, Avogadro's number) and ρ is solvent density. The L^2 cancel out and for water ρ is ~ 1.

$$W_T = 2 \int_0^\infty du \, \frac{\beta(u)}{(2+u)^2} \exp(U_T(h)/k_B T) \qquad (4)$$

and denominator

$$W_V = 2 \int_0^\infty du \, \frac{\beta(u)}{(2+u)^2} \exp(U_V(h)/k_B T) \qquad (5)$$

represent the respective collision probabilities. Given the potentials from Equations 1 and 3, the stability ratio obtained from the ratio of Equations 4 and 5, is given by Equation 14 of Honig et al. [14]. The Equation 4 and 5 integrals are over the reduced separation distance $u = h/a$ and the hydrodynamic factor:

$$\beta(u) = \frac{6u^2 + 13u + 2}{6u^2 + 4u} \qquad (6)$$

Equation 6 approximates the deviation from Stokes law viscous affects on the particle motion as particle surfaces approach at close distances.

Table I. Definition of Colloid Potential Energy for Stability Calculations

Model Parameter	Colloid Type			
	Montmorillonite	Hematite	ZrO$_2$	Meta-Autunite
a (nm) radius	300	60	30	250
$^\dagger \zeta_{eff}$	ζ	ζ	ζ	$\frac{4k_B T}{e} \tanh \frac{e\zeta}{4k_B T}$
A_H (J)	5×10^{-20}	5×10^{-20}	1×10^{-20}	2.06×10^{-20}
δ (nm) ‡	0.66	∞	∞	3.0
Ref.	Tombácz et al. [17], Figure 3, pg 77 Figure 4, pg 78, legend (Hamaker constant (A_H) reported in Joules)	Gunnarsson et al. [18], Figure 7, pg 456 pg 454, Section 3.3, line 14.	*Bitea et al. [19], Pg 59, Figure 3	Zheng et al. [16], Pg 53, Figure 7b Pg 53, section 3.2.4, line 4

†Fits of the surface potentials ζ, contained in the model set for ZrO$_2$ is shown in Figure 1
*Parameters determined from fits to data given in reference.
‡Reported interaction depth (or layer thickness) value in Tombácz et al. [17] (pg 79) was 1.3 nm but may also be between 1.55 and 1.85 nm depending on the ionic strength. Because of the nature of the charge distribution the plate thickness used in the van der Waals term is half this value at ~0.66 nm.

Table I presents the specific information to define the inter-particle potential used in the stability calculations for each colloid type. Two different values for the Hamaker constant for hematite were documented in Gunnarsson et al. [18] (section 3.3, pg 454). The selection of the lower energy value is conservative for model application as it results in greater stability for the colloids. Values for the fundamental physical constants reported in spreadsheets were obtained from Lide [15]. The values for the potential parameters can be found in each of the references in Table I with the exception of the ZrO_2 colloid. For the ZrO_2 case, the values were selected based on the stability ratio data (Figure 1) given by Bitea et al. [19]. The stability ratios for each colloid (see Equations 4 and 5) were calculated numerically.

RESULTS

The required input to the stability model is the ionic strength and pH of the in-package fluid resulting from reaction of water with the waste and ionic strength and pH of the seepage water entering the drift. The value set as the threshold position was at $W = 10$. The selection of $W = 10$ is discussed later. Experimental data was used to estimate colloid mass concentration formed from waste form degradation and the concentrations of radionuclides generated from degradation of the waste. The experimental points shown in Figure 1 were obtained by visual inspection of Figure 3, pg 59 from [19]. The points were then fit to a polynomial expression.

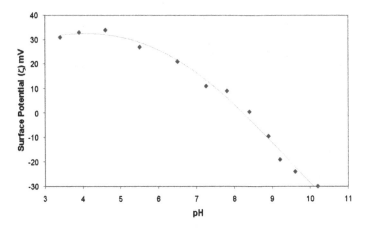

Figure 1. Measured variation in zeta potential with pH at an ionic strength of 0.001 mol/kg for ZrO_2 [adapted from ref [19].

The stability of the spent fuel derived colloid ("ZrO_2") can be described by two curves for pH regions above and below the zero point of charge. The ionic strength threshold to pH relationship was generated with a polynomial fit to the DLVO theoretical curve. It is important to note that these curves are pure mathematically fits for rapid evaluation of stability over various assumed conditions. If the reported ionic strength in the waste package (I_{wp}) is less than the value ($I_{threshold}$) calculated from input pH in the waste package (pH_{wp}) using the polynomial relationships described, the irreversible spent fuel derived colloids will be defined as stable. The plutonium concentrations for the spent fuel colloids were obtained from data reported in ref [11]. The plutonium concentrations ranged from 1×10^{-12} to 1×10^{-6} mol/L. Because of the potential

513

unknown interference of metallic components in the tests, the range was biased towards the bare fuel tests which resulted in a range of concentrations from 1×10^{-10} to 5×10^{-6} mol/L. Calculation of the mass concentration (mg/L) of waste form colloids from spent fuel was obtained by sampling a value from a uniform distribution that was based on the experimental data (see Figure 2) and then scaling to a mass of colloids.

The minimum concentration was a fixed value non-zero value for modeling purposes. When the pH was between 4 and 7, the following relationship was used based on the fits shown (see Figure 3, $W=10$ curve):

$$I_{threshold} = (0.0089\times pH^3) - (0.1466\times pH^2) + (0.7462\times pH) -1.092 \qquad (7a)$$

When the pH > 9.3, Equation 7b was used:

$$I_{threshold} = (0.087362\times pH^3) - (2.4078\times pH^2) + (22.126\times pH) - 67.791 \qquad (7b)$$

If the pH was < 4.0, the $I_{threshold}$ value that was calculated at pH 4.0 was used and if the pH > 10.6 the $I_{threshold}$ value calculated at pH 10.6 was used.

In Figure 2, a cumulative distribution plot of plutonium concentrations was adapted from data taken from ref [11], Appendix A. The plutonium colloid concentration values were from the >1.8-nm to <400-nm size fractions from several low burn-up spent nuclear fuels tests. All the tests were conducted at 85°C except HBR/BF-25 that was conducted at 25°C. In the figure the following acronyms apply: HBR= H.B. Robinson fuel (30.2 MWd/KgM), TP= Turkey Point fuel (27.7 MWd/KgM), BF= bare fuel, SD = Slit defect, HD = Hole defect, and UD = no defect.

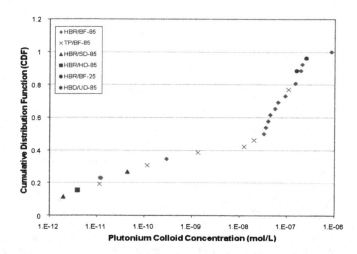

Figure 2. Cumulative distribution of concentrations of plutonium colloids from spent nuclear fuel corrosion tests.

Figure 3 shows stability curves for stability ratios (W) of 10, 100, 1×10^4, and 1×10^6. These calculations were performed in MS Excel using the 'Solver' function to obtain the ionic strength threshold at selected points. Curves were then fitted through the calculated values. In the example shown in Figure 3, the particle diameter was set at 60 nm. These calculated DLVO stability plots were fitted to polynomials (see Equations 7a and 7b) for ZrO_2 colloids separately for low pH high pH regions at $W=10$.

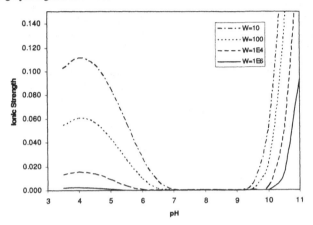

Figure 3. Stability of the spent fuel derived colloids with pH with various stability factors.

DISCUSSION

Small-scale laboratory tests may not duplicate the environment of a disposal environment. The experiments used in this model development were conducted with fragments of spent fuel solids [11] Other tests, notably [1], were also conducted with fragments of spent fuel solids, which though very useful for understanding corrosion behavior, may have been prone to physical filtration owing to the tortuous path from specimen surface to the collection vessel. As the colloid concentration distributions developed from the glass tests were from experiments that had minimal physical filtration of potential colloids, similar experiments were selected for the development of the colloidal plutonium distributions from spent fuel.

In contrast to the case of the spent fuel derived colloids, glass corrosion tests have indicated steady state colloid concentrations. There appears to be no effective zero point of charge for smectite but it is reasonable to suggest a value between 1 and 1.5, below which clay colloids would be unlikely to exist. Tombácz and co-workers [17] have investigated the stability of montmorillonite clay colloid suspensions as a function of pH and ionic strength in NaCl solution. Suspensions became unstable and flocculated at pH 2, 4, and 8 in 0.1 M (100 mmol/dm^3), ~0.225 M (200 to 250 mmol/dm^3), and ~0.375 M (350 to 400 mmol/dm^3) NaCl solutions, respectively. Results of calculations in this study, show high clay colloid stabilities over a range of pH in agreement with the experimental data reported in ref [17]. These results indicate that plutonium concentrations associated with glass-derived colloids could remain constant, if solution conditions remain constant. Smectite colloids are predicted to be highly stable under most conditions; and, indeed, clay particles have been observed in sampled groundwaters from the NTS [2].

The spent fuel corrosion tests that were used to establish plutonium concentrations were obtained over 1 to 174 days. The alteration products from these tests and other spent fuel corrosion tests do closely resemble secondary minerals found at natural uraninite deposits that have been contacted by silica-bearing solutions, such as uranophane and boltwoodite. The growth pattern of the secondary minerals under accelerated conditions may alter the particles colloidal behavior, as slow growth might result in the formation of large crystals that would be unlikely to become colloidal. In a further study, the data set from [11] was used to determine a stability ratio of the measured colloids (see Figure 4). The decrease in colloids with time was in agreement with a W value of $\sim 1 \times 10^6$. Even, under these conditions, the colloids would be predicted to remain stable for only 1 year. Particles are only stable under very high pH conditions and are predicted to be unstable under most disposal conditions. These predictions are in good agreement with experiments that show limited colloid concentrations at short times. These results suggest that high colloid concentrations can only be sustainable when the stability ratio is extremely high and that in natural environments colloids may only be present at trace concentrations. These relationships can explain the observed colloid concentrations in some short term experiments and go some way to explaining the trace quantities of colloids observed at the NTS.

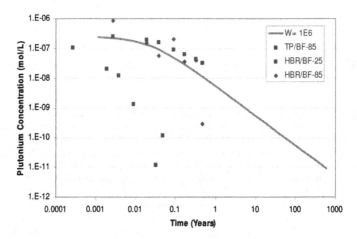

Figure 4. Predicted stability of plutonium colloids from corroded spent fuel based on data from [11] with $W = 1 \times 10^6$.

In this model, the stability ratio boundary was determined to divide the pH vs. ionic strength plane into regions where the colloid should be considered stable or unstable with respect to aggregation and gravity settling. Colloid aggregation kinetics can be approximated by considering only binary collisions represented by a second order rate equation:

$$\frac{dn}{dt} = -k_f\, n^2 \tag{8}$$

where n is the colloid number density and k_f is a rate constant that equals k_{smol}/W. The Smoluchowski rate constant, k_{smol}, is set to 6.16×10^{-18} m^{3}s^{-1} at room temperature[2]. The solution of Equation 8 shows that larger values of W effectively inhibit aggregation and increase the stability half-life. The characteristic stability time is then seen to be proportional to the stability ratio and inversely proportional to the initial concentration. Therefore, given a fixed initial concentration, the value of the stability ratio W is selected to best represent the stability boundary. Because the probability of collision is inhibited exponentially by electrostatic repulsion, it is possible to determine a region in the pH vs. ionic strength plane where W begins to change very rapidly. It is in this region where colloid stability transitions from rapid aggregation (a few seconds) to very stable.

CONCLUSIONS

By using theoretical calculations and laboratory data, the ionic strength at which a spent fuel derived colloid will be stable, can be predicted. Even though the true nature of colloid composition and heterogeneity, generation, and flocculation is complex and cannot be fully captured in this model, this conservatively bound treatment correctly predicted the reduction in colloid concentrations in the cited spent fuel corrosion tests. The representative colloids are designed to capture the behavior of the spent fuel derived colloids has been described in detail. Similar treatments to the other particles have generated a model that is better suited to handle more varied conditions than testing data. However, if the electrical surface properties of specific waste form derived colloidal particles were available much more accurate models could be developed.

REFERENCES

1. C.J. Mertz, R.J. Finch, J.A. Fortner, J.L. Jerden, T. Yifen, J.C. Cunnane and P.A. Finn, *Characterization of Colloids Generated from Commercial Spent Nuclear Fuel Corrosion*, Argonne National Laboratory Report (2003).

2. A. B. Kersting, D.W. Efurd, D.L. Finnegan, D.J. Rokop, D.K. Smith and J.L. Thompson, Nature **56**, 397 (1999).

3. A.P. Novikov, S.N. Kalmykov, S. Utsunomiya, R.C. Ewing, F. Horreard, A. Merkulov, S.B. Clark, V.V. Tkachev and B.F. Myasoedov, Science **314**, 638 (2006).

4. L. Soderholm, P.M. Almond, S. Skanthakumar, R.E. Wilson and P.C. Burns, Ang. Chemie Inter. Ed. **47**, 298 (2008).

5. J.K. Bates, J.P. Bradley, A. Teetsov, C. R. Bradley and M. Buchholtz ten Brink, Science **256**, 649 (1992).

---

[2] $k_{smol} = 4k_BT/3\eta$, where viscosity, η, is ~0.0009 Pa-s for water, k_B is the Boltzmann constant, T is the temperature in Kelvin. Therefore, $4k_BT/3\eta = 1.33 \times 1.38 \times 10^{-23} / 0.0009 = 6.16 \times 10^{-18}$ m^{3}s^{-1} (see Ref. [15].

6. J.A. Fortner, J. K. Bates, T. J. Gerding, *Analysis of Components from Drip Tests with ATM-10 Glass*, Report ANL-96/16 (Argonne National Laboratory, 1997).

7. E. C. Buck, P.A. Finn and J.K. Bates, Micron **35**, 235 (2004).

8. X. Feng, E.C. Buck, C. Mertz, J.K. Bates, J.C. Cunnane and D.J. Chaiko, Radiochim. Acta **66/67**, 197 (1994).

9. E.C. Buck and J.K. Bates, Appl. Geochem. **14**, 635 (1999).

10. P.A. Finn, E.C. Buck, M. Gong, J.C. Hoh, J.W. Emery, L.D. Hafenrichter and J.K. Bates, Radiochim. Acta **66/67**, 197 (1994).

11. C.N. Wilson, *Results from NNWSI Series 3 Spent Fuel Dissolution Tests*, Report PNL-7170 (Pacific Northwest Laboratory, 1990).

12. B. McNamara, B. Hanson, E. Buck and C. Soderquist, Radiochim. Acta **93**, 169 (2005).

13. M.D. Kaminski, N.M. Dimitrijevic, C.J. Mertz and M.M. Goldberg, J. Nucl. Mater. **347**, 77 (2005).

14. E.P. Honig, G.J. Roebersen and P.H. Wiersema, J. Coll. Inter. Sci. **36**, 97 (1971).

15. D.R. Lide, ed. *CRC Handbook of Chemistry and Physics,* 72nd Edition (CRC Press, 1995).

16. Z. Zheng, J. Wan, X. Song and T.K. Tokunaga, Coll. Surf. **A274**, 48 (2006).

17. E. Tombácz, I. Abraham, M. Gilde and F. Szanto, Coll. Surf. **49**, 71 (1990).

18. M. Gunnarsson, M. Rasmusson, S. Wall, E. Ahlberg and J. Ennis, J. Coll. Inter. Sci. **240**, 448 (2001).

19. C. Bitea, C. Walther, J.I. Kim, H. Geckeis, T. Rabung, F.J. Scherbaum and D.G. Cacuci, Coll. Surf. **A215**, 55 (2003).

Mater. Res. Soc. Symp. Proc. Vol. 1124 © 2009 Materials Research Society 1124-Q11-03

Evaluation of Colloid Transport Experiments in a Quarried Block

Hua Cheng and Vladimir Cvetkovic
Land and Water Resources Engineering, Royal Institute of Technology, Brinellvägen 32, SE-100 44 Stockholm, Sweden.

ABSTRACT

Colloid tracer experiments were performed in a single, heterogeneous fracture contained in a quarried block (QB) under the configuration of dipole tracer tests. The experiments were first performed using bentonite and 100 nm latex colloids, as well as conservative tracer iodide and bromide, under conditions of different flow rates in order to identify the flow rates that favour colloid transport. The tracer experiments were later expanded to include experiments with different colloid sizes and longer transport distances.

The aims of the present study are to identify the processes that affect colloid transport in the QB fracture and to estimate the retention parameters for the different sized colloids. We model the measured breakthrough curves (BTCs) using an advection-retention approach. The key feature of the advection-retention model is that advective transport and retention processes are related in a dynamic manner through the flow equation. Two Lagrangian random variables, τ and β, that depend solely on flow conditions, control the retention processes. Here τ is the nonreactive travel time and β is related to τ but also depends on the local aperture value.

We assume the water residence time distribution $g(\tau)$ to be inverse-gaussian. The first two moments of $g(\tau)$ were obtained by calibrating the measured BTCs of conservative tracers. We then model the colloid BTCs using $g(\tau)$ and take into account the retention processes. The modelling results indicate that dominating retention processes include first-order linear kinetic attachment/detachment on the fracture surface, and mass loss (removal) by filtration/sedimentation. Diffusion into the rock matrix is of a much lesser importance.

INTRODUCTION

In Sweden as well as in many other countries, the spent nuclear fuel from nuclear power plants will be deposited in deep geological formations. Suitable geological formations include crystalline granitic rocks. In a repository of spent fuel, the metal canister and the engineered buffer (compact bentonite clay and backfill materials) form man-made barriers for the retention of radionuclides. The buffer and backfill materials in contact with dilute groundwater may release bentonite colloids which may, in the case of a leaching canister, enhance the transport of radionuclides.

A study of colloid formation and transport is being carried out by the Swedish Nuclear Fuel and Waste Management Co. (SKB) at the Äspö Hard Rock laboratory (HRL). The objectives of the Colloid Project are to study the stability and mobility of colloids, the generation of colloids from bentonite clay, and the potential enhancement of radionuclide transport by colloids, and to monitor the background concentration of colloids in the groundwater at the Äspö HRL.

In support of the Äspö Colloid Project, Ontario Power Generation has initialized a laboratory experimental program to improve the understanding of the retention processes that affect colloid transport. The laboratory experiments are being undertaken at AECL's Whiteshell

Research Laboratory using a Quarried Block (QB) sample—a 1×1×0.7 m block of granite containing a single, heterogeneous fracture with well characterized aperture. The QB experiments include tracer tests using both bentonite and latex sphere colloids under the configuration of equal dipole tracer tests. The objectives of the QB experiments are to evaluate the effects of varying flow velocities, solution compositions, and particle size on transport of bentonite and latex colloids.

The tracer experiments were first performed using bentonite and 100 nm latex colloids under conditions of different flow rates in order to identify the flow rates that favour bentonite colloid transport. The tracer experiments were later expanded to include experiments with different colloid sizes and longer transport distances. The main purposes of the experiments are to improve the understanding of retention processes that affect colloid mobility, and to provide additional information that cannot be obtained in tests on the field-scale concerning transport of bentonite and latex sphere colloids.

In the present study, we evaluate the measured breakthrough curves (BTCs) of bentonite and latex colloids of different sizes under three flow conditions (low, medium and high flow rates). The aims of the present study are to identify the retention processes that affect colloid transport and to estimate the retention parameters.

Similar experiments have been conducted in saturated, naturally discrete fractured chalk cores using negatively charged fluorescent latex microspherical colloids, with Li+ and Br- as soluble tracers [1]. Even though we focus on evaluation of results only from the QB experiments, a brief comparison of the results in the different rock types (granite versus chalk stone) will also be discussed.

QUARRIED BLOCK

The QB is a 1m x 1m x 0.7 m block of granite containing a single, variable-aperture fracture (Figure 1, hereafter denoted as QB fracture). Further description of the QB sample can be found in Vilks and Miller [2]. The QB sample was excavated at the 240 level of AECL's Underground Research Laboratory (URL). The QB fracture is the only hydraulically active fracture observed at the 240-level.

The QB fracture was well characterised using a white-light imaging technique [3]. It is highly heterogeneous with apertures ranging from 0 to 27 mm (Figure 1). The mean aperture is approximately 2.5 mm with a standard deviation of 2.6 mm. Several boreholes were drilled in the upper part of the QB, including 13 boreholes of 6.4 mm (small circles in Figure 1) and 5 boreholes of 38 mm (large circles in Figure 1). These boreholes were used for tracer tests and subsequent sampling.

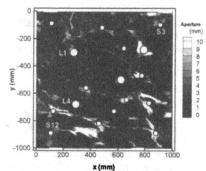

Figure 1. Aperture distribution and borehole locations for the QB fracture.

QB EXPERIMENTS

The tracer experiments were first performed along a shorter flow path of 380 mm (L1→ L4, Figure 1) using polydisperse bentonite and 100 nm latex colloids under conditions of different flow rates in order to identify the flow rates that favour bentonite colloid transport. The tracer tests have been performed using three injection/pumping rates: low rate (6 ml/h), medium rate (50 ml/h) and high rate (500 ml/h). The second phase of the tracer tests focused on the effects of particle size in monodisperse and polydisperse suspensions of colloids on the transport of bentonite and latex colloids using a medium injection/pumping rate. The tracers used in the second phase include conservative tracers (bromide, iodide), latex colloids of three sizes (20 nm, 100 nm, and 1 μm), and bentonite colloids of three size ranges (5-15 nm, 100 – 200 nm, and > 500 nm). The third phase of the experiments investigated latex and bentonite colloid transport along a significantly longer transport distance of 1100 mm (S3 → S12, Figure 1), using both a high flow rate and a low flow rate. Phase 4 consisted of opening the fracture for post-test analyses that would be used to visualize and quantify the extent of colloid deposition within the fracture. Detailed descriptions of the tracer tests are given in Vilks and Miller [2, 4].

EVALUATION MODEL AND STRATEGY

The QB fracture is considered as a single, two-dimensional planar fracture. The fracture aperture varies from point to point in the fracture plane (Figure 1), resulting in two-dimensional heterogeneous flow fields. We use a Lagrangian modeling approach in which the flow and transport simulations can be decoupled [5]. The flow field is assumed to be at steady state. Advection is the dominant hydrodynamic mode of transport inside the fracture. Injected tracers are advected and dispersed, and are subject to various retention processes with surrounding fracture surfaces. The retention processes considered in this study include:

For conservative tracers:
- Diffusion into the rock matrix characterized by a parameter group κ [5].

For colloids:
- Attachment/detachment on fracture surfaces described by a first-order, linear and reversible kinetic model (characterized by attachment rate coefficient k_1 [T^{-1}] and detachment rate coefficient k_2 [T^{-1}]).
- Filtration/sedimentation (characterized by irreversible first order mass removal rate coefficient λ[T^{-1}]).

Colloids and conservative tracers is assumed to not react/interact with each other. The evaluation procedure and strategy is as follows:
- To determine the water residence time distribution $g(\tau)$ by calibrating on measured BTCs of conservative tracers. $g(\tau)$ is assumed to be inverse-gaussian distributed. The first two temporal moments are calibrated.
- To use $g(\tau)$ to model the colloid BTCs by accounting for retention processes and to estimate the retention parameters for colloids, i.e., k_1, k_2 and λ.

EVALUATION RESULTS

We use the TT3 test from the QB experiments as an illustrative example of our evaluation. The measured BTCs of the TT3 test are presented in Figure 2. The total mass recoveries were 97% for Br, and 66%, 78% and 25% for 20 nm, 100 nm and 1 μm latex, respectively. Most measured BTCs for Br and for colloids of all sizes have no tailing parts. The lack of tailing implies that there is little diffusion into the rock matrix of any tracer, and that transport is advection-dispersion dominated. Matrix diffusion is thus neglected in this evaluation.

Calibrations on the measured BTCs of bromide give an average water residence time $<\tau>$ of 14 h and a standard deviation of 14 h (Figure 2a). Using these temporal moments and taking into account the retention processes, the evaluated BTCs of the colloids are compared with the measured BTCs in Figures 2b, 2c and 2d for 20 nm, 100 nm and 1 μm latex colloids. In this particular test, we have considered the mass loss for 20 nm, 100 nm and 1 μm latex colloids, and reversible attachment/detachment for the 100 nm latex colloid. The estimated mass loss rate coefficient λ for 20 nm latex is higher than that for 100 nm latex, which is consistent with the recovery data of these two colloids.

Similar evaluations as for the TT3 test were performed for the other tests. The estimated retention properties are shown in Figure 3 for the latex and bentonite colloids of different sizes as: a dimensionless parameter that is the product of attachment rate coefficient and the mean water residence time, $k_1<\tau>$ (Figure 3a); the product of detachment rate coefficient and the mean water residence time, $k_2<\tau>$ (Figure 3b); and the product of first order mass removal rate coefficient and the mean water residence time, $\lambda<\tau>$ (Figure 3c). In the figures, BTR1 and BTR2 denote the polydisperse bentonite colloids with an average size of approximately 200 nm (ranging from 5 nm to more than 2 μm). Figure 3 indicates no clear correlation between colloid size and colloid retention, but it still reveals that the medium sized colloids tend to have lower retention with lower values for all three retention parameters, while the small and larger sized colloids tend to have slightly higher retention.

DISCUSSION

For conservative tracers, two controlling parameters for matrix diffusion are fracture aperture and matrix porosity [5]. The fracture in the QB experiments we have evaluated is relatively large (on the average of 2.5 mm), which leads to a small extent of matrix diffusion. Since the QB fracture surfaces have been repeatedly washed and drained, any possible loose material or gouge material in the rim zone close to the fracture surfaces with larger porosity may have already been removed. Therefore, matrix diffusion in the QB experiments occurs mainly in the intact rock matrix. The very low porosity of the intact matrix (about 0.3%) also leads to a small extent of matrix diffusion. This is consistent with the experimental results (lack of trailing in the measured BTCs, Figure 2) in the QB experiments and our model assumptions. In other tests where the rock is more porous, matrix diffusion of conservative tracers is more significant [1].

In most of the QB tests, the colloids usually have lower recoveries compared to the conservative tracers under the same test conditions. This implies that retention of the colloids is stronger than the conservative tracers in the QB tests. The recoveries of the small colloids (10 - 20 nm) tend to be close to or slightly lower than the recoveries of the medium colloids (100 -200 nm), while the lowest recoveries were fond for the largest colloids (0.5 -1 μm). A similar trend

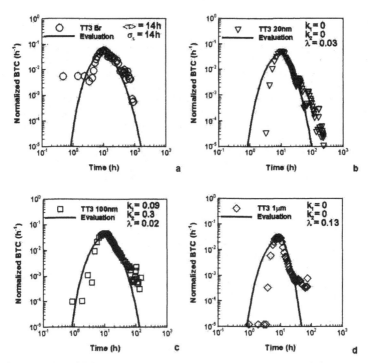

Figure 2. The evaluated BTCs are compared to the measured BTCs for the TT3 test. Symbols denote the measured BTCs. Lines denote the evaluations.

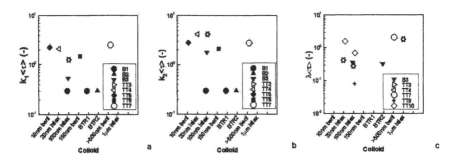

Figure 3. Retention parameters for different sized colloids. (a) Dimensionless attachment rate coefficient $k_1<\tau>$, (b) Dimensionless detachment rate coefficient $k_1<\tau>$, (c) Dimensionless mass removal rate coefficient $\lambda<\tau>$.

of retention of colloids in different sizes has also been observed in experiments in other rock types [1].

The slightly lower recovery of the smallest colloids compared with the medium sized colloids is probably due to the small colloids being trapped in cavities on the fracture surfaces, while the medium sized colloids travel in the middle of the fracture where the groundwater velocity is highest. The difference in recoveries between the small sized colloids and the medium sized colloids in the QB experiments is much smaller than those of the experiments in the chalk cores [1]. The reason could be that in the QB tests there is little loose or gouge material left, as has been discussed earlier. For the small sized colloids, the slightly low recovery may also imply the hydrodynamic chromatography effect.

The large sized colloids were estimated to have higher retention parameters, particularly the highest mass removal rate coefficient λ (Figure 3c). The possible mechanism for mass removal could be that the large sized colloids are relatively less stable and can possibly undergo the processes of irreversible filtration/sedimentation.

CONCLUSIONS

This study shows that the dominant transport process in the QB experiments is hydrodynamic dispersion. Matrix diffusion is of less significance. For small and medium sized colloids, the main retention processes can be represented by kinetic attachment/detachment onto fracture surfaces. For large sized colloids the dominant retention processes can be ascribed to filtration/sedimentation.

ACKNOWLEDGMENTS

This study has been supported by SKB, Stockholm, Sweden. Discussion and comments from Susanna Wold are greatly acknowledged.

REFERENCES

1. O. Zvikelsky and N. Weisbrod, *Water Resour. Res.* **42**, W12S08, doi:10.1029/2006WR004873 (2006.).
2. P. Vilks and N.H. Miller, *Laboratory bentonite colloid migration experiments to support the Äspö colloid project,* Report No: 06819-REP-01300-10123-R00 (Atomic Energy of Canada Limited, 2007).
3. D.J. Brush, *Quarried Block experiment: Numerical simulations of flow and transport experiments in a nature fracture*, Supporting Technical Report 06819-REP-01300-10075-R00 (Ontario Power Generation Nuclear Waste Management, 2003).
4. P. Vilks and N.H. Miller, *Laboratory bentonite and latex colloid migration experiments in a quarried block,* Report No: NWMO TR-2007-xx (Atomic Energy of Canada Limited, 2008).
5. V. Cvetkovic, H. Cheng and J.-O. Selroos, *Evaluation of Tracer Retention Understanding Experiments (first stage) at Äspö*, International Cooperation Report, ICR-00-01 (SKB, Äspö Hard Rock Laboratory, 2000).

Mater. Res. Soc. Symp. Proc. Vol. 1124 © 2009 Materials Research Society   1124-Q10-14

Release of Colloids From Injection Grout Silica Sol

Pirkko Hölttä[1], Martti Hakanen[1], Mari Lahtinen[1], Anumaija Leskinen[1], Jukka Lehto[1] and Piia Juhola[2]
[1]University of Helsinki, Department of Chemistry, P.O. Box 55, 00014 University of Helsinki, Finland.
[2]Posiva Oy, Olkiluoto, 27160 Eurajoki, Finland.

ABSTRACT

Non-cementitious grouts have been tested in Olkiluoto for the sealing of fractures with the small hydraulic apertures. A promising non-cementitious inorganic grout material for sealing the fractures with the apertures less than 0.05 mm is commercial colloidal silica called silica sol. The objective of this work was to determine colloid release from the silica sol gel and stability of silica colloids in different groundwater conditions. To use silica sol as a grout, the injected colloids have to aggregate and form a gel within a predictable time by using a saline solution as an accelerator. Silica sol gel samples were stored in contact with saline and low–salinity groundwater simulates. Release of silica colloids and colloid stability was followed by analyzing the colloid concentration, particle size distribution, concentration of reactive silica, solution pH and zeta potential after one month, half a year and one year.

The release and stability of silica colloids were found to be dependent significantly on groundwater salinity. Zeta potential values near zero and an increase in particle size at first and then the disappearance of large particles indicated particle flocculation or coagulation and instable colloidal dispersion in a saline groundwater simulate. In low–salinity groundwater simulate high negative zeta potential, small particle size, and constant size distribution indicate the existence of stable silica colloids. Under prevailing saline groundwater conditions in Olkiluoto no significant release of colloids from silica sol is expected but the possible influence of low–salinity glacial melt waters has to be considered.

INTRODUCTION

In Finland the final disposal of spent nuclear fuel is investigated by Posiva Oy in Olkiluoto utilizing the underground rock characterization facility ONKALO. The project seeks to obtain information about the bedrock at the site planned for the final disposal repository and assess its safety and test final disposal technology in actual deep underground conditions. Cement is typically used for permeation grouting in hard rock. Because high pH can be harmful for the Engineered Barrier System and cement-based grouts may have limited penetration, non-cementitious grouts have been tested for the sealing of fractures with a small hydraulic aperture. A promising non-cementitious inorganic grout material for sealing fractures with apertures of 0.05 mm or less is silica sol, which is commercial colloidal silica manufactured by Eka Chemicals in Bohus, Sweden [1]. The main experiences with colloidal silica come from geo-technical applications like the grouting of soil to increase its liquefaction resistance [2] and to seal narrow fractures in low permeable rock to prevent water leakage into tunnels [3, 4]. Silica sol gel is sufficiently stable to limit water ingress during the operational phase. The requirement that the pH of groundwater in the repository should not exceed 11 is fulfilled and the compatibility with EBS materials is expected to be good [5, 6].

Colloid-facilitated transport of radionuclides may significantly contribute to the long–term performance of a spent nuclear fuel repository. Several studies have indicated radionuclide (especially actinide) sorption on colloids and the mobility of radiocolloids. Field-scale studies at hazardous waste sites have evidenced that colloid transport can enhance actinide migration [7–9]. Laboratory experiments confirmed that colloids can accelerate the transport of cationic and anionic metals through porous and fractured media [10–12]. In Olkiluoto, the determined natural inorganic and organic colloid contents in groundwater are low – less than 1 ppm – but the bentonite buffer used in the EBS system is assumed to be a potential source of colloids [13, 14]. The use of colloidal material has to be considered in the long–term safety assessment of a spent nuclear fuel repository. The objective of this work was to determine colloid release from the silica sol gel and stability of silica colloids in different groundwater conditions.

EXPERIMENTAL

The silica sol and NaCl accelerator proposed to be used by Posiva has the brand name "MEYCO® MP320" [1]. Silica sol is manufactured from quartz and NaCl. It is a stable suspension of amorphous particles of silica [SiO_2], which builds randomly distributed [SiO_4]$^{4-}$ tetrahedra [15]. The particles have hydroxylated surfaces and are insoluble in water. The colloidal particle size is 5–100 nm. The average pH of the solution is 9.4 and viscosity of the sol is similar to water. To use silica sol as a grout, the particles have to aggregate and form a gel within a predictable time by using saline solution. The desired concentration of the accelerator depends on the properties of the bedrock to be stabilized, groundwater flow rate, water temperature, the salinity of the surrounding water, and the time required for the sol to gel. Higher salt concentration and temperature result in a shorter gelling time for the sol. In the experiments the proportion of the NaCl accelerator was 20%, corresponding to 30 min initial setting and 45 min final setting times. Groundwater simulates used as leaching solutions were saline OLSO [16] and low–salinity Allard [17] with adjusted pH values 7–11. In addition, groundwater simulate OLSO (pH 7) prepared without silica and deionized MilliQ water were used as leaching solutes.

Silica sol gel samples were made by mixing 4 mL silica sol and 1 mL accelerator in a 15 mL, 17 mm–diameter plastic centrifuge tube. After 45 min final setting time, 10 mL of groundwater simulate was added on top of silica sol gel. The contact area for water and silica sol gel was 2.3 cm². Silica sol gel samples were stored in contact with groundwater simulates and colloid release and colloid stability were followed by taking samples after about 0.1, 0.5 and 1 year. The first and second series of samples were taken by pouring the solution and replacing it with corresponding new solution. In the last case only the needed volume of solution was taken from the top of solution and the same solution was returned to the container after it was analyzed. Release and stability of colloids from silica sol gel were measured by analyzing the particle size distribution, particle concentration and zeta potential.

Colloid particle size distribution and zeta potential were determined using Malvern Zetasizer Nano ZS equipment applying dynamic light scattering (DLS) based on the analysis of the temporal fluctuation of the scattered laser light intensity originating from the Brownian movement of dispersed particles [18, 19]. Colloidal silica concentration was estimated by applying the Zetasizer count rate and calibration using a standard series prepared from silica sol solution. The derived count rate is roughly proportional to the concentration of particles and can be used to estimate colloid concentrations. The calibration curve was constructed by measuring known concentrations of particles with the same size as the colloids in the samples. Aerosil® EG50 high purity hydrophilic silica was used as a reference material. Colloid concentration estimated by DLS is not very accurate. Colloidal silica concentration was also determined from total silica concentration minus dissolved reactive silica concentration, i.e., monomers and lower oligomers. Dissolved reactive silica concentration was determined using molybdate blue (MoO_4) method [20]. Total silica concentrations were determined using ICP–MS spectrometry.

RESULTS AND DISCUSSION

Solution pH, particle diameter and particle concentration of released colloids and zeta potential are given in Table I. Results are given as a mean value and a standard deviation of five parallel samples. After one month, pH values differed remarkably from initial pH values (7–11). At the beginning, pH was stabilized predominantly due to atmospheric CO_2. After one year, measured pH values increased, likely due to hydroxide ions produced in the dissolution of silica. Differences in pH values between OLSO and Allard are explained by calcium and sodium ions present in saline OLSO in which calcium and sodium hydroxide is produced, resulting in lower pH values in OLSO. Mean particle size is presented as a function of measured pH after about 0.1, 0.5 and 1 year storage time in OLSO and Allard in Figure 1. Mean particle size distribution in OLSO samples was wide, from nanometers to thousands of nanometers after one month to half a year. After one year, small particle sizes indicated that reversible flocculation or irreversible coagulation had taken place. Because the samples were taken from the top of solution without agitation it is possible that flocculated colloids were not stirred up.

Released particle concentration was normalized to the contact area for water and silica sol gel. The accuracy of DLS determinations is dependent on the size and geometrical shape of the colloids. The colloids used in the calibration curve are spherical, which is not appropriate for aggregated silica colloids in OLSO groundwater simulate. For OLSO samples, reactive silica concentrations remained at the same level over the experimental time period, whereas for Allard samples dissolved reactive silica concentrations increased. The increase was probably due to the dissolution of silica from the silica gel besides the dissolution of silica colloids. ICP-MS measurements were performed for samples taken after half a year. Mean particle concentration calculated from total silica minus dissolved reactive silica concentration for OLSO samples was 6.7 ± 1.7 ppm·cm^{-2}, which was higher than Zetasizer measurements shown in Table I.

Table I. Release and stability of silica colloids after 0.1, 0.5 and 1 year storage in saline OLSO and low–salinity Allard groundwater simulates and deionized MilliQ water.

	Solution pH	Particle diameter (nm)	Concentration (ppm·cm^{-2})	Zeta potential (mV)
OLSO				
0.1 y	8.1 ± 0.1	$1 - 135$	2.5 ± 0.7	-4 ± 2
0.5 y	7.6 ± 0.1	$30 - 820$	4.8 ± 1.5	-2 ± 1
1 y	8.3 ± 0.2	0.70 ± 0.03	3.6 ± 0.7	-1 ± 1
Allard				
0.1 y	8.8 ± 0.1	35 ± 12	8.7 ± 1.4	-15 ± 2
0.5 y	8.5 ± 0.1	20 ± 1	4.3 ± 0.3	-10 ± 3
1 y	9.3 ± 0.1	19 ± 1	5.7 ± 1.3	-52 ± 2
MilliQ				
0.1 y	8.7 ± 0.1	$30 - 180$	5.8 ± 1.8	-21 ± 7
0.5 y	8.2 ± 0.1	25 ± 6	6.8 ± 3.9	-13 ± 11
1 y	9.5 ± 0.1	22 ± 3	9.0 ± 4.9	-52 ± 2

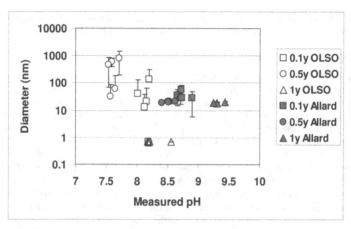

Figure 1. Mean particle diameter as a function of measured pH after 0.1, 0.5 and 1 year storage in OLSO and Allard groundwater simulates.

Zeta potential is presented as a function of measured pH in Figure 2 and particle size in Figure 3 after about 0.1, 0.5 and 1 year storage time in groundwater simulates OLSO and Allard. In a stable dispersion all suspended particles have a large negative or positive zeta potential, and they will tend to repel each other. The colloidal system is least stable near the isoelectric point where zeta potential is near zero and there is no force to prevent the particles from aggregating. In high salt concentrations, initial aggregation (flocculation) can be reversible and weak flocs may dissociate under an externally applied force such as vigorous agitation. The particles in colloidal dispersion may adhere to one another and form aggregates of successively increasing size that separates out by sedimentation. For the saline OLSO simulate, zeta potential values near zero and an initial increase in particle size, followed by disappearance of large particles later, indicate particle flocculation or coagulation and unstable colloidal dispersion. Because after one year the samples were taken from the top of solution without agitation, reversible flocculation cannot be excluded. For the low–salinity Allard simulate, highly negative zeta potential values, small particle size, and constant size distribution indicate the existence of stable silica colloids. The experiments were carried out under ambient air oxygen conditions at ca. 22°C. Relevant conditions in a repository are anoxic and temperature is expected to be between 10°C and 50°C. The appearance and kinetics of reactive, colloidal and precipitated silica under geologically relevant aqueous conditions is very complicated and dependent also on temperature. Compared to ionic strength and pH of the solution the effect of temperature is insignificant.

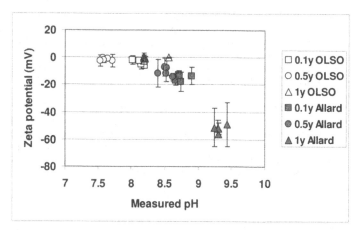

Figure 2. Mean zeta potential as a function of measured pH after 0.1, 0.5 and 1 year storage in OLSO and Allard groundwater simulates.

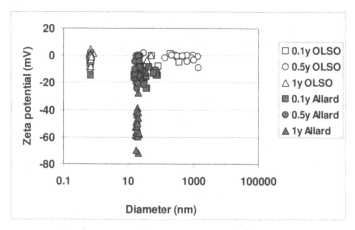

Figure 3. Mean zeta potential as a function of particle size after 0.1, 0.5 and 1 year storage in OLSO and Allard groundwater simulates.

CONCLUSIONS

The use of colloidal silica sol for permeation grouting must be considered in the long-term safety assessment of a spent nuclear fuel repository. In this work, silica colloid release from silica sol gel and stability of colloids in different groundwater conditions was determined. The release and stability of silica colloids was found to be dependent significantly on groundwater salinity. In the saline OLSO groundwater simulate, particle size distribution was at first wide from nanometers to thousands of nanometers. After one year flocculation or coagulation took place. Because sampling at one year avoided agitation, reversible flocculation could not be excluded. Zeta potential values of colloids were around zero, indicating favorable conditions for particle aggregation and instability of colloids. In the low–salinity Allard groundwater simulate, particle diameter remained small and particle size distribution was constant through time. Highly negative zeta potential values indicated the existence of stable silica colloids. The concentrations of the released silica colloids were slightly higher than concentrations of natural colloids determined in granitic groundwater. The potential relevance of colloid-mediated radionuclide transport is dependent on colloid stability and mobility in different geochemical environments. Under the prevailing medium saline to saline groundwater conditions in Olkiluoto, no significant release of colloids from silica sol is expected, but the possible influence of glacial melt waters has to be considered.

REFERENCES

1. EKA Chemicals, Eka gel MEYCO MP320 http://www.colloidalsilica.com/eka.asp.
2. P. Persoff, J. Apps, G. Moridis and J. M. Whang, *J. Alloys Compounds*, **125**, 461 (1999).
3. M. Axelsson and G. Gustafson, *Tunneling and Underground Space Tech.* **21**, 499 (2006).
4. J. Funehag and G. Gustafson, *Tunneling and Underground Space Tech.* **23**, 9 (2008).
5. A. Boden and U. Sievänen, SKB R-05-40/Posiva WR 2005-24 (2005).
6. B. Torstenfelt, M. Jansson and M. Atienza, SKB Arbetsrapport TU-05-04 (2005).
7. R. W. Buddemeier and J. R. Hunt, *Appl. Geochem.* **3**, 535 (1988).
8. A. B. Kersting, D. W. Efurd, D. L. Finnegan, D. J. Rokop, D. K. Smith and J. L. Thompson, *Nature* **397**, 56 (1999).
9. A. P. Novikov, S. N. Kalmykov, S. Utsunomiya, R. C. Ewing, F. Horreard, A. Merkulov, S. B. Clark, V. V. Tkachev and B. F. Myasoedov, *Science* **314**, 638 (2006).
10. R. W. Puls and R. M. Powell, *Environ. Sci. Technol.* **26**, 614 (1992).
11. P. Vilks and M. Baik, *J. Contam. Hydrol.* **47**, 197 (2001).
12. T. Yamaguchi, S. Nakayama, T. T. Vandergraaf, D. J. Drew and P. Vilks, *J. Power and Energy Systems* **2**, 186 (2008).
13. U. Vuorinen and H. Hirvonen, Posiva WR 2005-03 (2005).
14. M. Takala and P. Manninen, Posiva WR-2006-98 (2006).
15. R. K. Iler, *The Chemistry of Silica* (John Wiley & Sons, New York, 1979).
16. U. Vuorinen and M. Snellman, Posiva WR-1998-61 (1998).
17. B. Allard and J. Beall, *J. Environ. Sci. Health* **6**, 507 (1979).
18. Malvern Instruments Ltd., http://www.malvern.co.uk/LabEng/industry/colloids/colloids_home.htm.
19. M. Filella, J. Zhang, M. E. Newman and J. Buffle, *Colloids Surf. A* **120**, 27 (1997).
20. M. J. Fishman and L. C. Friedman (Eds.), *Techniques of Water-Resources Investigations of the U.S. Geological Survey,* USGS–TWRI Book 5, Chapter A1 (1989).

Mater. Res. Soc. Symp. Proc. Vol. 1124 © 2009 Materials Research Society   1124-Q11-04

Ni(II) Sorption on the Fracture Filling Mineral Chlorite

Åsa Zazzi[1] and Susanna Wold[2]
[1]School of Chemical Science and Engineering, Inorganic Chemistry, Royal Institute of Technology (KTH), SE-100 44 Stockholm, Sweden.
[2] School of Chemical Science and Engineering, Nuclear Chemistry, Royal Institute of Technology (KTH), SE-100 44 Stockholm, Sweden.

ABSTRACT

Abundance of the mineral chlorite as fracture filling material in granitic bedrock motivates sorption studies for quantification of the retention of radionuclides on this mineral. The activation product ^{63}Ni is an important component in spent nuclear fuel, accounting for a large contribution to the high activity level, and further motivates sorption studies of Ni on chlorite. Earlier sorption studies have been performed on larger mineral pieces; however, it is questionable if these data are representative for fracture filling material. Chlorite from a borehole core from Oskarshamn, Sweden, from the depth of 944 m has been characterized and the chemical composition determined prior to the experiments. The thickness of the fracture filling mineral is at maximum a couple of millimeters; therefore, careful removal of the thin chlorite layer from the core was performed with a carbide tool. The fraction was sieved to a size distribution of 63-118 μm and ultrasonically washed. The major oxides were found to be 34.4 % SiO_2, 21.6 % MgO, 15.4 % Al_2O_3 and 12.9 % Fe_2O_3 determined by Scanning Electron Microscopy-Energy Dispersive Spectroscopy (SEM-EDS) analysis. Sorption of Ni(II) to the sample was performed in 0.1 M $NaClO_4$ with a batch technique inside a glove-box with Ar-atmosphere, using ^{63}Ni as a tracer with the Ni carrier concentration of 10^{-6} and 10^{-8} M. The percent sorption in these experiments was 57% at pH 5.5, 85 % at pH 6.5 and 92 % at pH 8.3, with the last value corresponding to maximum sorption. No significant differences between the experiments performed in the two concentrations of 10^{-6} and 10^{-8} M were found. The degree of sorption on this fracture filling material is of the same magnitude as earlier studies of Ni sorption on chlorite provided from larger mineral pieces, using the same size fraction of the material. Sorption of Ni is, as expected, strongly pH dependent and ^{63}Ni escaping a breached canister will, reaching chlorite within the surrounding bedrock, be retarded due to strong sorption.

INTRODUCTION

Sweden is one of the countries planning a deep bedrock repository for spent nuclear fuel [1]. One of the barriers is the surrounding bedrock, which has been characterized in terms of minerals and fractures in different regions of Sweden, for example Forsmark, Äspö, Laxemar and Simpevarp [2-6]. The minerals that have been identified in the bedrock fracture zones include calcite, pyrite, biotite, haematite, chlorite and different clay minerals [2]. Among these minerals, chlorite is one that is potentially an important sorber of radionuclides. Laboratory experiments have been performed on chlorite taken from museum specimens that were collected from mineral finds, e.g. Taberg and Karlsborg, where dissolution rates [7-12] and sorption of cations to the surface have been determined [13-15]. One way to gain information about the Swedish bedrock is to examine drill cores such as the KOV:01 core, which was collected to evaluate the functionality of the drilling process. The KOV:01 core has been mapped [personal communication Fredrik Hartz] and the flow rate of flowing fractures was characterized at several depths [16]. Studies have also been performed on pieces from the KOV:01 core to investigate Fe-oxide fracture fillings [17].

Nickel as [63]Ni is an activation product in spent nuclear fuel with a half-life of 101.2 years [18], and therefore [63]Ni is one of the radionuclides that are included in safety assessment calculations [19, 20]. Since both general information on how different minerals act as sorbents and site-specific knowledge of samples are of interest, this study evaluated whether chlorites from mineral findings are representative of chlorite originating from fractures in granitic bedrock. This was achieved by determining [63]Ni sorption to different chlorite samples.

EXPERIMENTAL

The KOV:01 borehole core sample originates from the harbor area in Oskarshamn in south-east Sweden. This is the first vertical 1000 m-deep borehole core extracted with triple tubing in Sweden, and was drilled with an inclination of -78°, a bearing of 205° and a diameter of 76 mm [personal communication Eva-Lena Tullborg]. The core pieces used were extracted from a fracture zone at a depth of 943.78 to 944.10 m. Fracture coatings identified as chlorite, confirmed by SEM-EDS analysis, were scraped off from surfaces using a carbide blade (iron-free) provided by Sandvik Coromat. The mineral powder was sieved and ultrasonically washed, and the 63-118 μm fraction was used for batch sorption experiments. Chlorite samples from Taberg and Karlsborg were treated similarly to the KOV:01 sample and fractions of 63-118 μm and 120-200 μm were used for the experiments. Major element concentrations of KOV:01 were determined using a Zeiss DSM 940 scanning electron microscope (SEM) equipped with an Oxford Instruments Link Energy Dispersive X-Ray Spectrometer (EDS). The analyses were carried out with a working distance of 24 mm, an accelerating voltage of 25 kV, a specimen current of ~0.7 nA and a counting time of 40 s. The instrument is calibrated with natural minerals and simple oxide standards. Cobalt was used as a reference standard which was measured every 20 min to avoid drift. Raw counts were reduced using ZAF matrix correction on an online LINK ISIS computer system. Precision and accuracy for the major element analyses are better than 5 % with detection limits of ~0.1 wt% (0.3 % for Na_2O). A thin section of 30 μm, prepared at the Department of Earth Sciences Centre, University of Gothenburg, Sweden [21], coated with carbon for electron conductivity was used for the KOV:01 sample. Standard SEM images were collected on material released from the core surface. The specific surface area was measured by the Brunauer-Emmet-Teller (BET) method using a Micrometrics Flow Sorb II 2300 apparatus with N_2 as absorbing gas. The Fe(II)/Fe(III) ratio was determined by Mössbauer spectroscopy (Uppsala University, Uppsala, Sweden) at room temperature using [57]Fe. For all experimental solutions, Milli-Q deionised water (Millipore) and chemicals of analytical grade were used and sodium perchlorate, used as a background electrolyte, was prepared from $NaClO_4 \cdot H_2O$ (Merck p.a.).

Batch experiments were performed to investigate the degree of sorption of Ni to the chlorite samples. Within these sorption experiments, [63]Ni (PerkinElmer LifeScience, Inc.) was used as the radioactive tracer in a Ni(II) solution with a total Ni concentration of 10^{-8} M or 10^{-6} M. The experiments were performed in SARSTEDT polypropylene tubes and the chlorite content was 5g/L in 0.1 M $NaClO_4$; pH was adjusted with 0.1 M NaOH and 0.1 M $HClO_4$. Experiments were performed not more than in duplicate due to the small volume of the sample. After 7 days of slow shaking inside a glove-box with argon atmosphere, the samples were centrifuged and pH measured, and samples were collected for determination of ^{63}Ni remaining in the aqueous phase. ^{63}Ni was determined using β-liquid scintillation counting (LSC) on a PACKARD Tri-Carb Liquid Scintillation Analyser Model 1500 instrument.

The quantity of Ni sorbed to the chlorite surface can be described by:

$$\text{Ni-sorbed } (\%) = ((C_i-C_f)/C_i) \times 100 \qquad (1)$$

where C_i and C_f are the initial and final ^{63}Ni-activity in solution. The difference between the initial total ^{63}Ni activity in the solution and the ^{63}Ni-activity in the supernatant after centrifugation was taken to represent the amount of nickel sorbed to the solid phase. The distribution coefficient (K_d) also describes the amount of Ni sorbed to the chlorite surface:

$$K_d = ((C_i-C_f)/C_f) \times V / m \qquad (2)$$

where C_i and C_f are the initial and final ^{63}Ni-activity in solution, V is the volume of solution (cm^3) and m is the mass of dry chlorite (g). The values of K_d calculated in this way therefore have units cm^3/g. To measure Ni sorption onto the tube walls, reference samples without chlorite were analysed for each series at the corresponding pH values of the measured samples. The activity of the samples without chlorite was measured directly after the addition of Ni-solution and again after 7 days. The differences in activity corresponding to wall sorption lay within the uncertainty range of the LSC method.

RESULTS AND DISCUSSION

The mineral coating on the surface of fractures in the KOV:01 core was identified using SEM-EDS analysis and the coatings did not include any other major minerals. This coating was just a few millimeters thick, which is similar to findings in other drill cores [2]. The composition of the sample and its correlation with other chlorites are shown in Table I [11, 15].

Table I. Chemical composition of the different chlorite samples. (n.d.= below not detectable).

% oxide	SiO$_2$	Al$_2$O$_3$	Fe$_2$O$_3$	MgO	CaO	K$_2$O	MnO$_2$	Na$_2$O
Taberg [11]	33.6	13.7	5.95	32.9	<0.099	0.68	0.134	n.d.
Karlsborg [15]	30.0	19.3	16.1	19.8	0.26	3.84	0.327	n.d.
KOV:01	**34.4**	**15.4**	**12.9**	**21.6**	**0.55**	**0.12**	**0.51**	n.d.

The BET-area for the 120-200 µm fraction of the chlorites from Taberg and Karlsborg has been determined to be 6.73 m^2/g and 0.51 m^2/g, respectively [11, 15]. For the smaller fraction of 63 -118 µm the BET-area was determined to be 7.95 m^2/g and 0.59 m^2/g respectively. For KOV:01, there was insufficient sample available for N$_2$ BET area determinations. Mössbauer spectroscopy showed that the Fe(III)/Fe(III+II) ratio in the KOV:01 core was 0.41, thereby establishing the specific chemical formula to:

$$\text{KOV: } (Mg_{6.34}Fe^{II}_{1.12}Fe^{III}_{0.78}Al_{2.47}\bullet_{1.29})(Si_{6.91}Al_{1.09})O_{20}(OH)_{16} \qquad (3)$$

where • stands for uncertainties in the determination. The corresponding value of the Fe(III)/Fe(III+II) ratio for the Taberg chlorite was 0.29 and for the Karlsborg chlorite 0.16. Plate-like particles were found when the SEM images of untreated KOV:01 material were analyzed (Figure 1). The size of the particles was found to be approximately 50 µm to 150 µm before sieving. Similar particle shapes have been reported for other chlorites [10, 11] and at other depths in the KOV:01 core [17].

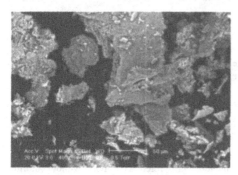

Figure 1. SEM image of plate-like particles (50 to ~150 µm) scraped from the surface of the KOV:01 core at 943.85 m depth.

Sorption of Ni(II) to chlorite from the KOV:01 core was clearly pH dependent, with a maximum Ni sorption of 92 % at pH 8.3 (Table II). Figure 2 shows the pH dependence when sorption is expressed as K_d (some data points are not visible because they overlap with other symbols). Our earlier Ni(II) sorption studies on Karlsborg chlorite showed no dependence of sorption on ionic strength [15, 22]. The strong pH dependency of sorption, together with the non-dependence on ionic strength, indicates that sorption occurs through surface complexation at specific surface sites on the chlorite, rather than through ionic exchange reactions. No difference in the degree of sorption was found between the two Ni concentrations investigated (10^{-8} and 10^{-6} M). These sorption results are in agreement with our previous results [15, 22]. The sorption data clearly demonstrated that Ni sorption to KOV:01 fracture chlorite is of the same magnitude as sorption to chlorite from larger pieces, regardless of particle size (Figure 2) and Table II and III

Table II. Degree of Ni sorption for the 63-118 µm fraction of different chlorite samples with initial Ni(II) concentration of 10^{-6} M.

pH	Taberg % sorbed	Karlsborg % sorbed	KOV:01 % sorbed
5.5	58.2 ± 0.2	56.3 ± 0.3	57.0 ± 0.3
6.5	85.2 ± 0.3	85.4± 0.3	85.0 ± 0.3
7.2	88.6 ± 0.1	86.8 ± 0.2	87.0 ± 0.2
8.3	90.1 ± 0.3	90.4 ± 0.2	92.0 ± 0.2

Table III. Degree of Ni sorption for the 120-200 µm fraction of Taberg and Karlsborg with initial Ni(II) concentration of 10^{-6} M, values used for comparision.

pH	Taberg Log K_d	Karlsborg Log K_d
7.2	88.6 ± 0.1	86.8 ± 0.2
8.3	90.1 ± 0.3	90.4 ± 0.2

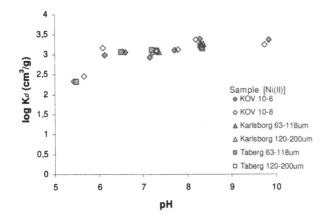

Figure 2. Log K_d from batch experiments as a function of pH for the different chlorite samples. Uncertainty ranges for K_d values are somewhat smaller than the size of the symbols.

CONCLUSIONS

The amount of Ni sorbed onto the chlorite sample from the fracture zone in grantic bedrock core KOV:01 is in agreement with Ni sorbed onto the chlorites taken from mineral collections. As expected the sorption of Ni to chlorite surfaces was pH-dependent, with a log K_d value at the sorption maxima of 3.4 ± 0.2. This is typical for chlorites with a low to medium iron content, such as those found within the Swedish bedrock.

ACKNOWLEDGMENTS

Thanks to Eva-Lena Tullborg for valuable information and answers and help with the sample, Björn Sandström for help with SEM measurements and Ali Firoozan for preparation of the thin section, Hans Annersten at Uppsala University for Mössbauer determinations and Fredrik Hartz, SKB, for providing a map of the Oskarshamn area .This work was funded by the Swedish Nuclear Fuel and Waste Management Co. (SKB).

REFERENCES

1. SKBF/KBS, *Final Storage of Spent Nuclear Fuel*, KBS-3, I-V, Art 716 1-5 (SKBF/KBS, Stockholm, 1983).
2. H. Drake, B. Sandström, and E.-L. Tullborg, *Mineralogy and geochemistry of rocks and fracture fillings from Forsmark and Oskarshamn: Compilation of data for SR-Can*, R-06-109 (SKB, Stockholm, 2006).

3. H. Drake and E.-L. Tullborg, *Oskarshamn site investigation - Fracture mineralogy and wall rock alteration Results from drill cores KAS04, KA1755A and KLX02*, P-05-174 (SKB, Stockholm, 2005).
4. B. Sandström and E.-L. Tullborg, *Forsmark site investigation - Fracture mineralogy. Results from fracture minerals and wall rock alteration in boreholes KFM01B, KFM04A, KFM05A and KFM06A*, P-05-197 (SKB, Stockholm, 2005).
5. B. Sandström and E.-L. Tullborg, *Forsmark site investigation Mineralogy, geochemistry, porosity and redox capacity of altered rock adjacent to fractures*, P-06-209 (SKB, Stockholm, 2006).
6. E.-L. Tullborg, *Mineralogical/Geochemical investigations in the fracture zone*, PR-25-95-06 (SKB, Stockholm, 1995).
7. C.A. Rochelle, K. Bateman, R. MacGregor, J.M. Pearce, D. Savage, and P.D. Wetton in *Scientific Basis for Nuclear Waste Management XVIII*, edited by T. Murakami and R.C. Ewing (*Mater. Res. Soc. Proc.* **353**, Pittsburgh, PA, 1995) pp. 149-156.
8. G.J. Ross, *Clays & Clay Minerals* **17**, 347 (1969).
9. H.M. May, J.G. Acker, J.R. Smyth, O.P. Bricker, and M.D. Dyar in *Proc. 32nd Annual Meeting Clay Minerals Society* (Baltimore, MD, 1995) pp. 88 (abstract).
10. F. Brandt, D. Bosbach, E. Krawczyk-Bärsch, T. Arnold, and G. Bernhard, *Geochim. Cosmochim. Acta* **67**, 1451 (2003).
11. Å. Gustafsson and I. Puigdomenech. *Scientific Basis for Nuclear Waste Management XXVI*, edited by R.J. Finch and D.B. Bullen (*Mater. Res. Soc. Proc.* **757**, Warrendale, PA, 2003) pp. 649-655.
12. R. Lowson, M. Comarmond, G. Rajaratnam, and P. Brown,, *Geochim. Cosmochim. Acta* **69**, 1687 (2005).
13. T. Shahwan and H.N. Erten, *J. Radioanal. Nucl. Chem.* **241**, 151 (1999).
14. M.H. Koppelman and J.G. Dillard, *Clays & Clay Minerals* **25**, 457 (1977).
15. Å. Gustafsson, M. Molera, and I. Puigdomenech. *Scientific Basis for Nuclear Waste Management XXVIII*, edited by J.M. Hanchar, S. Stroes-Gascoyne, and L. Browning (*Mater. Res. Soc. Proc.* **824**, Warrendale, PA, 2004) pp. 373-378.
16. J. Pöllänen and P. Rouhiainen, *Difference flow measurements in borehole KOV01 at Oskarshamn*, R-01-59 (Stockholm, SKB, 2001).
17. K. Dideriksen, B.C. Christansen, J.A. Baker, C. Frandsen, T. Balic-Zunic, E.-L. Tullborg, S. Mörp, and S.L.S. Stipp, *Chemical Geology* **244**, 330 (2007).
18. R. Collé, B.E. Zimmerman, P. Cassette, and L. Laureano-Perez, *Applied Radiation and Isotopes* **66**, 60 (2008).
19. A. Hedin, *Nucl. Technol.* **138**, 179 (2002).
20. M. Lindgren and F. Lindström, *SR 97 Radionuclide transport calculations*, TR-99-23 (SKB, Stockholm, 1999).
21. H. Drake and E.-L. Tullborg, *Oskarshamn site investigations. Fracture mineralogy. Results from drill core KSH03A+B*, P-06-03 (SKB, Stockholm, 2006).
22. Å. Zazzi, A.-M. Jakobsson, and S. Wold, unpublished manuscript (2009).

Mater. Res. Soc. Symp. Proc. Vol. 1124 © 2009 Materials Research Society   1124-Q11-06

Performance Assessment for Depleted Uranium Disposal in a Near-Surface Disposal Facility

Karen E. Pinkston, David W. Esh, and Christopher J. Grossman
U.S. Nuclear Regulatory Commission, 11545 Rockville Pike,
Rockville, MD 20852, U.S.A.

ABSTRACT

The U.S. Nuclear Regulatory Commission (NRC) staff has conducted a technical analysis to assess the potential impacts of disposal of large quantities of depleted uranium in a near-surface disposal facility. The nature of the radiological hazards associated with depleted uranium presents challenges to the estimation of long-term effects from its disposal – namely that its radiological hazard gradually increases over time due to the in-growth of decay products. In addition, these decay products include a daughter in gaseous form (Rn-222), which has significantly different mobility in the environment than the parent radionuclides. NRC staff developed a screening performance assessment model of a reference low-level radioactive waste (LLRW) disposal facility to evaluate the risk and uncertainties associated with the disposal of depleted uranium as low-level waste. The model was constructed with the dynamic simulation software package GoldSim®, a Monte Carlo simulation software solution for dynamically modeling complex systems. The depleted uranium source is modeled as releasing to a backfill assumed to surround the depleted uranium in the disposal cells. Radionuclides released to the backfill are vertically transported via advection through unsaturated zone cells to an underlying aquifer, where they are transported to a receptor well. Radon that emanates from radium present in the depleted uranium is modeled as diffusing through an engineered cap into the interior of a residence placed over the disposal area or to the external environment. The model evaluates the radiological risk to future residents and intruders (acute or chronic exposures) near or on the land overlying the disposal facility. Calculations were performed probabilistically to represent the impact of variability and uncertainty on the results. Key variables evaluated included: disposal configurations, performance periods, institutional control periods, wasteforms, site conditions, pathways, and scenarios. The impact of these variables on projected radiological risk can be significant. For example, estimated risks are very sensitive to the performance period, and estimated disposal facility performance is strongly dependent on site specific hydrologic and geochemical conditions. In addition, radon fluxes to the environment are very sensitive to the long-term moisture state of the system and to the disposal depth.

INTRODUCTION

The NRC staff conducted a technical analysis to assess the potential impacts of disposal of large quantities of depleted uranium (DU) in a generic near-surface disposal facility. DU is produced in the enrichment process as a waste product or byproduct. The source term is described as "depleted" because the enrichment process concentrates both the U-235 and U-234 in the product, and therefore, these radionuclides are depleted in the waste or byproduct, which primarily consists of U-238. Staff developed a screening model to evaluate the radiological risk to potential future residents and intruders (acute or chronic exposures) near or on the land overlying a hypothetical disposal facility for DU and to understand the impacts of key variables on the risks. Key variables evaluated were: disposal configurations, performance periods, institutional control periods, waste forms, site conditions, pathways, and scenarios. The impact of these variables on projected radiological risk can be significant. The model was constructed with the dynamic simulation software package GoldSim®, developed by GoldSim Technology Group of Issaquah, WA, USA, and calculations were performed probabilistically to represent the impact of variability and uncertainty on the results. The model contains 3,252 GoldSim elements, and stochastic inputs are specified for over 400 variables. Conditions such as infiltration rates, liquid saturation, hydraulic gradient, unsaturated zone thickness, hydraulic conductivities, and geochemical conditions were represented in the analysis as epistemic uncertainty to account for a range of sites. In reality, many of these parameters can be constrained for a particular site and disposal system. Because site-specific waste management decisions or other variables can strongly influence whether performance objectives can be met, the results should not be taken out of the analysis context.

THEORY

The nature of the radiological hazards associated with DU presents challenges to the estimation of long-term effects from its disposal because of unique characteristics of the source term. Metallic DU initially contains approximately 99.75% U-238, 0.25% U-235, and 0.002% U-234 [1]. The activity for DU would be expected to remain relatively constant initially, but begin increasing at around 1,000 years as the parent radionuclides decay through the uranium series decay chains. Peak activity, assuming no release from the source, would not be attained until after one million years following disposal. In addition, the activity of some risk significant radionuclides (e.g., Rn-222, Pb-210) increase by a much more significant amount than the overall activity. Because different elements can have different mobility and radiotoxicity, total activity cannot be directly translated to risk (dose). The most prevalent forms of DU for disposal resulting from fuel cycle activities are depleted uranium hexafluoride (UF_6) and depleted uranium oxide (UO_2 or U_3O_8), which results from deconversion of fluoride forms. Both UO_2 and U_3O_8 are solids that are significantly more stable than UF_6 over common disposal conditions, making the oxide forms more suitable for long-term storage or disposal.

DISCUSSION

Evaluation of releases of radioactivity from the disposal of DU was performed for leaching of contaminants to a water pathway and diffusion of radon to the atmosphere. The population was assumed to reside offsite during the institutional control period, and then outside a buffer zone surrounding the disposal area boundary after the institutional control period of 100 years. The protection of individuals from inadvertent intrusion was evaluated with acute and chronic exposure scenarios following either excavation into the waste, excavation above the waste but not into the waste, or drilling through the waste. The particular intruder scenario evaluated was based on the depth to waste. Below a disposal depth of 3 m (9.8 ft), disruption of the waste via excavation was not believed to be credible for a resident-intruder scenario (Figure 1). The DU source is modeled as releasing to a backfill assumed to surround the DU in the disposal cells. Radon can partition between the gas and liquid phases, and diffuse in the gas phase through clay, soil, and basement foundation layers, as applicable. Radionuclides released to the backfill are vertically transported via advection through unsaturated zone cells to an underlying aquifer, where they are transported to a receptor well. Contaminated water is then extracted and used for farming or domestic purposes. The dose was calculated using the probabilistic dose model BDOSE™ developed for the NRC by the Center for Nuclear Waste Regulatory Analyses [2]. Exposure pathways in this model include external exposure from surface, air, and water; internal exposure from inhalation of air; and internal exposure from ingestion of drinking water, vegetables/fruits, milk, beef, game, fish, and soil.

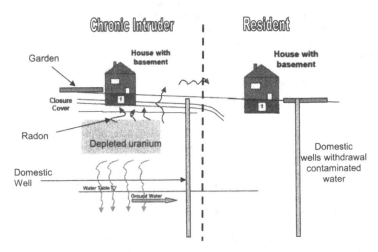

Figure 1. Conceptual Model Showing the Primary Scenarios.

Source term

The source term submodel represents the failure of waste containers over time as well as the gradual degradation of the waste form. The source term submodel applies distribution coefficients, based on material type, to partition radionuclides between solid and liquid phases. Solubility limits are also applied, in addition to partitioning, to estimate liquid phase concentrations of radionuclides. Partition coefficients and solubility values are selected with multi-dimensional lookup tables based on sampled values for pH and carbonate concentration. Numerous references were used to develop the lookup tables [3-8]. It was assumed that depleted uranium would be disposed of in an oxide form. The quantities assumed were 700,000 metric tons (770,000 tons) from DOE and 700,000 metric tons (770,000 tons) from operation of commercial uranium enrichment facilities [9-11]. The disposal system was assumed to have an engineered cover that would limit infiltration. The infiltration cover was assumed to lose its effectiveness a few hundred years after site closure. For arid sites, the long-term infiltration rate was assumed to be on the order of a few millimeters per year. For humid sites the long-term infiltration rate was assumed to be on the order of tens of centimeters per year.

Groundwater transport

The groundwater transport modeling was relatively simple from the perspective of temporal and spatial variability because the assessment was designed to evaluate a range of sites. Transport through the unsaturated zone was assumed to be vertical to the saturated zone; and transport through the saturated zone was assumed to be horizontal or lateral to a receptors well. Groundwater transport through the unsaturated zone is represented with a series of mixing cells. Advection, partitioning between liquid and solid phases, solubility limits, and decay and in-growth are included in the mathematical representation of a cell. Groundwater transport through the saturated zone is represented with GoldSim pipe elements. Pipes are modeled as reactive columns and include advection, partitioning between liquid and solid phases, decay and in-growth, and dispersion. Because the analysis was generic and hydrologic systems can have widely variable properties, the input distributions were fairly wide, resulting in hydraulic residence times in the pipe from less than ten to greater than 1,000 years.

Radon

Radon that emanates from radium present in the DU is modeled as diffusing to the surface through an engineered cap. The engineered cap contains a clay layer as well as a soil layer. Modeling of radon transport in partially saturated media is subject to a high degree of uncertainty because the gas phase diffusion of radon in partially saturated porous media is highly dependent on the saturation of the media. To take this into account, the tortuosity used in the diffusion calculations is corrected for the saturation of the pore space in the soil and the clay. The outdoor concentration of radon is calculated by modeling the air above the site as a mixing cell in which the radon is diluted and removed by wind. If a residence is located over the DU disposal area, the radon is also modeled as diffusing through the foundation of the house and into the house. Barometric pumping was not included. The validity of this assumption is questionable for shallow disposal depths in arid environments in particular. However, under those conditions, the doses were sufficiently large that the primary output metric of whether the system could meet the performance objectives would not be impacted (i.e., the results already exceeded the performance objectives).

Modeling results

Table I provides the percent of realizations that meet doses of either 0.25 mSv/yr (25 mrem/yr) to the public or 5 mSv/yr (500 mrem/yr) to the intruder for a variety of scenarios and configurations. The results in Table I demonstrate that performance period (i.e., the length of time the analysis was run for), disposal depth at arid sites, and site conditions are important variables to consider for the disposal of DU. Most of the dose at a humid site comes from the groundwater pathway, and the dose at arid sites is mainly due to inhalation of radon. The dose from radon increases greatly at low moisture contents and at shallow disposal depths. With a short performance period, many sites and disposal configurations would be able to meet the performance objectives. For an arid site, radon has not ingrown sufficiently when the performance period is relatively short (e.g., 1,000 years). For both arid and humid sites, the delay in transport may be sufficient to achieve the performance objectives, except for shallow disposal. Disposal of large quantities of DU at depths less than 3 m (9.8 ft) results in projected chronic intruder doses much in excess of 5 mSv/yr (500 mrem/yr). Grouting of the waste may improve the likelihood of an arid site meeting the performance objectives, though the performance of grout over long periods of time is very uncertain.

An uncertainty analysis was performed using genetic variable selection algorithms using a neural network software product, Neuralware NeuralWorks Predict® [12]. For the water dependent pathways at an arid site, important parameters were the hydraulic conductivity and gradient of the aquifer, the infiltration rate, and geochemical conditions that determine sorption and solubilities. For radon at an arid site, the liquid saturation of the materials and properties of the residence and scenario, such as house height, foundation porosity, air exchange rate in the house, and fraction of time spent indoors, were most significant.

Table I. Percent of Probabilistic Realizations that Meet the Performance Objectives

Scenario	Performance Period (yr)	Resident[1]			Chronic Intruder[2]
		Total dose	Drinking water	Inhalation	Total dose
Arid, 1 m (3.3 ft) disposal depth	1,000	100	100	100	<2
	10,000	40	90	50	0
	100,000	10	60	20	0
	1,000,000	<1	40	8	0
Arid, 3 m (9.8 ft) disposal depth	1,000	100	100	100	2
	10,000	80	90	100	0
	100,000	50	60	80	0
	1,000,000	20	40	70	0
Arid, 5 m (16 ft) disposal depth	1,000	100	100	100	100
	10,000	80	90	100	100
	100,000	50	60	90	90
	1,000,000	30	40	90	70
Humid, 5 m (16 ft) disposal depth	1,000	70	70	100	100
	10,000	0	0	100	20
	100,000	0	0	100	0
	1,000,000	0	0	97	0
Arid,[3] 5 m (16 ft) disposal depth, Grout	1,000	100	100	100	100
	10,000	90	90	100	100
	100,000	80	80	100	90
	1,000,000	60	60	90	80

[1] Percent of realizations that are below 0.25 mSv/yr (25 mrem/yr) TEDE.
[2] Percent of realizations that are below 5 mSv/yr (500 mrem/yr) TEDE. When the waste depth is greater than 3 m (9.8 ft), the waste disruption process is through well drilling, not home excavation.

CONCLUSIONS

The model results indicate that the estimated risks from the disposal of DU are sensitive to the performance period and the disposal depth, and estimated disposal facility performance is strongly dependent on site-specific hydrologic and geochemical conditions. Radon fluxes to the environment are very sensitive to the long-term moisture state of the system. Radon is a major contributor to the dose at arid sites with shallow disposal. The groundwater pathway is the major contributor to the dose at humid sites. Grouting of the waste may improve the likelihood of an arid site meeting the performance objectives; however, grout may enhance the mobility of uranium in the groundwater pathway after the grout degrades. It is essential that the site hydrology and geochemistry be well-understood when determining the risk from disposal of large quantities of DU, because site-specific conditions are the primary determinant of the dose. Uranium and daughter radionuclide speciation and partitioning, as well as, radon transport in natural systems are complex processes; the analysis of the near-surface disposal of DU must adequately evaluate and manage this uncertainty. In particular, measurements of site-specific infiltration rates, radionuclide sorption and solubilities, radon diffusion and emanation rates, waste release rates, and soil-to-plant transfer factors can greatly reduce the uncertainty in the estimated future performance of a disposal site.

REFERENCES

1. Kozak, M.W., T.A Feeney, C.D. Leigh, and H.W. Stockman, 'Performance Assessment of the Proposed Disposal of Depleted Uranium as Class A Low-Level Waste,' Sandia National Laboratories, Albuquerque, NM. 1992.
2. Simpkins, A.A., et al, 'Description of Methodology for Biosphere Dose Model BDOSE.' Center for Nuclear Waste Regulatory Analyses, Southwest Research Institute, San Antonio, TX. 2007.
3. Allard, B., 'Sorption of Cs, I, and Actinides in Concrete Systems.' SKB Technical Report 84-15, Sweden. 1984.
4. Allard, B., 'Chemical Properties of Radionuclides in a Cementitious Environment.' SKB Progress Report 86-09, Sweden. 1987.
5. BSC, 'Dissolved Concentration Limits of Radioactive Elements.' ANL-WIS-MD-000010 Rev 3, Bechtel SAIC Company, Las Vegas, NV. 2004.
6. EPA, 'Understanding Variation in Partition Coefficient, Kd, Values.' EPA-402-R-99-004A. 1999.
7. EPA 'Understanding Variation in Partition Coefficient, Kd, Values. Volume III: Review of Geochemistry and Available Kd Values for Americium, Arsenic, Curium, Iodine, Neptunium, Radium, and Technetium' EPA-402-R-04-002C. 2004.
8. Sheppard, M.I. and D.H. Thibault. *Default Soil Solid/Liquid Partition Coefficients, Kds, for Four Major Soil Types: A Compendium. Health Physics.* Vol. 59. pp. 471–482. 1990.
9. U.S. Department of Energy (DOE). 'Draft Supplement Analysis for Location(s) to Dispose of Depleted Uranium Oxide Conversion Product Generated from DOE's Inventory of Depleted Uranium Hexaflouride.' DOE/EIS-0359-SA1. Office of Environmental Management. 2007.
10. NRC, 'Environmental Impact Statement for the Proposed National Enrichment Facility in Lea County, New Mexico, Final Report.' NUREG-1790, June 2005.
11. NRC, 'Environmental Impact Statement for the Proposed American Centrifuge Plant in Piketon, Ohio, Final Report.' NUREG-1834, 2006.
12. Neuralware, NeuralWorks Predict® Product Version 2.40, Carnegie, PA. 2001.

Mater. Res. Soc. Symp. Proc. Vol. 1124 © 2009 Materials Research Society   1124-Q07-01

3D Random Walk Simulation of Migration Behavior of Radionuclide in a Granite Core

Yoshimi Seida[1,2], Hiroyasu Takase[3], Hiroaki Takahashi[1] and Yuichi Niibori [4]
[1] Nuclear Chem. Eng. Center, Inst. Res. & Innovation, Takada 1201, Kashiwa, 277-0861 Japan
[2] Dept. Env. Chem. & Tech., Tokyo Inst. of Tech., Nagatsuta 4259, Yokohama, Japan
[3] Quintessa Japan Co. Ltd., Yokohama, 220-6007 Japan
[4] Dept. Quantum Sci. & Energy Eng., Tohoku Univ., 6-6-01-2 Sendai, 980-8579 Japan

ABSTRACT

A 3D random walk simulation (RWS) of a heterogeneous intact rock system was performed to examine the influences of heterogeneity of the rock, such as spatial distribution of minerals and/or sorption sites and migration path, on its retardation behavior for diffusants. A cube based 3D granite model with heterogeneous mineral distribution was constructed based on the mineral composition and distribution of the Inada granite of Japan. Impulse responses of the 3D granite model for both sorptive and non-sorptive walker particles were obtained by the RWS and were evaluated in comparison with the breakthrough patterns of a simple 1D diffusion model in a homogeneous system with various retardation parameters.

INTRODUCTION

Migration of radionuclides in natural barriers is a key process that should be clarified for the safe and reliable deep geological disposal of high level radioactive wastes in Japan [1]. The general approach to obtain rock retardation parameters is diffusion and batch sorption experiments performed within an acceptable time period using small rock samples. However, even at laboratory scale, the experimental method to obtain reliable retardation parameters of intact rock has not been established well due to the complexity and heterogeneity of the physico-chemical structure of natural rocks (sorption and diffusion properties of each constituent mineral and their space distribution) and little knowledge of migration behavior in the heterogeneous media. Representativeness and reliability of the retardation parameters obtained experimentally using small rock specimens should be carefully examined if a significant heterogeneity exists in the rock sample [2]. The representative retardation parameter of heterogeneous rock can be explained by a component additive model of retardation parameters of each constituent mineral. But additivity has not been examined well so far, especially in the intact system. Random walk simulation (RWS) analysis is one way by which the influences of heterogeneity on the experimentally observed retardation behavior of rock can be examined.

In the present study, RWS of sorptive and non-sorptive diffusants in heterogeneous small intact rock was performed to investigate the influence of heterogeneity on the retardation behavior of rock and the overall diffusion and partition coefficients. A cube-based three-dimensional (3D) granite model with heterogeneous mineral distribution was constructed based on the mineral composition and distribution in granite. Pulse-injected breakthrough patterns for sorptive and non-sorptive walker particles were obtained by the RWS in the heterogeneous system. Retardation parameters were examined through fitting with superposition of breakthrough patterns obtained from a one-dimensional (1D) diffusion equation with various retardation parameters in a homogeneous system. The effectiveness of the 3D model-based simulation was discussed.

CONSTRUCTION OF MODEL GRANITE

Mineral distribution in granite

The internal mineral distribution in a granite core (Inada granite of Japan, 32mmϕx 25mmL) was used to produce a model granite. The internal mineral distribution was observed by means of SEM/EDX and X-ray CT with tomographic resolution of 1024x1024 pixels for each cross sectional image collection. 3D structure of the mineral distribution was constructed by volume rendering of 653 sheets of the slice images. Mineral phases were identified from X-ray absorption and elemental distribution. Details of the analysis and the image treatment are shown elsewhere [3].

Modeling of granite based on cube model

Using the obtained composition and geometric distribution of constituent minerals in the granite, a 3D granite model was constructed using a cubic approximation of minerals proposed in principle by Niibori et al. [4]. A cubic space (20x20x20 mm) was first divided into iso-cubic 2 mm small cells (represents elemental grain size of minerals) with mineral ID number (N=1: biotite; 2:quartz; 3:feldspar) referring the mineral distribution observed by the X-ray CT. The small cells were coarsely divided into 0.2 mm mesh cells. The boundaries between different minerals were interpreted as grain boundaries. The small cell size roughly corresponds to the grain size of biotite observed in the Inada granite by microscope. The other minerals were produced by clustering small cells. The cells were distributed to have the volume ratio of each mineral consistent with a point counted mineral composition (biotite:quartz:feldspar = 4:36:60). Details of the cell construction are reported elsewhere [3].

MODEL OF MIGRATION AND SIMULATION PRCEDURE

The migration of radionuclides in porous media without fluid convection is expressed by the following diffusion equation under the assumption of Henry-type equilibrium sorption.

$$R_n \frac{\partial C_n}{\partial t} - \frac{\partial}{\partial x} D_e \frac{\partial C_n}{\partial x} = 0 \qquad R_n = 1 + \frac{\rho K_d}{\theta} \qquad (1)$$

The D_e, K_d, ρ, θ and R_n indicate diffusion coefficient, partition coefficient, density, porosity and retardation coefficient, respectively. The discrete form of Equation 1 for the RWS gives Equation 2.

$$\frac{\partial \bar{r}(p)}{\partial t} = \bar{v}_D(\bar{r}, p) \qquad (2)$$

where the $\bar{v}_D(\bar{r}, p)$ represents the random variable (diffusive velocity in a unit time step) at position $\bar{r}$. The p represents the probability of both direction and position of walkers with Gaussian distribution functions. Diffusion behavior with Gaussian distribution can be approximated by a binomial distribution with a travel distance $\sqrt{2D\Delta t}$ in a single walk step under the law of large numbers. The R_n is included in the diffusion coefficient. The diffusion term in Equation 2 was approximated by a binomial distribution for each direction. Displacement of walkers can be given by Equation 3.

$$\Delta \bar{r} = \frac{\bar{D}(\bar{r})}{|D_i(\bar{r})|} \sqrt{2D_i(\bar{r})\Delta t} \qquad (3)$$

where $\bar{D}(\bar{r})$ represents the vector of transport (diffusion vector) and $D_i(\bar{r})$ is the diffusion coefficient at position $\bar{r}$. The diffusion behaves isotropically so that $\bar{D}(\bar{r})/|D_i(\bar{r})| = (\pm 1, \pm 1, \pm 1)$. To check the statistical reliability of the RWS, the fluxes of walker particles at the exit plane obtained by the RWS were compared in advance to those obtained by numerical solutions of Equation 1 with a diffuse source at the center of the cube. The flux profiles obtained by both methods for three-phase diffusion with respective D_e, θ, K_d and ρ values corresponded fairly well [3]. 100,000 walkers were found to be sufficient for reliable RWS in this study.

The effective diffusion coefficient, D_e and K_d were assigned to each sub-divided cell. The D_e and K_d values used in the present RWS were collected from reported papers and listed in Table I [3]. Width of the grain boundaries was set to 1 μm. D_e at the cells containing boundaries was determined to be 1.26×10^{-12} m^2/s for the cells of biotite and quartz and 2.55×10^{-12} m^2/s for feldspar, based on a flux-weighted average of D_e between the boundary and the mineral area. The diffusion coefficient at grain boundaries is equivalent to that in pore water. The walker particles were introduced by pulse-style from the bottom plane of the cube ($z=0$). Zero concentration of particles was set at the opposite outer plane boundary ($z=20$ mm). A periodic boundary condition was used at the other outer plane boundaries. Effective and apparent and diffusion coefficients, D_e, D_a ($=\theta + \rho K_d$) can be determined from the fluxes for the sorptive and non-sorptive particles, respectively.

Table I The parameters used in the RWS

	Biotite	Quartz	Feldspar	Grain boundary
K_d [m^3/kg]	50	0	0.05	
D_e [m^2/s]	1×10^{-14}	1×10^{-14}	1×10^{-12}	1×10^{-9}
θ [%]	1	1	1	10

RESULTS

Spatial distribution of each mineral was visualized by CT analysis as shown in Figures 1(a) and (b). Heterogeneous distributions of biotite and alkaline feldspar were identified. Mineral grains were found to include many micro fractures separating the mineral domains into smaller sub particles. The fractures were introduced as porosity of the mineral phase in the cube model. The domain sizes of quartz and

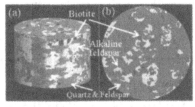

Figure 1. Distributions of minerals in the Inada granite obtained by X-ray CT analysis (a) 3D image and (b) cross sectional image

feldspar were confirmed to be comparative by microscope observation [3]. Figures 2(a)-(d) show the constructed granite cube model and the distribution of D_e. The heterogeneous granite structure was successfully simulated in the model.

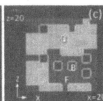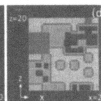

Figure 2. Mineral (a, c) and D_e (b, d) model distributions. (a), (b): 3D views. (c), (d): cross sections. Colors correspond to the mineral phase and D_e value. B=biotite, F=feldspar, Q=quartz.

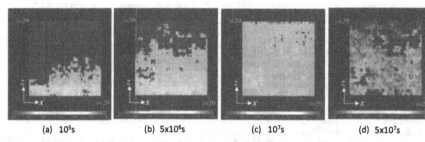

| (a) 10^5s | (b) $5x10^6$s | (c) 10^7s | (d) $5x10^7$s |

Figure 3. Snapshots of the particle distribution for the series of travel times (non-sorptive particles). The shades represent the concentrations normalized to inlet concentration.

Non-sorptive particles

Figures 3(a)-(d) are snapshots of particle migration in the RWS at a cross section perpendicular to the injection plane. The concentration of particles shown by shading was normalized to the initial concentration at the injection plane ($z=0$). The grid lines indicate grain boundaries. As the D_e values in feldspar and at grain boundaries are larger than those in biotite and quartz, particles migrate faster in those areas preferentially. Figure 4 shows the time sequence of effluent flux [1/yr] for unit pulse input obtained from the RWS and its fittings by numerical solutions of Equation 1. The major fitted curve ("single" dashed line) was obtained with single $\theta = 0.036$ and $D_e = 1.7x10^{-11}$ m^2/s, but lacks early stage fitting of the flux. The simulated flux was then fitted by the weighted superposition of numerical solutions of Equation 1 with various retardation parameters ("multi" dotted line). The parameters used for the fitting are listed in Table II. Fitting coefficients in Table II indicate the weight in the superposition. From the fitting coefficients, the grain boundaries of feldspar and

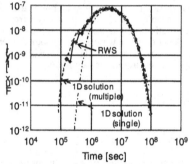

Figure 4. Time sequence of effluent flux obtained from the RWS and its fittings by the numerical solutions of 1D diffusion equation with single and multiple retardation parameters.

Table II. The fitting parameters for non-sorptive migration

Assumed pathway	Inside feldspar	Feldspar boundary	Quartz/biotite boundary	Inside quartz and biotite
D_e [m^2/s]	$1.0x10^{-10}$	$2.3x10^{-11}$	$1.3x10^{-11}$	$1.0x10^{-12}$
Fitting coef.	0.027	0.527	0.429	0.016

those between quartz and biotite were found to be major pathways of migration. 95% of the particles traveled along the grain boundary. In addition, the representative D_e value of the simulated system was a middle value between the D_e of feldspar grain boundaries ($2.3x10^{-11}$ m^2/s) and that of quartz/biotite grain boundaries ($1.3x10^{-11}$ m^2/s). The representative D_e estimated by the fitting is smaller than the volume averaged D_e ($5x10^{-11}$ m^2/s).

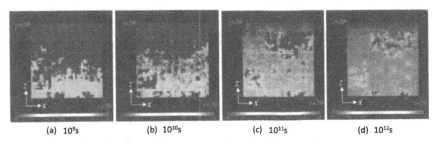

(a) 10^9s   (b) 10^{10}s   (c) 10^{11}s   (d) 10^{12}s

Figure 5. Snapshots of the particle distribution for the series of travel times (sorptive particles). The shades represent the concentrations normalized to inlet concentration.

Sorptive particles

Slow particle transport due to sorption onto sorptive minerals was observed (Figure 5) The times in Figure 5 are much longer than those in Figure 3. It is noteworthy that the concentration inside quartz, the small D_e mineral, is higher than in other areas. Once transported into the mineral, it will be difficult for particles to escape due to the small D_e. The effluent flux in the RWS revealed a significant peak in the early stage of breakthrough (Figure 6; note the longer time scale compared to Figure 4). This indicates the existence of a fast migration pathway even for sorptive particles due to the heterogeneous structure, indicating that diffusivity will be influenced significantly by the heterogeneity of the transport medium. The breakthrough curve was not a simple single curve. The result also indicates the importance of the representative minimum volume of sample specimens with physico-chemical heterogeneous structure, which will be used to obtain retardation

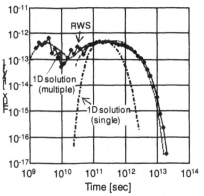

Figure 6. Time sequence of effluent flux obtained from the RWS and its fittings by the numerical solutions of 1D diffusion equation with single and multiple retardation parameters.

Table III. The retardation parameters used in the curve fitting

Fitted area	1st peak	Post edge of 2nd peak	Pre-Center of 2nd peak	Center of 2nd peak	Post center of 2nd peak
$D_a = (\times D_e$ [m^2/s]	2.0×10^{-14}	1.0×10^{-15}	4.0×10^{-16}	2.0×10^{-16}	1.0×10^{-16}
D_e [m^2/s]	2.3×10^{-13}	1.5×10^{-13}	7.3×10^{-14}	6.2×10^{-14}	1.9×10^{-13}
K_d [m^3/kg]	0.012	0.250	0.620	1.240	2.470
Fitting coef.	0.004	0.059	0.072	0.121	0.743

parameters experimentally. Attention should also be paid to the use of a homogeneous diffusion model in the case of small test samples with potentially heterogeneous structure. The best fitted superposition flux profile is shown by the dotted line in Figure 6; the fitting coefficients are shown in Table III. Migration of 75% of the particles was characterized by $D_a=10^{-16}$ m^2/s and appeared as the second elution peak in the breakthrough. This result corresponds to migration through biotite grain boundaries and intra-particle feldspar. The steep and narrow first peak of the flux (fitted by $D_a=2.0 \times 10^{-14}$ m^2/s) indicates the existence of migration through the major minerals with small sorption capacity (quartz and feldspar) but not contact with the minor mineral with large sorption capacity (biotite). A harmonic average diffusivity estimated from the fractional content of minerals and grain boundaries (pore volume) and diffusion coefficients at each region was very small (8.8×10^{-18} m^2/s) compared with the D_e estimated by the fitting of second flux peak. This level of migration will be hard to observe by conventional continuous injection experiments due to its very slow process.

DISCUSSION

The results indicate the potential significance of minerals with large volume fractions and accessible pathways even if the mineral K_d is not large. Connectivity of the grain boundaries and the fraction of the percolated pathway are important parameters in the estimation of the retardation parameter in the rock with heterogeneous structure. The minerals along the accessible migration path with large sorption capacity will actually contribute significantly to the retardation performance of the rock. When minerals with high K_d exist but at small abundance in the rock, we may need to pay attention to the representative elemental volume (REV). The retardation behavior of heterogeneous rock with various spatial migration pathways, diffusivities and partition coefficients can be simulated by the simple additive model of the conventional migration equation with an adequate fitting weight. Representativeness of the retardation parameter increases with size of the test sample since probability of nuclide contact with heterogeneously distributed minerals will increase with the size of the test sample. We will report the size effect in a companion paper.

CONCLUSIONS

The heterogeneity of a rock sample influences breakthrough behavior when the heterogeneity is significant in comparison with the size of sample. To determine reliable retardation parameters for heterogeneous rocks with the diffusion experiment within an acceptable time period, the influence of physico-chemical heterogeneity of the test sample on the experimentally obtained parameters should be considered. The simulation model examined in the present study supplies a simple tool to estimate the REV of the heterogeneous sample specimen that will be used in the diffusion experiment, along with the migration behavior of radionuclides in the media.

ACKNOWLEDGMENTS
This work was partly supported by Ministry of Economy, Trade and Industry of Japan.

REFERENCES
1. Japan Nuclear Cycle (JNC) Development Institute report, *Technical reliability of deep geological repository of HLW in Japan*, JNC TN1400 99-023(1999); JNC, *Technical examination of TRU waste treatment*, JNC TY1450 2005-001(2), (2005).
2. E. Yoshida and Y.Seida, submitted to *Eng. Geol.* (2008).
3. Institute of Research and Innovation of Japan (IRI), *H17 Report of Advanced Evaluation of Rradionuclides Migration in the Barrier System* (IRI, 2007).
4. Y. Niibori *et al.* in *Scientific Basis for Nucl. Waste Management XXVII*, edited by V. Oversby and L. Werme (Mater. Res. Soc. Proc. **807**, Warrendale, PA, 2004) pp.645-650.
5. Y. Seida *et al.*, *Proc. of Fall Meeting of Atomic Energy Society of Japan (AESJ) 2007*, AESJ, I13 (AESJ, 2007) p. 492.

Mater. Res. Soc. Symp. Proc. Vol. 1124 © 2009 Materials Research Society 1124-Q07-02

Sorption and Retardation Processes of Cs in Granite Under Groundwater Conditions

Yoshimi Seida[1,2*], Takahiro Kikuchi[1], Hiroaki Takahashi[1], Hisao Sato[3], Akira Ueda[3] and Hidekazu Yoshida[4]

1. Institute of Research and Innovation, 1201 Takada, Kashiwa, 277-0861 Japan
2. Tokyo Institute of Technology, Nagatsuta 4259, Yokohama 226-8502 Japan
3. Mitsubishi Materials Co. Ltd., Naka, 311-0102, Japan
4. Nagoya University Museum, Chikusa, Nagoya, 464-8601 Japan
* seida.y.aa@m.titech.ac.jp

ABSTRACT

The dynamics of Cs^+ sorption and migration behavior in granite under groundwater conditions were investigated using a granite sample collected from Inada, Japan. A Cs^+ sorption isotherm for the Inada granite was obtained originally by batch sorption experiment with a powdered rock sample. The extended Langmuir model with two sorption sites fit the adsorption isotherm fairly well. The Cs^+ sorption behavior of the dominant sorption mineral, biotite, was observed by means of phase-shift interferometry (PSI) as an *in-situ* probe. Sorption of Cs^+ from the edge to the interlayer of biotite, but not deep inside, was observed under groundwater conditions. Cs^+ sorption was considered to occur near the edge of the biotite layer, where K^+ ions in the layer were replaced by Na^+ ions in the groundwater first, followed by exchange with Cs^+. Breakthrough behavior of intact granite (32 mm in diameter x 25 mm length) for Cs^+ was observed over six months in a flow-through experiment using a centrifuge system. The Cs^+ breakthrough curve showed a plateau. The migration behavior of Cs^+ in the granite was modeled based on the series of results observed in the flow-through experiment, PSI analysis and the batch sorption experiment. A migration model with cascade-type dual modes of kinetic sorption was examined to interpret the Cs^+ breakthrough data.

INTRODUCTION

Migration of radionuclides in natural barriers is a key process that should be clarified for the safe and reliable deep geological disposal of high level radioactive wastes considered in Japan. The general approach to obtain rock retardation parameters is diffusion and batch sorption experiments performed within an acceptable time period using rock samples [1,2]. However, even at laboratory scale, it will take a long time to obtain useful retardation data (breakthrough curve and/or diffusion and distribution coefficients) in an intact rock system. As a result, retardation data such as diffusivity, distribution coefficient and retardation mechanism have not been frequently collected in intact systems. In the present study, sorption and migration behaviors of Cs^+ in an intact granite core sample were investigated using a flow-through diffusion method with centrifugal acceleration of transport and phase shift interferometry (PSI). The batch sorption behavior of Cs^+ onto powdered granite was obtained independently to assist modeling of Cs^+ migration in the granite core. A sorption isotherm was obtained as a function of granite particle size. Dynamics of Cs^+ sorption on biotite, the major sorption mineral, were analyzed using PSI in terms of layer height/thickness of biotite during the sorption process in various solutions. The Cs^+ breakthrough curve obtained by the flow-through experiment was simulated using a dual-modes kinetic sorption model developed in the present study according to the results observed by the series of analyses.

EXPERIMENTS AND MODELING

Batch and flow-through diffusion experiments
The granite sample used in the present study was collected from Inada, Japan. Powdered and core (32 mm in diameter x 25 mm length, L) samples were used in the batch sorption and flow-through experiment, respectively. Three different particle sizes were used in the batch sorption experiment (2.5-1.5, 1.5-0.5 and < 0.5 mm in diameter). The batch sorption experiment was performed based on Atomic Energy Society of Japan guidelines [3].

Breakthrough behavior of Cs^+ in the intact granite was observed over six months by a flow-through experiment using the centrifuge system developed by UFA Inc., USA [4]. The centrifuge system accelerates access of nuclides to sorption sites along with transport of pore water by body force in the intact rock without losing retardation capacity. Details of the flow-through experiment were shown elsewhere [5]. The pseudo-K_d defined by Equation 1 was calculated from the concentration of Cs^+ in effluent to evaluate a contribution of sorption to retardation in the intact system.

$$\text{pseudo - K}_d = \frac{q_{ap}}{C_{av}} = \frac{\sum (C_{in} - C_{out})V_i / w_{rock}}{\sum C_i V_i / \sum V_i} \tag{1}$$

where q^{ap}, C_{av}, C_{in}, C_{out}, C_i, V_i and w_{rock} are concentration of adsorbed Cs^+, average concentration in the effluent, inlet concentration, outlet concentration, concentration in the effluent at each time, respectively and V is the product of flow velocity and time.

The composition of groundwater used in the present study is shown in Table I. The concentration of Cs^+ in each liquid phase was analyzed by ICP-MS. The amount of sorption, represented by q_{ap} in Equation 1, was calculated based on mass balance.

PSI observation of the Cs^+ sorption process
The sorption process of Cs^+ onto the dominant sorption mineral biotite at its near edge was observed by means of PSI as an *in-situ* probe for sorption dynamics, followed by observation of Cs^+ distribution in the biotite using electron beam probe analysis (EPMA). Details of the PSI and EPMA analysis are reported elsewhere [5].

Table I. Composition of synthetic groundwater used and conditions during the batch sorption experiment.

	[mol/dm³]	
NaHCO₃	3.6	10³
CaSO₄ + 5H₂O	1.1	10⁴
KCl	6.2	10⁵
MgSO₄	5	10⁵
Solution pH	8.5	
Liquid/Solid ratio	10 cm³/g	
Sorption period	7 to 109 days with shaking	
Temperature	298 K	

Kinetic migration model
The basic equation of Cs^+ transport in the granite core under steady-state centrifugal fluid flow can be shown by Equation 2.

$$\frac{\partial C}{\partial t} = D \frac{\partial^2 C}{\partial x^2} - u \frac{\partial C}{\partial x} - \frac{(1-\varepsilon)}{\varepsilon} \rho_r \frac{\partial q}{\partial t} \tag{2}$$

where C, D, u ρ_r, q, ε and t are concentration in pore water, dispersion coefficient, flow velocity, density of rock, concentration of adsorbed Cs^+, porosity, and time, respectively.

The second term on the right side of the equation accounts for convective transport by the centrifuge fluid flow. Biotite was considered the dominant Cs^+ sorber in the rock. Dual-modes kinetic sorption was assumed according to the sorption isotherm, breakthrough behavior in the flow-through experiment, and the PSI and EPMA observations. Sorption of Cs^+ on the biotite occurs first at the edge near fractures and/or accessible biotite grain boundaries in the granite, followed by diffusion into interlayers of biotite. The sorption capacity of granite was expressed by the extended Langmuir equation with two sorption sites: edge and interlayer of biotite. The linear driving force model was applied for the rates of sorption from pore water to layer edges and from the edges to interlayers (cascade type dual modes sorption) and is presented in Equations 3 to 5 [2].

$$\frac{\partial q}{\partial t} = \frac{\partial q_{edge}}{\partial t} + \frac{\partial q_{layer}}{\partial t} \tag{3}$$

$$\frac{\partial q_{edge}(x)}{\partial t} = h_{edge}(q_{edge}^{eq}(C(x)) - q_{edge}(x)) \tag{4}$$

$$\frac{\partial q_{layer}(x)}{\partial t} = h_{layer}\frac{q_{edge}(x)}{q_{edge}^{max}}(q_{layer}^{eq}(C(x)) - q_{layer}(x)) \tag{5}$$

where q_i, q_i^{eq}, q^{max} and h are concentration of Cs^+ adsorbed at i-site, equilibrium sorption capacity at i-site, maximum adsorption capacity and apparent rate constant, respectively. The set of non-linear Equations 2 to 5 was solved numerically using a finite difference method with the initial and boundary conditions corresponding to the experimental ones (Equations 6 and 7).

$$t = 0; C = 0, q_i = 0 \ at \ 0 < x < L \tag{6}$$
$$t > 0; C = C_{inlet} \ at \ x = 0 \tag{7}$$

The parameters used in the simulation are listed in Table II. The apparent rate constants, h, were fitting parameters in the present study. The fitting was performed roughly by varying the h_i from viewpoints of breakthrough point and the plateau height [2]. The pseudo-K_d defined by Equation 1 was also calculated from the simulated breakthrough curve.

Table II. The parameters used in the simulation.

Length, L	2.5 cm	Flow velocity, u	2.0×10^{-6} m/s
Porosity, ε	0.01 [-]	Dispersion coef., D	1.0×10^{-10} m²/s
Langmuir const., k_{edge}	5.0×10^8 cm³/mol	q_{edge}^{max}	4.0×10^{-8} mol/g
Langmuir const., k_{layer}	1.5×10^6 cm³/mol	q_{layer}^{max}	9.0×10^{-6} mol/g
Apparent rate const., h_{edge}	3.0×10^{-8} s⁻¹	Density of rock, ϱ	2.55 g/cm³
Apparent rate const., h_{layer}	1.0×10^{-9} s⁻¹		

RESULTS

Batch sorption experiment

Figure 1 shows the sorption isotherm for Cs^+ (109 days). The amount of sorption was higher for smaller grain size due to the increase of contact area with solution. This suggests that the intact system will take a long time to use the full sorption capacity due to the small contact area, as well as small diffusion coefficient, in the rock. The sorption isotherm for the fine powder was used in sorption modeling. The line in the figure is the best fit by the extended Langmuir model with two sorption sites (Equation 8).

$$q^{eq} = \sum_i \frac{k_i q_i^{max} C}{1 + k_i C} \qquad (8)$$

Bradbury and Baeyens [6] showed that clay minerals sorb Cs^+ at three different sites: fray-edge, planar and type II sorption sites. Nevertheless, we are able to fit our sorption isotherm using a two sites sorption model without losing the trend of the obtained data; this also allows simplicity in dual-modes cascade sorption modeling.

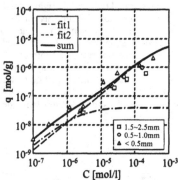

Figure 1. Cs^+ sorption isotherms for Inada granite for different particle sizes.

Flow-through experiment

Figures 2(a) and 2(b) show the Cs^+ breakthrough curve for the granite core and the pseudo-K_d calculated from the breakthrough data. Migration of Cs^+ was retarded in the intact system even in the forced flow condition. The breakthrough curve shows a plateau at long times [Figure 2(a)]. This result indicates multi-mode sorption, by which the plateau breakthrough would appear for the long period.

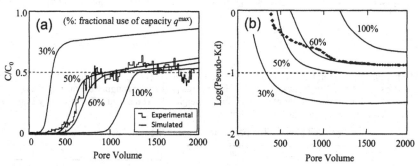

Figure 2. (a)Breakthrough curves of Cs^+ obtained from the flow-through experiment and from the model simulation as a function of effective sorption capacity and (b) pseudo-K_ds calculated from the experimental breakthrough data (diamonds) and the model simulation.

PSI and EPMA analysis

Figure 3(a) shows a PSI image around a layer edge and an etch pit of biotite. Observation of edge swelling due to contact with $0.1M$ Na^+ (NaCl) solution was performed first by PSI [Figure 3(b)]. The increase of layer height due to sorption of Na^+ ions was observed as shown the cross sectional height distribution in Figure 3(d). Then, sorption of Cs^+ (from a $0.1M$ NaCl solution with $0.01M$ Cs^+) in the interlayer of biotite from the edge (but not deep inside) was confirmed from the decrease of height at the edge [Figures 3(c) and 3(d)] and the EPMA image of Cs^+ distribution in Figure 3(e). In groundwater experiments, similar swelling behaviors of biotite were obtained [5]. Cs^+ sorption in groundwater conditions was confirmed to occur near

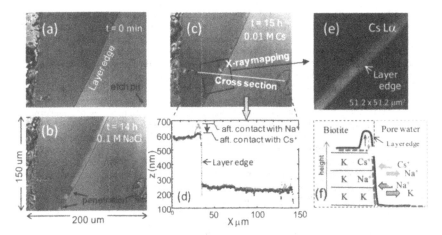

Figure 3. (a) PSI image of granite surface with layer edge and etch pit (t=0 min), (b) the image of the surface in contact with 0.1M NaCl for 14 h, (c) the image of the surface in contact with the 0.1M NaCl + 0.01M CsCl solution for 15 h, (d) height distributions from images b and c along the cross section in image c, (e) EPMA image of Cs$^+$ distribution in the square area shown in image c, and (f) schematic diagram of the dynamics of Cs sorption supposed from the PSI observation.

the layer edge at the interface between biotite and solution where K$^+$ ions in the layer were replaced by Na$^+$ ions in the solution in advance. The area of sorption propagated inside the layer but not deep enough during the observed duration (1 month). This indicates that Cs$^+$ will not readily sorb in deep biotite interlayers during short-term experiments. Distribution of Cs$^+$ in micro fractures (near edge) and grain boundaries of biotite and feldspar was also observed by the EPMA analysis of the cross-sections of granite core from the flow-through experiment.

Model simulation

The breakthrough curves and pseudo-K_ds simulated based on the model with cascade type kinetic sorption are shown in Figures 2(a) and 2(b). An apparent sorption capacity of rock, which is indicated in the figure by % of maximum sorption capacity of rock (which is obtained in the batch sorption experiment), was assumed in the simulation to examine the influence of effective capacity on the breakthrough behavior in the intact system. The developed model successfully simulated the breakthrough behavior with long plateau region. According to the results of the simulation, the model with half the maximum sorption capacity fit the breakthrough data and pseudo-K_d profiles fairly well.

DISCUSSION

From the results of this study, sorption of cesium in the interlayer of biotite under groundwater conditions was estimated to occur by the following process [Figure 3(f)]. First, Na$^+$ ions in the groundwater exchange with K$^+$ ions in the interlayer near edge, resulting in the relaxation of the layer structure to enable exchange by larger cations. The layer distance of biotite changes from 0.99 nm (interlayer cation is K$^+$) to 1.05 and 1.20 nm by ion exchange with

Cs$^+$ and Na$^+$, respectively. The local swelling and/or increase of layer height at the surface near edge of biotite indicated the sorption of Na$^+$ ions through cation exchange with the original K$^+$ ions [Figure 3(b)]. When in contact with the groundwater with Cs$^+$, the near-edge height decreased due to Cs$^+$ sorption. In our extra experiments, Cs$^+$ was not exchanged by K$^+$ in the interlayer in Na$^+$-free conditions. Sorption of Na$^+$ ions prior to the Cs$^+$ sorption may be an important process to receive the larger size Cs$^+$ ions in the interlayer.

Retardation behavior in heterogeneous media such as the granite core can be difficult to describe with a single retardation parameter due to the existence of multiple sorption sites and/or modes of migration. Steefel et al. [7] reported that a kinetic sorption model with dual sorption sites fit Cs$^+$ breakthrough data and a partition coefficient calculated with the amount of sorption in a Hanford site soil column fairly well. In this case, rocks composed of heterogeneous minerals indeed have multiple sorption sites, and migration behavior in the media cannot be simply simulated by a single mode sorption model with a single retardation parameter. The granite core involves both strong sorption affinity, but with a small capacity, and weak sorption affinity, with a large capacity, as can be understood from the sorption isotherm in Figure 1. The model simulation fit fairly well the breakthrough behavior and pseudo-K_d of Cs$^+$ by taking into account multiple sorption sites and modes. The % of effective sorption capacity in the core sample indicates the potential sorption sites accessible in the intact system. The breakthrough data for the intact system fit fairly well the case of 50 to 60 % sorption capacity measured for the powdered samples. However, the simulated breakthrough curve shows a slight increase with time which has not been confirmed experimentally at the present time. Further observation should be done to confirm the reliability of the model.

CONCLUSIONS

The dynamics of Cs$^+$ retardation behavior in granite in groundwater conditions was investigated in an intact system. The retardation behavior of the granite core can be simulated by the dual modes kinetic sorption model developed in the present study fairly well.

ACKNOWLEDGMENTS

This work was supported in part by Ministry of Economy, Trade and Industry of Japan.

REFERENCES

1. J.P. Absakim, N.M. Crout and S.D. Young, *Environ. Sci. Tech.* **30**, 2735 (1996).
2. Y. Seida, H.Takahashi, M. Yuki and K.Suzuki, *Proc. of Fall Meeting of Atomic Energy Society of Japan (AESJ) 2007,* AESJ I13 (AESJ, Tokyo, 2007) p. 492.
3. AESJ standard, *Measurement Method of the Distribution Coefficient on the Sorption Process,* AESJ-SC-TR001 (AESJ, Tokyo, 2006).
4. ASTM Standard Test Procedure, *Test Method for Determining Hydraulic Conductivity in porous Media by Open-Flow Centrifugation,* D6527-00 (ASTM, 2008).
5. Institute of Research and Innovation of Japan (IRI), *H17 Report of Advanced Evaluation of Radionuclide Migration in the Barrier System* (IRI, Japan, 2007).
6. M. H. Bradbury and B. J. Baeyens, *J. Contam. Hydrol.* **42**, 141 (2000).
7. C. I. Steefel, S. Carroll, P. Zhao, S. Roberts, *J. Contam. Hydrol.* **67**, 219 (2003).

Mater. Res. Soc. Symp. Proc. Vol. 1124 © 2009 Materials Research Society   1124-Q07-03

Physical Rock Matrix Characterization:
Structural and Mineralogical Heterogeneities in Granite

M. Voutilainen*, S. Lamminmäki**, J. Timonen*, M. Siitari-Kauppi** and D Breitner***

*Department of Physics, University of Jyväskylä, P.O.Box 35, FIN-40351 Jyväskylä, Finland
**Laboratory of Radiochemistry, University of Helsinki, P.O. Box 55, FIN-00014 University of Helsinki, Finland
***Lithosphere Fluid Research Lab, Institute of Geography and Earth Sciences, Eötvös University, Budapest, 1117, Hungary

ABSTRACT

Evaluation of the transport and retardation properties of rock matrices that serve as host rock for nuclear waste repositories necessitates their thorough pore-space characterization. Relevant properties to be quantified include the diffusion depth and volume adjacent to water conducting features. The bulk values of these quantities are not sufficient due to the heterogeneity of mineral structure on the scale of the expected transport/interaction distances.

In this work the 3D pore structure of altered granite samples with porosities of 5 to 15%, taken next to water conducting fractures at 180–200 m depth in Sievi, Finland, was studied. Characterization of diffusion pathways and porosity were based on quantitative autoradiography of rock sections impregnated with ^{14}C-labelled polymethylmethacrylate (PMMA). Construction of 3D structure from PMMA autoradiographs was tested. The PMMA method was augmented by field emission scanning electron microscopy and energy-dispersive X-ray analyses (FESEM/EDAX) in order to study small pore-aperture regions in more detail and to identify the corresponding minerals. The 3D distribution of minerals and their abundances were determined by X-ray microtomography. Combining the mineral specific porosity found by the PMMA method with these distributions provided us with a 3D porosity distribution in the rock matrix.

INTRODUCTION

Over extended periods, long-lived radionuclides within geologic disposal sites may be released from the spent fuel and migrate to the geo/biosphere. In the bedrock, contaminants will be transported along fractures by advection and retarded by sorption on mineral surfaces and by molecular diffusion into stagnant pore water in the matrix along a connected system of pores and micro-fissures. Because chemical interactions of groundwater and transported components with inner mineral surfaces play a major role in the retardation process, the effectiveness of the rock matrix as a natural barrier is influenced by the size, shape and spatial arrangement of the effective rock porosity network. A reliable picture of the pore-space structure in different mineral phases of rock can be determined using different complementary methods of characterization.

For many years the ^{14}C-polymethylmethacrylate method (PMMA) [1, 2] has been widely used as a pore-structure analysis tool for low-porosity granitic rock. PMMA porosity results have been compared with those of confocal laser scanning microscopy [3,4] and X-ray micro computed tomography (μCT) [5]. These non-destructive methods have proved to be very useful

[6, 7] because of their ability to provide three-dimensional (3D) visualizations of the structures of cm-sized samples.

The pore structure of an altered tonalite sample (SY KR7) from Sievi Syyry, Finland, was studied by the PMMA method so as to determine the spatial distribution of its porosity from a sub-μm to cm scale. Field emission scanning electron microscopy and electron dispersive X-ray spectroscopy (FESEM/EDAX) allowed detailed determination of pore apertures down to 1 μm. The mineralogical and pore structure was analysed by μCT with 14 μm resolution.

EXPERIMENTAL

The samples studied in this work were taken from Sievi Syyry, Finland. Geologically, Sievi is situated in the Central Finland Granitoid Complex (age about 1900-1800 Ma) near the Svecofennian schist. It was one of the five areas selected for preliminary investigations for the nuclear waste disposal site in Finland [8], and a large number of drill holes were made there. The Sievi area, approximately 8 km^2, is surrounded by major fracture zones. The rocks present in the region comprise volcanic and sedimentary schist, gneisses, and basic and acidic plutonic rocks. The main rock types are tonalite, granodiorite and quartz diorite, all acidic plutonic rocks. The samples used here (SY KR7 176 m, 182 m) represent strongly altered tonalites that consist of sericite (~50%), plagioclase (20-30%), biotite (10-20%), hornblende (1-2%) and quartz (1-2%). Kfeldspar (1-3%) is probably a result of later secondary alteration, and occurs only rarely. Sericite, muscovite, chlorite, epidote, carbonate, sphene, apatite, and opaque minerals may also occur.

Rock core SY7 A1 (176m) was impregnated with ^{14}C-labeled PMMA for porosity determination. Sample SY7 A1 was also studied by electron microscopy. A parallel non-impregnated sample, SY7 B2 (182m), was studied with X-ray μCT.

The ^{14}C-PMMA method involves impregnation of cm-scale rock cores with ^{14}C-MMA (Russian Scientific Centre for Applied Chemistry, St. Petersburg, Russia) in vacuum, followed by irradiation polymerization after impregnation of ^{14}C-MMA, sample partitioning and autoradiography. Irradiation initiates the polymerization reaction in monomeric MMA forming ^{14}C-PMMA inside the rock, fixing the resin into the pore space of the rock. The tracer activity was 280 kBq/ml. PMMA-impregnated rock cores were sawn and polished for autoradiography. Distributions of porous minerals, grain boundaries and micro fissures were imaged on X-ray film (Kodak X-omat MA) for 2-5 days. These autoradiographs were digitized with a scanner (CanoScan 9900F, Canon, optical resolution 2400 dpi), and analyzed with Matlab 7.6 using the Image Processing Toolbox (The MathWorks, MA, USA).

The electron microscopy and energy dispersive X-ray analyses were performed on polished sections of the PMMA impregnated samples to study pore apertures of porous regions in greater detail, and to determine the corresponding minerals. These analyses were performed on a Hitachi FE-SEM S-4800 field emission scanning electron microscope, and the elemental composition of minerals was analyzed quantitatively on carbon-coated samples using an INCA350 EDS system with a UTW Si(Li) detector operated in the backscattered electron mode. In these measurements, 20 kV acceleration voltage and 20 μA beam current with a spot diameter of about 0.5 μm were used. Magnification by 250 - 500 times was used to reveal apertures down to a size of 1 μm.

In computed tomography (CT), cross-sectional images of the sample are determined by solving an inverse problem from transmission data, shadowgrams in the absorptive-contrast X-ray CT, taken in many directions. CT can thus be used to make a 3D image of the internal X-ray absorption distribution of the sample. As absorption is proportional to material density, this image also represents its 3D density distribution.

CT imaging was done at the University of Jyväskylä with a SkyScan 1172 µCT tabletop scanner that uses an X-ray tube with a spot size of less than 5 µm. The Feldkamp algorithm was used for the inversion problem because of the conical geometry of the X-ray beam. Conical beam has an advantage of varying scanning resolution with varying sample size. In SkyScan 1172, scanning resolutions of 0.9 µm to 20 µm can be used for maximal sample sizes of 2 mm to 68 mm, respectively. In this work a 14.4 µm voxel size was used to scan the 25 mm-diameter sample. After the 3D reconstruction, a non-linear variance-based filter was used to reduce noise in the images [9], so as to get better information about the 3D structures of pores and mineral distributions. The abundances of main minerals were also determined from these results.

Figure 1 shows the partitioning of the PMMA-impregnated samples for autoradiography and the dimensions of the µCT sample.

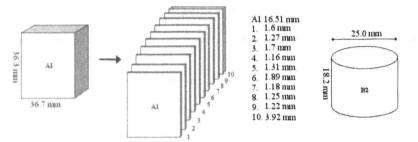

Figure 1. Left: Partitioning of sample SY7 A1 for autoradiography. Right: Tomographic imaging was done on a parallel sample SY7 B2.

RESULTS AND DISCUSSION

The PMMA porosity of SY7 A1 samples varied from 9 % to 13 % with an average porosity of 11.2 %. A 3D pore-structure construction of the (3.6x3.7x1.7) cm^3-sized sample A1 is shown in Figure 2. The structure of heterogeneous porosity is visualized by PMMA autoradiographs of the ten A1 subsamples of Figure 1. They cover a rock volume of 21 cm^3. By superimposing an autoradiograph with the respective rock surface, three different categories of porosity pattern can be distinguished (Figure 3 left). Porosities of 30 to 100% are congruent with sericitized plagioclase, and those of 5 to 10% are congruent mainly with dark minerals. Finally, porosities below 5% belong to plagioclase. The porosity distribution of subsample $A_1(10)$ of Figure 1 is shown in Figure 3 (right). The porosity of each pixel is determined, and the height of a column represents the relative proportion of the respective porosity among the pixel-based porosities.

Figure 4 shows two backscattered-electron images of the PMMA-impregnated SY7 A1 sample. These images include pores that are closed, i.e., not filled with PMMA (left), and pores that are connected, i.e., filled with PMMA (right). The pore apertures in all studied minerals varied from 1 to 300 µm. The main mineral components in altered tonalite are plagioclase, its alteration product sericite (>75%), and dark minerals: biotite and its alteration product chlorite (>20%). The accessory minerals are zircon, apatite, and sphene. According to SEM analyses the largest holes/pores, filled or unfilled by PMMA, were observed in the chlorite and sericitized plagioclase minerals.

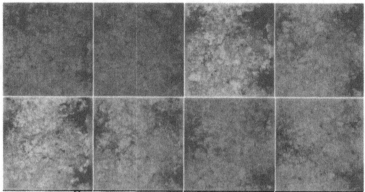

Figure 2. Eight ^{14}C-PMMA autoradiographs from SY7 A1 illuminate the 3D pore structure with an average depth resolution of ~ 1.5 mm. Subsample $A_1(1)$ is in the top left corner, followed horizontally by subsequent subsamples. White corresponds to 0 % and black to 100 % porosity.

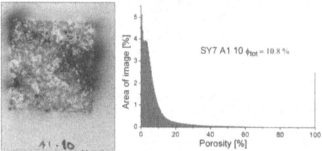

Figure 3. Left: A ^{14}C-PMMA autoradiograph of the subsample $A_1(10)$. Right: Porosity histogram of this area shows porosities up to 100%.

Figure 4. SEM image of albite plagioclase showing a cavity of 200 μm (left), and PMMA mixed with crushed mineral grains in a highly porous weathered phase (right). Width of images is circa 0.5 mm.

A 3D tomographic reconstruction of sample SY7 B2 helps to visualize its whole 3D structure. The μCT image in Figure 5 consists of about 1200 cross-sectional images like the one in Figure 6 (left). In such images we have a continuous distribution of grayscale values, but after segmentation of minerals we get a homogeneous coloring in each segmented region. In this case the sample was segmented visually into four regions, corresponding to pores and three different mineral components depending on the material density (Figure 6 right). Based on these images it is clear that most of the pores are located in plagioclase and its weathered product sercite, and that most of the denser minerals are surrounded by biotite and its weathered product chlorite. The SEM analyses corroborate μCT results: the largest pore openings are in sericite.

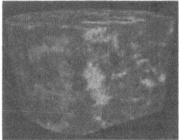

Figure 5. A 3D visualization of the μCT sample showing the void space (dark gray) and minerals divided into three main components. Light grayscale value corresponds to high density. The diameter of the sample is 25.0 mm

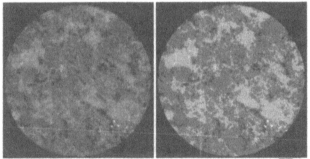

Figure 6. A typical μCT cross-section (left) and its segmented (right) image. Black corresponds to void and white to the densest mineral.

A detailed comparison of PMMA and tomography results shows that unweathered plagioclase (albite) corresponds to the lowest porosity phases (0-2%), biotite corresponds to porosities of 2-10%, and sericite appears in fracture zones showing porosities up to 100%. The total porosity determined from μCT images (5.3 %; Table I) is only about a half of PMMA porosity (11.2 %, on average), because the μCT method can only distinguish pores that are bigger than the scanning resolution, while the PMMA method detects even small scale porosity but averages it over pixel-sized areas. Based on this we can conclude that one half (six percent units) of the porosity is in pores whose apertures are under 14.4 μm.

Table I. Relative abundances of pores and three main minerals in the SY7 B2 sample determined from μCT images.

	Color	Relative abundance
Pores	Dark gray	5.3 %
Plagioclase and sercite	Gray	70.0 %
Biotite and chlorite	Light gray	24.4 %
Zircon, apatite and sphene	White	0.3 %

CONCLUSIONS

The PMMA method can resolve the mineral porosities, but pore apertures cannot be measured by PMMA. Pores down to 1μm aperture are easily seen by SEM, and EDAX allows detection of corresponding minerals. The benefit of combining μCT with PMMA + SEM is that the real 3D distribution of porosity can thus be determined. The best combinations of methods for tight granitic rock and for rock matrix adjacent to water conducting fractures are (1) mineral-specific porosity in 3D by combining μCT results for mineral composition with mineral specific porosity by PMMA, and (2) 3D mineral composition by μCT combined with mineralogy based on SEM, polarization microscopy and other petrographic analysis. Such combinations offer an excellent opportunity to study the structure and mineral composition of rock from micrometer to centimeter scales.

REFERENCES

1. K. H. Hellmuth, M. Siitari-Kauppi, and A. Lindberg, *J. Contam. Hydrol.* **13**, 403 (1993).
2. K. H. Hellmuth, S. Lukkarinen, and M. Siitari-Kauppi, *Isotopes Environ. Health Stud.* **30**, 47 (1994).
3. T. Lähdemäki, M. Kelokaski, M. Siitari-Kauppi, M. Voutilainen, M. Myllys, T. Turpeinen, J. Timonen, F. Mateos and M. Montoto in *Scientific Basis for Nuclear Waste Management XXX*, edited by D.S. Dunn, C. Poinssot, and B. Begg (Mater. Res. Soc. Proc. **985**,Warrendale, PA 2007) pp. 587-592
4. M. Montoto, A. Martínez-Nistal, A. Rodriguez-Rey, N. Fernández-Merayo, and P. Soriano, *J. Microscopy* **177** (2), 138 (1995).
5. E. Rosenberg, R. Ferreira De Paiva, P. Guéroult and J. Lynch, *International Symposium on Computerized Tomography for Industrial Applications and Image Processing in Radiology*, DGZfP-Proc. BB67-CD (1999).
6. K.-H. Hellmuth, M. Siitari-Kauppi, P. Klobes, K. Meyer, J. Goebbels, *Phys. Chem. Earth (A)* **24(7)**, 569 (1999).
7. V. Cnudde, B. Masschaele, M. Dierick, J. Vlassenbroeck, L. Van Hoorebeke, and P. Jacobs, *Appl. Geochem.* **21**, 826 (2006).
8. P. Anttila, A. Kuivamäki, A. Lindberg, M. Kurimo, M. Paananen, K. Front, P. Pitkänen, A. Kärki, Nuclear Waste Commission of Finnish Power Companies, Report YJT-93-19 (1993).
9. R. Gonzales and R. Woods, in *Digital Image Processing*, 2nd ed., edited by M:J. Horton (Prentice-Hall, New Jersey, 2002), p.239.

Mater. Res. Soc. Symp. Proc. Vol. 1124 © 2009 Materials Research Society   1124-Q07-05

Experimental Study and Modeling of Uranium (VI) Sorption Onto a Spanish Smectite

T. Missana, U. Alonso, M. Garcia-Gutierrez, N. Albarran, T. Lopez
CIEMAT, Departamento de Medioambiente, Avenida Complutense 22, 28040 Madrid (SPAIN)

ABSTRACT

Adsorption of uranium onto a Spanish smectite was studied, analyzing the effects of the most important parameters such as pH, ionic strength, radionuclide concentration and solid to liquid ratio. Batch sorption studies, in anoxic conditions under N_2 atmosphere, were carried out on the bentonite previously purified and converted into the homoionic Na-form.

In the sorption edges, two regions could be clearly distinguished. At pH lower than 5, sorption depended strongly on the ionic strength, possibly indicating the predominance of the uranyl ionic exchange process. At higher pH, sorption did not depend on the ionic strength but only on pH. The sorption behavior in this region suggested the predominance of a surface complexation mechanism. Sorption isotherms showed a non linear behavior in the concentration range used.

Sorption data were interpreted using a non electrostatic standard model combining surface complexation, with the weak and strong SOH sites of the clay, and ionic exchange. The acid – base properties of the weak SOH sites were determined by potentiometric titrations. The model used was able to reproduce, in a very satisfactory way, all the data in a wide range of experimental conditions.

INTRODUCTION

The sorption and transport properties of radionuclides in the near- and far-field barriers of a deep geological radioactive waste repository are amongst the principal aspects to be evaluated for the performance assessment (PA) of such disposal systems. The study of the clay materials is of particular importance since the backfill material is compacted clay in most designs and also because argillaceous formations are particularly suitable as host rock formations. In these systems, the transport of radionuclides is mainly a diffusion-controlled process retarded by sorption. For the description of sorption processes, PA models use the Kd-approach, which lumps together all the solid/water interactions. A mechanistic approach to sorption processes, which is based on a thermodynamic description of the radionuclide/solid interactions, is more appropriate to predict sorption behavior under variable conditions, and its implementation in performance assessment calculations would be of great interest instead of the usual "K_d-approach". For this reason, it is important to analyze in detail radionuclide sorption processes in clayey materials and to use appropriate mechanistic models for their description.

The Spanish FEBEX bentonite has been used over the last decade in different international research projects (FEBEX I and II, CRR, CFM, FUNMIG) as reference backfill material and its sorption properties were analyzed. Previous works studied the sorption, by a mechanistic approach, of Cs [1], Eu [2], Se [3] and divalent elements as Sr, Co and Ca [4]. No uranium sorption data on this clay were published before. Therefore, the main objective of the present work is to describe the sorption behavior of uranyl (U^{VI}) in the FEBEX clay and to use the simplest possible mechanistic model to interpret sorption data.

The FEBEX clay is mainly formed by smectite, a 2:1 layer type clay. In these clays, the main expected sorption mechanisms are cationic exchange and/or surface complexation that have to be independently evaluated by means of different sorption tests.

Previous studies on the sorption of U^{VI} onto different smectites exist [5, 6, and references therein] where, in fact, different sorption mechanisms were identified, including spectroscopic techniques.

In this work, several tests in a wide range of experimental conditions were carried out: pH varied from 3 to 10; ionic strengths from $3 \cdot 10^{-3}$ to $1 \cdot 10^{-1}$ M in $NaClO_4$; solid to liquid ratios from 1 to 6 g/L and radionuclide concentration $[U^{VI}]$ from $1 \cdot 10^{-8}$ to $1 \cdot 10^{-3}$ M. This approach provides a very large experimental data set for uranyl sorption onto smectite. The ultimate aim is to show that a simple model, combining surface complexation and ionic exchange processes, is capable of adequately describing the uranium sorption and can be used in PA.

EXPERIMENTAL

Materials

The smectite used in these experiments (FEBEX bentonite) comes from the Spanish deposit of Cortijo de Archidona. This clay contains mainly smectite (93±2%), with quartz (2±1%), plagioclase (3±1%), cristobalite (2±1%), potassium feldspar, calcite and trydimite as accessory minerals. The cation exchange capacity (CEC) of the FEBEX clay is 102 ± 4 meq/100g, the BET surface area is 33 m^2g^{-1}. Further details on this clay can be found elsewhere [7]. Prior to sorption experiments, the FEBEX clay was purified and homoionised in Na form. The colloidal clay fraction (size lower than 0.5 μm) was obtained by centrifuging (600xg, 10 min) and was used for sorption experiments. The suspensions were measured by photon correlation spectrometry (PCS), which is a technique that allows measuring the size and concentration of colloids. The hydrodynamic diameter of bentonite particle is around 300 nm.

The smectite content in this fraction is higher than 98 %. The homoionisation process did not significantly affect the main properties of the clay (CEC and surface area). The concentration of the clay in the suspensions was determined by gravimetry after drying the solid and the contact electrolyte in the oven 2 days at 105 °C.

The radionuclide used in this study was ^{233}U from a solution of $UO_2(NO_3)_2$ in 1M HNO_3. ^{233}U has a half-life of $1.59 \cdot 10^5$ years. Its alpha activity in solution was measured by liquid scintillation counting using a 2700 TR Packard liquid scintillation analyzer. The scintillation cocktail used is Beckman Ready Gel™. The volume of the Ready Gel is the same as that of the liquid sample. Typically, 2 mL of supernatant were used. In the case of low level activity samples (when high sorption existed), the volumes were doubled.

Sorption Experiments

Sorption experiments were carried out at room temperature under N_2 atmosphere in an anoxic glove box. Suspensions at different ionic strengths were used for the experiments. Sorption edges were carried out changing the pH from pH 3 to 11 with NaOH or HCl 0.1 M. Sorption isotherms (22 ± 2 °C) were carried out at a fixed pH and fixed background electrolyte concentrations and varying the radionuclide concentration. The highest concentrations (> $1 \cdot 10^{-6}$ M) were achieved by adding, in addition to the radiotracer, $UO_2(NO)_3$ salts of high purity (Merck). The pH was readjusted, if necessary, after the radionuclide addition.

The suspension at the selected pH was introduced in 12.5 ml ultracentrifuge tubes which were sealed and maintained in continuous stirring for 7 days. This contact time was selected after kinetic studies. The separation of the clay from the supernatant was carried out by ultra-centrifuging (645000xg, 30 min) with a Beckman L90-Ultra ultracentrifuge. PCS measurements indicated that after centrifuging the concentration of colloids is under the detection limit (0.1 ppm).

After the solid separation, three aliquots of the supernatant from each tube were extracted for the analysis of the final activity; the remaining solution was used to check the final pH.

Mean distribution coefficients (K_d) were calculated from the three aliquots of the supernatant with this formula:

$$K_d = \frac{C_i - C_f}{C_f} \cdot \frac{V}{m} \tag{1}$$

where C_i is the initial activity (counts/ml), C_f the final activity in the supernatant (counts/ml), m the mass of the solid (gr) and V the liquid volume (ml). Sorption on the ultracentrifuge tubes was checked after the sorption tests; uranium sorption was lower than 1 %, and it was not accounted for in K_d calculations.

MODELING

As previously mentioned, in clays, both ionic exchange and surface complexation may contribute to different extents to the sorption of radionuclides; thus, it is important to differentiate these contributions. The ionic exchange reactions between a cation B with charge z_B which exists in the aqueous phase, and a cation A, with charge z_A, at the clay surface ($\equiv X$) is defined by:

$$z_B A \equiv X + z_A B \Leftrightarrow z_A B \equiv X + z_B A \tag{2}$$

The cation exchange reactions can be described in terms of selectivity coefficients, following the Gaines and Thomas definition [8].

At the edge sites (SOH), the pH-dependent charge is determined by the following protonation / deprotonation reactions:

$$SOH_2^+ \Leftrightarrow SOH + H^+ \qquad K_{a1} \tag{3}$$

$$SOH \Leftrightarrow SO^- + H^+ \qquad K_{a2} \tag{4}$$

where SOH_2^+, SOH and SO^- represent the positively charged, neutral and negatively charged surface sites, respectively, and K_{a1} and K_{a2} are the intrinsic equilibrium acidity constants. The specific adsorption of cations at these surface functional groups can be described by reactions of the type:

$$SOH + M^{z+} \Leftrightarrow SOM^{z-1} + H^+ \tag{5}$$

The acid-base properties of the clay edge sites are generally determined by potentiometric titrations. The acid-base characteristics of the Na-smectite used here were previously analyzed using a non electrostatic model [1]. Also sorption modeling was carried out using a non electrostatic model. The pKs of protonation / deprotonation obtained for the FEBEX Na-smectite are included in Table I.

RESULTS AND DISCUSSION

Figure 1 (left) shows the sorption edges obtained at the investigated different ionic strengths using a similar solid to liquid ratio (~ 1g/L and $[U^{VI}]=4.4 \cdot 10^{-7}$ M). Two regions can be clearly distinguished: the first one, at acidic pH, where sorption strongly depends on the ionic strength and a second one, at basic pH where sorption does not depend on the ionic strength but only on pH. These results indicate that two mechanisms are taking place; ionic exchange might be the predominant one, at acid pH, whereas surface complexation seems to be the main sorption mechanism at high pH. In general, there was no evidence that the solid to liquid ratio affected sorption, at least at this U concentration.

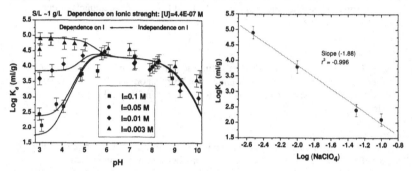

Figure 1: <u>Left:</u> Sorption edges at different ionic strengths. Curves superimposed on the experimental points correspond to the model calculated using the parameters of Table I. <u>Right:</u> dependence of K_d values in the acidic pH region (pH~3) with the ionic strength.

Table I: Parameters used for modeling of uranyl sorption onto the FEBEX smectite.

BET: 33 m^2/g ; CEC: 102 meq/100 g = 30.91 µeq/m^2		
S_WOH density: 1.82 µeq/m^2	S_SOH density: 0.061 µeq/m^2	
SURFACE ACIDITY		
Specie	**Composition**	**LogK**
S_WO^-	S_WOH, -H^+	-8.4 [4]
$S_WOH_2^+$	S_WOH, +H^+	5.3 [4]
S_SO^-	S_SOH, -H^+	-9.9 [4]
$S_SOH_2^+$	S_SOH, +H^+	4.8 [4]
SURFACE COMPLEXATION MODEL		
Specie	**Composition**	**LogK**
$S_WOUO_2^+$	S_WOH, -H^+, 1UO_2^{2+}	-1.5 (this work)
S_WOUO_2OH	S_WOH, -2H^+, 1UO_2^{2+}, 1H_2O	-5.8 (this work)
$S_SOUO_2^+$	S_SOH, -H^+, 1UO_2^{2+}	2.5 (this work)
S_SOUO_2OH	S_SOH, -2H^+, 1UO_2^{2+}, 1H_2O	-3.3 (this work)
IONIC EXCHANGE		
Specie	**Composition**	**LogK**
=X_2UO_2	2XNa, -2Na^+, 1UO_2^{2+}	0.28 (this work)

The logarithm of the K_d values obtained at acidic pH (~pH 3) were plotted as a function of the [Na] concentration in solution as shown in Figure 1 (right). A linear relation with a slope of -1.88 is found. The slope of the curve is approximately -2, in agreement with an exchange process at the bentonite surface between Na and a bivalent cation. The small deviation to the theoretical behavior might be due to the presence of some ions from bentonite dissolution (for example Ca^{2+}) competing for sorption [2]. From these experimental data the selectivity constant for UO_2^{2+} in the reaction $2XNa + UO_2^{2+} \Leftrightarrow X_2UO_2 + 2Na^+$ could be determined, following the methodology described in [9] and is indicated in Table I.

Figure 2 (left) shows the sorption isotherms obtained at $1 \cdot 10^{-1}$ and $1 \cdot 10^{-2}$ M in $NaClO_4$ and two different pH. In Figure 2 (left) data are expressed as $Log[U]_{ads}$ vs. $Log[U]_{eq}$ the straight lines that fit the experimental points always have a slope less than one indicating that sorption is not linear. Figure 2 (right) shows only the isotherms obtained at 0.1 M and the data are more clearly expressed as the logarithm of K_d vs. $Log[U]_{eq}$. The non linearity of sorption is evidenced by the fact that K_d values are decreasing as the concentration increase. The fact that sorption is not linear over the entire range of concentrations investigated might be caused by the existence of two different types of complexation sorption sites for uranium, as observed in the FEBEX clay for other elements [2]. In the sorption model, two complexation sites will thus be considered (S_WOH, for weak and S_SOH, for strong sites). The acid-base properties of the weak sites can be determined by potentiometric titrations (Table I), whereas the properties of strong sites, whose density is very small, are usually determined by model fit [4]. The density of S_SOH in the FEBEX Na-smectite was previously experimentally determined to be 0.061 $\mu mol \cdot m^{-2}$ [2]. The pK of protonation/deprotonation of these strong sites, obtained by the best fit of experimental data in [2], is included in Table I.

It is generally accepted that the dominant charged aqueous species are those that most likely are involved in the formation of surface complexes. Aqueous speciation as a function of the pH for uranium concentration of $4.4 \cdot 10^{-7}$ M and ionic strength $1 \cdot 10^{-1}$ M was calculated.

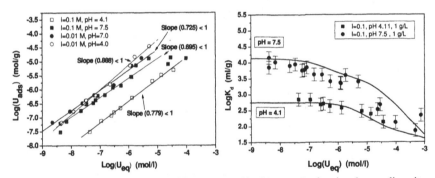

Figure 2: <u>Left</u>: sorption isotherms at different pH and ionic strength, showing the non-linearity <u>Right</u>: Sorption isotherms (K_d vs U final concentration) at I=0.1 M and pH 4.1 and 7.5; the curves correspond to modeling performed with the parameters in Table I.

A zero partial pressure of CO_2 was supposed and the sets of constants of Grenthe [10] included in the EQ3/6 data base of the CHESS code [11] was used. CHESS was used for thermodynamic calculations and sorption modeling. The dominant species at pH lower than 5 are UO_2^{2+} and UO_2OH^+. At pH between 5.5 and 8.5, $UO_2(OH)_2^0$ becomes the major dissolved species, whereas at higher pH the negatively charged $UO_2(OH)_3^-$ predominates.

For surface complexation, the two major species UO_2^{2+} and UO_2OH^+, were assumed to react with the surface sites, SOH (weak and strong):

$$SOH + UO_2^{2+} \Leftrightarrow SOUO_2^+ + H^+ \qquad\qquad K_c^1 \qquad\qquad (6)$$

$$SOH + UO_2OH^+ \Leftrightarrow SOUO_2OH + H^+ \qquad\qquad K_c^2 \qquad\qquad (7)$$

The parameters used for the modeling are all summarized in Table I. The model fit is shown superimposed to the experimental points in Figure 1 (left, sorption edges) and Figure 2 (right, sorption isotherms).

CONCLUSIONS

Uranium (VI) adsorption onto the FEBEX Na-smectite colloids was studied under a wide range of pH, ionic strengths and uranium concentrations. A simple model combining both surface complexation and ionic exchange was applied which succeeded in reproducing satisfactorily all the experimental data. The mean selectivity coefficient determined for the Na-UO_2^{2+} exchange was ($Log\,_{Na}^{U}K_{SEL} = 0.28$). Sorption by surface complexation was explained by a non-electrostatic model with the formation of the complexes $SO\text{-}UO_2^+$ and $SO\text{-}UO_2OH$ with both weak and strong SOH surface sites.

ACKNOWLEDGMENTS

This work has been supported within the frame of CIEMAT-ENRESA association, and by the Spanish Ministry of Education and Science under the grant CGL2005-01482/BTE (PROMICOL).

REFERENCES

1. T. Missana, M. Garcia-Gutierrez, V. Fernandez and P. Gil, *Application of mechanistic models for the interpretation of radionuclides sorption in clays. Part 1.* CIEMAT/DIAE/54610/03, (Madrid-Spain, 2002), pp. 1-47.
2. T. Missana, U. Alonso, M. Garcia-Gutierrez, N. Albarran and T. Lopez. *Experimental study and modeling of europium adsorption onto smectite clay colloids.* CIEMAT /DMA /2G102 /05/08 (Madrid-Spain, 2008), pp. 1-28.
3. T. Missana, U. Alonso and M. Garcia-Gutierrez, *J. of Colloid Interface Sci.*, Submitted.
4. T. Missana and M. Garcia-Gutierrez, *Phys. Chem. Earth*, **32**, 559 (2007).
5. G.D Turner, J.M. Zachara, J.P. McKinley and S.C. Smith, *Geochim. Cosmochim. Acta*, **60(18)**, 3399 (1996).
6. A. Kowal-Fouchard, R. Drot, E. Simoni and J.J. Ehrhardt, *Environ. Sci. Technol.*, **38**, 1399 (2004).
7. A. M. Fernandez, B. Baeyens, M.Bradbury, P. Rivas, *Phys.Chem. Earth*, **29(1)**, 105 (2004).
8. G.I. Gaines and H.C. Thomas, *J. Chem. Phys.*, **21,** 714 (1953).
9. M.H. Bradbury and B. Baeyens, *Sorption by Cation Exchange: Incorporation of a cation exchange model into geochemical computer codes.* PSI Bericht 94-07 (Villingen-Switzerland, 1994).
10. I. Grenthe, *Chemical thermodynamics of uranium*, North Holland Elsevier (1992).
11. J. van der Lee and L. DeWindt, *CHESS tutorial and cookbook.* Technical Report École des Mines, LHM/RD/99/05 (Paris-France,1999).

Mater. Res. Soc. Symp. Proc. Vol. 1124 © 2009 Materials Research Society   1124-Q07-12

Influence of Humic Acid on Se and Th Sorption on Sedimentary Rock

Y. Seida[1], Y. Tachi[1], A. Kitamura[1,] T. Nakazawa[2] and N. Yamada[2]
[1] Geological Isolation Research and Development Directorate, Japan Atomic Energy Agency, Tokai, 319-1194, Japan
[2] Mitsubishi Materials Corporation, Naka, 311-0102, Japan

ABSTRACT

The influence of humic acid (HA) on solubility of radioactive elements [Se(IV), Th(IV)] and on their retardation parameters in sedimentary rock under groundwater conditions was studied using sedimentary rocks obtained from Horonobe URL site and commercially supplied humic acid as a model organic substance. The retardation parameters were evaluated with sorption and diffusion experiments as a function of HA concentration. The distribution coefficient, K_d, of the rock for Se(IV) in the presence of HA was around 2 ml/g and little influence of the HA on the K_d was observed. On the other hand, the K_d for Th(IV) was estimated to be around 1000 ml/g below 100 mg/l of HA. Influence of HA on the K_d was observed when the concentration of HA was 100 mg/l. Sorption of Th(IV) was largely hindered due to complexation with the dissolved HA and the partition of HA when the concentration of HA is high. The through-diffusion experiment showed that the presence of HA did not affect the diffusion of Se(IV).

INTRODUCTION

Organic substances dissolved in groundwater or sorbed onto natural rocks are among the uncertain factors those may influence migration of radioactive elements and the retardation capacity of the rocks in deep geological disposal of high level radioactive waste considered in Japan. Understanding the influence on the retardation of radioactive elements in the barrier system is important to reducing uncertainties for reliable performance assessment. The influence on the retardation capacity of barrier materials such as natural rocks will depend on characteristics of the organic substances, partition of the radioactive elements in the solute-organic substance-rock ternary system, and pore structure of rock. However, details of them have not been well understood, at least for Se(IV) and Th(IV) in sedimentary rock [1-3].

In the present study, the influence of humic acid on sorption and diffusion of Se(IV) and Th(IV) in sedimentary rock under groundwater conditions was examined. Distribution coefficients and diffusion coefficients for Se(IV) and Th(IV) were measured by batch sorption and through-diffusion experiments as a function of HA concentration.

EXPERIMENTAL

The sedimentary rock used in the present study was obtained from Horonobe URL site in Hokkaido, Japan (Borehole HDB-10, 500m depth in Wakkanai formation, CEC = 20 meq/100g, specific surface area = 90 m^2/g). Major minerals are quartz, smectite, illite and pyrite. Powdered (below 200 μm in diameter) and core (60 mmϕ x 5 mm long) samples of the rock were used in the sorption and diffusion experiments, respectively. Synthetic groundwater with an ionic composition similar to that of Horonobe URL site's was used as a pore water (Table I). The synthetic groundwater was a high salinity solution with high NaCl content [ionic strength (IS) = 0.23]. Humic acid (HA) supplied from Aldrich (~150 kDa, sodium salt) was used as a model organic substance dissolved in the groundwater. The HA was purified before use according to a conventional procedure. ThO_2 and SeO_2 were used in the preparation of test solutions of Th(IV) and Se(IV), respectively. The metal oxides were first dissolved by an aliquot of nitric acid and then mixed with the synthetic groundwater with pH adjustment by HCl and NaOH. All chemicals used in the present study were reagent grade.

Table I. Composition of synthetic groundwater

pH	8.2
Element	[mg/l]
Na^+	4761
K^+	98
Li^+	17
Ca^{2+}	84
Mg^{2+}	140
SO_4^{2-}	110
Cl^-	7420
TIC	194

Stability of test solutions

Due to the large ionic strength of the synthetic groundwater used, the prepared test solutions were monitored in advance for over one month in terms of solution pH, total organic carbon (TOC) and the concentrations of Se(IV) and Th(IV). The TOC and the concentration of each element were analyzed by means of TOC analyzer (Shimazu co. ltd., TOC-5000A) and ICP-MS (Yokogawa HP-4500). The pH, TOC and concentration of Se(IV) and Th(IV) remained constant for more than one month, indicating they are stable in the synthetic groundwater from a viewpoint of dissolution. Se(IV) and Th(IV) do not precipitate nor are consumed by sorption on container material, indicating that the test solutions will be stable over the periods of sorption and diffusion experiments performed in the present study. Se(IV) was considered to be dissolved as SeO_3^{2-} in the system under study. Th(IV) was estimated to be dissolved mainly as $Th^{IV}(OH)_3CO_3^-$, which was estimated from pH dependent speciation diagrams obtained by geochemical equilibrium calculation using PHREEQC with the JNC-TDB thermodynamic database.

pH dependence of HA-nuclide interaction

Interaction between the HA and each element in the synthetic groundwater was evaluated from the viewpoint of molecular size by means of filtration of their mixtures and analysis of the filtrates using 0.45μm pore membrane (ADVANTEC DISMIC) and ultrafiltration membrane [ADVANTEC USY-1, molecular weight cut off (MWCO) = 10000 (2 nm)]. The series of pH adjusted test solutions containing 100 mg/l HA and 10^{-4} M Se(IV) or 10^{-6} M Th(IV) were kept for 18 days at room temperature with shaking. The concentration of each solute in the filtrates was measured by UV-VIS spectrophotometry for HA (Shimazu UV-2400, λ=400nm) and ICP-MS for Se and Th after filtration by the respective membranes. The radioactive elements that interact with the HA via complexation will be excluded with HA by the filtration, depending on its size, by which the complexation/interaction of Se(IV) and Th(IV) with HA was evaluated.

Sorption and Diffusion experiments

Se(IV) and Th(IV) sorption behavior on the sedimentary rock in the presence of HA was observed by batch sorption experiments using the powdered rock sample according to a standard test method [4]. Liquid to solid ratio of the test solutions was 10 ml/g. The concentration of HA was varied from 1 to 100 mg/l [Se(IV)=10^{-4}M, Th(IV)=10^{-6}M]. Each test solution with the powdered rock was kept at room temperature for 7 days with shaking. The concentrations of HA and radioactive elements in the solution were analyzed by UV-VIS spectrophotometry and ICP-MS. The distribution coefficient, K_d of the rock for the elements was calculated based on mass balance.

The through-diffusion experiment was based on a standard test method [4]. The test solution containing 100 mg/l HA and each radioactive element [Se(IV)=10^{-4}M and Th(IV)=10^{-6}M] was introduced in the upper reservoir of a diffusion cell. From the obtained breakthrough curve, the D_e of diffusant was determined by steady-state slope analysis and curve fitting with an analytical model calculation of unsteady-state diffusion developed by Zhang et al. [5].

RESULTS

Figure 1 shows the concentration of HA in the filtrates obtained by filtration of the series of test solutions containing HA (initial content = 100 mg/l) and each nuclide. The concentration of HA in each filtrate treated by 0.45 µm membrane filtration (MF) and ultra-filtration (UF) is shown for the pH series of solutions. The concentration of HA in the absence of the nuclides is also shown in the figure by dotted lines. Around 60% of HA was excluded by the ultra-filtration (UF), irrespective of the coexistence of nuclides. This means that 60 % of the HA has molecular sizes between 0.45 and 0.002 µm in the solution. The influence of coexistence of Se(IV) or Th(IV) on the size/dissolution of HA was not observed by the present resolution of filtration.

Figures 2(a) and (b) show the concentrations of Se(IV) and Th(IV) for the series of filtrates of test solutions. Concentration of Se(IV) in the filtrates (in both MF and UF treatments) was not changed for the series of solution pH examined, indicating little influence of HA on the solubility of Se(IV). On the contrary, Th(IV) concentration was reduced by the UF treatment, indicating an interaction/complexation of Th(IV) with HA and/or aggregation of Th(IV). Almost one tenth of Th(IV) was excluded by the UF treatment at each solution pH, indicating that most of Th(IV) formed complexes with HA.

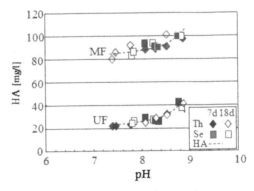

Figure 1 Concentration of HA in the filtrates of each test sample
with Se(IV) and Th. (HA:100 mg/l, Se(IV):10^{-4} M, Th:10^{-6} M)

Figures 3(a) and (b) indicate the influence of HA on the K_ds of each nuclide for the rock. In the case of Se(IV), the K_d was small, at 2 ml/g, and the influence of HA coexistence on the K_d seems to be very small. The K_d for Th(IV) was more than 1000 in the dilute HA system. Due to detection limit of the ICP-MS, the Kd was derived based on the detection limit of the detector. The influence of HA on the K_d for Th(IV) was not observed clearly below 100 mg/l of HA. Below the 100 mg/l of HA, almost all of the introduced HA could be sorbed onto the sedimentary rock. 1 mg/l of detected HA in the initial HA range below 10 mg/l as shown in Figure 3 will be due to organic substances originated from the rock. Most of the Th(IV) interacting with the added HA sorbs to the rock along with the sorption of introduced HA. In the case of 100 mg/l HA, some amount of HA remains in the solution due to its partition equilibrium and the Th(IV) complexed with the HA also remains in the solution. The result indicates that partitioning of Th(IV) that complexed strongly with HA will be controlled by the sorption behavior of HA. The K_d of Th(IV) in the absence of HA was also more than 1000 ml/g [6].

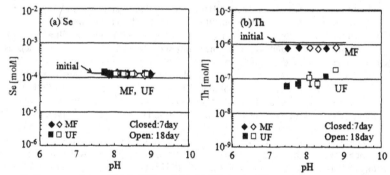

Figure 2 Concentrations of (a) Se(IV) and (b) Th(IV) in the filtrates of each test solution. (HA:100 mg/l, initial Se=10^{-4} M, initial Th=10^{-6} M)

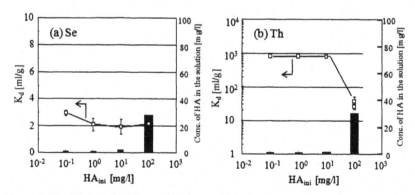

Figure 3 Distribution coefficient of (a) Se(IV) and (b) Th(IV), and the concentration of HA in the solution after the sorption equilibrium, as a functions of initial HA.

Figure 4 shows the D_e of Se(IV) as a function of HA concentration. Consistent with the results obtained by the above mentioned stability and sorption experiments [little interaction between Se(IV) and HA and little influence of HA on the K_d], the D_e of Se(IV) was not influenced by the HA over the range of HA concentration examined. Carbonate ions affect the solubility, sorption, and diffusion behavior of some actinide and lanthanide metal ions. In the case of a NaHCO₃ solution instead of the synthetic groundwater with common carbonate content but small ionic strength (IS = 0.016), decrease of D_e was observed (Figure 4). This phenomena was considered to be due to the increase of the electric double layer (Debye length) of clay minerals with anion exclusion in the rock, indicating a diffusion pathway of Se(IV) in the clay minerals in the rock.

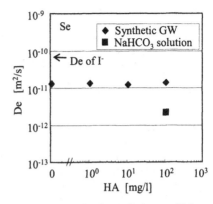

Figure 4 The effective diffusion coefficient, De, of Se(IV) in the sedimentary rock in the presence of HA

Carbonate ions will not affect the De in the present system. The breakthrough of Th (IV) has not been observed yet (after 1 year), although the breakthrough of HA was observed 8 months into the experiment. The mechanism of HA effects on Th retardation is now under study and the diffusion experiment has been continued for Th(IV)

DISCUSSION

Both elements exist as anions in the solutions used in the present study. Se(IV) does not interact with the HA, but Th(IV) does. The exclusion of Th(IV) by the UF in the presence of HA obviously indicates the complexation of Th(IV) with the HA as reported elsewhere. Th(IV) complexation with HA through ligand-exchange with carboxylate in the HA was proposed by Schild [7]. The formation of Th(IV)-HA complex will result in slow diffusion of Th(IV) due to the size of the complex. The sedimentary rock used in the present study includes connected macropores with sub-micron meter size and 95% of several nanometer size micropores [8]. The size of macropores in the rock is rather large relative to the molecular size of HA (60% of HA was above 2 nm in diameter). The HA fraction with small molecular weight will migrate preferentially and will breakthrough faster than the other HA [9]. Th(IV) may not form complex with the small molecular weight HA. At least, significant reduction of the retardation function of the rock in the presence of <100mg/l HA has not been observed under the present condition for Th(IV). The HA used in the present study was a model HA and is not the HA expected *in-situ*. The introduced HA may reveal sorption processes independent of the *in-situ* HA sorbed on the rock originally. This situation will be important to consider when examining the influence of HA on the retardation function of rock using foreign HA (HA obtained from other source but not in-situ). The foreign HA will form complexes with nuclide elements similar to the *in-situ* HA, but stability constants of the foreign HA will be different from those of *in-situ* HA. Those factors could cause over- or under- estimation of the influence of HA on the retardation function of rock.

It is generally accepted that *in-situ* HA is better for the study to reduce any uncertainties caused by the use of foreign HA. According to our results, sorption of Th(IV) that forms complexes with HA is mostly controlled by the sorption behavior of the HA onto the rock. Namely, when the HA distributes between solution and rock, the Th(IV) also distributes between solution and rock depending on the sorption behavior of HA, resulting in the decrease of retardation due to the distribution. We are continuing the Th(IV) diffusion experiment and the influence of HA on the diffusion will be reported in a later paper.

CONCLUSIONS

Existence of HA in groundwater does not affect the solubility, K_d and D_e of Se(IV) significantly under atmospheric conditions due to little interaction of Se(IV) with the HA used in the present study. Th(IV) interacts with the HA through complexation, resulting in the increase of molecular size and the decrease of K_d at higher HA concentration due to the partition of HA sorption onto the rock. The complexation of Th(IV) with HA is a potential factor that influences the retardation performance of rock for Th(IV).

ACKNOWLEDGMENTS

This study was financed by the Ministry of Economy, Trade and Industry of Japan. The authors appreciate it very much to Prof. T. Sasaki, K. Iijima and M.Terashima for valuable discussion.

REFERENCES

1. P. Reiller, V. Moulin, F. Casanova and C. Dautel, *Radiochim Acta*, **91**, 513 (2003).
2. T. Zuyi, C. Taiwei, D. Jinzhou, D. XiongXin, G. Yingjie, *Appl. Geochem.*, **15**, 133 (2000).
3. K. L. Nash. and G. R. Choppin, *J. Inorg. Nucl. Chem.*, **42**, 1045(1980).
4. AESJ, *Measurement Method of the Distribution Coefficient on the Sorption Process*, Standard AESJ-SC-TR001 (Atomic Energy Society of Japan, Tokyo, 2006).
5. M. Zhang, M. Takeda and H. Nakajima, in Scientific Basis for Nucler Waste Management XXIX, ed by P. V. Iseghem (*Mater. Res. Soc. Symp. Proc.*, **932**, Warrendale, PA, 2006) pp.135-142.
6. H20 report of "Project for Assessment methodology Development of Chemical Effects on Geological Disposal System", funded by Ministry of Economy, Trade and Industry of Japan, (Japan Atomic Energy Agency, 2009), in press.
7. D. Schild and C. M. Marquardt, *Radiochim. Acta*, **88**, 587(2000).
8. H. Takahashi and Y. Seida, *Proc. 2008 Fall Meeting of Atomic Energy Society of Japan*, L47, CD-ROM (Atomic Energy Society of Japan, 2008).
9. K. Iijima, S. Kurosawa, M. tobita, S. Kibe and Y. Ouchi, *Proc. 2008 Fall Meeting of the Atomic Energy Society of Japan*, L54, CD-ROM, (Atomic Energy Society of Japan, 2008).

Mater. Res. Soc. Symp. Proc. Vol. 1124 © 2009 Materials Research Society

Sorption and Diffusion of Cs in Horonobe-URL's Sedimentary Rock: Comparison and Model Prediction of Retardation Parameters From Sorption and Diffusion Experiments

Yukio Tachi, Yoshimi Seida, Reisuke Doi, Xiaobin Xia and Mikazu Yui
Geological Isolation Research and Development Directorate, Japan Atomic Energy Agency,
4-33 Muramatsu, Tokai-mura, Ibaraki 319-1194, Japan

ABSTRACT

Sorption and diffusion of Cs in the sedimentary rock from the Horonobe generic URL were studied from the viewpoints of reliability of experimental evaluation and model prediction, focusing on its applicability to intact systems. The distribution coefficient, K_d, for Cs was measured under the same chemical conditions by both batch sorption and diffusion experiments. To obtain reliable parameters of sorbing species for intact rock, through-diffusion experiments coupled with multiple curve analysis including tracer depletion, breakthrough and depth-concentration curves were examined, and resulted in good agreement with those predicted by conventional transport models using only one set of retardation parameters. The K_d values obtained by the diffusion tests using intact rock were consistent with those obtained by the batch tests with crushed rock. The sorption behavior was modeled by considering additive ion-exchange reactions for illite and smectite, which were assumed to be dominant sorption minerals based on microscopic observation and their known sorption mechanism. The model predicted the K_d values obtained by the series of experiments reasonably well.

INTRODUCTION

Sorption and diffusion of radionuclides in the deep geological environment are key processes to be understood for the safe geological disposal of high-level radioactive waste (HLW). To set reliable parameters for the safety assessment, it is necessary to establish a reliable method for experimental evaluation and model prediction of sorption and diffusion. One of the key challenges for this issue is the applicability and up-scaling of data from dispersed batch systems to intact/compacted systems, the details of which have been discussed by many authors [e.g., 1-3].

The aim of the present study is to provide a comprehensive experimental and modeling approach to evaluate the retardation parameter for the sedimentary rock in the Horonobe generic underground research laboratory (URL) in the northernmost region of Hokkaido, Japan. Cesium was selected because it is a key radionuclide in the safety assessment of HLW, a simple and moderately sorbing species, and many experimental and modeling studies have been previously undertaken with Cs [e.g., 3-7]. To examine the reliability of experimental evaluation and the consistency between batch and intact systems, distribution coefficients, K_d, were measured under the same chemical conditions by both batch sorption and diffusion experiments. In the diffusion experiment, the effective diffusion coefficient, D_e, and K_d were derived from through-diffusion tests with variable concentrations in the reservoirs by coupling the multiple curve analysis. The sorption parameters obtained from batch sorption and diffusion methods were compared, and were also interpreted under the assumption of dominant sorption minerals based on microscopic observation and their known sorption mechanism.

EXPERIMENTAL

Materials

Two rock samples were extracted from cores of borehole HDB-6 at two different depths (-512 m and -596 m) in the Wakkanai formation. Both samples are siliceous mudstone and are similar in appearance and characteristics. The mineral assemblage identified by XRD and chemical composition data are qualitatively evaluated to be opal A, opal CT, quartz, feldspars, clay minerals and pyrite, and are quantitatively analyzed to contain 7% illite and 11% smectite [8]. The CEC measured by the ammonium acetate method is 19.6 meq/100g. The dry density and porosity measured by the water saturation method are 1.55 g cm^{-3} and 37.5 %, respectively [9]. The pore size distribution by Hg porosimetry indicates a unimodal curve with a large percentage of the pore sizes below 100 nm.

The synthetic groundwater used in the experiments was prepared as in the previous study [5], which is based on a combination of measured chemical composition and geochemical calculations. It was prepared at ionic strength of 0.41 M and a pH of 8.3 by the dissolution of salts in distilled water as follows; 0.31 M NaCl, 0.06 M NaHCO$_3$, 0.03M NH$_4$Cl, and 0.01M KBr.

Batch sorption experiment

Batch sorption experiments were conducted with a crushed rock sample, sieved to collect particles < 250 μm, at room temperature under aerobic conditions. The samples were contacted with synthetic groundwater at a liquid to solid ratio of 30 ml g^{-1} in polypropylene vessels. The initial concentrations of Cs ranged between 5x10^{-6} and 2x10^{-4} mol L^{-1}. The vessels were shaken frequently and pH values of the solutions were periodically monitored to be around 8.3. Cs sorption as a function of time was measured to confirm sorption kinetics and equilibrium. After the experimental period of 14 days, which kinetics tests indicated was sufficient to reach sorption equilibrium, the solution was filtered through a 0.45 μm membrane filter and analyzed by ICP-MS. All tests were conducted in duplicate in parallel with blank tests without rocks.

Diffusion experiment

The diffusion experiments were conducted by using a through-diffusion type cell as shown schematically in Figure 1. The conventional through-diffusion experiment ideally needs to maintain a constant concentration in the reservoirs and to reach steady-state diffusion. It is, however, difficult for sorbing tracers to attain these requirements because of the complex experimental procedure and longer experimental runtimes [10, 11]. In this study, the through-diffusion experiment with variable concentrations in the reservoirs was performed. The tracer concentrations in both reservoirs were allowed to change and monitored continuously, and multiple

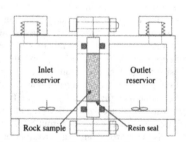

Figure 1. Through-diffusion cell.

curve analysis was applied to obtain reliable retardation parameters. The sample (32 mm in diameter and 10 mm thick) was cut from the core in the direction parallel to the core axis and was sandwiched between two continuously stirred reservoirs, each containing 190 ml of the synthetic groundwater. The initial concentration of Cs in the inlet reservoir was 3.7×10^{-5} mol L^{-1}. Diffusion tests were conducted in triplicate for each rock sample. The Cs concentrations in both reservoirs were periodically sampled and measured by ICP-MS until the Cs concentration in the outlet reservoir reached a near linear increase. After an additional diffusion time, the diffusion cell was disassembled and the rock sample was successively ground to powder from the surface at 1-2 mm intervals. Then each powder was immersed in 3 M HNO_3 solution, and analysed by the ICP-MS. The depth profile of Cs concentration was obtained by correcting for extraction efficiency.

Microscopic observations

To identify the dominant minerals for Cs sorption, microscopic observations were conducted for a rock sample from borehole HDB-3 located near HDB-6, which has a similar mineralogy to HDB-6. Electron probe microanalysis (EPMA) was used for morphological observations and quantitative chemical analysis. After reaction of crushed rock samples with diluted groundwater including Cs, the rock samples were separated from the liquid phase and dried, embedded in resin, and gold coated for EPMA analysis.

RESULTS AND DISCUSSION

Sorption kinetics and isotherms

The results of the sorption kinetics test showed that one week was sufficient time to achieve sorption equilibrium. From batch experiments, K_d was determined as the ratio of equilibrated Cs concentrations sorbed in the solid phase and dissolved in the liquid phase. The K_d values are shown in Table I for different dissolved Cs concentrations. At lower Cs concentration, K_d values are nearly linear, showing constant sorption, and K_d values decrease at higher concentration. There was no significant difference in the K_d values between the two rock samples.

Diffusion analysis and consistency of retardation parameters

The diffusion and retardation parameters for the rock were evaluated by the conventional diffusion model in porous media. Fick's diffusion equation combined with a linear sorption isotherm was applied, given by:

$$\frac{\partial C_p}{\partial t} = \frac{D_e}{\alpha} \frac{\partial^2 C_p}{\partial x^2}, \quad \alpha = \varepsilon + \rho K_d \quad (1)$$

where C_p is the concentration in pore water (mol L^{-1}), D_e is the effective diffusion coefficient ($m^2 s^{-1}$), α is the rock capacity factor, ε is the porosity, ρ is the dry density (g cm^{-3}), and K_d is the distribution coefficient (ml g^{-1}). An example set of the tracer-depletion curve in the inlet reservoir and break-through curve in the outlet reservoir is shown in Figure 2(a). These curves are analyzed by the rigorous analytical solution of equation (1) derived by [11] for through-diffusion experiments with variable concentrations in the reservoirs. Both curves are fitted

575

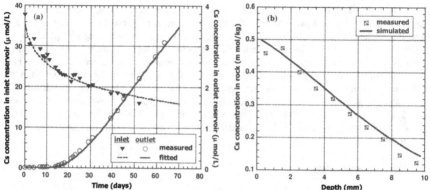

Figure 2. Typical results of Cs through-diffusion tests; (a) reservoir-depletion (inlet reservoir) and breakthrough curves (outlet reservoir), (b) depth profile in rock samples (No.596-2).

simultaneously by the numerical method proposed in [11], and D_e and α are determined. The depth profile obtained for the rock sample showed a linear profile and good agreement with the curve calculated by the analytical solution with the same parameters derived from the above fitting as shown in Figure 2(b). The tracer-depletion, breakthrough curves and depth profile were all simulated simultaneously by a conventional transport model with one set of diffusion and linear sorption parameters. The proposed approach using through-diffusion with variable concentrations in the reservoirs and multiple curve analysis for both reservoirs can provide reliable retardation parameters in intact rock samples.

The diffusion and sorption parameters determined from the diffusion experiments are summarized in Table I with the results of the batch sorption experiment. The distribution coefficients were calculated from the rock capacity factor, porosity (37.5 %) and dry density (1.55 g cm^{-3}). There was no significant difference in the D_e and α values between the two samples, and between triplicate tests for each sample. This indicates diffusion and sorption of Cs in this sedimentary rock is quite homogeneous. The effective diffusion coefficients are relatively large—2×10^{-10} m^2 s^{-1}—and reflect the large porosity of the rock samples.

As shown in Table I, the K_d values from the two different methods, the diffusion test using intact rock and batch sorption test using crushed rock, are fairly consistent. This consistency is attributed to the same experimental conditions in both sorption and diffusion experiments. In addition, this is because the Cs concentration in the diffusion test is set in the range corresponding to linear sorption, for which K_d values are relatively constant. The latter point was justified from the fact that diffusion curves can be simulated with the conventional diffusion model with a linear sorption isotherm. The consistency of K_d values between diffusion and batch sorption experiments has been observed in other studies of simple cations in sedimentary rocks [e.g., 2, 3]: however, this comparative evaluation should also be studied in other systems with different elements and rock types.

Table I. Retardation parameters obtained from diffusion and batch sorption experiments.

Sample No. (HDB-6)	from diffusion experiment			from batch sorption experiment
	D_e (m^2 s^{-1})	α (-)	K_d (ml g^{-1})	K_d (ml g^{-1}) (at equilibrated [Cs]$_{aq}$)
512-1	2.1 x 10^{-10}	52	33	45±4 (2.1x10^{-6}M), 39±3 (4.4x10^{-6}M),
-2	2.2 x 10^{-10}	52	33	38±5 (1.7x10^{-5}M), 42±5 (2.4x10^{-5}M),
-3	2.1 x 10^{-10}	49	31	26±5 (1.0x10^{-4}M)
596-1	2.1 x 10^{-10}	44	28	45±5 (2.1x10^{-6}M), 39±4 (4.4x10^{-6}M),
-2	2.1 x 10^{-10}	45	29	36±4 (1.7x10^{-5}M), 41±4 (2.4x10^{-5}M),
-3	2.2 x 10^{-10}	47	30	25±5 (1.0x10^{-4}M)

Sorption mechanism and model prediction

In sedimentary rock, Cs is strongly and selectively sorbed by ion-exchange reactions at fixed charge sites on clay minerals such as illite and smectite. Many studies have revealed the importance of high-affinity frayed edge sites for illite and low-affinity planar sites for smectite [e.g., 4, 6, 12]. The results of microscopic observation indicated selective sorption of Cs on the sedimentary rock studied (Figure 3). EPMA analysis showed that Cs was concentrated in specific areas related to the occurrence of illite particles; the identification of illite was based on morphological and chemical information such as potassium distribution.

Following on from the above results and observations, Cs sorption on the rock samples was modeled by competitive ion-exchange reactions on illite and smectite. Model parameters, including site types, site capacities, and equilibrium constants of ion-exchange reactions with Na$^+$, K$^+$, and NH$_4^+$ based on the equivalent fraction formalism, were taken directly from the published data as shown in Table II. The three-site ion-exchange model for illite [6] and one-site model for smectite [7, 13] were applied. The K_d values were calculated using the geochemical code PHREEQC [16]. In the model calculations, the clay contents of illite and smectite in the rock sample are key parameters that influence sensitively K_d values. The clay contents of the rock sample were assumed to be 10 % for illite and 16 % for smectite, based on measured mineral compositions and CEC. As shown in Figure 4, the range of measured K_d values show a reasonable agreement with those predicted by the sorption model considering additive contributions of illite and smectite.

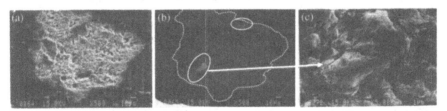

Figure 3. A typical result of EPMA analysis; (a) rock fragment, (b) element map for Cs, (c) illite particle.

Table II. Site capacity and reaction parameters of ion-exchange model for illite and smectite.

Clay minerals		Illite model		Smectite model
CEC [eq/kg]		0.2		1.1[*a]
site type	FES	Type-II	PS	-
site capacity [eq/kg] (site distribution)	5x10⁻⁴ (0.2%)	4x10⁻² (20%)	0.16 (80%)	1.1[*a]
log K of reaction	X-Na	20.0	20.0	20.0	20.0
	X-K	22.4	22.1	21.1	20.42[*a]
	X-NH₄	23.5	-	-	20.46[*b]
	X-Cs	27.0	23.6	21.6	21.6[*c]
reference		[6]		*a:[14], *b:[15], *c:[7]
clay contents of rock sample[*1]		10%		16%

[*1]; evaluated by measured mineral composition and CEC.

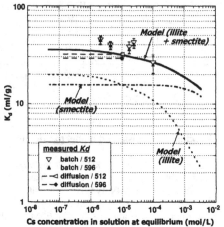

Figure 4. K_d values measured by batch and diffusion tests, and predicted by models.

CONCLUSIONS

Cs retardation parameters for the sedimentary rock from Horonobe-URL were comprehensively investigated by batch sorption and diffusion experiments. Through-diffusion coupled with multiple curve analysis including tracer depletion, breakthrough, and depth concentration curves is proposed to produce reliable retardation parameters. The K_d values obtained from the diffusion tests using intact rock were consistent with those obtained from the batch tests using crushed rock. The sorption behavior was interpreted using the additive ion-exchange model for illite and smectite, assuming dominant sorption minerals based on microscopic observation and their known sorption mechanism. The model predicted the K_d values obtained by the series of experiments reasonably well. Further study is underway to confirm the applicability of the proposed approaches to more sorbing species.

ACKNOWLEDGMENTS

This study was partly funded by the Ministry of Economy, Trade and Industry of Japan. The experiments were partly performed by the Tokyo Nuclear Service Inc. and UI Science Inc. This study was supported by the Horonobe Underground Research Center, JAEA, by providing rock samples and valuable discussions. The authors also thank Mr. M. Kubota, Inspection Development Corporation, by supporting the EPMA analysis.

REFERENCES

1. Organisation for Economic Co-operation and Development (OECD)/Nuclear Energy Agency (NEA), *NEA Sorption Project Phase II* (OECD Publications, Paris, 2005).
2. L.R. Van Loon, B. Baeyens and M.H. Bradbury, *Appl. Geochem.* **20**, 2351 (2005).
3. T. Melkior, S. Yahiaoui, S. Motellier, D. Thoby and E. Tevissen, *Appl. Clay Sci.* **29**, 172 (2005).
4. J.M. Zachara, S.C. Smith, C. Liu, J.P. McKinley, R.J. Serne and P.L. Gassman, *Geochim. Cosmochim. Acta* **66(2)**, 193 (2002).
5. X. Xia, K. Iijima, G. Kamei and M. Shibata, *Radiochim. Acta* **94**, 683 (2006).
6. M.H. Bradbury and B. Baeyens, *J. Contam. Hydrol.* **42**, 141 (2000).
7. H. Wanner, Y. Albinsson and E. Wieland, *Fresenius' J. Anal. Chem.* **354**, 763 (1994).
8. M. Hiraga and E. Ishii, *Mineral and chemical composition of rock core and surface gas composition in Horonobe Underground Research Laboratory Project (Phase 1)*, JAEA-Data/Code 2007-022 (in Japanese) (Japan Atomic Energy Agency, 2007).
9. Japan Nuclear Cycle Development Institute (JNC), TJ5400 2005-004 (in Japanese) (JNC, 2004).
10. C.D. Shackelford, *J. Contam. Hydrol.* **7**, 177 (1991).
11. M. Zhang, M. Takeda and H. Nakajima in *Scientific Basis for Nuclear Waste Management XXIX*, edited by P. Van Iseghem, (Mater. Res. Soc. Proc. **932**, Warrendale, PA, 2006) pp. 135-142.
12. R.W. Cornell, *J. Radioanal. Nucl. Chem.* **171**, 483 (1993).
13. M. Ochs, B. Lothenbach, M. Shibata, H. Sato and M. Yui, *J. Contam. Hydrol.* **61**, 313 (2003).
14. C. Oda and M. Shibata, JNC TN8400 99-032 (in Japanese) (JNC, 1999).
15. P. Fletcher and G. Sposito, *Clay Miner.* **24**, 375 (1989).
16. D.L. Purkhurst and C.A.J. Appelo, *User's Guide to PHREEQC (ver.2)*, USGS Report 99-4259 (U.S. Geological Survey, Washington, DC, 1999).

Mater. Res. Soc. Symp. Proc. Vol. 1124 © 2009 Materials Research Society 1124-Q07-14

Migration Behavior of Bentonite Colloids Through a Fractured Rock

Yoshio Kuno and Hiroshi Sasamoto
Geological Isolation Research and Development Directorate, Japan Atomic Energy Agency, Tokai-mura, Ibaraki, 319-1194, JAPAN

ABSTRACT

Bentonite colloids released into groundwater from the buffer in a geological repository for high-level radioactive waste may influence the migration of radioactive elements in fractures of the host rock. In the present study, column experiments were carried out to investigate the migration behavior of bentonite colloids in artificially fractured granite. The objective of the study is to determine whether the colloids are filtered by the fracture, considering the effects of the fracture length and the ionic strength of transporting solutions. Results indicate that bentonite colloids are not filtered if the transporting solution is dilute (i.e., distilled water or 10^{-4} M NaCl solution). This is evidenced by the observed tendency for normalized colloid concentrations in the column effluent (C/C_0) to rapidly approach 1. An initial increase in the breakthrough curves is also observed at early stages of migration experiments involving more concentrated solutions (10^{-3} M NaCl solution), but the curves later evolve to an approximate steady state with $C/C_0 < 1$. The amount of filtered colloids tends to increase with increasing fracture length. These results suggest that bentonite colloids can be filtered by interactions between the colloids and fracture surfaces, and that these interactions are affected by the ionic strength of the transporting solution. This migration behavior is simulated by a model of colloid migration, taking into consideration the effects of colloid filtration by fracture surfaces. The information on the filtration of bentonite colloids would contribute to a more realistic assessment of colloid effects on radionuclide migration.

INTRODUCTION

Bentonite is planned for use as a buffer material in many geological repository concepts for the permanent disposal of high-level radioactive waste (HLW). The buffer is expected to swell into open fractures in the repository's host rock, which raises the possibility that colloidal particles of bentonite could form at the intruding front and be migrated away from the repository in flowing groundwater [1]. Such clay colloids are known to be stable in solutions of low ionic strength [2,3], and to be strong sorbents for many radionuclides [4,5]. Colloidal migration of sorbed radionuclides could enhance the migration of these elements from a HLW repository.

In order to investigate these colloid effects, *in-situ* experiments were carried out at the Grimsel Test Site in Switzerland [6]. The results indicate that the migration of bentonite colloids in fractured crystalline rock is not retarded if the groundwater flow rate is high (i.e., much higher than is expected in a deep underground repository for HLW). On the other hand, significant amounts of bentonite colloids have been observed to be filtered in similar laboratory experiments in which the flow rate of water through the fractured rocks was relatively low [7,8]. A possible explanation for this difference in behavior is that clay colloids are filtered by interactions involving fracture surfaces when the groundwater flow rate is sufficiently low.

The migration behavior of bentonite colloids in artificially fractured granite was examined under slow-flow conditions in the present study. The possibility of colloid filtration by the fractured rock was investigated by varying such experimental conditions as the column length and the ionic concentration of the transporting solutions. Experimental results are compared with a model of colloid migration to assess the applicability of numerical analyses, taking into consideration the effects of colloid filtration by fracture surfaces.

EXPERIMENTAL

Preparation of experimental solutions

We used an industrially refined Na-type bentonite, Kunipia-F®, manufactured by Kunimine Industries Co. Ltd. This bentonite contains more than 99 % montmorillonite. Colloidal suspensions were prepared by mixing 100 mg of Kunipia-F® in 1 liter of distilled water, 10^{-4} M NaCl or 10^{-3} M NaCl solutions. Bentonite colloids are known to be stable under these conditions [3]. The suspensions were stirred and filtered using membrane filters with 0.8 μm pore size to exclude larger particles. Following filtration, new colloidal solutions were prepared with a concentration of 70 mg/L.

Column experiments

Column experiments were carried out using cylindrical samples of a granite core containing a single artificial fracture, as shown in Figure 1. Two lengths of the core, 15 cm and 30 cm, were used in separate experiments to investigate the effects of fracture length on colloid migration behavior. To form a simple parallel-plate fracture, the aperture width was maintained at 0.1 cm by installing spacers between the two half-cylinders of granite. Prior to the start of each experiment, a colloid-free solution having the same ionic concentration as the colloid solution was pumped through the column to condition the granite fracture. The colloid solution was then pumped through the fracture at a low flow velocity of 2.5 cm/hr. Effluent solutions exiting from granite column were sampled periodically. Silicon concentrations in the effluent samples were measured using an ICP-AES analyzer. A portion of each sample was filtered using an ultrafilter with an effective molecular weight cutoff (MWCO) of 10,000. The Si concentration of the filtrate was also measured in order to deduct the amount of dissolved ions. The colloid concentration in the sample was determined by the colloid content of Si, which was the major element in the bentonite. After each experiment, the fracture was cleaned with fresh water, which was pumped through the column at a high flow rate. A new experiment was not started until it was determined that the fresh water effluent contained no residual colloids.

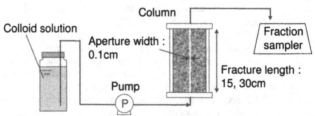

Figure 1. Setup of column experiments using bentonite colloids.

RESULTS AND DISCUSSION

Column experiments

Breakthrough curves showing the time evolution (i.e., in terms of total effluent volume) of normalized concentrations of bentonite colloids in solutions exiting from the 15-cm and 30-cm experimental columns are shown in Figures 2 (a) and (b), respectively. The normalized concentrations, (C/C_0), relate the effluent colloid concentration, C, to the influent colloid concentration, C_0. As shown in Figure 2 (a), C/C_0 rapidly approaches 1 and then remains at this value when the transporting solution is either distilled water or 10^{-4} M NaCl solution, which indicates that bentonite colloids are not filtered from these dilute solutions as they flow through the 15-cm fracture. A rapid increase in the breakthrough curves is also observed initially in experiments involving 10^{-3} M NaCl solutions (Figure 2 (a)), but the curves later evolve to an approximate steady state with C/C_0 between 0.8 - 0.9. The column experiments using 10^{-3} M NaCl solutions were therefore carried out twice to confirm the experimental reproducibility of migration behavior in the fractured granite.

Similar behavior is overall observed in the experiments involving the 30-cm fracture (Figure 2 (b)), but steady-state C/C_0 values in the experiments involving 10^{-3} M NaCl solutions are between 0.75 - 0.8. This range is slightly lower than that observed in the corresponding 15-cm experiment, suggesting that the amount of filtered colloids increases with increasing fracture length. These results indicate that bentonite colloids can be filtered by granite fractures due to an interaction between the colloids and rock surfaces, and that this interaction is affected by the ionic strength of the transporting solution. Based on these results it may be reasonable to expect that bentonite colloids will be filtered by fractured granites if the colloidal solution has a relatively high ionic strength (i.e., equivalent to about 10^{-3} M NaCl, or higher).

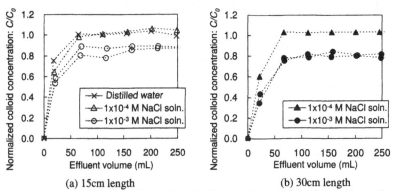

(a) 15cm length   (b) 30cm length

Figure 2. Migration behavior of bentonite colloids through an artificial fracture in granite. Column experiments using 10^{-3} M NaCl solutions were carried out twice.

Numerical analysis of colloid migration

The migration behavior of colloids was simulated in the present study using a numerical analysis that takes into account colloid filtration by fractures. The relevant governing equation, based on the one-dimensional advection-dispersion equation, is given by [9]:

$$\frac{\partial}{\partial t}(C + \frac{\sigma}{b}) + v\frac{\partial C}{\partial x} - D\frac{\partial^2 C}{\partial x^2} = 0 \ , (1)$$

where C is the concentration of mobile colloids in the flowing water (kg/m^3), b is the half aperture of the fracture (m), v is the colloid velocity in the fracture (m/s), D is the colloid dispersion coefficient (m^2/s), t is time (s) and σ is the mass of filtered colloids on the fracture surface per unit area (kg/m^2). Recent experimental work suggests that trace amounts of bentonite colloids can diffuse into granite [10]. It has also been noted elsewhere, however, that the matrix diffusion coefficient of bentonite colloids is several orders of magnitude lower than would be required to fully account for colloid retention in breakthrough experiments, and that other mechanisms, such as colloid filtration by fractures, are likely responsible for colloid retention in fractured rocks [11]. Matrix diffusion is therefore ignored in the present study, where it is assumed that colloids are prevented from entering matrix pores.

The concentration change of the filtered colloids on the fracture surface is modeled using [9]:

$$\frac{\partial \sigma}{\partial t} = \lambda v C b, (2)$$

where λ is the filter coefficient (1/m). The rate of colloid filtration by the fracture is represented in proportion to this parameter and the flux of mobile colloids. It is assumed that colloid filtration is unlimited and that any detachment of colloids from the fracture surface (i.e., remobilization of the filtered colloids) can be neglected. These assumptions are considered to be relevant to the early stages of the colloid migration experiments because the effects of filtered colloids to block further colloid filtration, and colloid detachment from the fracture surface, would be unimportant under such conditions.

Table I summarizes selected parameter values for calculations using the model described above. The fracture aperture, colloid velocity and fracture lengths are based on experimental conditions. The colloid velocity in the fracture is assumed to be equal to the average velocity of flowing water, and the colloid dispersion coefficients were estimated by considering the appropriate dispersion lengths for each fracture. The filter coefficient was determined by fitting the experimental breakthrough curves.

Table I. Calculation parameters of colloid migration through a fracture.

Parameters	Symbol	Value
Half aperture of the fracture	b	5.0×10^{-4} m
Colloid velocity in the fracture	v	6.9×10^{-6} m/s
Colloid dispersion coefficient	D	1.0×10^{-7} m^2/s (Fracture length: 15 cm)
		2.0×10^{-7} m^2/s (Fracture length: 30 cm)
Filter coefficient of colloids	λ	7.5×10^{-1} 1/m

Calculated results based on the model described above are compared in Figure 3 with breakthrough curves determined in experiments involving 10^{-3} M NaCl solutions in both fracture lengths because sufficient colloids were filtered in these conditions. The model correctly predicts that mobile colloids are continuously eliminated from the flowing water and that the extent of colloid filtration is controlled by the filter coefficient. This parameter determines a constant C/C_0 value corresponding to the steady-state, plateau-like portion of the breakthrough curve. Colloid migration in both fracture lengths was simulated using an identical value of $\lambda = 7.5 \times 10^{-1}$ 1/m (Table I). The influence of fracture length on colloid migration behavior is explained by the filtration effect of colloids.

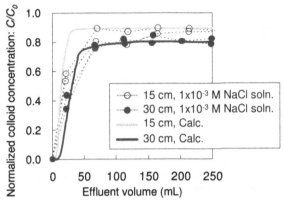

Figure 3. Comparison of experimental breakthrough curves with calculated results.

Estimation of filtration effect of colloids

The effect of colloid filtration on colloid migration over longer migration distances than were considered in the experiments described above was estimated in the present study. In these estimations, flow conditions are assumed to be similar to those in the column experiments. A sensitivity study of the effects of the filter coefficient was evaluated for hypothetical conditions of fracture lengths extrapolated over long distances. Correlations between the filter coefficient and colloid concentration at the steady-state, plateau-like portion of the breakthrough curve are shown in Figure 4, where it can be seen that C/C_0 tends to decrease with increasing values of the filter coefficient and with increasing migration distance. The steady-state C/C_0 values in the range of 0.75 - 0.9 at 15 cm and 30 cm lengths were obtained from our experiments. These C/C_0 values correspond to $\lambda = 7.5 \times 10^{-1}$ 1/m, as used in the numerical analysis described above. The results indicate that if the fracture length is increased to 10 m, the steady-state C/C_0 value in the effluent will be quite low. This suggests that sufficient colloids would become immobile over migration distances of about 10 m or greater if the colloids are effectively filtered by the fracture surfaces. This also raises the possibility that the filtered colloids could sorb radionuclides in the flowing groundwater and thereby restrict their migration from a HLW repository. This information about the filtration of bentonite colloids would contribute to the estimation of colloid effects on radionuclide migration. It is important to note, however, that these results are hypothetical and that they do not consider the durability of colloid filtration. The applicability of such filtration effects over longer periods of time than were considered in the present experiments should thus be evaluated in order to develop a more realistic assessment of long-term filtration effects on colloid migration.

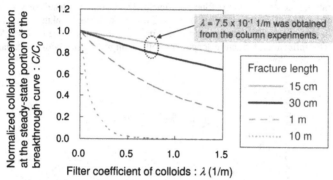

Figure 4. Estimation of colloid filtration effects for fracture lengths up to 10m.

CONCLUSIONS

Column experiments were carried out to investigate the migration behavior of bentonite colloids in a fractured granite. The study focuses on the effects of colloid filtration by fracture surfaces. It was found that bentonite colloids are not filtered if the transporting solutions are relatively dilute (distilled water or 10^{-4} M NaCl solution), but are filtered if the solutions are relatively concentrated (10^{-3} M NaCl solution). A characteristic decrease of steady-state C/C_0 values in effluent observed in the experiments involving 10^{-3} M NaCl solutions is consistent with the results of numerical simulations of the experimental system if the simulations include a filter coefficient for the fractured rock. Preliminarily estimates suggest that sufficient colloids would be filtered from groundwaters flowing through fractured granites over longer migration distances. This also raises the possibility that sorption of radionuclides dissolved in the groundwater by the filtered bentonite colloids could restrict the migration of these elements from a HLW repository.

ACKNOWLEDGMENTS

The authors would like to thank Dr. R. C. Arthur for editing English of the manuscript.

REFERENCES

1. R. Pusch, SKB Technical Report TR-99-31, (1999).
2. H. van Olphen, *An Introduction to Clay Colloid Chemistry*, 2nd ed., (Krieger Publishing Co., 1991), pp. 92-110.
3. Y. Kuno, G. Kamei and H. Ohtani, *Mat. Res. Soc. Proc.*, **713**, 841 (2002).
4. N. Lu and C. F. V. Mason, *Appl. Geochem.*, **16**, 1653 (2001).
5. S. Painter, V. Cvetkovic, D. Pickett and D. R. Turner, *Environ. Sci. Technol.*, **36**, 5369 (2002).
6. A. Möri, W. R. Alexander, H. Geckeis, W. Hauser, T. Schäfer, J. Eikenberg, T. Fierz, C. Degueldre and T. Missana, *Colloids Surf. A: Physicochem. Eng. Aspects*, **217**, 33 (2003).
7. T. Schäfer, H. Geckeis, M. Bouby and T. Fanghänel, *Radiochim. Acta*, **92**, 731 (2004).
8. T. Missana, Ú. Alonso, M. García-Gutiérrez and M. Mingarro, *Appl. Geochem.*, **23**, 1484 (2008).
9. M. Ibaraki and E. A. Sudicky, *Water Resour. Res.*, **31**, 2945 (1995).
10. Ú. Alonso, T. Missana, A. Patelli and V. Rigato, *Phys. Chem. Earth, Parts A/B/C*, **32**, 469 (2007).
11. G. Kosakowski, *J. Contam. Hydrol.*, **72**, 23 (2004).

AUTHOR INDEX

589

SUBJECT INDEX

reactivity, 111
Ru, 141
Rutherford backscattering (RBS), 339

scanning electron microscopy (SEM), 141, 333, 393
Se, 567
second phases, 399
Si, 319
simulation, 213, 407, 519, 543, 549
slurry, 181
sol-gel, 525
steel, 463
storage, 117, 251, 289, 295, 301
surface
    chemistry, 231, 289
    reaction, 99

transmission electron microscopy (TEM), 213, 237
tribology, 421

U, 105, 195, 387, 393, 537, 561

vapor pressure, 307

waste management, 3, 15, 29, 41, 53, 65, 77, 99, 111, 131, 141, 147, 153, 161, 167, 181, 187, 207, 231, 243, 251, 263, 277, 295, 301, 307, 313, 319, 327, 333, 339, 351, 357, 373, 379, 387, 407, 421, 427, 463, 525, 537, 555, 561, 581
water, 251, 307, 387

x-ray
    diffraction (XRD), 167
    fluorescence, 131, 257
    tomography, 555

Zr, 175, 481